TIPERs

Sensemaking Tasks for Introductory Physics

Curtis J. Hieggelke
Joliet Junior College

David P. Maloney
Indiana University-Purdue University Fort Wayne

Stephen E. Kanim
New Mexico State University

Thomas L. O'Kuma
Lee College

PEARSON

Boston Columbus Indianapolis New York San Francisco Upper Saddle River
Amsterdam Cape Town Dubai London Madrid Milan Munich Paris Montréal Toronto
Delhi Mexico City São Paulo Sydney Hong Kong Seoul Singapore Taipei Tokyo

Publisher: Jim Smith
Program Manager: Katie Conley
Editorial Assistant: Sarah Kaubisch
Managing Development Editor: Cathy Murphy
Cover Designer: Derek Bacchus
Manufacturing Buyer: Jeff Sargent
Senior Marketing Manager: Will Moore
Cover Photo Credit: Getty Images/Sola deo Gloria

33 2019

ISBN 10: 0-132-85458-9; ISBN 13: 978-0-132-85458-0

www.pearsonhighered.com

PEARSON SERIES IN EDUCATIONAL INNOVATION

Peer Instruction: A User's Manual
Eric Mazur

Tutorials in Introductory Physics
Lillian C. McDermott
Peter S. Shaffer
The Physics Education Group, University of Washington

Just-In-Time Teaching: Blending Active Learning with Web Technology
Gregor M. Novak
Evelyn T. Patterson
Andrew D. Gavrin
Wolfgang Christian

Ranking Tasks in Physics: Student Edition
Curtis J. Hieggelke
David P. Maloney
Thomas L. O'Kuma

E&M TIPERs: Electricity and Magnetism Tasks
Curtis J. Hieggelke
David P. Maloney
Thomas L. O'Kuma
Stephen E. Kanim

nTIPERs: Newtonian Tasks Inspired by Physics Education Research
Curtis J. Hieggelke
David P. Maloney
Stephen E. Kanim

About the Authors

Since 1991, Curtis Hieggelke, David Maloney, Thomas O'Kuma, and Stephen Kanim have led over 35 workshops in which educators learned how to use and develop TIPERs. Many of these workshops were part of the Two-Year College Physics Workshop Project (supported by seven grants from the National Science Foundation and co-directed by Hieggelke and O'Kuma), which has offered a series of more than 60 professional development workshops for over 1,200 participants of two-year college and high school physics teachers. Working with Alan Van Heuvelen and Thomas O'Kuma, Maloney and Hieggelke also developed the Conceptual Survey of Electricity and Magnetism, which has become a standard instrument for measuring electricity and magnetism conceptual gains in introductory physics courses. The American Physical Society gave the 2009 Excellence in Physics Education Award to Hieggelke, Maloney, and O'Kuma, in part for their work on TIPERs. This is the fifth book for Maloney and Hieggelke dealing with curriculum materials based on Physics Education Research (PER).

Curtis J. Hieggelke received a B.A. in physics and mathematics from Concordia College (Minnesota) and a Ph.D. in theoretical particle physics from the University of Nebraska-Lincoln. He spent his professional career as a physics teacher at Joliet Junior College until his retirement in 2003. Hieggelke has served as president and section representative of the Illinois Section of the American Association of Physics Teachers (AAPT) and received the Distinguished Service Citation in 1993. He was awarded the Distinguished Service Citation by AAPT in 1994, and he was elected to the Executive Board of AAPT for three years as the two-year college representative. He has served as the principal investigator or co-principal investigator for 12 National Science Foundation (NSF) grants for workshops and curriculum materials development based on PER.

David P. Maloney received a B.S. in physics from the University of Louisville and an M.S. in physics and Ph.D. in physics, geology, and education from Ohio University. A member of the Indiana University-Purdue University-Fort Wayne faculty since 1987, he has also taught at Wesleyan College and Creighton University. He was awarded the Distinguished Service Citation by AAPT in 2005. His main research interests concern the study of students' commonsense ideas about physics, how those ideas interact with physics instruction, and the study of problem solving in physics. Maloney has authored or coauthored two-dozen articles and has been the principal investigator or co–principal investigator for eight NSF grant projects.

Stephen E. Kanim received a B.S. in electrical engineering from UCLA and a Ph.D. in physics from the University of Washington and has been a member of the physics faculty at New Mexico State University since 1998. He has varied research interests related to the teaching and learning of physics, including student conceptual understanding of specific physics topics, reasoning about proportions, and experimental tests of models of student thinking. He has worked on several research-based curriculum development projects, including development of a set of introductory labs for mechanics. He previously taught high school physics in Las Cruces, New Mexico and in Palo Alto, California, and worked as an electrical engineer in Santa Clara, California.

Thomas L. O'Kuma received his B.S. and M.S. in physics and mathematics from Louisiana Tech University. Since 1976 he has been a full-time, two-year college physics instructor and has been at Lee College since 1989. He received the Distinguished Service Citation in 1994 and the Award for Excellence in Introductory College Physics Teaching in 2002 from the AAPT, and the Robert N. Little Award in 1994 for Outstanding Contributions to Physics in Higher Education in Texas. He was elected president of AAPT and served on the Executive Board of AAPT for four years. He has served as the principal investigator or co–principal Investigator for 19 NSF grants. O'Kuma's research interests are in how students learn physics and in developing tools to help students learn physics.

INTRODUCTION

A WORKBOOK TO HELP MAKE SENSE OF PHYSICS

One definition of science is that it is refined common sense. We humans use our common sense to help us navigate and cope with the world. The ideas we are capable of using in coping are those that make sense to us. If an idea does not make sense to us, it will be harder to remember, and it is harder to put the idea into practice.

Suppose you find the following paragraph in a book you've picked up about an unfamiliar subject:[1]

> *It is very important that you learn about traxoline. Traxoline is a new form of zionter. It is monotilled in Ceristanna. The Ceristannians gristerlate large amounts of fevon and then bracter it to quasel traxoline. Traxoline may well be one of our most lukised snezlaus in the future because of our zionter lescelidge.*

At the end of the chapter, you find the following questions:

> *1. What is traxoline?*
>
> *2. Where is traxoline monotilled?*
>
> *3. How is traxoline quaselled?*
>
> *4. Why is traxoline important?*

Even though you will have no trouble answering these questions, it doesn't mean that you understand anything about traxoline! The form of these questions is such that you are able to "make answers" without "making sense."

Often when people first start to study physics, they also fall into a habit of answer-making instead of sensemaking. Typical physics problems describe a situation and require that students use the information provided to solve for some unknown quantity. This can be a useful process, as you learn what the equations are for and what you can find out by using them. However, sometimes we can solve these problems by plugging values into equations without really understanding the underlying ideas. It is possible to learn these equations, and even to become good at using these equations to solve these types of problems, without really understanding their meaning. Students who have done this may become really good at making physics answers, but that doesn't mean that they understand physics.

The equations are mathematical expressions of general rules or ideas about the workings of the universe we live in. Understanding physics requires understanding these rules and ideas—where they apply (and where they don't), what factors affect the results (and what factors don't), and how one idea or rule is related to another. To really understand physics, you will need to connect underlying ideas to their mathematical representation—to make sense of the equations.

The exercises in this workbook are intended to promote sensemaking. The various formats of the questions have been chosen so that it is difficult to make answers by using physics equations as formulas. You will need to develop a solid qualitative understanding of the concepts, principles, and relationships in physics. In addition, you will have to decide what is relevant and what isn't, which equations apply and which don't, and what the equations tell us about physical situations.

Many students are not used to connecting mathematical relationships to concepts: This workbook is intended to provide opportunities to practice making these connections. It is not a textbook, and you won't get much out of it by just reading through it. Instead, you need to actually work through the exercises. With practice, when you are given a more typical physics problem where you are asked to find a value for an unknown quantity, you will be able to make sense of the problems in addition to making answers. We hope that learning physics will be much more rewarding as a result.

GENERAL INFORMATION AND CONVENTIONS

This workbook contains tasks in ten different formats. These formats are likely to be new to you, and it might take time to learn how to respond properly. We (the authors) believe the different formats will encourage you to think more productively about the ideas of physics and about how these ideas are described mathematically.

For almost all formats, you will be required to give an explanation as well as an answer. Writing out your

---

[1] "The Montillation of Traxoline" was written by Dr. Judith Lanier, professor and dean emeritus, Department of Education, Michigan State University.

explanations in a form that your classmates can understand is an extremely useful habit for several reasons. First, doing this will give you a second chance to think through your response to the task and will allow you to double-check your reasoning. It also encourages the habit of writing careful explanations that can concisely argue for a position—an indispensable skill in whatever line of work you choose to pursue.

TASK FORMATS

The ten task formats in this workbook are Bar Chart Tasks (BCT), Changing Representation Tasks (CRT), Comparison Tasks (CT), Linked Multiple Choice Tasks (LMCT), Qualitative Reasoning Tasks (QRT), Ranking Tasks (RT), Student Contention Tasks (SCT), Troubleshooting Tasks (TT), What's Wrong Tasks (WWT), and Working Backwards Tasks (WBT).

Several of these formats are reasonably self-explanatory. For example, troubleshooting tasks ask you to identify what is wrong in a situation and explain how to correct it. However, several of the formats are likely to be unfamiliar to you, and we will describe those here.

Ranking Tasks require you to compare four to six variations of a physical situation and to rank them on the basis of some property or quantity. In writing out a ranking sequence, it is possible that you might believe two, or more, of the variations are equal for the specified basis. If this happens, it is important that you explicitly identify which variations you believe are tied. Another possibility for the ranking tasks is that all of the variations have the same value for the specified basis; if that happens, you should choose the appropriate option from those at the right end of the answer choices. And in some tasks, all of the variations may have a zero value of the specified basis for ranking, so that option is also provided in each RT.

Comparison Tasks are similar to ranking tasks except that there are only two cases (labeled A and B) to compare, so the choices are always that A is *larger than, smaller than,* or *equal to* B on the specified basis.

Changing Representations Tasks will require you to go from one representation to another. To understand a physical situation and as a step toward representing a physical situation mathematically, physicists have developed a number of useful representations. These representations may be a particular form of drawing, a chart, a graph, or a map. In a way, representations are a little like models of the systems. Learning how to represent a physical situation in different ways will strengthen your understanding of the underlying physics, and this is the purpose of these tasks. For example, you may be given an acceleration versus time graph that describes a particular motion, and your task will be to make a strobe position diagram of that motion.

Student Contentions Tasks present a variety of student statements about a physical situation. You will be asked which of the statements (if any) you agree with. Often, the students' contentions will disagree so it will only make sense to choose at most one of the students to agree with, or to decide that none of their statements are presenting reasonable ideas. However, there will be other versions of these tasks where the students may be saying similar things in different ways and/or saying different things that are reasonable for the situation. In those cases you should agree with as many of the student statements as you believe are appropriate.

Bar Chart Tasks are tasks requiring you to represent physical quantities with bars whose length represents the amount of that quantity. Your instructor may want you to work with these as qualitative tasks, where the relative number of bars for each entry is what is important, or more quantitatively where a specific number of bars for each entry is needed. Bar chart tasks are particularly useful for understanding *conserved quantities.* As an example, if everyone in your classroom bought and sold things from each other, as long as no money came into the classroom or left, the total amount of money would be a conserved quantity. We might in this case make a bar chart showing how much money each person had at the beginning and how much each person had at the end. In physics, energy is an example of a conserved quantity, and bar chart tasks are especially useful for describing how energy changes form. If the task is worked qualitatively the total number of bars initially plus or minus the number during the process has to equal the total number of bars finally to satisfy the conservation of energy relation.

Next, we discuss several issues about some conventions used in physics that we have adopted in this workbook. Physics instructors tend to have strong opinions about conventions, and your instructor may have good reasons for disagreeing with our choices. For this reason, you may want to check with your instructor about what convention he/she will be using.

USE OF POSITIVE AND NEGATIVE SIGNS

Positive and negative signs are used in physics in at least four different ways, and sometimes (unfortunately) you need to infer the meaning from the context. Appropriate interpretation of how the signs are being used is critical if you want to be sure you are reading physics texts properly. Here we describe the various uses of positive and negative signs.

The first use is for directions of vectors when dealing with one-dimensional situations. For example, suppose we have cars moving along a straight East-West road. We would typically choose one of these directions—say, East—as positive and the opposite direction as negative. In other words, we are using the positive and negative signs to represent direction, *and only direction*. That means the signs say nothing about the magnitude (size) of the quantity, for example, displacement or velocity, we are dealing with, only which way they point.

As a result of this interpretation, a velocity of –8 m/s is considered a larger velocity than +3 m/s. You should be careful in this context not to use the analogy to the mathematical number line and treat positive values, of any size, as larger than negative values, of any size. In this workbook, we will often ask for the *magnitude* of quantities that have direction, which means that you should ignore their directions and consider only the size.

The place where the number line analogy is reasonable is with the second use of positive and negative in physics, when we are dealing with scalar quantities (i.e., quantities that do not have a direction) which can be "above" or "below" a zero value. Examples of such quantities are gravitational potential energy, electric potential energy, electric potential, work, and temperature (on the Celsius scale). There is another analogy that works for this situation, which is the "bank balance" model, or at least the classic bank balance where one could go a little "in the hole." A positive bank balance does mean you have more money than a negative balance, which would actually mean you owe the bank money. Here the signs do tell us something about the size of the quantity and (for example) it is perfectly reasonable to state that +2 J is larger than –7 J for gravitational potential energy.

A third way positive and negative are used in physics is to indicate polarity, primarily with electric charges. Here the sign indicates only that an object is a member of either category A (+) or category B (–) and again says nothing about how big a charge it is. However, when determining quantities involving electric charges such as force, electric field, electric potential, or electric potential energy, which category the object(s) are members of will affect the signs for those quantities and thus can affect their sizes or their directions.

A final place where positive and negative signs are used is with lenses and mirrors. In this case, signs are assigned by convention to the focal lengths to indicate whether the lenses or mirrors are converging or diverging. They are also used on magnification to indicate upright (positive) or inverted (negative) images. In addition, they are assigned to image distance to indicate location of the image relative to the object; for example, positive image distances mean the object and image are on opposite sides of the lens. And finally, object distances can be positive or negative, with positive values indicating real objects.

OTHER CONVENTIONS IN THIS BOOK

Often, physicists must use simplifications of physical situations in order to make sense of them. For example, it is often the case that we will ignore air resistance when we analyze the motion of a projectile. In this workbook, we will assume the following simplifications to physical situations unless otherwise noted:

(1) We will ignore air resistance.

(2) We will assume that surfaces and pulleys are frictionless.

(3) We will assume that pulleys and strings have no mass.

(4) We will assume that all fluids have the same density.

(5) We will assume that batteries, ammeters, and voltmeters are ideal. (Your instructor will explain what this means at the appropriate time.)

(6) We will assume that bulbs in a circuit are identical.

There is a standard convention for how graphs are labeled in science and math courses that will be used in this book. The convention is to label graphs by naming the quantity that is plotted on the vertical (y) axis followed by a dash and then the quantity that is plotted on the horizontal (x) axis. So the general format for the name/label of a graph is y-x. Specific examples are velocity-time or force-distance.

IDENTIFYING THE SYSTEM

When analyzing physical situations or solving physics problems, an important aspect of the analysis is deciding where and how to apply the appropriate concepts and principles. Normally, the first step in the process is identifying the object, or objects, that will be the focus of the analysis. The object (or objects) is called the system. Identifying the system is important because situations or problems can often be analyzed different ways depending on which system is chosen.

Let's take a simple example: Suppose you throw a ball straight up in the air. We will analyze changes in energy as the ball goes up to the top, stops instantaneously, and then accelerates down using two different systems. In the first case, we will take as the system just the ball itself. From this perspective the ball starts with kinetic energy and then the Earth does work on the ball as it travels up to the top and then falls back down. Potential energy is *not* involved here because potential energy is an interaction between two or more objects within the system, and our system has only one object—the ball.

If instead we take the ball and the Earth as the system, the analysis would be portrayed differently. First, we would have to choose a zero level for the gravitational potential energy of the system. Usually this would be the initial height of the ball. With that choice the system initially has kinetic energy but zero potential energy. As the ball travels up to the top, the kinetic energy of the system decreases and the gravitational potential energy increases. At the top the system has zero kinetic energy and maximum gravitational potential energy (equal to the initial kinetic energy). Then, as the ball falls back down, the kinetic energy increases and the gravitational potential energy decreases back to zero. Notice that there is no work involved with this choice of system because work is an energy exchange involving an agent outside the system.

In a number of the situations in physics, what constitutes the system will be obvious, but there will be times when it will be necessary to explicitly identify what you are going to take as your system. Getting in the habit of explicitly identifying the system will make your life easier in the long run.

Acknowledgments

First, we want to acknowledge the support of our families and spouses (Estelle Hieggelke, Irene Maloney, Kathe Kanim, and Kathy O'Kuma) and thank them for all their assistance.

An endeavor like this book requires input from many people in addition to the authors. These tasks were extensively reviewed and improved by Martha Lietz of Niles West High School (Skokie, IL), Robert Morse of St. Albans School (Washington, DC), and William P. Hogan (Joliet Junior College, IL). We also valued the feedback from the many teachers who have attended our workshops or tested our materials in their classes.

Since this book contains Tasks Inspired by Physics Education Research, we would like to acknowledge several of our most important inspirations. First, Alan Van Heuvelen deserves mention because his ALPS manual, which used bar charts and changing representation tasks, provided an early and strong guiding idea, and he was the co-developer of the Working Backwards (Jeopardy) format. Second, the University of Washington Physics Education Group developed the Conflicting Contentions format, which we modified to produce our Student Contentions format. Third, we thank the many Physics Education Research investigators whose names we unfortunately cannot mention individually. Finally, we want to recognize and thank Arnold B. Arons, Robert Karplus, Lillian C. McDermott, and Paul G. Hewitt for their pioneering work in promoting conceptual understanding in physics courses.

We would like to acknowledge and thank the National Science Foundation and Duncan McBride in particular, for supporting the development of these materials and the Physics Workshop Projects. This material is based in part upon work supported by the Division of Undergraduate Education of the National Science Foundation under grant numbers USE 9150334, USE 9154271, DUE 9255466, DUE 9353998, DUE 9554683, DUE 9952735, DUE 0125831, DUE 0632963, and DUE 0633010. Any opinions, findings, and conclusions or recommendations are those of the authors and do not necessarily reflect the views of the National Science Foundation.

Finally, we thank James Smith, our editor, and the staff at Pearson for their assistance and their willingness to publish a unique type of book.

Legend for Task Names	
BCT	Bar Chart Tasks
CRT	Changing Representation Tasks
CT	Comparison Tasks
LMCT	Linked Multiple Choice Tasks
QRT	Qualitative Reasoning Tasks
RT	Ranking Tasks
SCT	Student Contention Tasks
TT	Troubleshooting Tasks
WWT	What's Wrong Tasks
WBT	Working Backwards Tasks

TABLE OF CONTENTS

Answering Ranking Tasks

Example-RT1: Stacked Blocks—Mass of Stack

Shown below are stacks of various blocks. All masses are given in the diagram in terms of M, the mass of the smallest block.

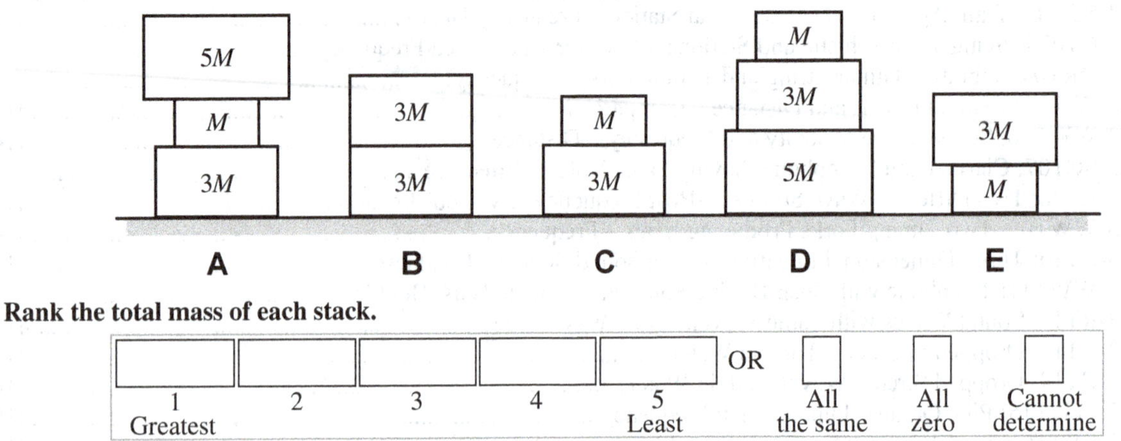

Rank the total mass of each stack.

Explain your reasoning.

Stacks A and D have a total mass of 9M, C and E have a mass of 4M, and B has a mass of 6M.

Example answer formats

One way the ranking can be expressed is as "A = D > B > C = E" based on the reasoning listed above.

Using the ranking chart, this answer could be expressed either as

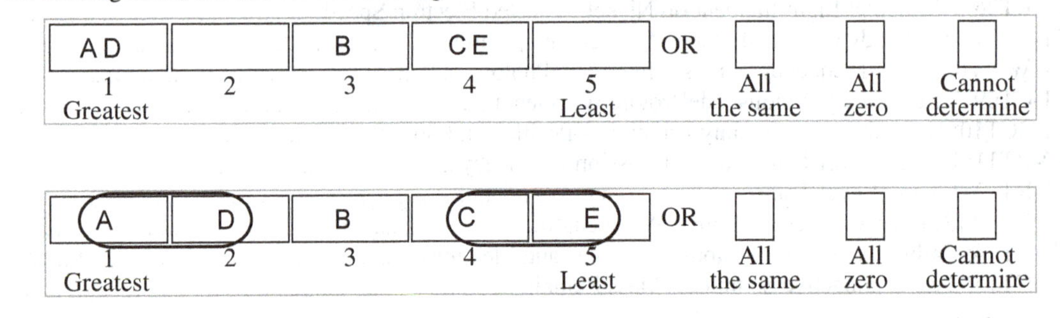

or as

where the ovals around the letters indicate a tie in the ranking. Note that the order of two items ranked as equal is not important, but some instructors encourage students to use alphabetical order.

PRACTICE-RT2: STACKED BLOCKS—NUMBER OF BLOCKS

Shown below are stacks of various blocks. All masses are given in the diagram in terms of M, the mass of the smallest block.

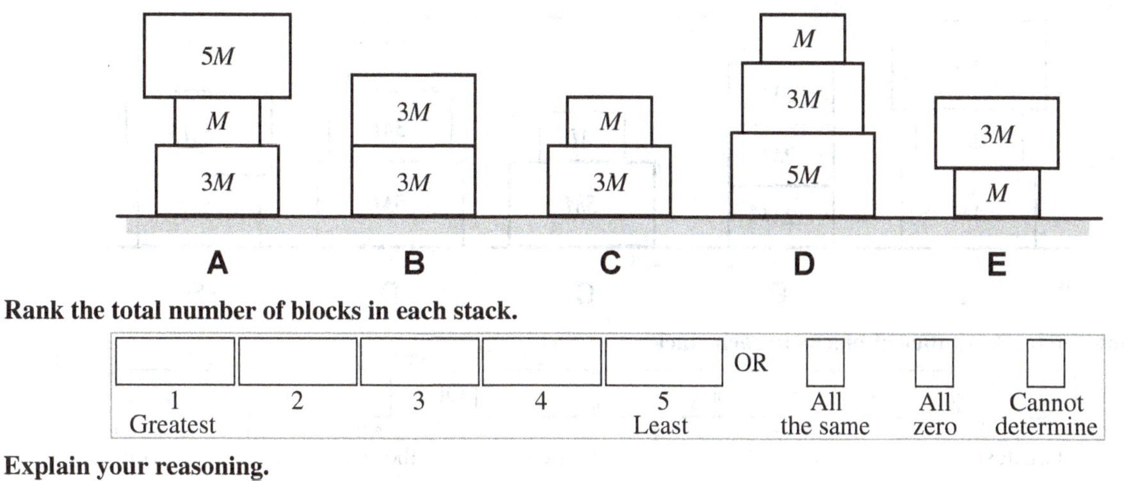

Rank the total number of blocks in each stack.

1 Greatest	2	3	4	5 Least	OR	All the same	All zero	Cannot determine

Explain your reasoning.

PRACTICE-RT3: STACKED BLOCKS—AVERAGE MASS

Blocks are stacked on a table. Masses are given in terms of M, the mass of the smallest block.

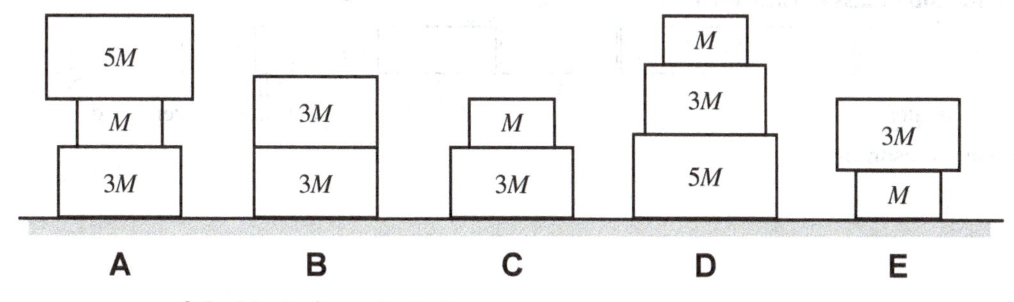

Rank the average mass of the blocks in each stack.

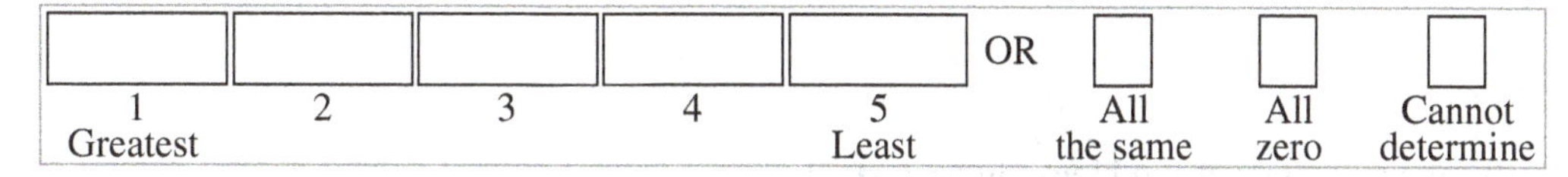

Explain your reasoning.

PRACTICE-RT4: STACKED BLOCKS—NUMBER, TOTAL MASS, AND AVERAGE MASS

Shown below are stacks of various blocks. All masses are given in the diagram in terms of M, the mass of the smallest block.

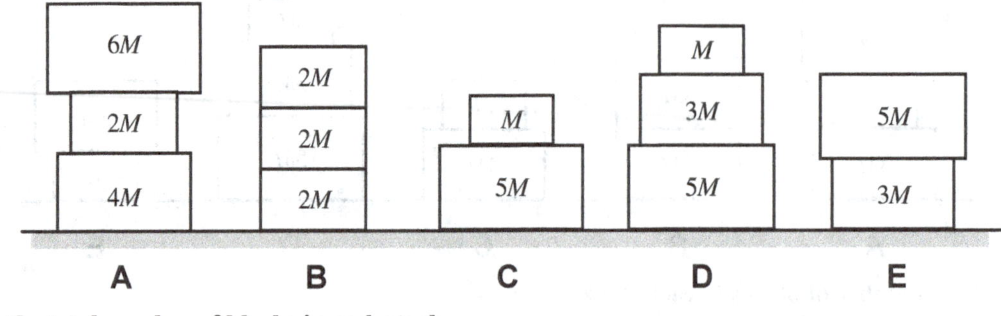

(a) Rank the total number of blocks in each stack.

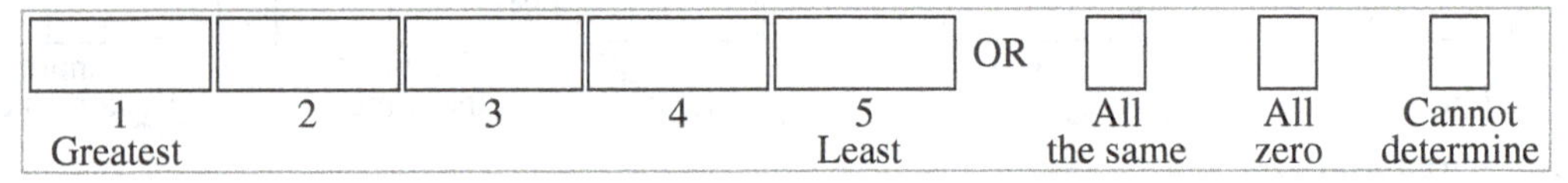

Explain your reasoning.

(b) Rank the total mass of each stack.

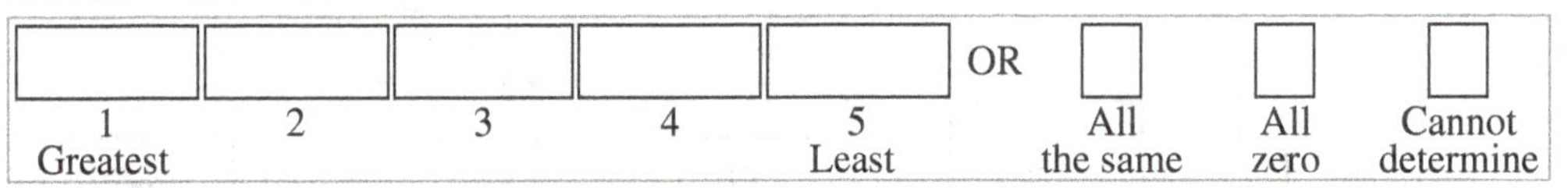

Explain your reasoning.

(c) Rank the average mass of the blocks in each stack.

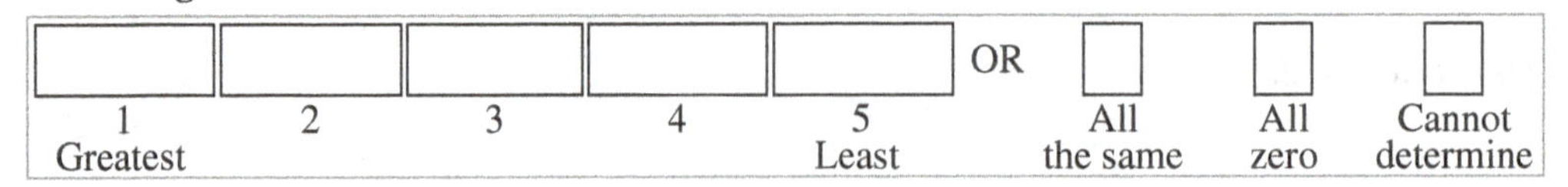

Explain your reasoning.

A1 PRELIMINARIES

A1-RT01: LINE GRAPH I—SLOPE

Four points are labeled on a line.

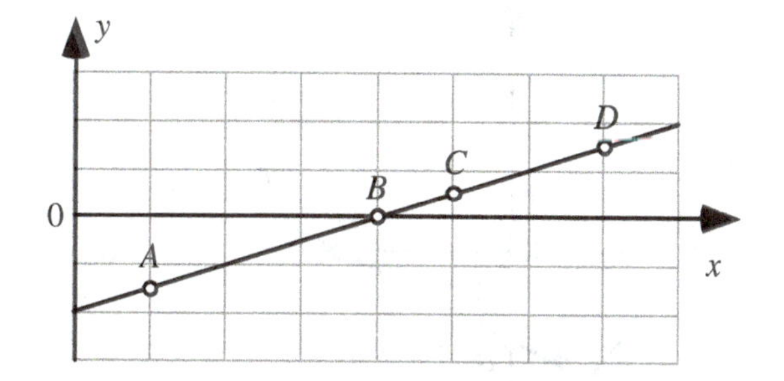

Rank the magnitudes (sizes) of the slopes of the line at the labeled points.

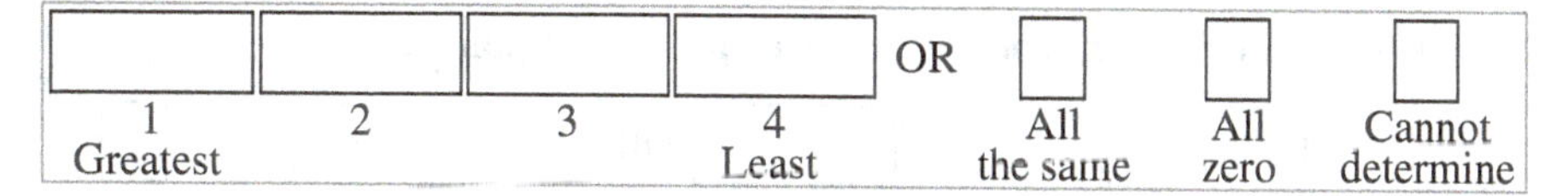

Explain your reasoning.

A1-RT02: *Y-X* GRAPH LINES—SLOPE

Shown are several lines on a graph.

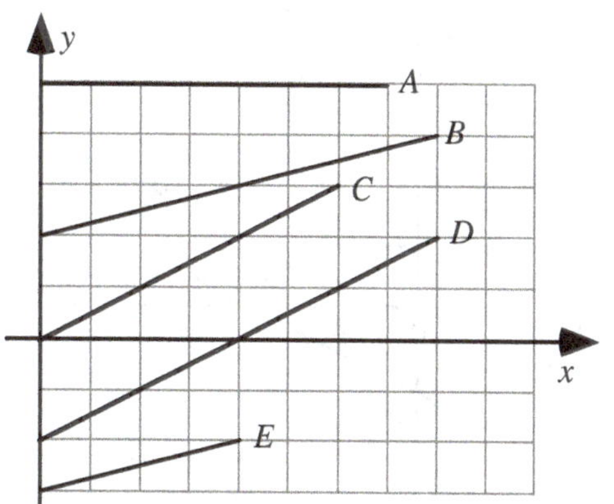

Rank the slopes of the lines in this graph.

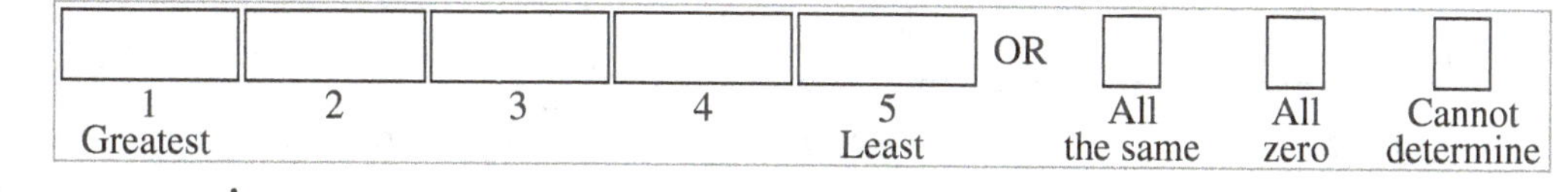

Explain your reasoning.

A1-RT03: Complex Line Graph—Magnitude of Slope

Four points are labeled on a graph.

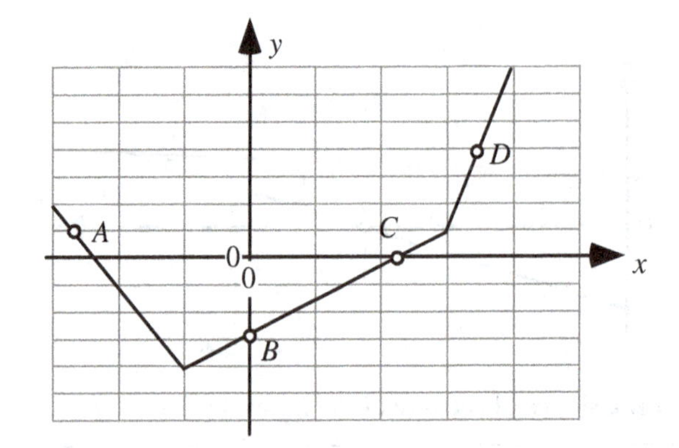

Rank the magnitudes (sizes) of the slopes of the graph at the labeled points.

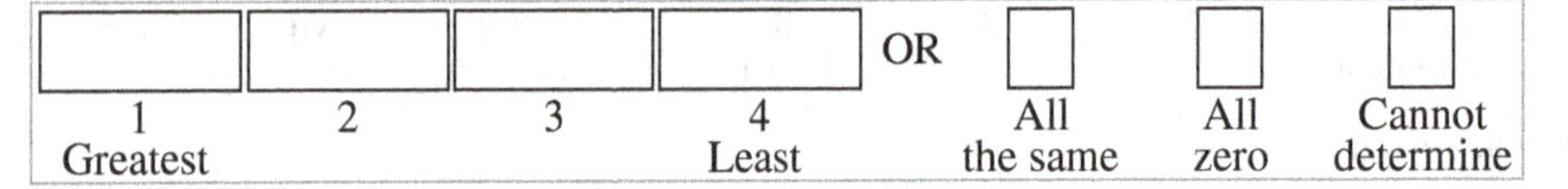

Explain your reasoning.

A1-RT04: Complex Line Graph—Slope

Four points are labeled on a graph.

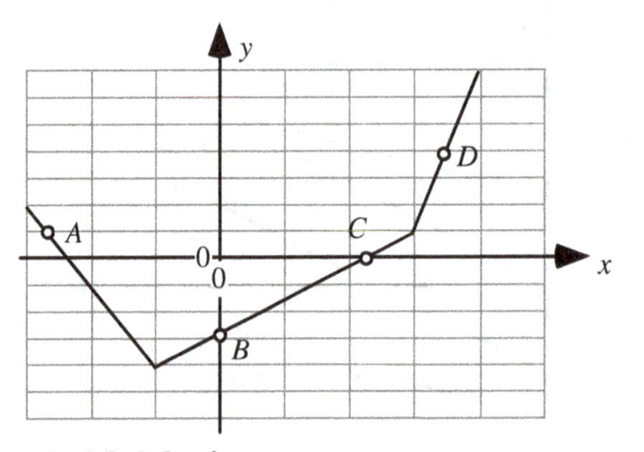

Rank the slopes of the graph at the labeled points.

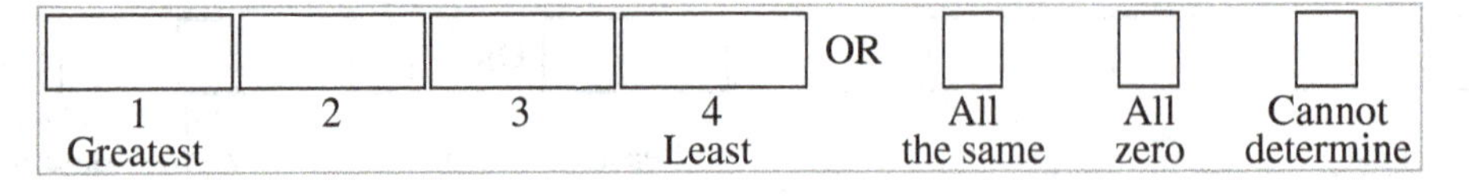

Explain your reasoning.

A1-RT05: FOUR RECTANGLES—SLOPE OF DIAGONALS

In each case, a rectangle is drawn on a grid.

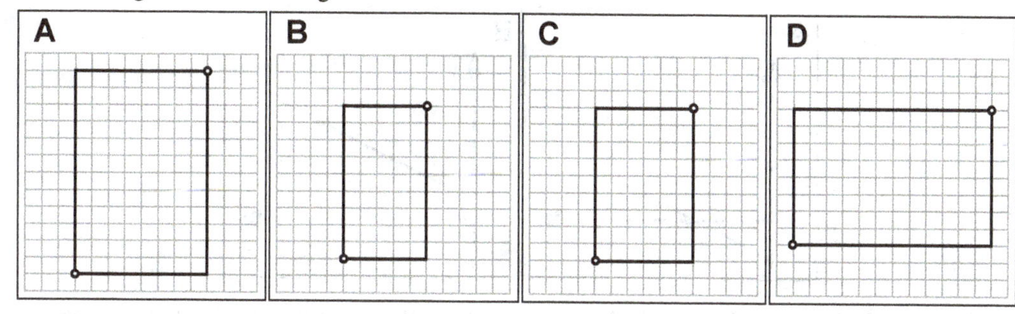

Rank the slopes of the diagonals between the points marked with dots for these rectangles.

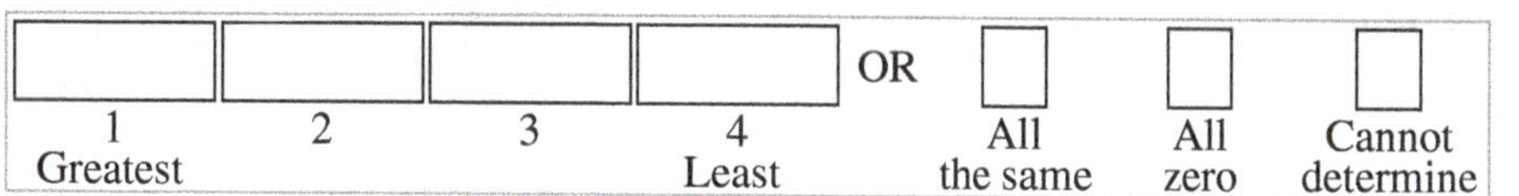

Explain your reasoning.

A1-WWT06: RECTANGLE—SLOPE OF DIAGONALS

In each case, a rectangle is drawn on a grid. A student makes the following statement in comparing the slopes of the diagonal lines connecting the corners marked by dots:

> *"The steepness of a line depends on how much the line rises compared to its run. For Case A the rise is 9, and the run is 6, and the difference between rise and run is 3. For Case B, the rise is 8 and the run is 12 and the difference is minus 4. Case B has a smaller slope than Case A, and in Case B the slope is negative."*

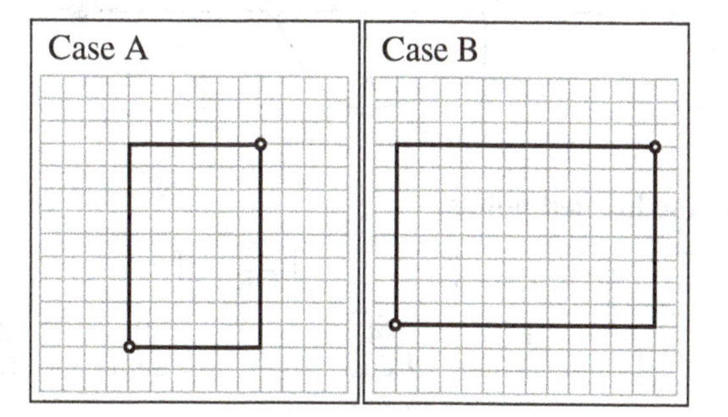

What, if anything, is wrong with this student's statement? If something is wrong, identify and explain how to correct all errors. If this statement is correct, explain why.

TIPERs

A1-CT07: Line Graph II—Slope

Shown are two graphs.

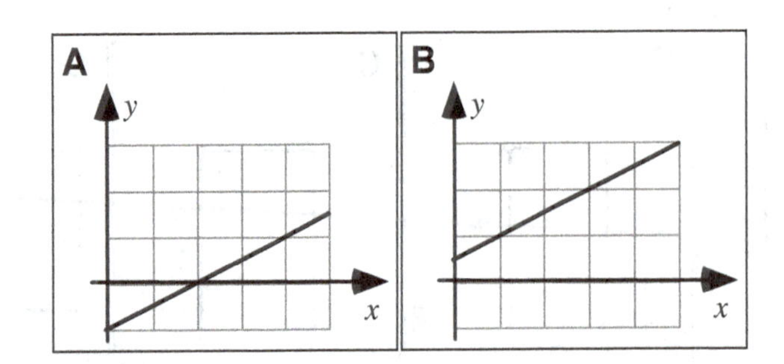

Is the slope of the graph (i) *greater in Case A*, (ii) *greater in Case B*, or (iii) *the same in both cases?* _____

Explain your reasoning.

A1-RT08: Curved Line Graph—Slope

Four points are labeled on a graph.

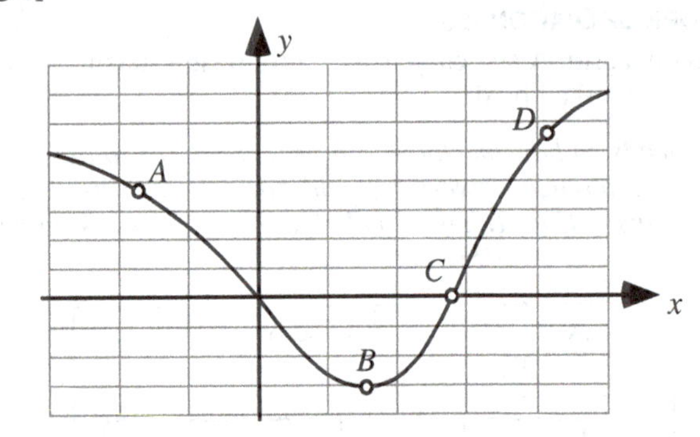

Rank the slopes of the graph at the labeled points.

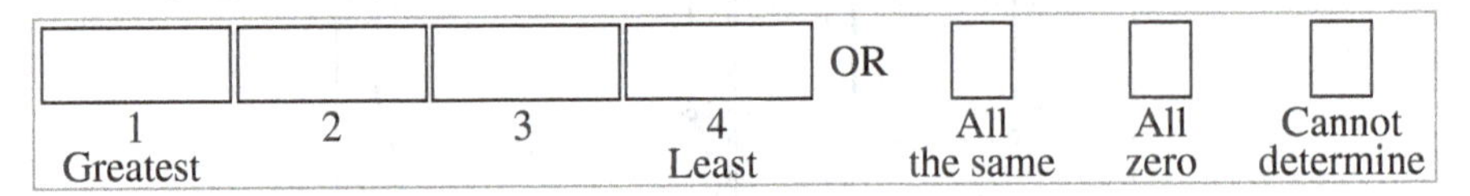

Explain your reasoning.

A1-WWT09: TWO COLUMNS OF DATA—DATA GRAPH

A student uses data from a table to make a graph as shown.

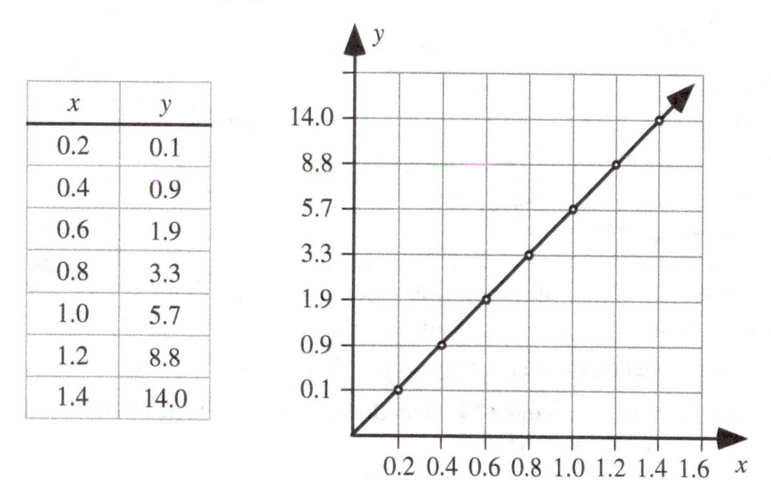

x	y
0.2	0.1
0.4	0.9
0.6	1.9
0.8	3.3
1.0	5.7
1.2	8.8
1.4	14.0

What, if anything, is wrong with this graph? If something is wrong, identify and explain how to correct all errors. If this statement is correct, explain why.

A1-WWT10: Monthly Website Visits Graph—Interpretation

A website posts the following graph of the number of monthly visits during the past year.

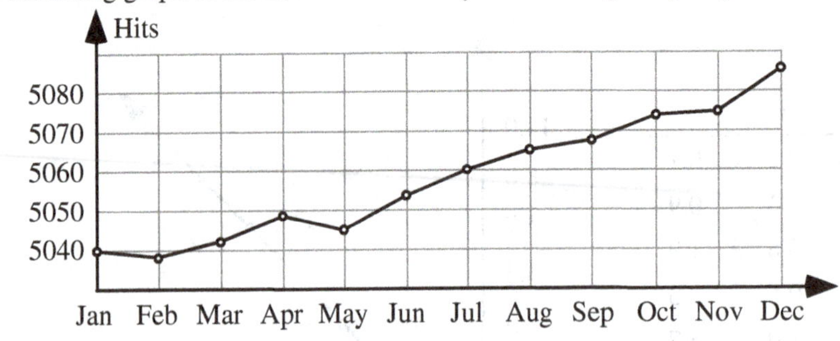

The website owner makes the following statement about this graph:

"As you can see, this year our popularity has grown dramatically, and we look forward to continued success."

What, if anything, is wrong with this statement? If something is wrong, identify and explain how to correct all errors. If this statement is correct, explain why.

A1-WWT11: Cat Moving Away from a Dog Graph—Cat Speed

A cat is moving away from a sleeping dog along a hallway. A graph of the distance of the cat from the dog as a function of time is shown. A student uses the equation rate times time equals distance to calculate the speed of the cat at time 6 seconds:

Rate = distance/time = 16 m/6 s = 2.667 meters per second.

What, if anything, is wrong with this calculation? If something is wrong, identify and explain how to correct all errors. If this is correct, explain why.

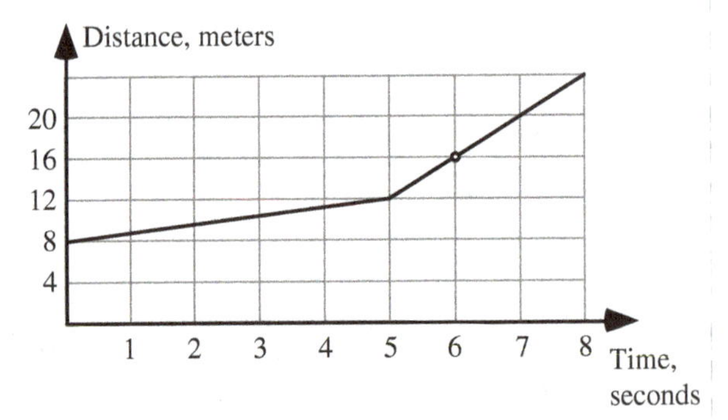

A1-WWT12: FILLING A CYLINDRICAL GLASS—HEIGHT OF WATER-TIME GRAPH

A cylindrical glass is filled using a tap with a constant flow rate of 4 ml per second. A student graphs the height of the water in the glass as a function of time as shown:

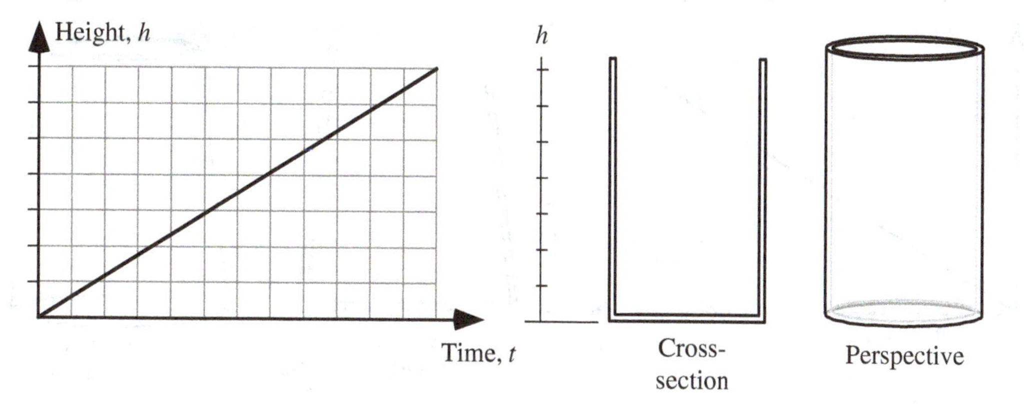

What, if anything, is wrong with this graph? If something is wrong, identify and explain how to correct all errors. If this is correct, explain why.

A1-WWT13: FILLING TAPERED GLASS—HEIGHT OF WATER-TIME GRAPH

A glass is tapered so that it is wider at the top than at the bottom. The glass is filled using a tap with a constant flow rate of 4 ml per second. A student graphs the height of the water in the glass as a function of time as shown:

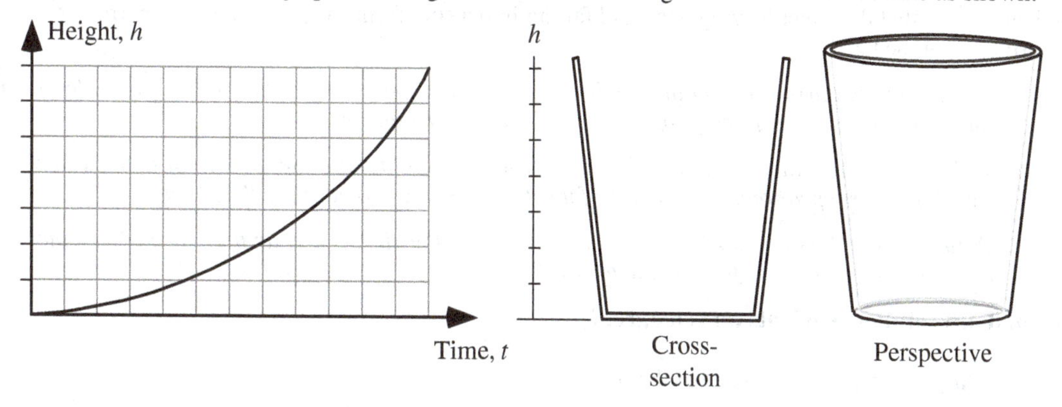

What, if anything, is wrong with this graph? If something is wrong, identify and explain how to correct all errors. If this is correct, explain why.

A1-WWT14: Filling Inverted Tapered Glass—Water Height-Time Graph

A glass is tapered so that it is narrower at the top than at the bottom. The glass is filled using a hose with a constant flow rate of 4 ml per second. A student graphs the height of the water in the glass as a function of time as shown:

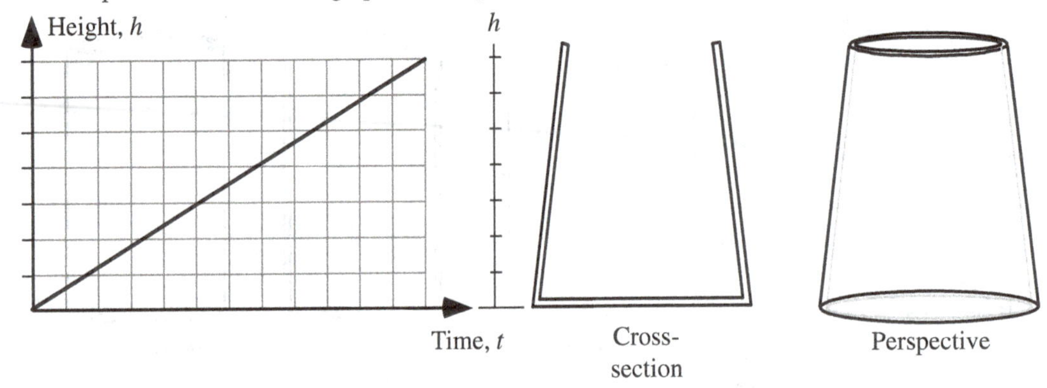

What, if anything, is wrong with this graph? If something is wrong, identify and explain how to correct all errors. If this is correct, explain why.

A1-SCT15: Water Stream—Time to Fill Glass

Three students notice that the water leaving a tap and falling into a sink forms a stream that gets narrower as it gets farther from the tap.

Andre: *"The stream is narrower, and there's less water as you get closer to the sink. That's why if you want to fill a glass quickly you should hold it near the faucet."*

Bela: *"It doesn't matter where you hold the glass, it will fill up in the same amount of time. Water doesn't just disappear once it leaves the faucet—it has to go somewhere."*

Carl: *"Actually, the glass fills up faster if you hold it near the sink, not near the faucet. Near the sink is where the water is flowing the fastest."*

With which, if any, of these students do you agree?

Andre _____ Bela _____ Carl _____ None of them_____

Explain your reasoning.

A1-WWT16: FILLING COMPLEX FLASK AT CONSTANT RATE—HEIGHT OF WATER-TIME GRAPH

A flask has the complicated shape shown. The flask is filled using a hose with a constant flow rate of 4 ml per second.

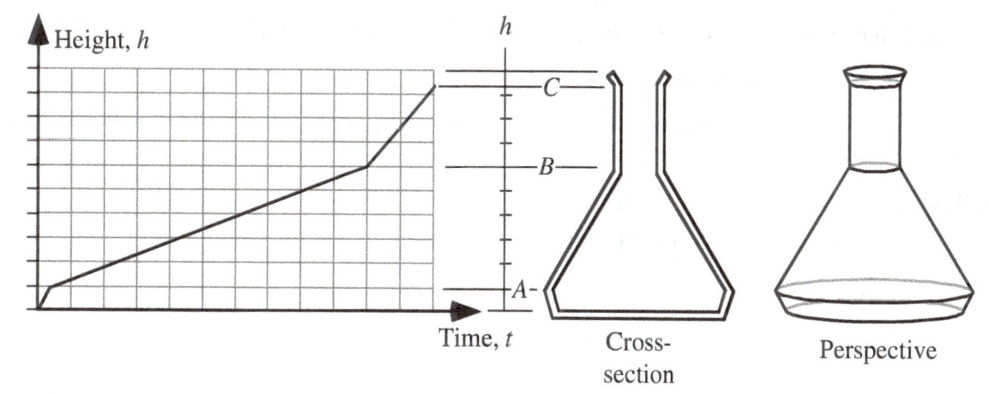

A student graphs the height of the liquid in the flask as a function of time as shown above and makes the following statement:

> *"At first the flask is getting wider, so the graph increases quickly, then it gets narrower, so the height doesn't increase as quickly. Then, when the water reaches the neck, the flask stays a constant width, so the height increases at a constant rate."*

What, if anything, is wrong with this graph? If something is wrong, identify and explain how to correct all errors. If this is correct, explain why.

A1-WWT17: CARS AND TRUCKS IN PARKING LOT—STATEMENT

A student is told that the equation $3y = x$ represents the statement:

"There are three times as many cars as pickup trucks in the parking lot." She says, "The letter x represents the cars, and the letter y represents the pickups."

What, if anything, is wrong? If something is wrong, identify and explain how to correct all errors. If this is correct, explain why.

A1-SCT18: Inches and Feet—Equation

To express the relationship between inches and feet, someone writes "$12I = 1F$." Three students discussing this relation state:

Amy: *"The letter I is used as a unit in this case. There aren't any variables in this equation."*

Bea: *"The letter I is a variable and represents the number of inches."*

Cari: *"That can't be right, because if I let I equal 12 inches, then I get 144 inches equals 1F. And there are only 12 inches in a foot."*

With which, if any, of these students do you agree?

Amy _____ Bea _____ Cari _____ None of them_____

Explain your reasoning.

A1-WWT19: Boys and Girls on Dance Floor—Equation

A student is asked to represent the statement *"For every 5 girls on the dance floor there are 3 boys"* using G to represent the number of girls and B to represent the number of boys. He writes $5G = 3B$.

What, if anything, is wrong? If something is wrong, identify and explain how to correct all errors. If this is correct, explain why.

A1-WWT20: Students and Teachers—Equation

A student is asked to represent the statement *"There are 42 more students than teachers in the classroom"* using S for the number of students and T for the number of teachers. She writes $S = T + 42$.

What, if anything, is wrong? If something is wrong, identify and explain how to correct all errors. If this is correct, explain why.

A1-WWT21: TEXTING AND COST—STATEMENT

A student is told that the equation $7y = 4x$ represents the statement *"For every 7 hours of texting I get charged 4 dollars."* She says, *"The letter* x *represents the number of hours of texting, and the letter* y *represents the number of dollars I am charged."*

What, if anything, is wrong? If something is wrong, identify and explain how to correct all errors. If this is correct, explain why.

A1-WWT22: LINE DATA GRAPH—INTERPRETATION

A student makes the following claim about some data that he and his lab partners have collected:

"Our data show that the value of y *decreases as* x *increases. We found that* y *is inversely proportional to* x.*"*

What, if anything, is wrong with this statement? If something is wrong, identify and explain how to correct all errors. If this statement is correct, explain why.

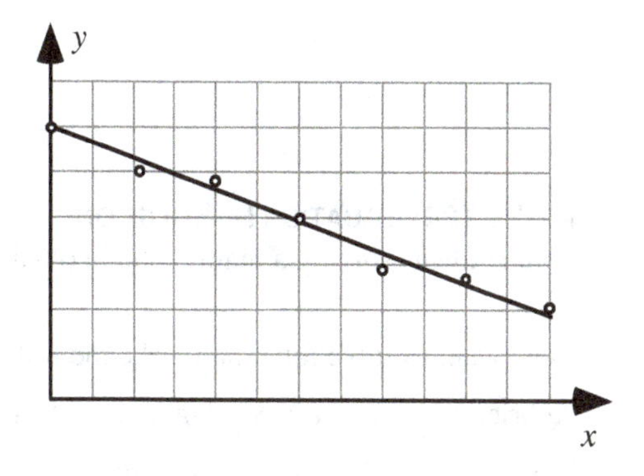

A1-QRT23: Statement about Y-X Graphs—Doubling Graph

Six y- versus x- graphs are shown on a single set of axes.

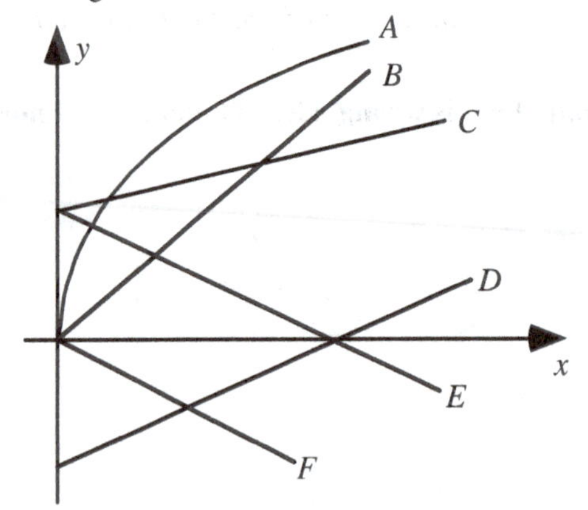

Which, if any, of these graphs is consistent with the statement *"If x doubles, then y also doubles?"* _____

Explain your reasoning.

A1-SCT24: Equation I—Solution

Four students solving a math problem obtain the following equation:

$$4abx = cax + 2b$$

They want to solve for x, and make the following statements about what to do next:

Aubrie: *"There's nothing we can do until we plug in numbers for a, b, and c."*

Bayan: *"We need to get all the* x *terms on one side, so we should subtract* cax *from both sides."*

Cherise: *"I agree, but first we can simplify. There's an* "a" *on both sides, and we can cancel them."*

Didier: *"I agree that we need to get all the* x *terms on one side, but to do that we should divide by* cax.*"*

With which, if any, of these students do you agree?

Aubrie _____ Bayan _____ Cherise _____ Didier _____ None of them_____

Explain your reasoning.

A1-SCT25: EQUATION II—UNITS ANALYSIS

Four students solving a math problem obtain the following equation:

$$4abx = cax + 2b$$

They are not sure the equation is reasonable, and make the following statements about the units:

Alan: *"I don't think this can work because the different variables should have different units, which means the equation would be inconsistent."*

Bri: *"I disagree; I think it could work if* a *and* x *are something like* m *and* 1/m.*"*

Chas: *"I agree with Bri, but only if* c *has the same units as* b *in addition to what Bri said."*

With which, if any, of these students do you agree?

Alan _____ Bri_____ Chas _____ None of them_____

Explain your reasoning.

A1-WWT26: INVERSE QUANTITIES—STATEMENT

A student is solving a problem using an equation that includes the variables *x*, *y*, and *z*. She says,

"We need to simplify the equation $\frac{1}{x} = \frac{1}{y} + \frac{1}{z}$ *and then solve for* z*. First we invert, which gives* $x = y + z$*. Then we solve for* z *by subtracting* y *from both sides, and we get* $z = x - y$*."*

What, if anything, is wrong? If something is wrong, identify and explain how to correct all errors. If this is correct, explain why.

A1-QRT27: Statement—Doubling Equation

A. $y = 2x$

B. $y = 3x$

C. $y = 2x + 7$

D. $y = -4x$

E. $y = x^2$

Which, if any, of these equations is consistent with the statement *"If* x *doubles, then* y *also doubles?"*

Explain your reasoning.

A1-RT28: Lemonade from Concentrate—Flavor Strength

Four students are mixing lemonade using a lemonade concentrate and water. They all have different recipes. In the diagrams, the darker cans represent lemonade concentrate and the lighter cans represent water.

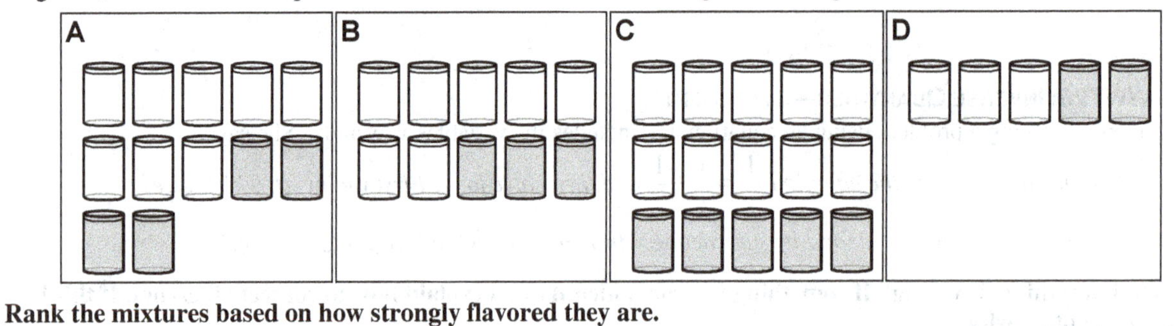

Rank the mixtures based on how strongly flavored they are.

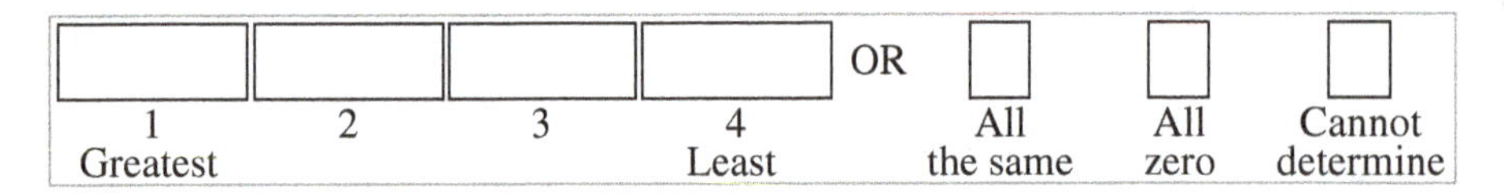

Explain your reasoning.

A1-RT29: FOUR BASKETBALL PLAYERS—FREE-THROW SKILLS

Four basketball players have the following statistics for free-throws:

	Name	Baskets Made	Baskets Missed
A	Aliza	13	6
B	Berta	14	3
C	Claudia	7	3
D	Diana	6	3

Based only on this small amount of data, rank the free-throw skills of the players.

☐	☐	☐	☐	OR	☐	☐	☐
1 Greatest	2	3	4 Least		All the same	All zero	Cannot determine

Explain your reasoning.

A1-QRT30: FOUR BASKETBALL PLAYERS—MAKING TEAMS

Four basketball players have the following statistics for free-throws:

	Name	Baskets Made	Baskets Missed
A	Amalie	43	12
B	Beth	77	18
C	Cami	61	19
D	Diethra	58	11

(a) Based only on their free-throw statistics, choose teams for a 2-on-2 basketball game that is as evenly matched as possible.

Explain your reasoning.

(b) For the teams you have chosen (and again based only on their free-throw statistics), which team is likely to win?

Explain your reasoning.

A1-SCT31: Large and Small Picture—Size

Picture B at the right was made by enlarging picture A on a photocopying machine. The distance between point *x* and point *y* measures 1.4 times as large in picture B as in picture A, and the distance between point *x* and point *z* also measures 1.4 times as large in picture B.

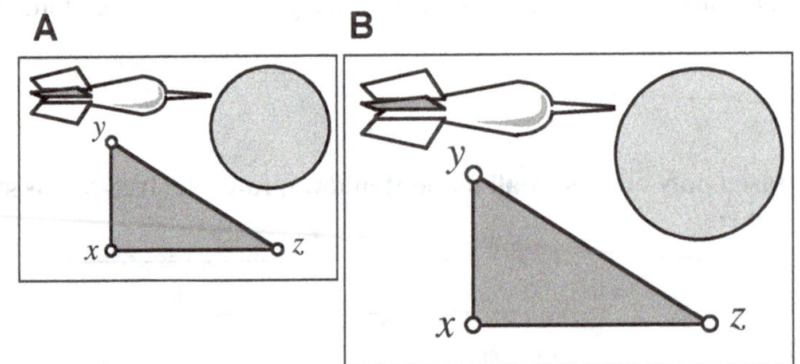

Consider the following claims made by students about the two pictures:

Andres: *"The distance from* y *to* z *is also 1.4 times as large in picture B as in picture A. Picture B is 1.4 times as large in any direction."*

Blas: *"The area of the triangle* xyz *will be 1.4 times as large in picture B as in picture A. Picture B is 1.4 times as large in any direction."*

Cervita: *"The angle* yzx *is also 1.4 times as large in picture B as in picture A."*

Daniel: *"The circumference of the circle in the upper-right-hand corner of the picture is 1.4 times as large in picture B as in picture A."*

Esther: *"The area of the circle in the upper right hand corner of the picture is 1.4 times as large in picture B as in picture A. Anything you can measure in picture B will be 1.4 times as large as what you measure in A."*

Freddie: *"The length of the dart is 1.4 times as large in picture B as in picture A."*

Genaro: *"The area of the dart is 1.4 times as large in picture B as in picture A."*

With which, if any, of these students do you agree?

Andres ____ Blas ____ Cervita _____ Daniel _____ Esther ____ Freddie _____ Genaro _____ None of them_____

Explain your reasoning.

A1-CT32: SCALE MODEL PLANES—SURFACE AREA AND WEIGHT

A woodworker has made four small airplanes and one large airplane. All airplanes are exactly the same shape, and all are made from the same kind of wood. The larger plane is twice as large in every dimension as one of the smaller planes. The planes are to be painted and then shipped as gifts.

(**a**) The amount of paint required to paint the planes is directly proportional to the surface area. **Will the amount of paint required for the single plane in Case A be (i) *greater than,* (ii) *less than,* or (iii) *equal to* the total amount of paint required for all four planes in Case B? _____**

Explain your reasoning.

(**b**) The shipping cost for the planes is proportional to the weight which is related directly to the volume. **Will the weight of the single plane in Case A be (i) *greater than,* (ii) *less than,* or (iii) *equal to* the total weight of all four planes in Case B? _____**

Explain your reasoning.

A1-WWT33: FIVE KILOGRAMS OF PENNIES—VALUE OF FIVE KILOGRAMS OF NICKELS

A student is told that 5 kg of pennies have a value of $20. The student has to find the value of the same mass of nickels, knowing that a nickel has twice the mass of a penny. The student carries out the following calculation:

> *"$20 is 2000 pennies, and they have a mass of 5 kg. If I have the same mass of nickels that are each twice the mass of a penny, then I will only have 1,000 nickels. That means I would have $50."*

What, if anything, is wrong with this student's analysis? If something is wrong, identify it and explain how to correct it. If nothing is wrong, complete the explanation.

A1-SCT34: Coal Mine Rescue Shaft—Time

Three students are working on the following problem:

> *"If it takes 70 hours to dig a coal mine rescue shaft 300 feet deep, how long should it take to dig another coal mine rescue shaft 1,500 feet deep?"*

The students make the following contentions:

Ally: *"Since it took 70 hours to dig 300 feet it will take five times as long to dig five times as deep, so it will take 350 hours."*

Bill: *"It takes them a little less than a quarter of an hour to dig one foot, so I get 349.5 hours, which almost agrees with Ally, but I am not sure who has the right answer."*

Clyde: *"The workers were able to dig at a rate of 4.29 feet per hour for the first shaft, so if they can do the same for the second it will take 349.7 hours. So I think we all agree and the numbers only differ because of rounding errors."*

With which, if any, of these students do you agree?

Ally _____ Bill _____ Clyde _____ None of them _____

Explain.

A1-QRT35: Pennies—Number of Pennies

Since 2,000 pennies have a mass of *n* kg, how many pennies would be needed to produce a mass of 1.6*n* kg?

Explain.

A1-TT36: Six Page Physics Test—Weight

A certain physics test, which is several pages long, is copied on 20-pound paper (500 sheet have a weight of 20 pounds) so the weight per test is 0.25 pounds. The class has 39 students, so the instructor makes 40 copies of this test. What is the weight of the paper that he brings to class on test day?

A student presented with this problem carries out the following analysis:

> *"To find the total weight of the paper in the tests I need to divide the 40 copies by the 0.25 test/pound, and that will give me 16 pounds."*

The student's analysis is wrong; identify the problem and explain how to correct it.

A2 VECTORS

A2-QRT01: VECTORS ON A GRID I—MAGNITUDES

Eight vectors are shown superimposed on a grid.

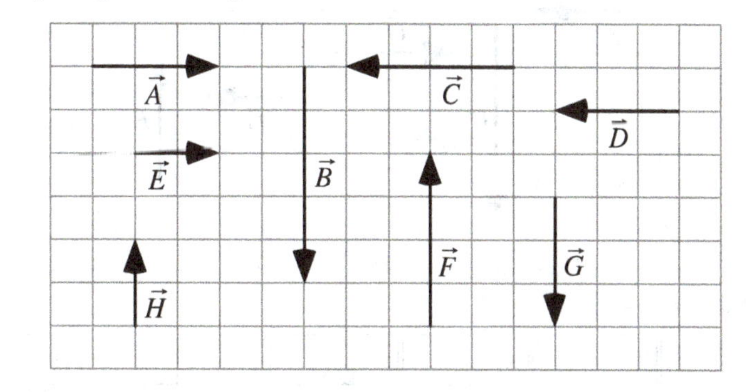

(a) List all of the vectors that have the same magnitude as vector $\vec{A}$.

(b) List all of the vectors that have the same magnitude as vector $\vec{B}$.

(c) List all of the vectors that have the same magnitude as vector $\vec{C}$.

(d) List all of the vectors that have the same magnitude as vector $-\vec{A}$.

(e) List all of the vectors that have the same magnitude as vector $-\vec{B}$.

(f) List all of the vectors that have the same magnitude as vector $-\vec{D}$.

A2-RT02: Vectors on a Grid II—Magnitudes

Five vectors are shown superimposed on a grid.

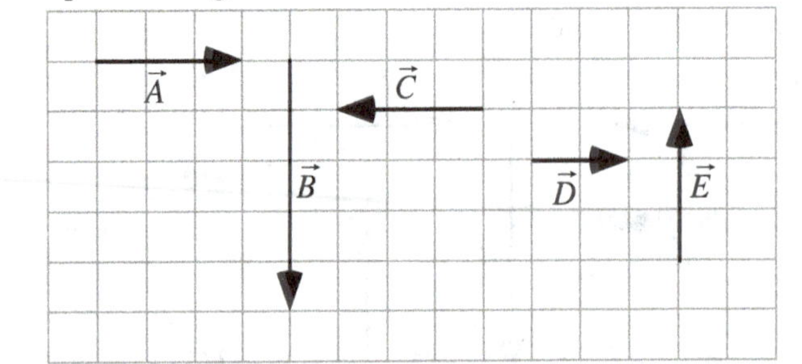

Rank the magnitudes of the vectors.

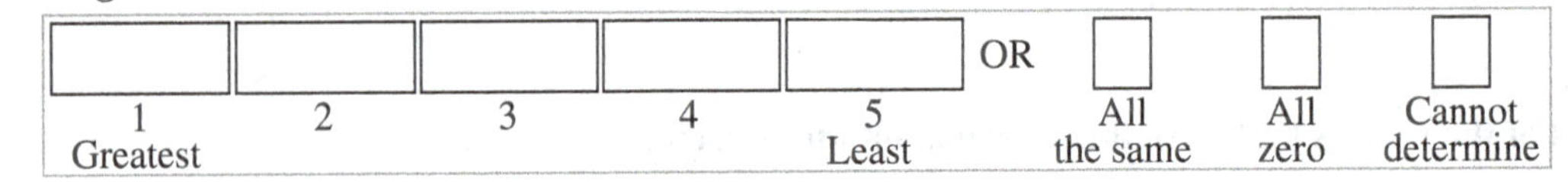

Explain your reasoning.

A2-QRT03: Vectors on a Grid III—Directions

Nine vectors are shown superimposed on a grid.

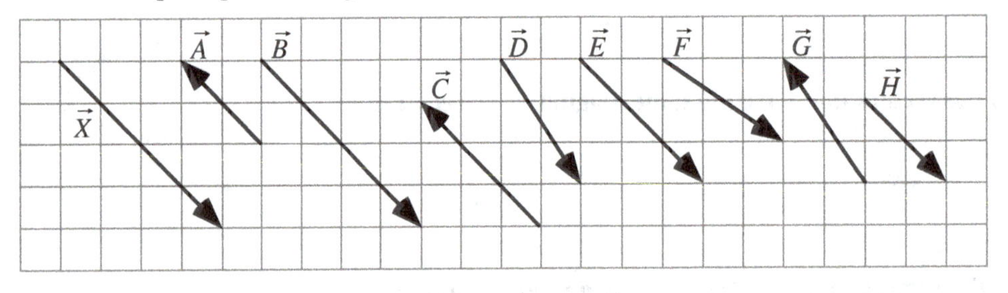

(a) List all of the vectors that have the same direction as vector $\vec{X}$.

(b) List all of the vectors that have the same direction as the vector $-\vec{X}$.

A2-QRT04: VECTOR GRAPHICAL ADDITION AND SUBTRACTION I—EXPRESSION

Three vectors, labeled $\vec{P}$, $\vec{Q}$, and $\vec{R}$, are shown below.

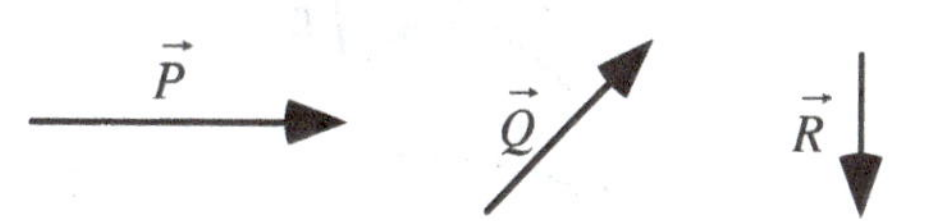

Write an expression using the vectors $\vec{P}$, $\vec{Q}$, and $\vec{R}$ for the resultant vectors shown.

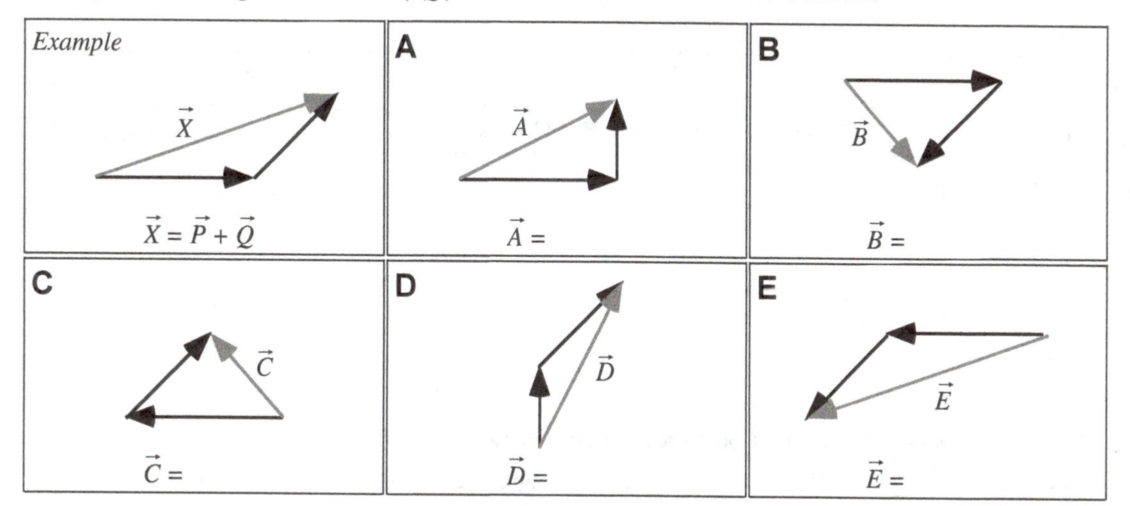

A2-QRT05: VECTOR GRAPHICAL ADDITION AND SUBTRACTION II—EXPRESSION

Three vectors, labeled $\vec{P}$, $\vec{Q}$, and $\vec{R}$, are shown below.

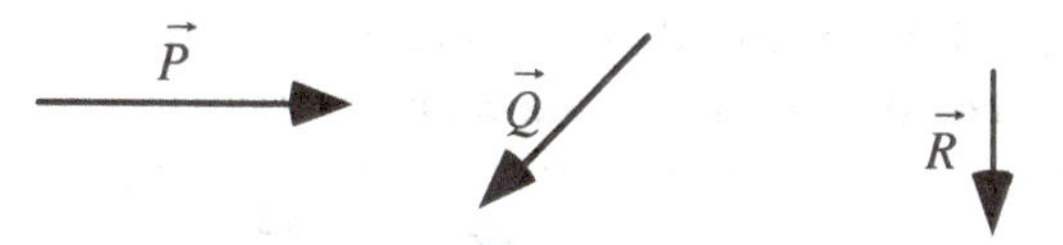

Write an expression using the vectors $\vec{P}$, $\vec{Q}$, and $\vec{R}$ for the resultant vectors shown.

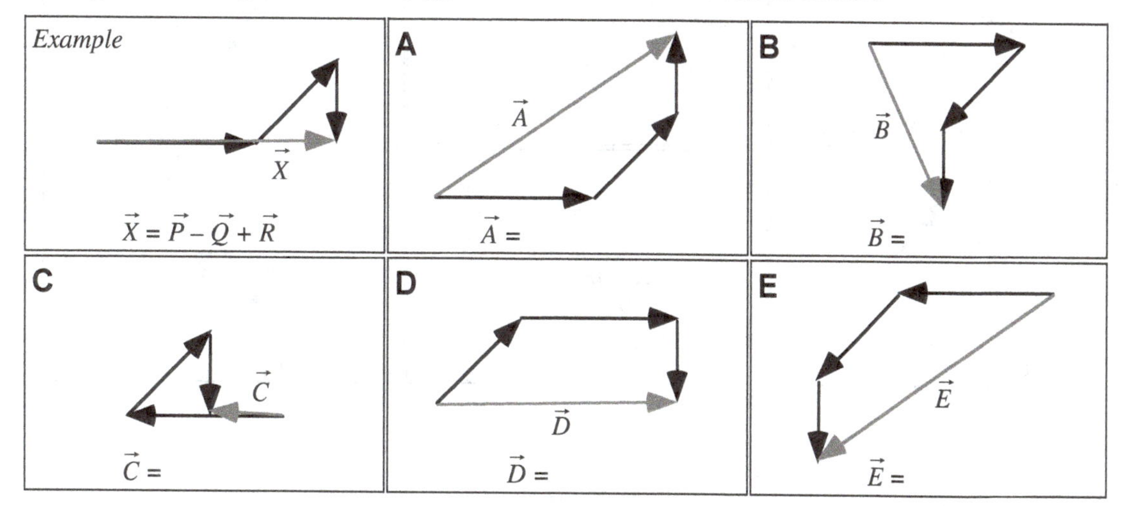

A2-QRT06: Vector Expression of Graphical Relationship—Equation

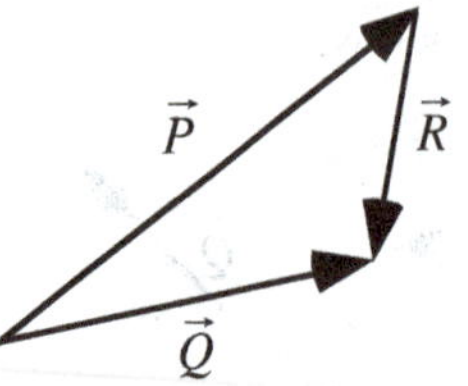

Which of the following vector equations correctly describes the relationship among the vectors shown in the figure?

A	B	C	D	E
$\vec{P} + \vec{Q} = \vec{R}$	$\vec{P} = \vec{Q} + \vec{R}$	$\vec{P} + \vec{R} = \vec{Q}$	$\vec{P} + \vec{Q} + \vec{R} = 0$	None of equations $A - D$ is correct.

Explain your reasoning.

A2-WBT07: Resultant—Graphical Addition or Subtraction

Three vectors, labeled $\vec{P}$, $\vec{Q}$, and $\vec{R}$, are shown below. The magnitude of each vector is given in arbitrary units.

$\vec{P}$ Length = 4 $\vec{Q}$ Length = 3 $\vec{R}$ Length = 2

Construct the vectors shown in cases A to E below by adding or subtracting two of the vectors $\vec{P}$, $\vec{Q}$, and $\vec{R}$ as shown in the example. Write a vector equation representing each process.

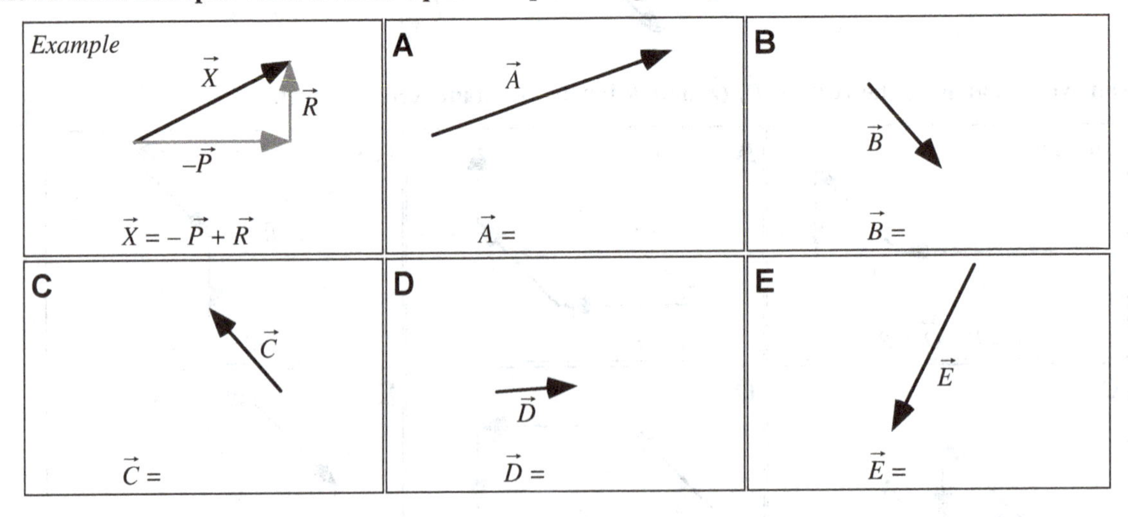

A2-SCT08: ADDING TWO VECTORS—MAGNITUDE OF THE RESULTANT

Three students are discussing the magnitude of the resultant of the addition of the vectors $\vec{A}$ and $\vec{B}$. Vector $\vec{A}$ has a magnitude of 5 cm, and vector $\vec{B}$ has a magnitude of 3 cm.

Alexis: *"We'd have to know the directions of the vectors to know how big the resultant is going to be."*

Bert: *"Since we are only asked about the magnitude, we don't have to worry about the directions. The magnitude is just the size, so to find the magnitude of the resultant we just have to add the sizes of the vectors. The magnitude of the resultant in this case is 8 cm."*

Cara: *"No, these are vectors, and to find the magnitude you have to use the Pythagorean theorem. In this case the magnitude is the square root of 34, a little less than 6 cm."*

Dacia: *"The resultant is the vector that you have to add to the first vector to get the second vector. In this case the resultant is 2."*

With which, if any, of these students do you agree?

Alexis _____ Bert _____ Cara _____ Dacia_____ None of them_____

Explain your reasoning.

A2-QRT09: VECTORS ON A GRID IV—GRAPHICAL REPRESENTATION OF SUM

Shown at left below are vectors labeled $\vec{K}$, $\vec{L}$, $\vec{M}$, and $\vec{N}$, with lengths in arbitrary units, superimposed on a grid.

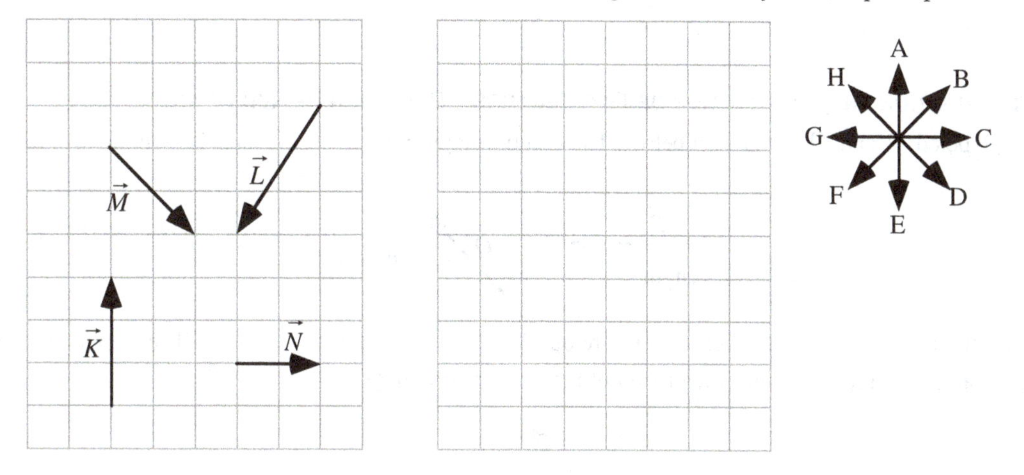

On the grid on the right, construct a graphical representation of $\vec{J} = \vec{K} + \vec{L} + \vec{M} + \vec{N}$ with labels for each vector, and indicate the direction of $\vec{J}$: __________ (closest to one of the directions listed in the direction rosette above).

A2-RT10: Vectors on Grid V—Components of Vectors

Shown below are vectors superimposed on a grid.

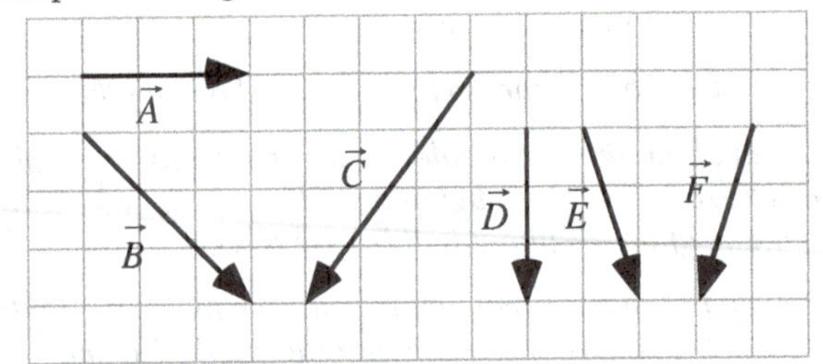

(a) Rank the magnitudes of the *x*-components of each vector.

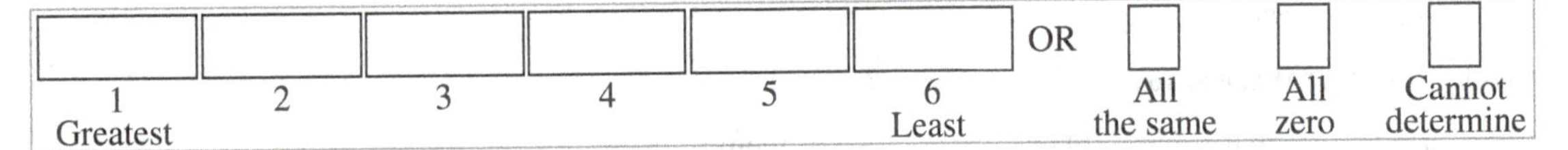

Explain your reasoning.

(b) Rank the magnitudes of the *y*-components of each vector.

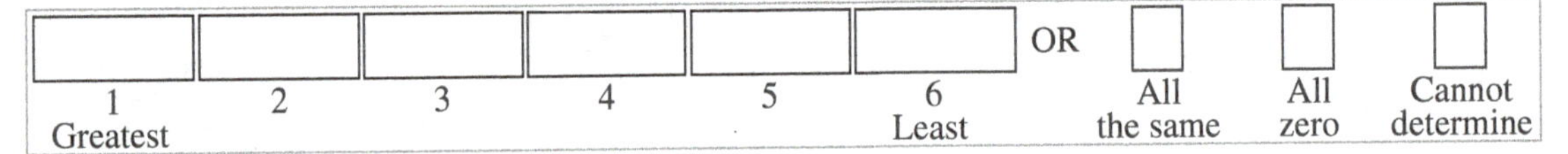

Explain your reasoning.

A2-WWT11: Addition and Subtraction of Two Vectors—Direction of Resultant

Two vectors, labeled $\vec{P}$ and $\vec{Q}$, are shown below. The length (magnitude) of each vector is given in arbitrary units.

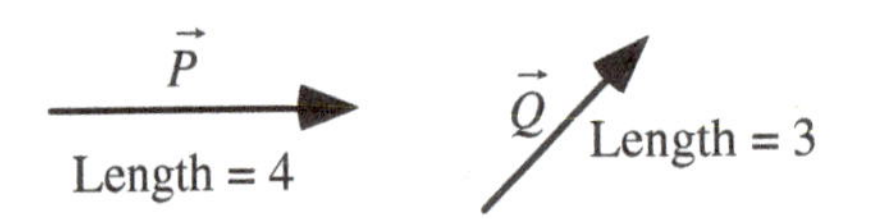

A student constructs the figure shown below to figure out the sum of the vectors $\vec{P}$ and $\vec{Q}$. The student contends that the lighter arrow $\vec{A}$ represents the vector sum of the vectors $\vec{P}$ and $\vec{Q}$.

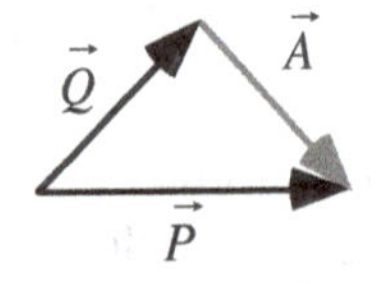

What, if anything, is wrong with the student's work? If something is wrong, identify it, and explain how to correct it. If his/her work is correct, explain why.

A2-RT12: VECTORS ON GRID VI—RESULTANT MAGNITUDES OF ADDING TWO VECTORS

Six vectors are shown superimposed on a grid.

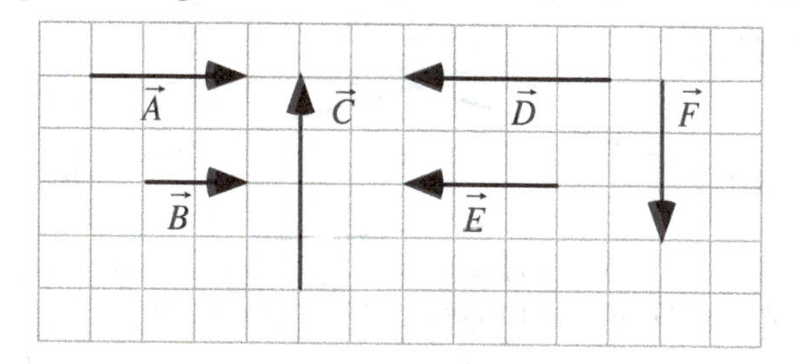

Rank the magnitude of the vector resulting from adding vector $\vec{A}$ to each vector ($\vec{A}+\vec{A}$, $\vec{B}+\vec{A}$, $\vec{C}+\vec{A}$, etc.).

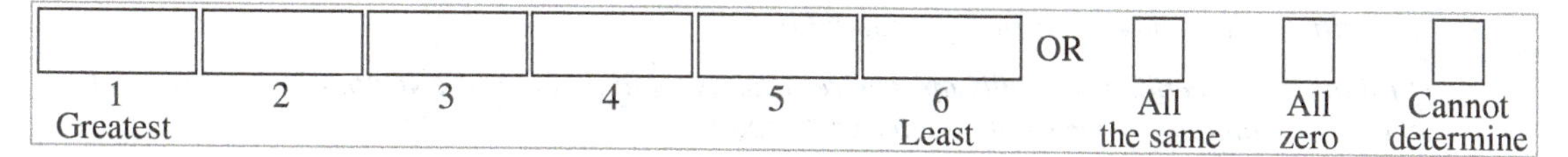

Explain your reasoning.

A2-RT13: VECTORS ON GRID VII—RESULTANT MAGNITUDES OF ADDING TWO VECTORS

Six vectors are shown superimposed on a grid.

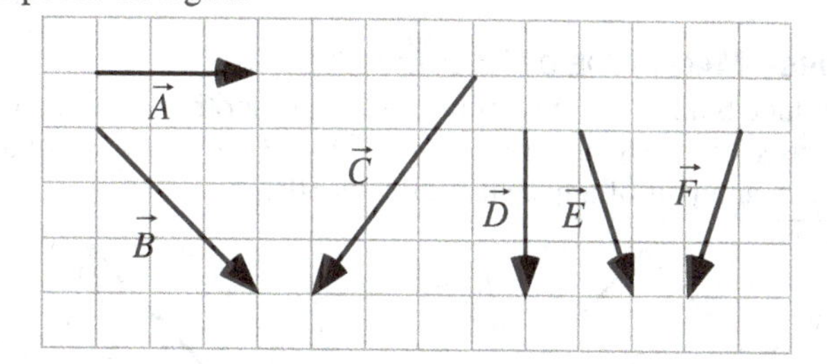

Rank the magnitude of the vector resulting from adding vector $\vec{A}$ to each vector ($\vec{A}+\vec{A}$, $\vec{B}+\vec{A}$, $\vec{C}+\vec{A}$, etc.).

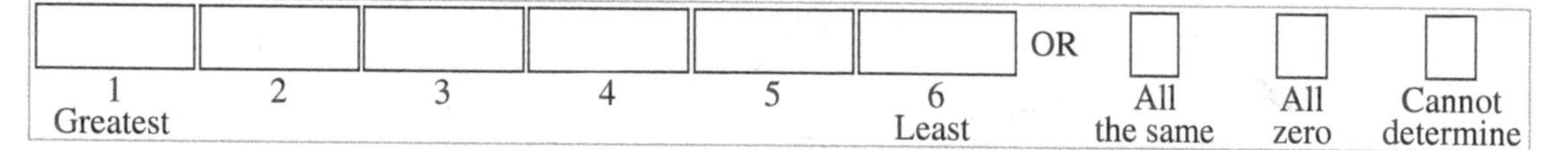

Explain your reasoning.

A2-SCT14: Combining Two Vectors—Resultant

Two vectors each have a magnitude of 6 units, and each makes a small angle α with the horizontal as shown. Four students are discussing the resultant vector obtained by adding these two vectors.

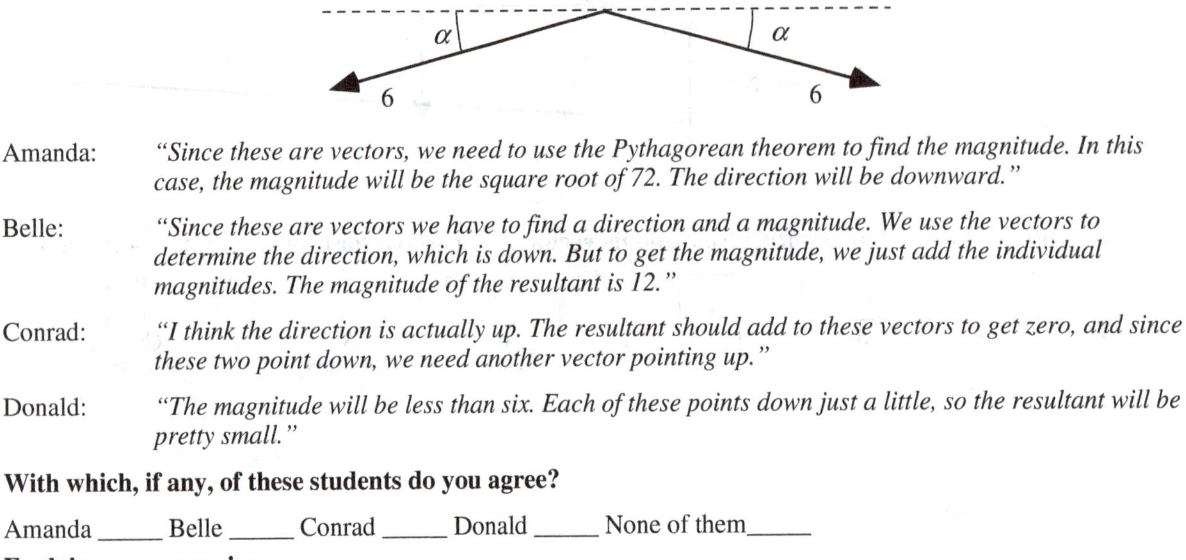

Amanda: *"Since these are vectors, we need to use the Pythagorean theorem to find the magnitude. In this case, the magnitude will be the square root of 72. The direction will be downward."*

Belle: *"Since these are vectors we have to find a direction and a magnitude. We use the vectors to determine the direction, which is down. But to get the magnitude, we just add the individual magnitudes. The magnitude of the resultant is 12."*

Conrad: *"I think the direction is actually up. The resultant should add to these vectors to get zero, and since these two point down, we need another vector pointing up."*

Donald: *"The magnitude will be less than six. Each of these points down just a little, so the resultant will be pretty small."*

With which, if any, of these students do you agree?

Amanda _____ Belle _____ Conrad _____ Donald _____ None of them_____

Explain your reasoning.

A2-CT15: Combining Vectors—Magnitude of Resultant

In Case A, two vectors of magnitude 6 units are at right angles to one another. In Case B, four vectors, each of magnitude 3 units, are arranged as shown. The outer vectors in Case B are also at right angles to one another, and the difference in direction between any pair of adjacent vectors is 30°.

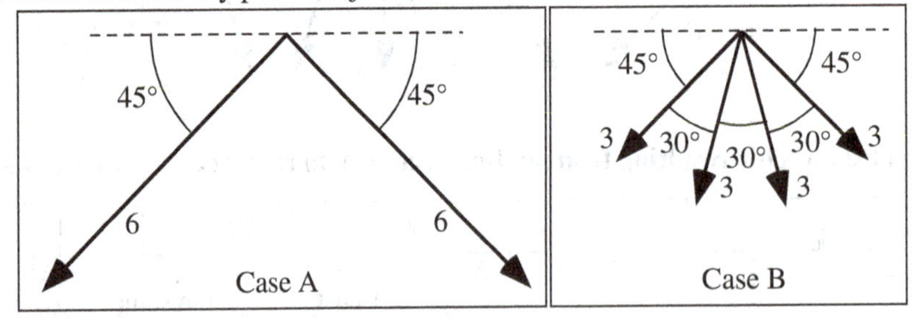

If all vectors in each case are added together, is the magnitude of the resultant in Case A (i) *greater than,* (ii) *less than,* or (iii) *equal to* the magnitude of the resultant in Case B? _____

Explain your reasoning.

A2-QRT16: VECTOR COMBINATION I—DIRECTION OF RESULTANT

For each situation below, combine the vectors as indicated and determine the direction of the resultant vector. Then select the closest direction to the resultant from the direction rosette at the right.

A, B, C, D, E, F, G, H

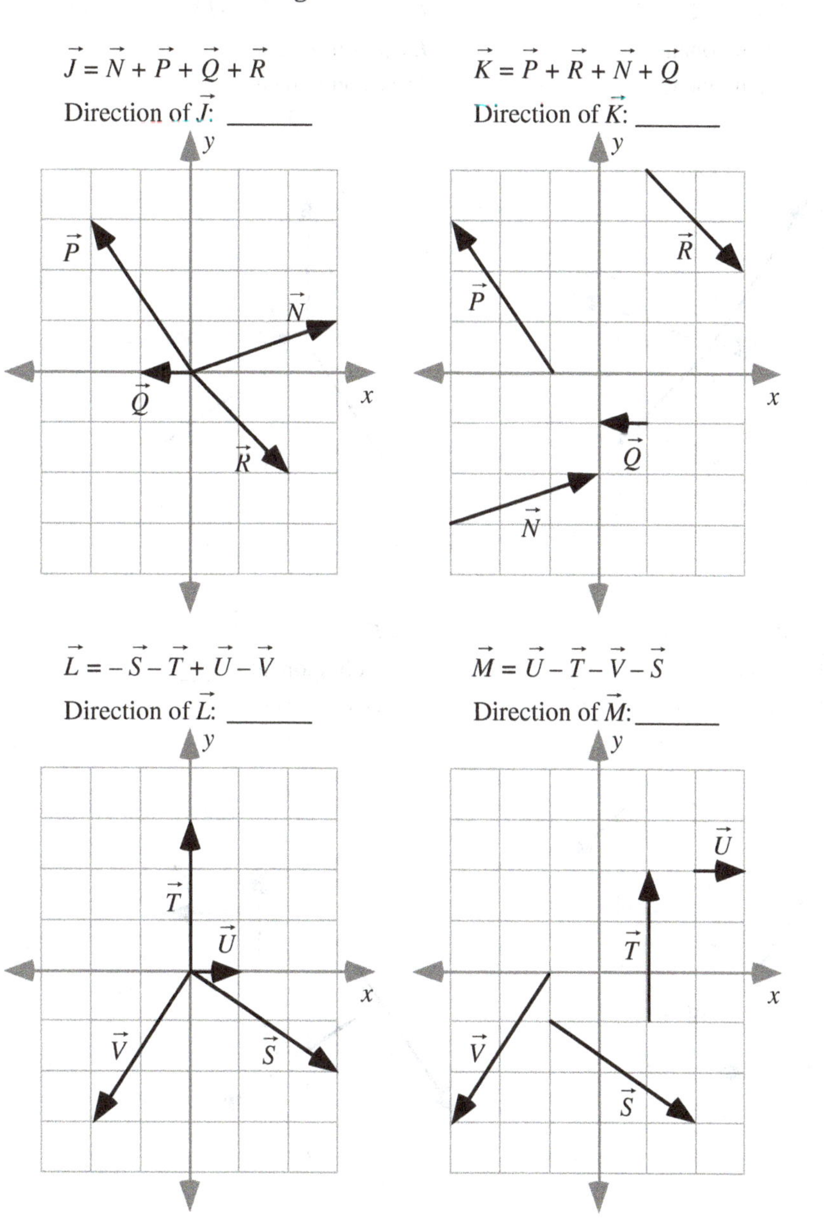

A2-QRT17: Vector Combinations II—Components of the Resultant Vector

For each situation below, determine the components of the resultant vectors.

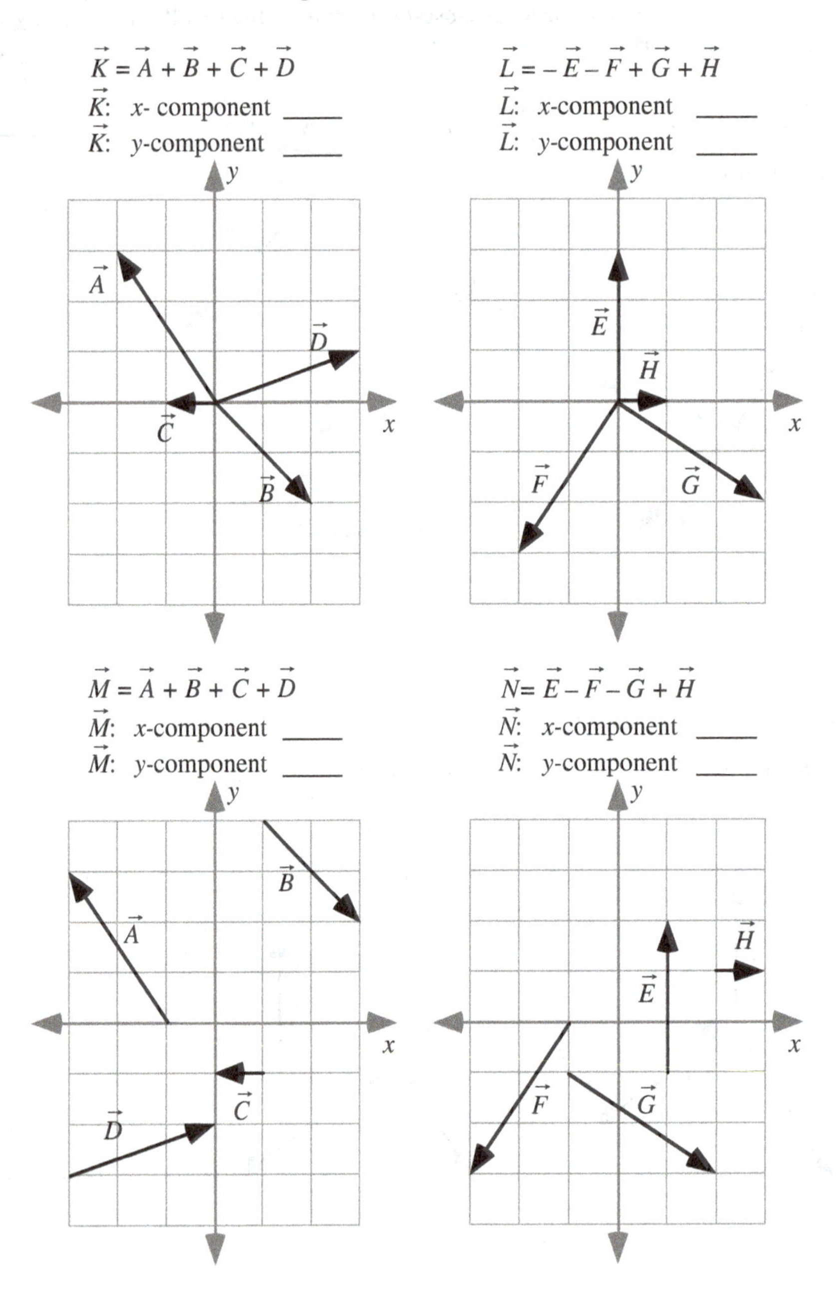

A2-QRT18: FORCE VECTORS—PROPERTIES OF COMPONENTS

Shown below are vector diagrams representing two sets of forces.

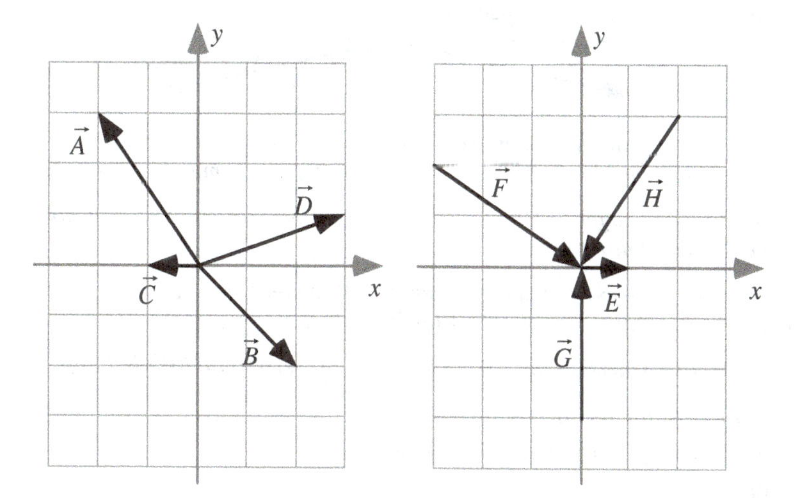

(a) List all the forces that have a zero x-component:

(b) List all the forces that have a zero y-component:

(c) List all the forces that have an x-component pointing in the positive x-direction:

(d) List all the forces that have a y-component pointing in the negative y-direction:

A2-QRT19: VELOCITY VECTORS—PROPERTIES OF COMPONENTS

Shown below are vector diagrams representing velocities.

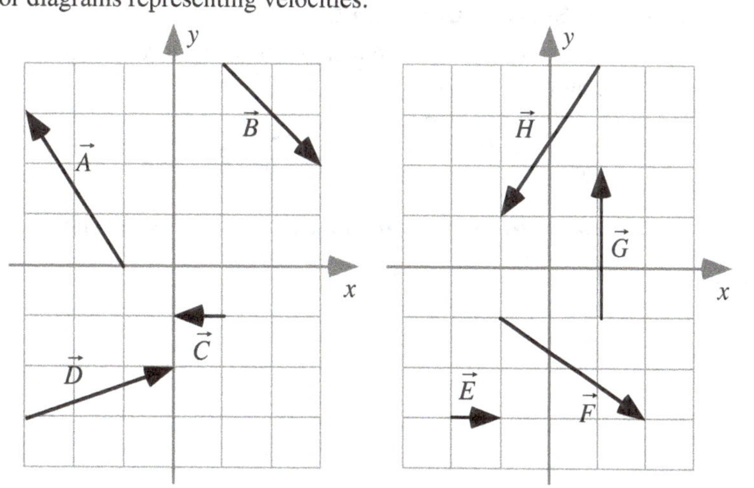

(a) List all the velocities that have a zero x-component:

(b) List all the velocities that have a zero y-component:

(c) List all the velocities that have an x-component pointing in the positive x-direction:

(d) List all the velocities that have a y-component pointing in the negative y-direction:

A2-SCT20: Two Vectors—Vector Difference

Two vectors labeled $\vec{A}$ and $\vec{B}$, each having a length of 6 meters, make a small angle α with the horizontal as shown. Four students are discussing the vector difference $\vec{C} = \vec{A} - \vec{B}$.

Arlo: *"Since we're subtracting vector B, we flip it around so it points in the same direction as vector A. The difference will be 12 meters long and will point in the same direction as vector A."*

Bob: *"We're subtracting, so the resultant will be smaller than six. Both vectors point down, so the difference will point down as well."*

Celine: *"When you flip vector B around to get negative B, it points up and to the left. Then we add it to vector A, we get a long vector pointing horizontally to the right."*

Delbert: *"Both vectors are 6 meters long, so the difference is zero. It doesn't point in any direction."*

With which, if any, of these students do you agree?

Arlo _____ Bob _____ Celine _____ Delbert _____ None of them_____

Explain your reasoning.

A2-CT21: Two Vectors—Vector Sum and Difference

Two vectors labeled $\vec{A}$ and $\vec{B}$ each have a magnitude of 6 meters, and each makes a small angle α with the horizontal as shown. Let $\vec{C} = \vec{A} + \vec{B}$ and $\vec{D} = \vec{A} - \vec{B}$.

Is the magnitude of $\vec{C}$ (i) *greater than,* (ii) *less than,* or (iii) *equal to* the magnitude of $\vec{D}$? _____

Explain your reasoning.

A2-RT22: ADDITION AND SUBTRACTION OF THREE VECTORS—DIRECTION OF RESULTANT

Three vectors, labeled $\vec{P}$, $\vec{Q}$, and $\vec{R}$, are shown below.

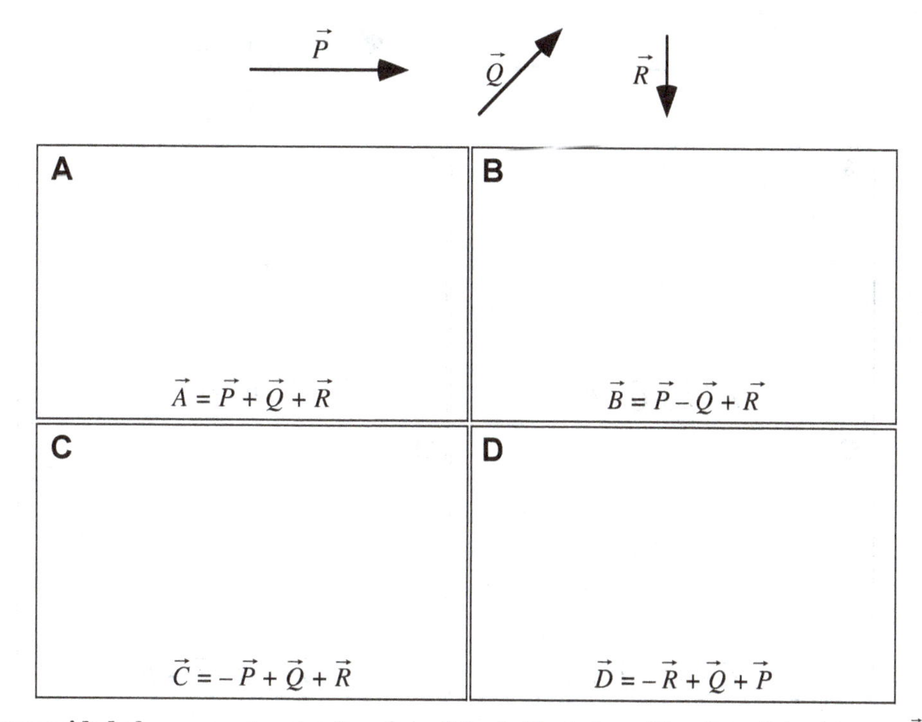

In each space provided above, construct a drawing of the indicated combination of the vectors $\vec{P}$, $\vec{Q}$, and $\vec{R}$, and then rank the magnitude of the angle that the resultant vector makes with the vector $\vec{P}$.

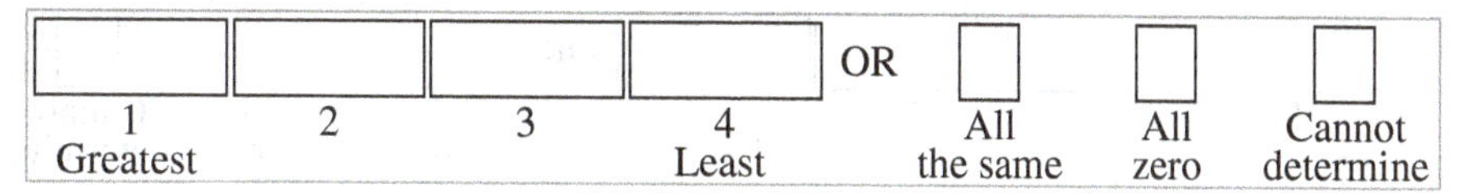

Explain your reasoning.

A2-RT23: Addition and Subtraction of Three Vectors—Magnitude of Resultant

Three vectors, labeled $\vec{P}$, $\vec{Q}$, and $\vec{R}$, are shown. The length of each vector is given in arbitrary units.

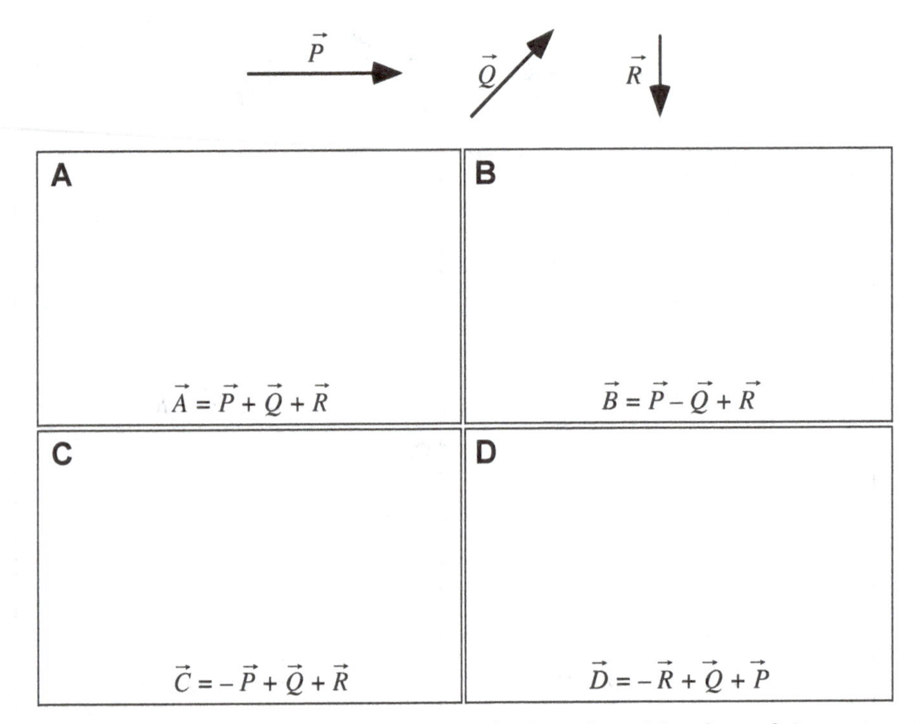

In each space provided above, construct a drawing of the indicated combinations of the vectors $\vec{P}$, $\vec{Q}$, and $\vec{R}$, and then rank the magnitude of these resultant vectors.

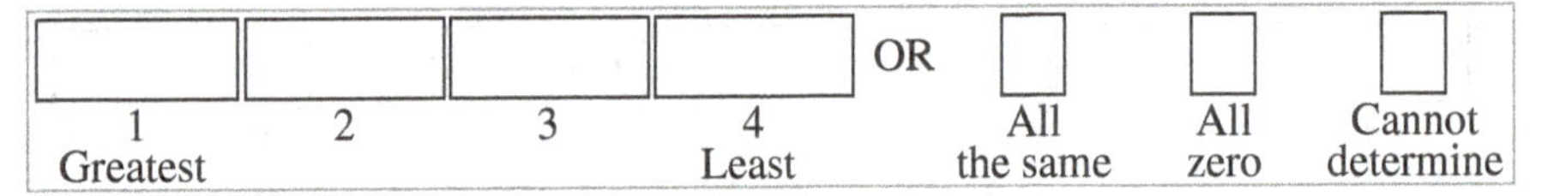

Explain your reasoning.

B1 MOTION IN ONE DIMENSION

B1-RT01: STROBE PHOTOGRAPHS OF SPHERES—DISPLACEMENT I

In each case, a sphere is moving from left to right next to a tape marked in meters. A strobe (flash) photograph is taken every second, and the location of the sphere is recorded. The total time intervals shown are not the same for all spheres.

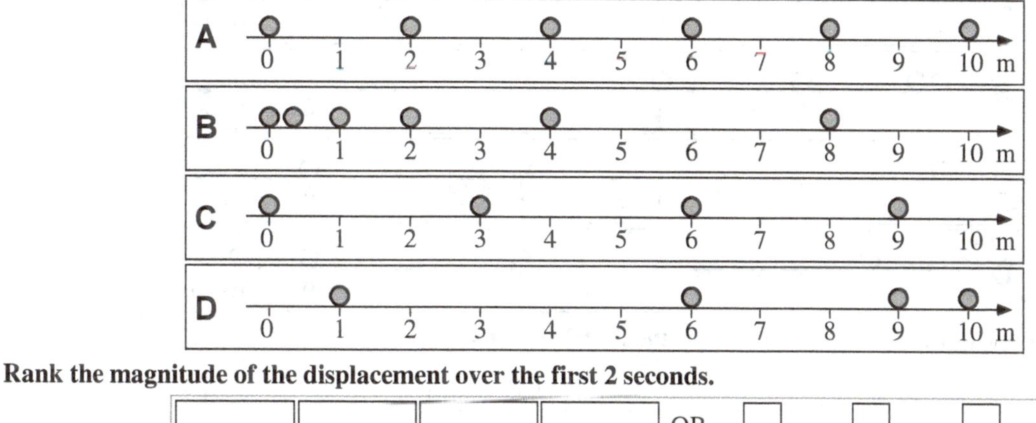

Rank the magnitude of the displacement over the first 2 seconds.

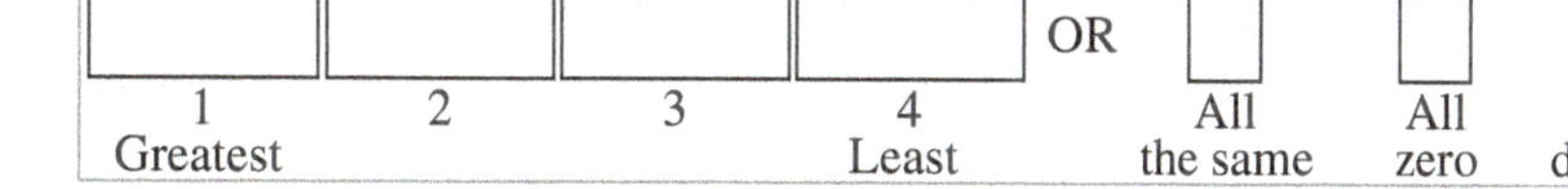

Explain your reasoning.

B1-RT02: STROBE PHOTOGRAPHS OF SPHERES—DISPLACEMENT II

In each case, a sphere is moving from left to right next to a tape marked in meters. A strobe (flash) photograph is taken every second, and the location of the sphere is recorded. The total time intervals shown are not the same for all spheres.

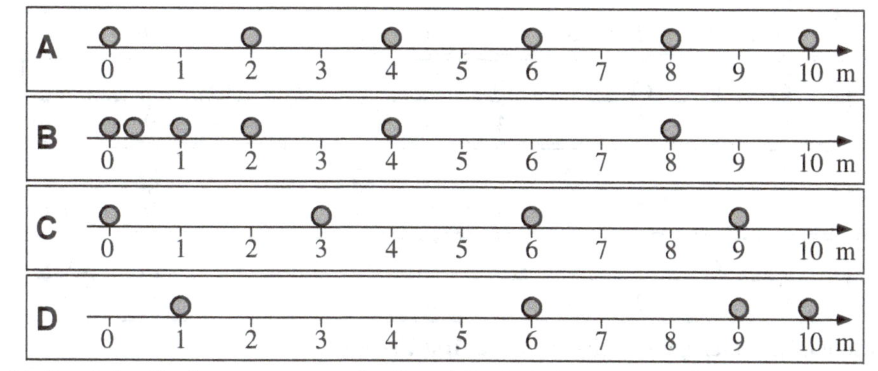

Rank the magnitude of the displacement over the first 3 seconds.

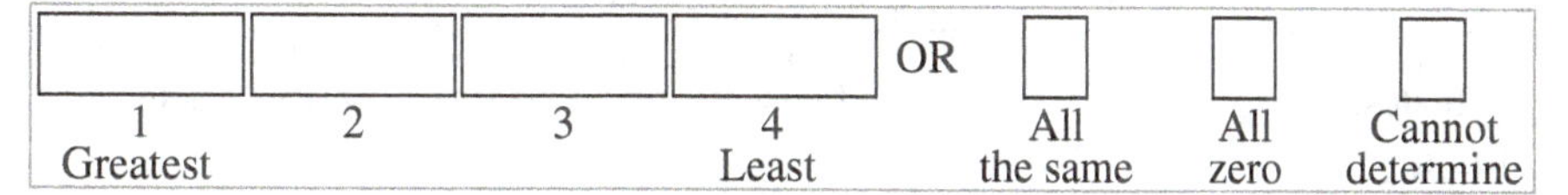

Explain your reasoning.

B1-RT03: Strobe Photographs of Spheres—Average Velocity I

In each case, a sphere is moving from left to right next to a tape marked in meters. A strobe (flash) photograph is taken every second, and the location of the sphere is recorded. The total time intervals shown are not the same for all spheres.

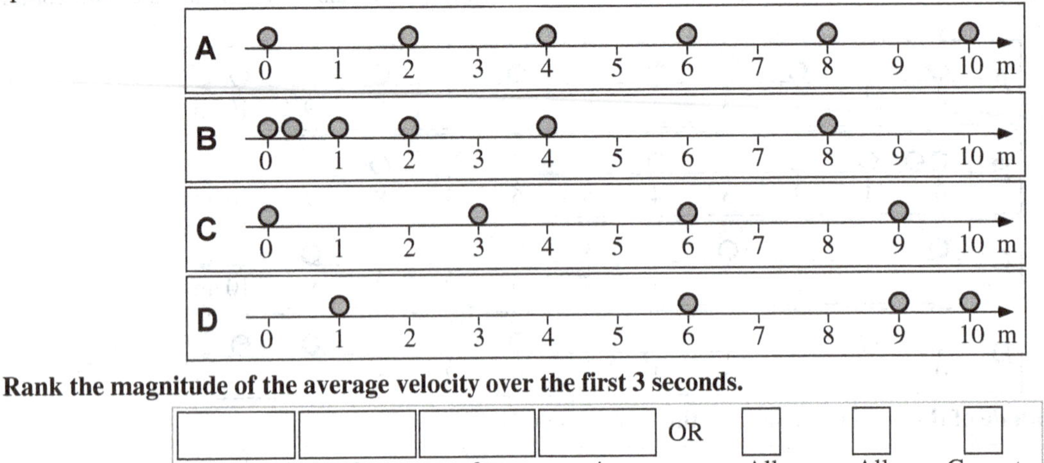

Rank the magnitude of the average velocity over the first 3 seconds.

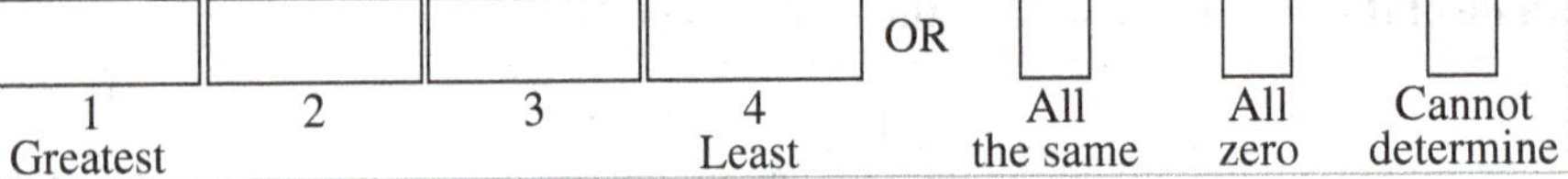

Explain your reasoning.

B1-RT04: Strobe Photographs of Spheres—Average Velocity II

In each case, a sphere is moving from left to right next to a tape marked in meters. A strobe (flash) photograph is taken every second, and the location of the sphere is recorded. The total time intervals shown are not the same for all spheres.

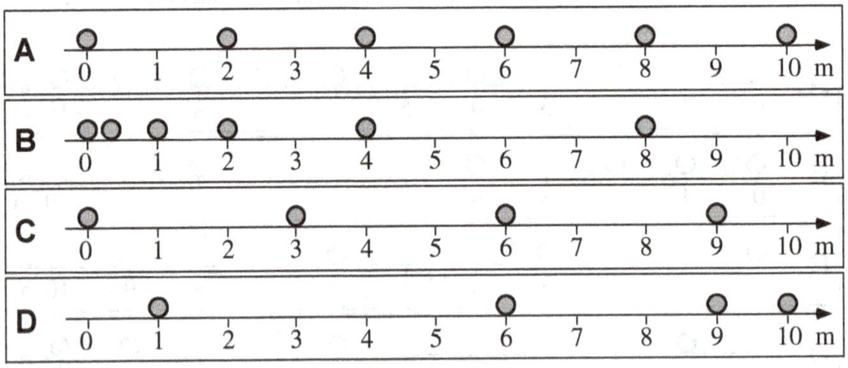

Rank the magnitude of the average velocity over the first 2 seconds.

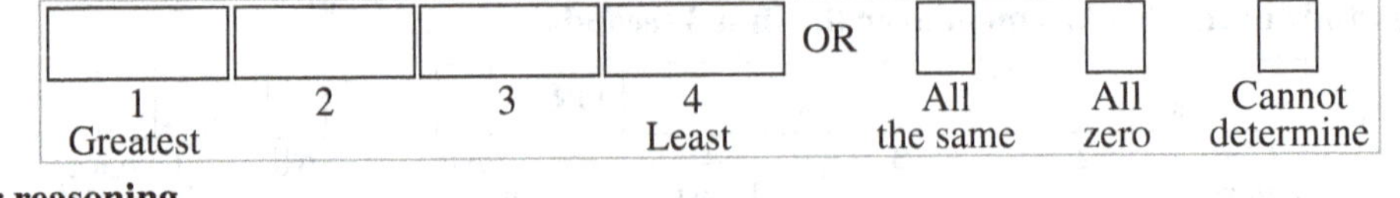

Explain your reasoning.

B1-RT05: Ball Strobe Diagrams—Average Velocity

The following drawings represent strobe (flash) photographs of a ball moving in the direction of the arrow. The circles represent the positions of the ball at succeeding instants of time. The time interval between successive positions is the same in all cases.

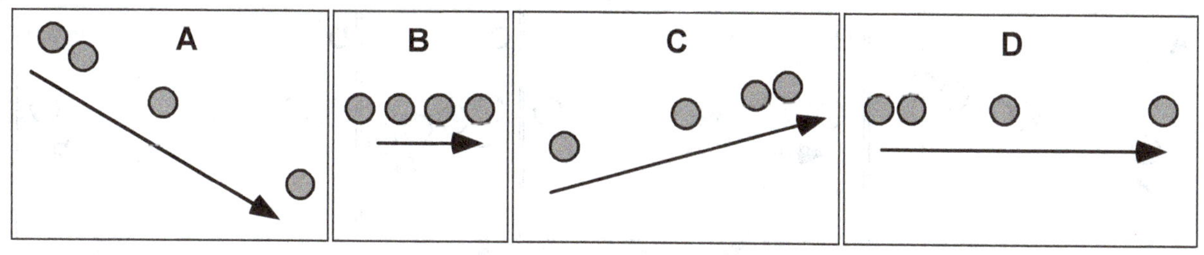

Rank the magnitude of the ball's average velocity in the last time interval.

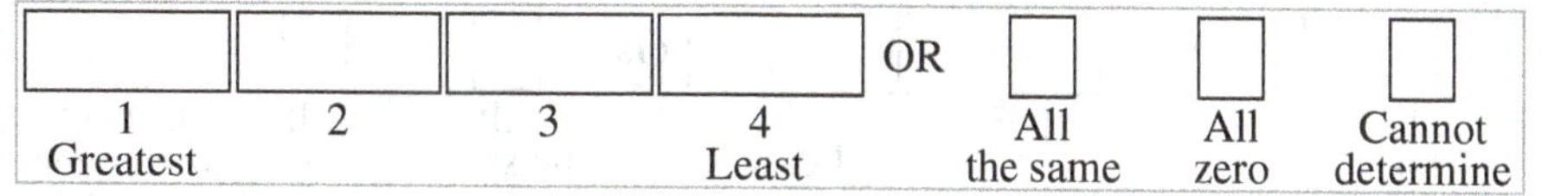

Explain your reasoning.

B1-RT06: Ball Strobe Diagrams—Acceleration

The following drawings represent strobe (flash) photographs of a ball moving in the direction of the arrow. The circles represent the positions of the ball at succeeding instants of time. The time interval between successive positions is the same in all cases. Assume all accelerations are constant.

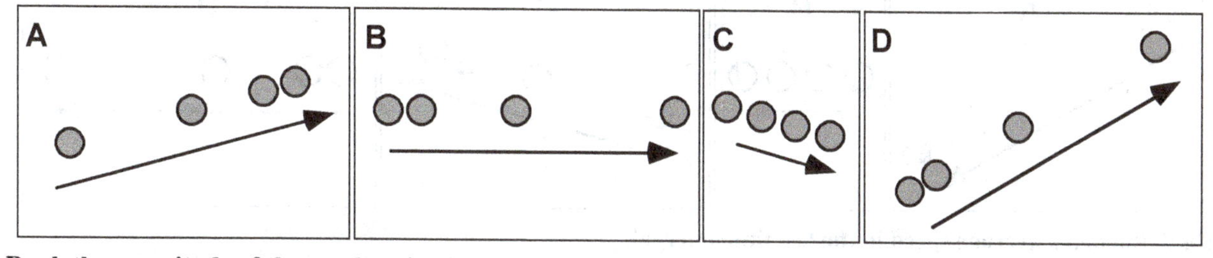

Rank the magnitude of the acceleration based on the drawings.

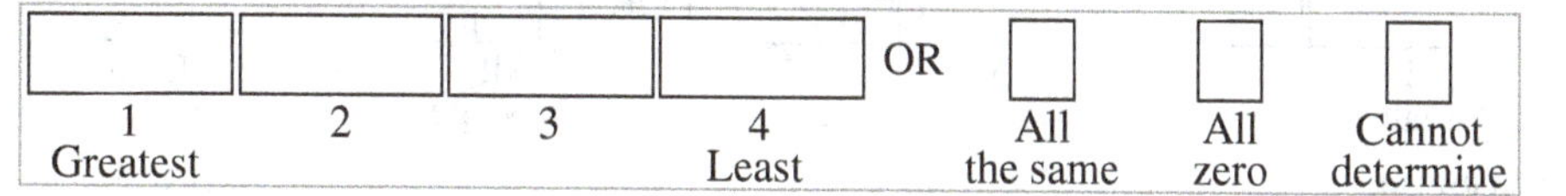

Explain your reasoning.

B1-RT07: Ball Strobe Diagrams—Acceleration

The following drawings represent strobe (flash) photographs of a ball moving in the direction of the arrow. The circles represent the positions of the ball at succeeding instants of time. The time interval between successive positions is the same in all cases. Assume all accelerations are constant.

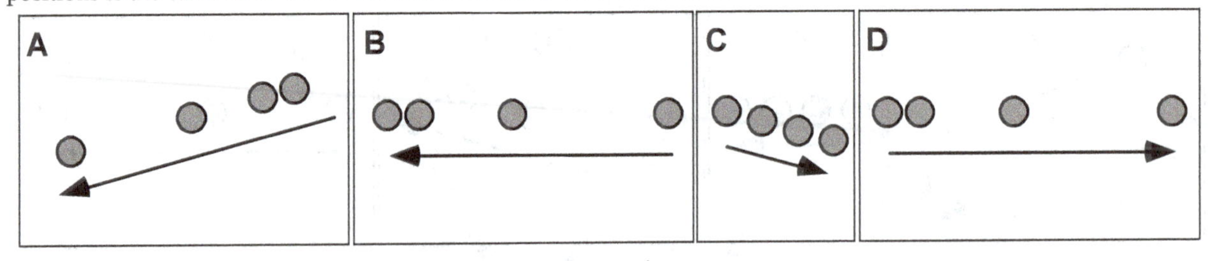

Rank the magnitude of the acceleration based on the drawings.

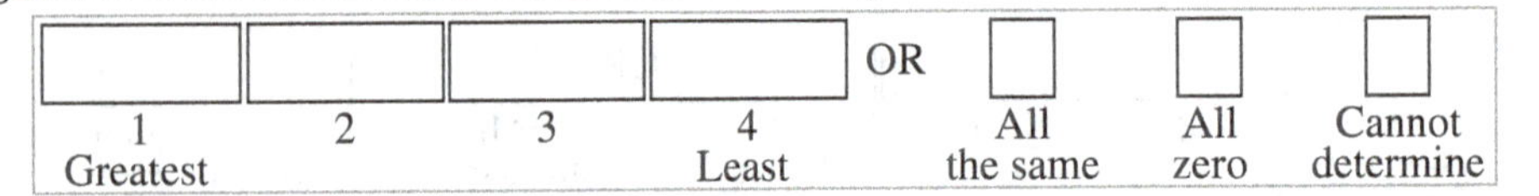

Explain your reasoning.

B1-RT08: Ball Strobe Diagrams—Speed

The following drawings represent strobe (flash) photographs of a ball moving in the direction of the arrow. The circles represent the positions of the ball at succeeding instants of time. The time interval between successive positions is the same in all cases.

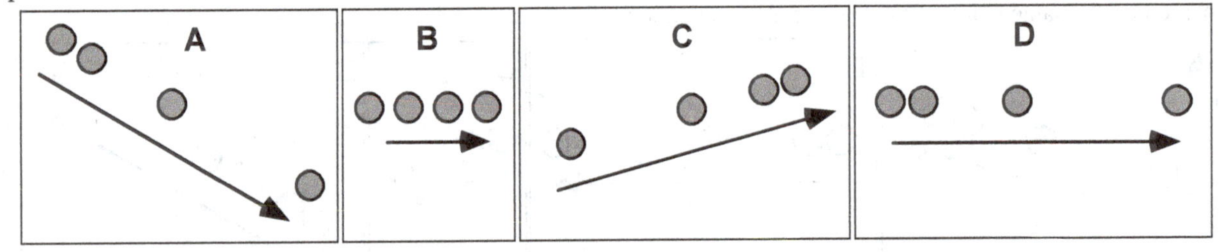

Rank the ball's average speed in the last time interval.

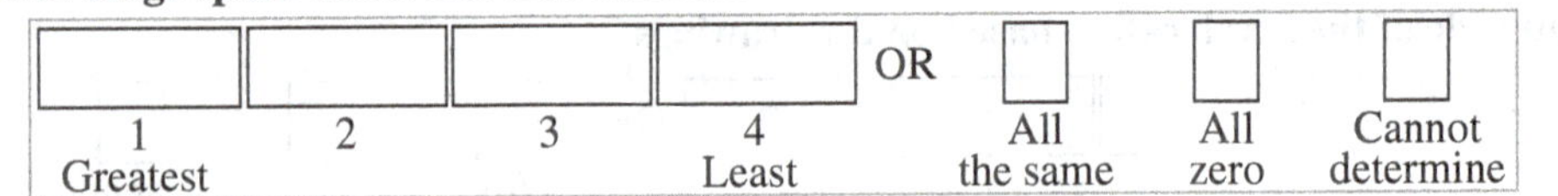

Explain your reasoning.

B1-CRT09: VELOCITY-TIME GRAPHS—SPEED-TIME GRAPHS

Velocity versus time graphs for six toy cars that are traveling straight along a hallway are shown. All graphs have the same time and velocity scales.

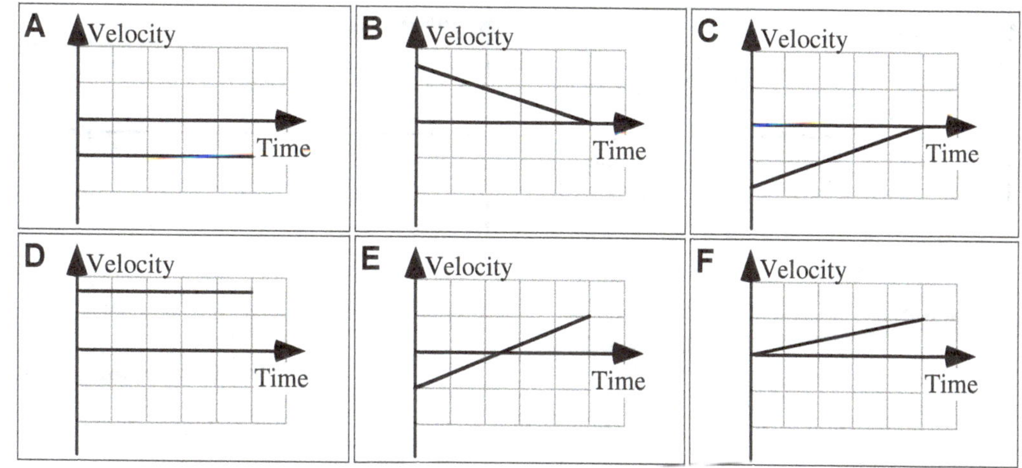

Draw below the speed versus time graphs for these graphs.

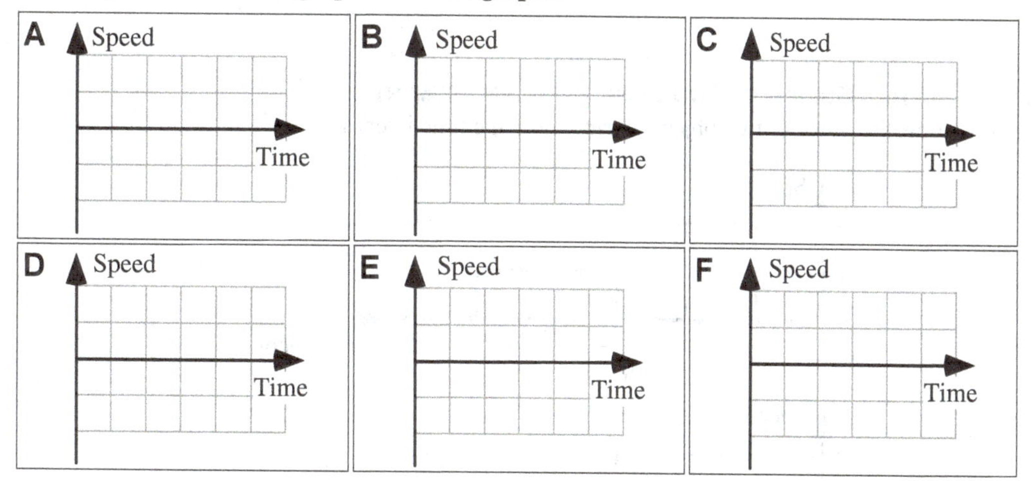

Explain your reasoning.

B1-CRT10: Velocity-Time Graph—Speed-Time Graphs

Given this velocity-time graph, draw the corresponding speed-time graph.

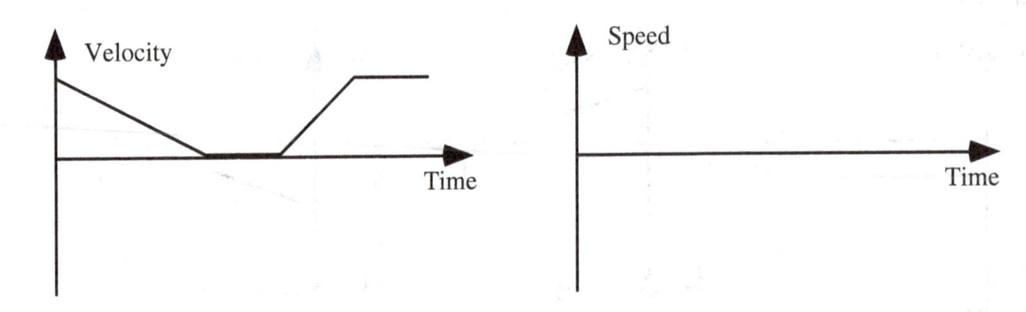

Explain your reasoning.

B1-WWT11: Speed-Time Graphs of Two Objects—Displacement

The graphs below show the speed of two objects during the same time interval.

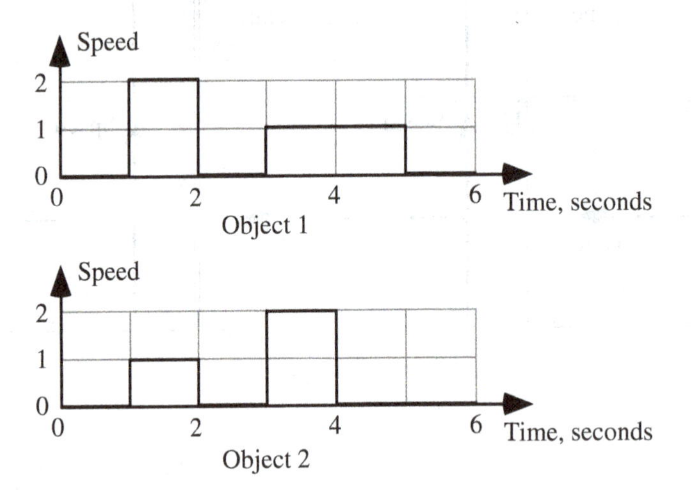

A student considering these two graphs states:

"Object 1 will be farther from its starting point after this 6-second interval than Object 2 because Object 1 had a larger displacement than Object 2."

What, if anything, is wrong with the student's statement? If something is wrong, explain the error and how to correct it. If the graph is correct, explain why.

B1-WWT12: BALL THROWN UPWARD AND COMES BACK DOWN—SPEED-TIME GRAPH

A ball is thrown straight upward and falls back to the same height. A student makes this graph of the speed of the ball as a function of time.

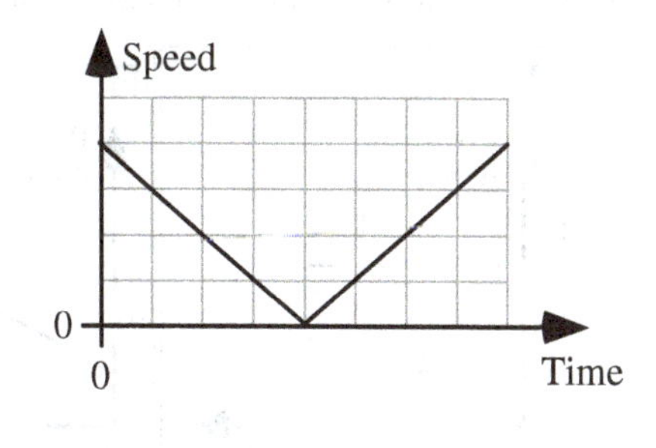

What, if anything, is wrong with the student's graph? If something is wrong, explain the error and how to correct it. If the graph is correct, explain why.

B1-RT13: POSITION-TIME GRAPHS—DISPLACEMENT

Each graph below shows the position of an object as a function of time.

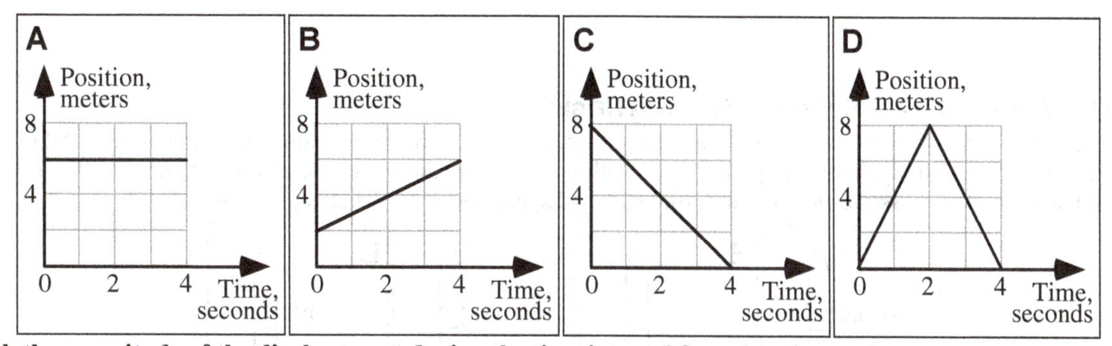

Rank the magnitude of the displacement during the time interval from 0 to 4 seconds.

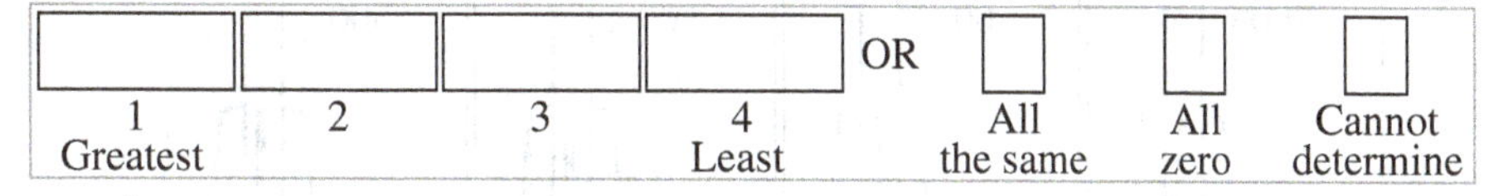

Explain your reasoning.

B1-RT14: Position-Time Graphs—Average Speed

Each graph below shows the position of an object as a function of time.

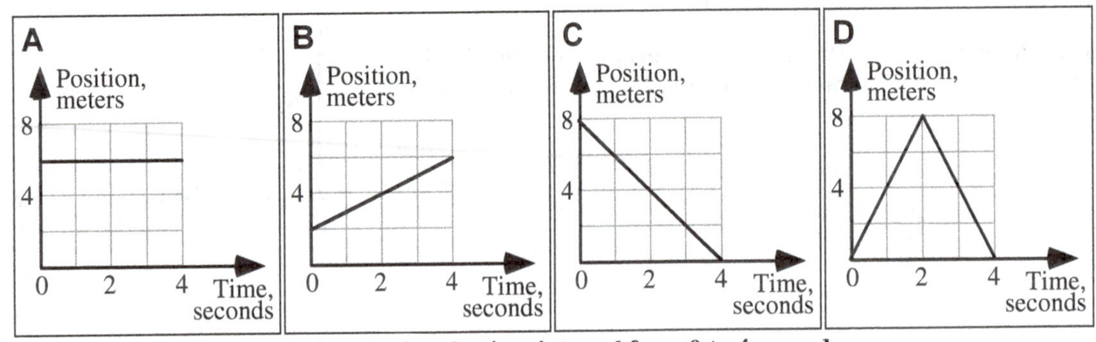

Rank the average speed of the object during the time interval from 0 to 4 seconds.

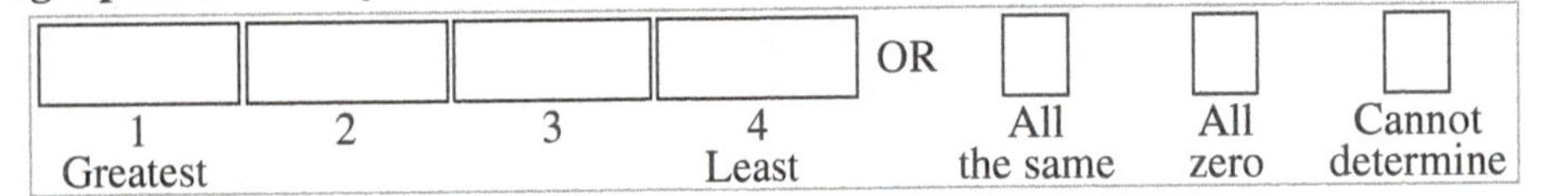

Explain your reasoning.

B1-RT15: Vertical Model Rockets—Maximum Height

The model rockets depicted below have just had their engines turned off when they are at the same height. All of the rockets are aimed straight up, but their speeds differ. Although they are the same size and shape, the rockets carry different loads, so their masses differ. The specific mass and speed for each rocket is given in each figure.

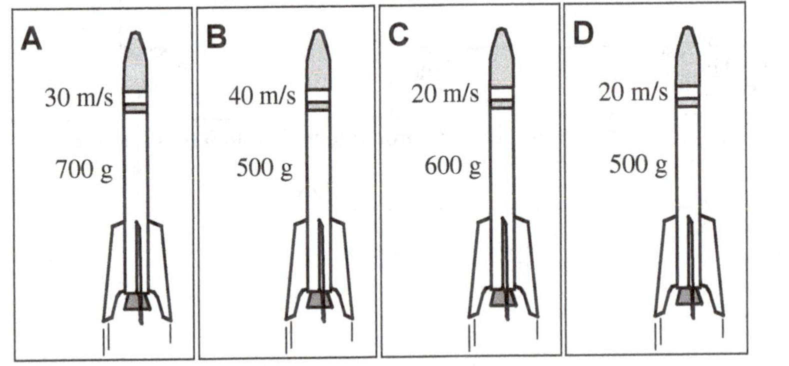

Rank the maximum height the model rockets will reach.

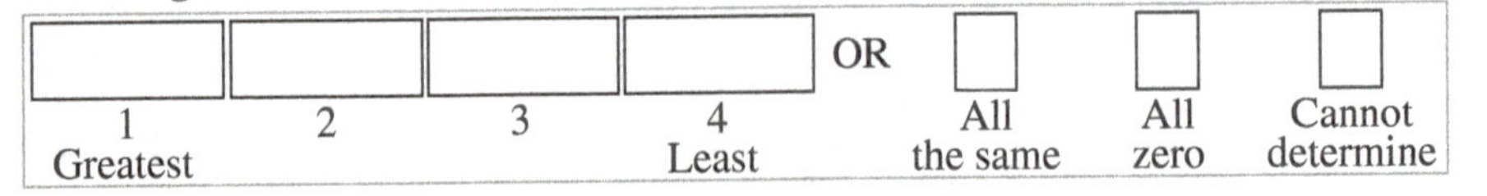

Explain your reasoning.

B1-RT16: CARS—CHANGE OF VELOCITY

In each figure below, a car's velocity is shown before and after a short time interval.

	Before	After
A	+10 m/s	+20 m/s
B	+10 m/s	0
C	+10 m/s	-10 m/s
D	+30 m/s	+20 m/s

Rank the magnitude of the change in velocity during the time interval.

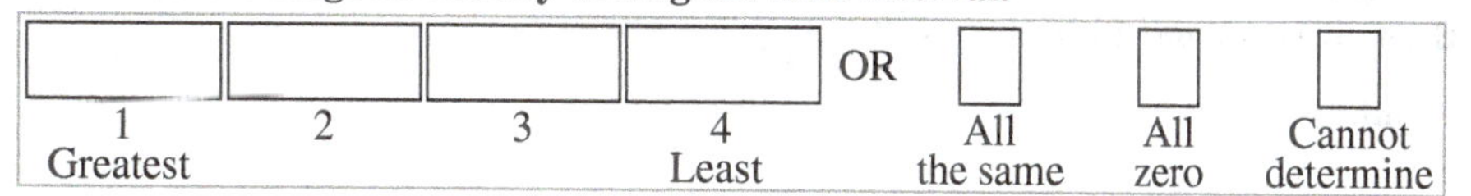

Explain your reasoning.

B1-WWT17: VELOCITY-TIME GRAPH—ACCELERATION-TIME GRAPH

A student obtains a graph of an object's velocity versus time and then draws the graph of the acceleration versus time for the same time interval.

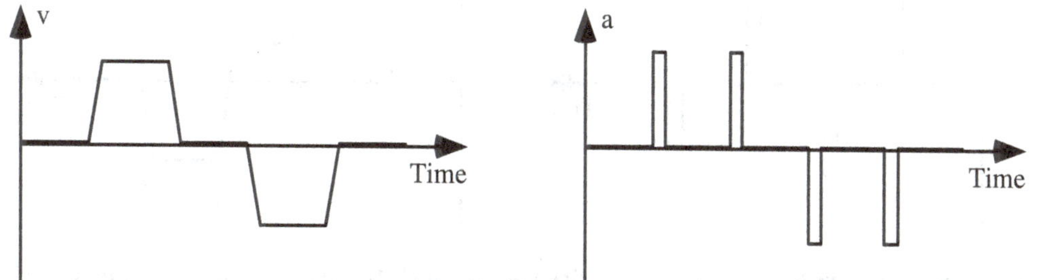

What, if anything, is wrong with the graph of the acceleration versus time? If something is wrong, identify it and explain how to correct it. If the graph is correct, explain why.

B1-CT18: Velocity-Time Graphs—Displacement

The graphs represent the velocity of two toy robots moving in one dimension for a particular time interval. Both graphs have the same time and velocity scales.

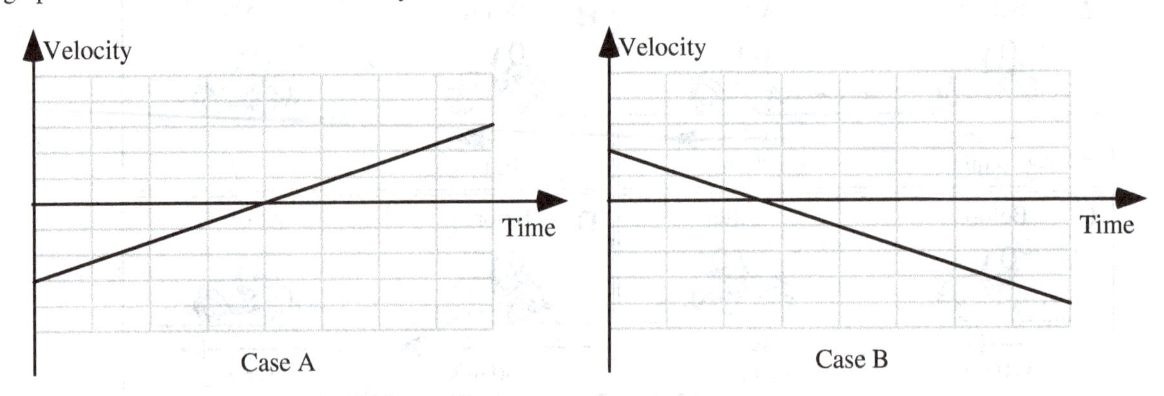

Is the magnitude of the displacement of the robot for the entire time interval shown (i) *greater* in Case A, (ii) *greater* in Case B, or (iii) *the same* in both cases? _____

Explain your reasoning.

B1-WWT19: Acceleration-Time Graph—Velocity-Time Graph

A student obtains a graph of an object's acceleration versus time and then draws the graph of the velocity versus time for the same time interval. The object starts from rest.

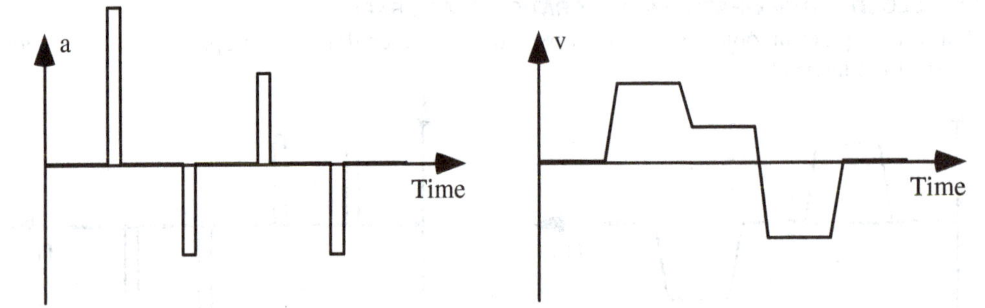

What, if anything, is wrong with the graph of velocity versus time? If something is wrong, identify it and explain how to correct it. If the graph is correct, explain why.

B1-WWT20: VELOCITY-TIME GRAPH—ACCELERATION-TIME GRAPH

A student obtains a graph of an object's velocity versus time and then draws the graph of the acceleration versus time for the same time interval.

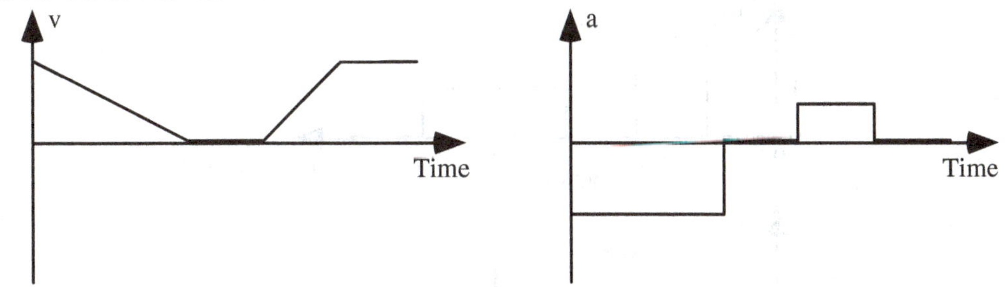

What, if anything, is wrong with the graph of the acceleration versus time? If something is wrong, identify it and explain how to correct it. If the graph is correct, explain why.

B1-CRT21: ACCELERATION-TIME GRAPH—VELOCITY-TIME GRAPH

Sketch a possible velocity versus time graph given the acceleration graph for the same time interval.

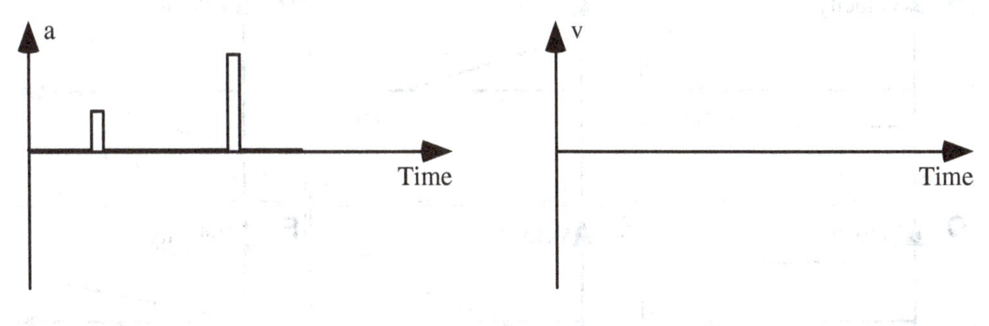

Explain your reasoning.

B1-CT22: Velocity-Time Graphs of Objects—Displacement

The graphs below show the velocity of two objects during the same time interval.

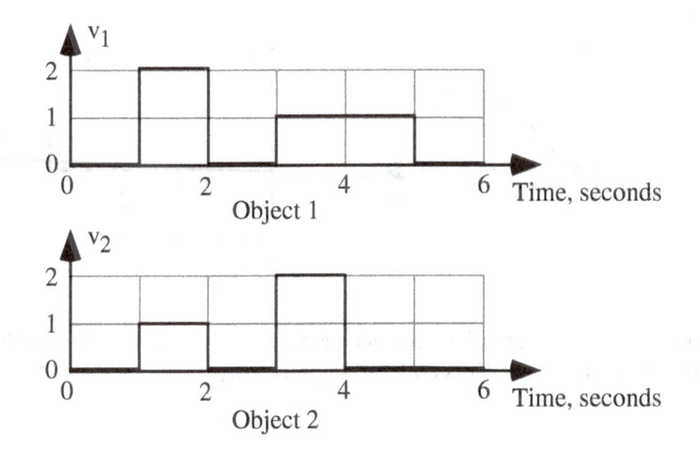

For the 6-second time interval shown, is the displacement of Object 1 in the upper graph (i) *greater than,* (ii) *equal to,* or (iii) *less than* the displacement of Object 2 in the lower graph? _____

Explain your reasoning.

B1-RT23: Velocity-Time Graphs—Displacement

Shown below are six velocity-time graphs for toy robots that are traveling along a straight hallway. All graphs have the same time and velocity scales.

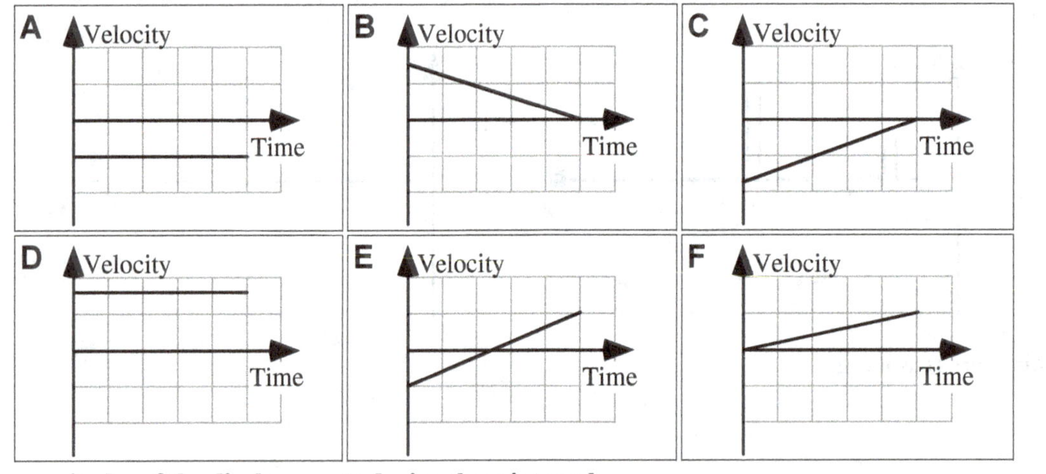

Rank the magnitudes of the displacements during these intervals.

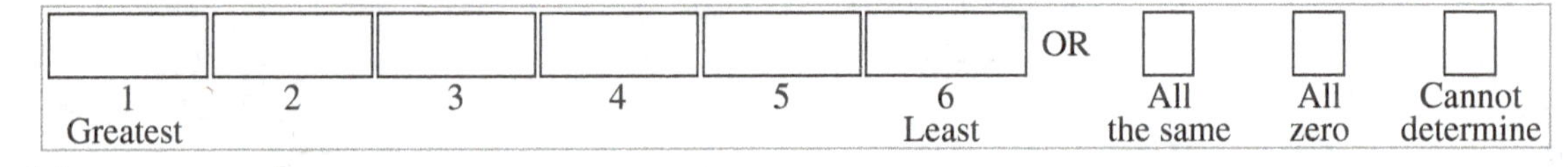

Explain your reasoning.

B1-WWT24: Ball Thrown Upward and Comes Back Down—Velocity-Time Graph

A ball is thrown straight upward and falls back to the same height. A student makes this graph of the velocity of the ball as a function of time.

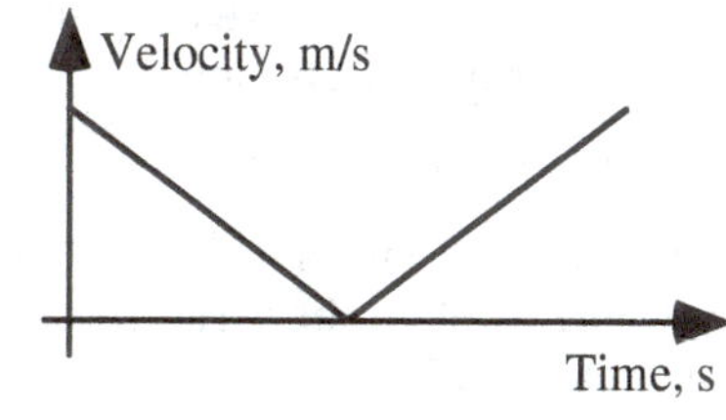

What, if anything, is wrong with the student's graph? If something is wrong, explain the error and how to correct it. If the graph is correct, explain why.

B1-SCT25: Ball Thrown Upward—Graph of Speed-Time

A ball is thrown straight upward and falls back to the same height. A student makes the graph of the speed of the ball as a function of time. Three students who are discussing this graph make the following contentions:

Speed, m/s

Time, s

Akira: *"I don't think this can be correct because the sign of the acceleration changes on this graph, but the acceleration on the ball will be constant."*

Burt: *"No, I think this is right because it is only showing what happens to the speed, which will decrease to zero at the top and then increase as the ball falls. Since the slopes for both segments are the same except for sign that means the acceleration is constant."*

Catalina: *"This graph makes sense to me because it shows the speed decreasing on the way up. But I disagree with Burt, because I think this means the acceleration is also decreasing until the ball gets to the top and stops. Then both the speed and acceleration increase as the ball falls down again."*

With which, if any, of these three students do you agree?

Akira _____ Burt _____ Catalina _____ None of them_____

Explain your reasoning.

B1-WWT26: Velocity-Time Graph of Two Objects—Fastest Object

A student is shown the velocity-time graphs for two objects and is asked to decide which object is moving faster. The student responds:

"B is faster because it has the steeper slope."

What, if anything, is wrong with the student's statement? If something is wrong, explain the error and how to correct it. If the statement is correct, explain why.

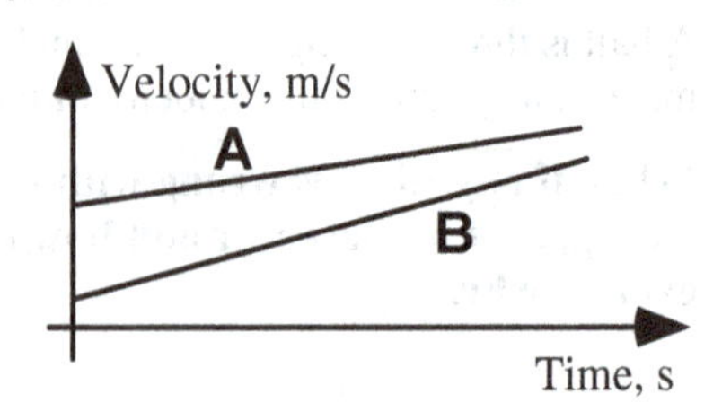

B1-CRT27: Traveling Students—Velocity-Time Graph

Carmela and Desi leave a parking lot separately and drive west. They both start from rest. Desi leaves first, traveling with an acceleration of 4 m/s^2 west for the first 6 seconds, and then driving at a constant velocity. Two seconds after Desi started, Carmela starts with an acceleration of 3 m/s^2 west for 10 seconds, and then she drives at a constant velocity.

Graph the velocity of both travelers as a function of time up to t = 16 seconds starting at time t = 0 when Desi leaves the classroom. Use a solid line for Desi's velocity and a dashed line for Carmela's velocity.

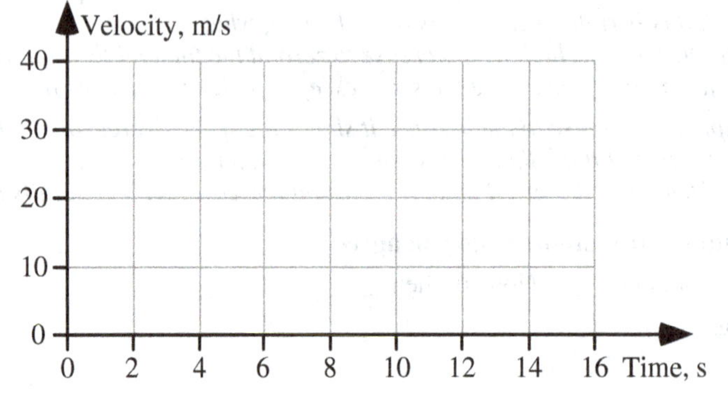

Explain your reasoning.

B1-WWT28: POSITION-TIME GRAPH OF TWO OBJECTS—FASTEST OBJECT

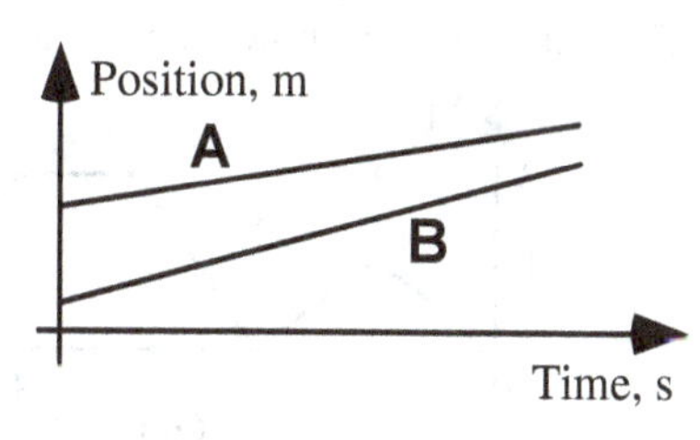

A student is shown the position-time graphs for two objects and is asked to decide which object is moving faster. The student responds:

"B is faster because it has the steeper slope."

What, if anything, is wrong with the student's statement? If something is wrong, explain the error and how to correct it. If the statement is correct, explain why.

B1-RT29: VELOCITY-TIME GRAPHS—DISTANCE TRAVELED

Velocity-time graphs for six toy robots that are traveling along a straight hallway are shown. All graphs have the same time and velocity scales.

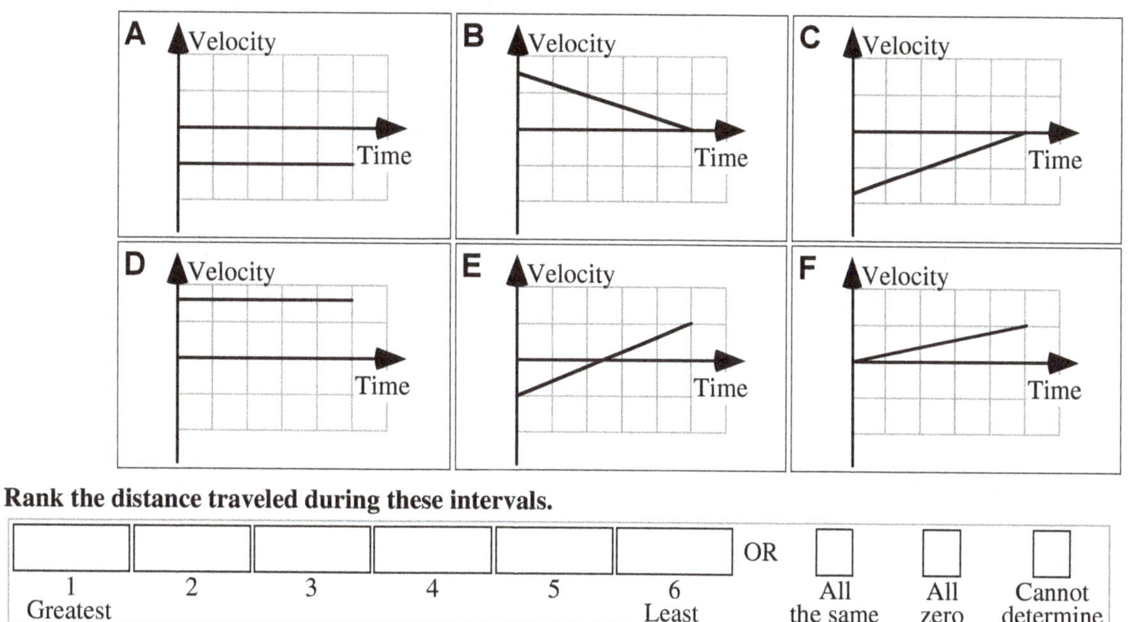

Rank the distance traveled during these intervals.

						OR	☐	☐	☐
1 Greatest	2	3	4	5	6 Least		All the same	All zero	Cannot determine

Explain your reasoning.

B1-SCT30: Velocity-Time Graphs of Objects—Displacement

The graphs below show the velocity of two objects during the same time interval.

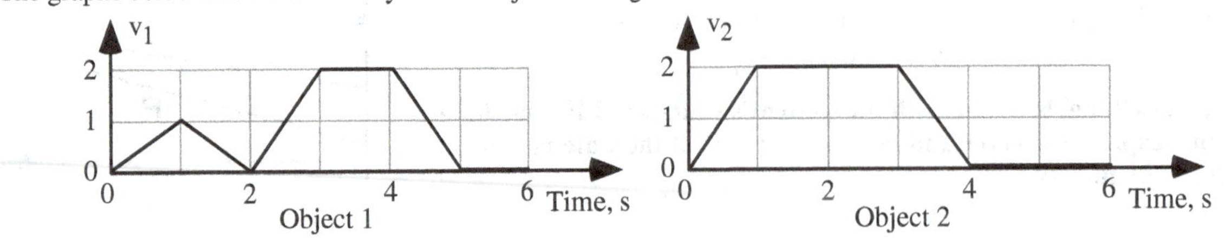

Three students are discussing the displacements of these objects for this interval.

Amos: *"I think Object 2 will have the greater displacement because it gets to a higher speed faster than Object 1."*

Badu: *"No, Object 1 will have the greater displacement because it travels for a longer time than Object 2."*

Candi: *"I agree with Amos, but for a different reason. Object 2 has the larger displacement because the area under the graph is greater."*

With which, if any, of these three students do you agree?

Amos_____ Badu _____ Candi _____ None of them_____

Explain your reasoning.

B1-QRT31: POSITION OR VELOCITY GRAPHS—CHANGE DIRECTION

The graph shown is for an object in one-dimensional motion.

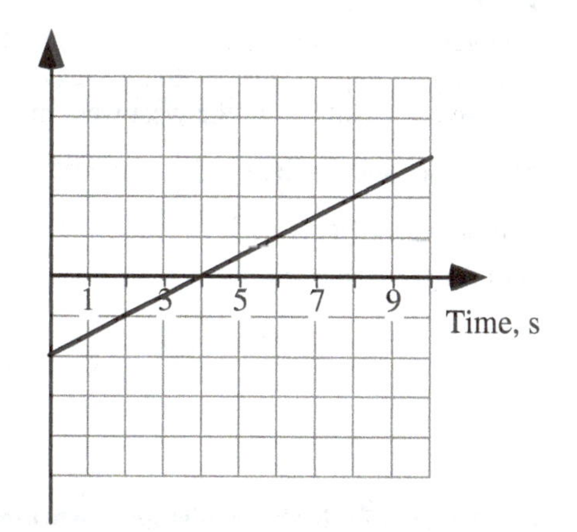

(a) If the vertical axis of the graph is position, does the object ever change direction?

If so, at what time or times does this change in direction occur?

Explain your reasoning.

(b) If the vertical axis of the graph is velocity, does the object ever change direction?

If so, at what time or times does this change in direction occur?

Explain your reasoning.

B1-WWT32: BALL THROWN STRAIGHT UPWARD—TIME TO REACH TOP

A student throws a ball straight upward. A friend times how long it takes the ball to reach its maximum height.

The student predicts:

"Faster things take less time. If I throw the ball faster, it will reach its highest point in less time."

What, if anything, is wrong with this statement? If something is wrong, identify it and explain how to correct it. If this statement is correct, explain why.

B1-QRT33: Position-Time Graph—Direction

A bicyclist is moving along a straight street oriented east—west. In drawing the graph, positions to the east of the origin were marked as positive and positions to the west were marked as negative.

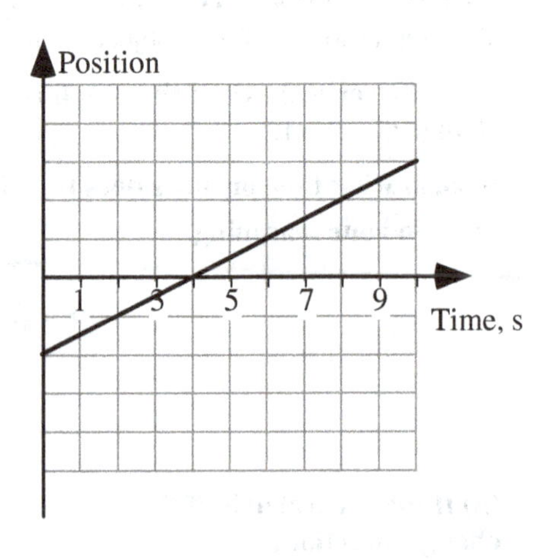

(a) At 1 second, is the cyclist moving?

If so, in what direction?

Explain your reasoning.

(b) At 1 second, is the cyclist accelerating?

If so, in what direction?

Explain your reasoning.

(c) At 9 seconds, is the cyclist moving?

If so, in what direction?

Explain your reasoning.

(d) At 9 seconds, is the cyclist accelerating?

If so, in what direction?

Explain your reasoning.

(e) At 4 seconds, is the cyclist moving?

If so, in what direction?

Explain your reasoning.

(f) At 4 seconds, is the cyclist accelerating?

If so, in what direction?

Explain your reasoning.

B1-QRT34: VELOCITY-TIME GRAPH—DIRECTION

A unicyclist, someone riding a single-wheel cycle, is moving along a straight street oriented east—west. In drawing the graph, east was taken as the positive direction and west the negative direction.

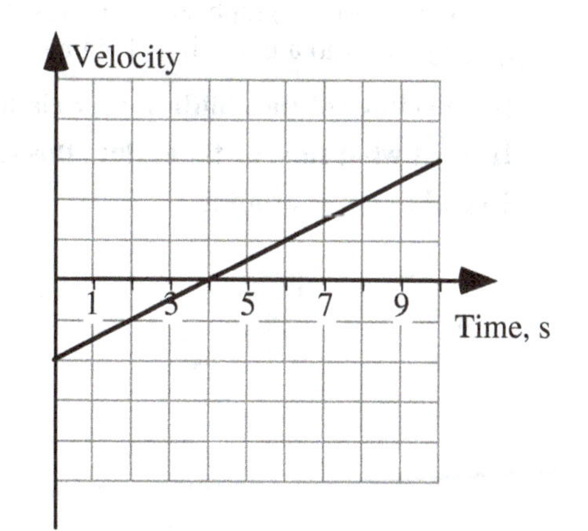

(a) At 1 second, is the cyclist moving?
If so, in what direction?
Explain your reasoning.

(b) At 1 second, is the cyclist accelerating?
If so, in what direction?
Explain your reasoning.

(c) At 9 seconds, is the cyclist moving?
If so, in what direction?
Explain your reasoning.

d) At 9 seconds, is the cyclist accelerating?
If so, in what direction?
Explain your reasoning.

(e) At 4 seconds, is the cyclist moving?
If so, in what direction?
Explain your reasoning.

(f) At 4 seconds, is the cyclist accelerating?
If so, in what direction?
Explain your reasoning.

B1-QRT35: Position-Time Graphs of Children—Kinematic Quantities

The position-time graph shown represents the motion of two children, Ariel and Byron, who are moving along a narrow, straight hallway.

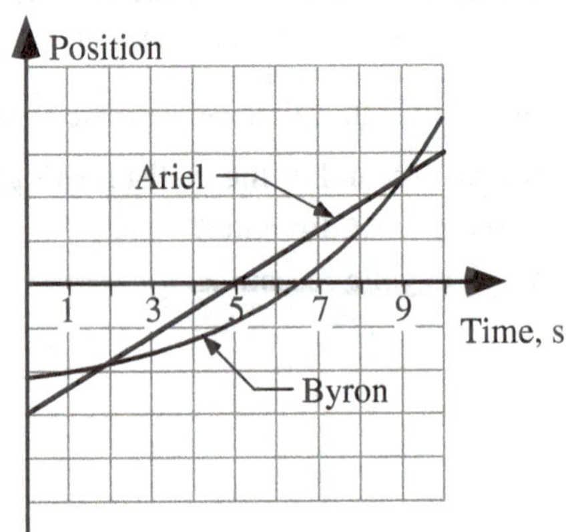

(a) Do either of the children ever change direction?

If so, at what time or times does this change in direction occur?

Explain your reasoning.

(b) Are the two children ever at the same position along the hallway?

If so, at what time or times does this happen?

Explain your reasoning.

(c) Do the two children ever have the same speed?

If so, at what time or times does this happen?

Explain your reasoning.

(d) Do the two children ever have the same acceleration?

If so, at what time or times does this happen?

Explain your reasoning.

B1-QRT36: VELOCITY-TIME GRAPHS OF CHILDREN—KINEMATIC QUANTITIES

The velocity-time graph shown represents the motion of two children, Ariel and Byron, who are moving along a narrow, straight hallway.

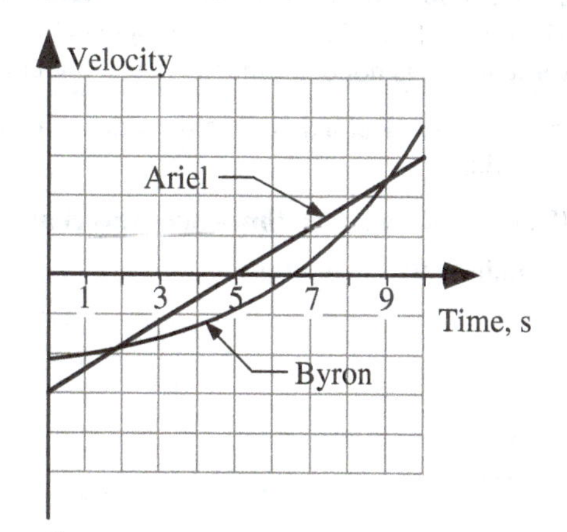

(a) Do either of the children ever change direction?
If so, at what time or times does this change in direction occur?
Explain your reasoning.

(b) Do the two children ever have the same velocity?
If so, at what time or times does this occur?
Explain your reasoning.

(c) Do the two children ever have the same acceleration?
If so, at what time or times does happen?
Explain your reasoning.

B1-WBT37: POSITION EQUATION—PHYSICAL SITUATION

Describe the motion of an object that is represented by the following equation:

$$x = 33.6 \text{ m} - (2.8 \text{ m/s})t$$

B1-QRT38: Position or Velocity Graphs—Change Direction

The graph shown is for an object in one-dimensional motion. The vertical axis is not determined, so it is not labeled.

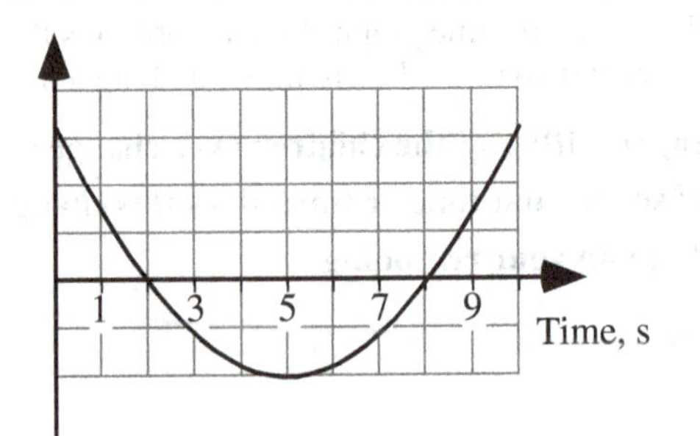

(a) If the vertical axis is position, does the object ever change direction?

If so, at what time or times does this change in direction occur?

Explain your reasoning.

(b) If the vertical axis is velocity, does the object ever change direction?

If so, at what time or times does this change in direction occur?

Explain your reasoning.

B1-SCT39: Bicyclist on a Straight Road—Average Speed

Three students are discussing the motion of a bicyclist who travels at a steady 18 m/s for 10 minutes, then at 6 m/s for 20 minutes, and finally at 12 m/s for 15 minutes along a straight, level road. Students make the following contentions about the bicyclist's average speed for the overall trip:

Aaron: *"The average speed is 12 m/s because you add the three velocities, but then you have to divide by three."*

Bessie: *"I think you have to take the amount of time at each speed into account. The total time for the trip is 45 minutes, so the cyclist is going at 12 m/s for one-third of the time, at 6 m/s for 20/45ths of the time, and at 18 m/s for 10/45ths of the time. You'd have to weight the speeds according to the times."*

Cesar: *"The average speed is 10.7 m/s because that is what you get when you divide 28,800 m, the total distance traveled on the straight road, by 2700 seconds, the total time it took."*

With which, if any, of these three students do you agree?

Aaron_____ Bessie _____ Cesar _____ None of them_____

Explain your reasoning.

B1-QRT40: POSITION EQUATIONS—MOTION CHARACTERISTICS

Equations for the position of four objects executing one-dimension motion in the x-direction as a function of time are given.

A	B
$x = 5\,m + (3\,m/s)t + (4\,m/s^2)t^2$	$x = -5\,m + (3\,m/s)t + (4\,m/s^2)t^2$
C	**D**
$x = -5\,m + (3\,m/s)t$	$x = 5\,m - (3\,m/s)t + (8\,m/s^2)t^2$

(a) What is the initial position for Object A?
Explain your reasoning.

(b) What is the initial position for Object B?
Explain your reasoning.

(c) What is the initial velocity for Object C?
Explain your reasoning.

(d) What is the initial velocity for Object D?
Explain your reasoning.

(e) What is the acceleration for Object A?
Explain your reasoning.

(f) What is the acceleration for Object D?
Explain your reasoning.

(g) What is the position for Object A at 1 second?
Explain your reasoning.

(h) What is the position for Object D at 1 second?
Explain your reasoning.

(i) Which, if any, of these objects is moving at a constant velocity and what is its velocity?
Explain your reasoning.

B1-WWT41: Acceleration-Time Graph—Interpretation

A student is given the following acceleration versus time graph for a motorcyclist traveling along a straight, level stretch of road.

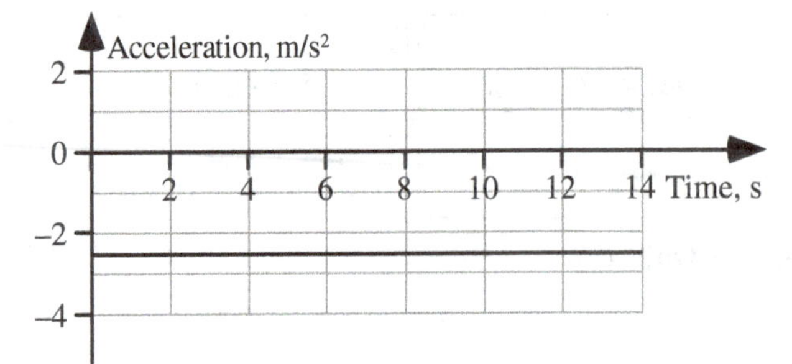

The student states:

"This motorcyclist was slowing down during the period up to 14 seconds because her acceleration was negative during this period."

What, if anything, is wrong with this student's contention? If something is wrong, identify it and explain how to correct it. If it is correct, explain why.

B1-RT42: Velocity-Time Graphs—Instantaneous Velocity

The graphs below show the velocity versus time for boats traveling along a straight, narrow channel. The scales on both axes are the same for all of these graphs. In each graph, a point is marked with a dot.

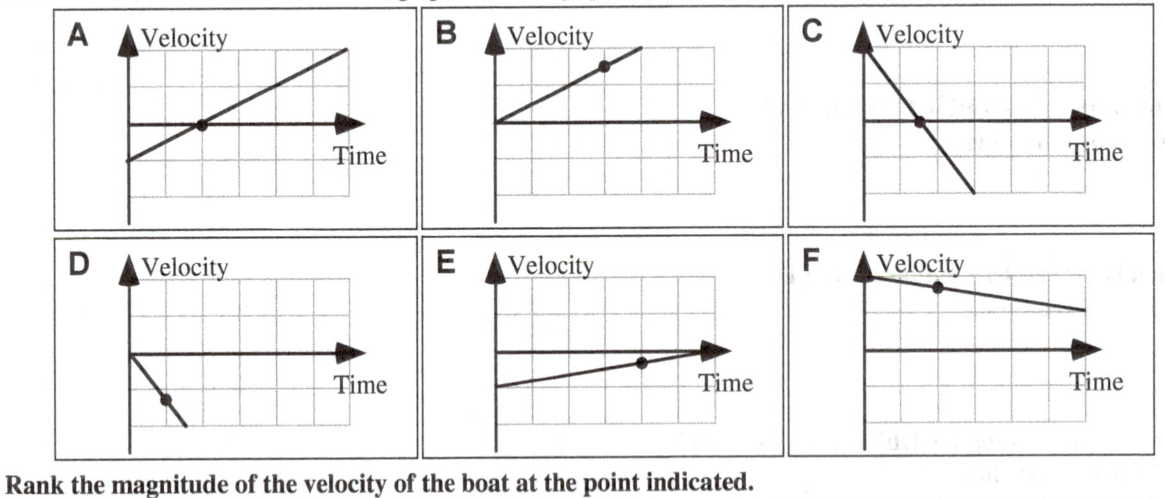

Rank the magnitude of the velocity of the boat at the point indicated.

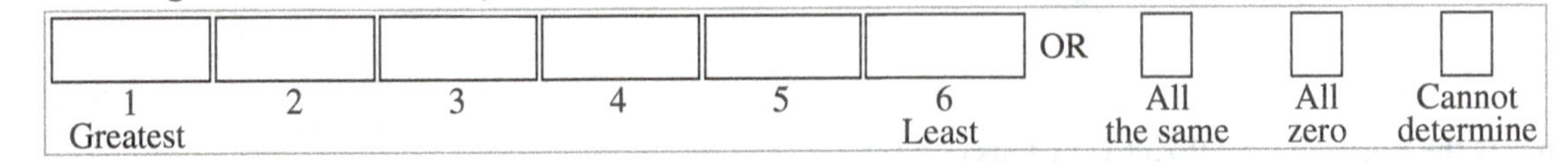

Explain your reasoning.

B1-RT43: VELOCITY-TIME GRAPHS—ACCELERATION

The graphs below show the velocity versus time for boats traveling along a straight, narrow channel. The scales on both axes are the same for all of these graphs. In each graph, a point is marked with a dot.

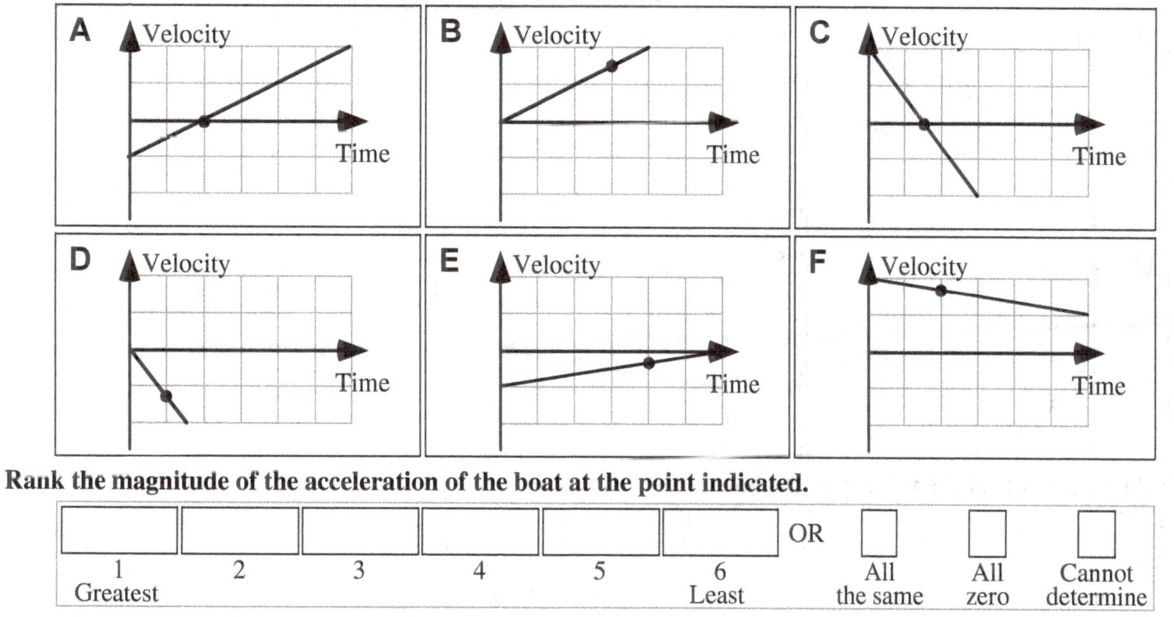

Rank the magnitude of the acceleration of the boat at the point indicated.

1 Greatest | 2 | 3 | 4 | 5 | 6 Least — OR — All the same | All zero | Cannot determine

Explain your reasoning.

B1-RT44: POSITION-TIME GRAPHS—INSTANTANEOUS SPEED

These graphs show position versus time for boats traveling along a straight, narrow channel. The scales on both axes are the same for all of these graphs. In each graph, a point is marked with a dot.

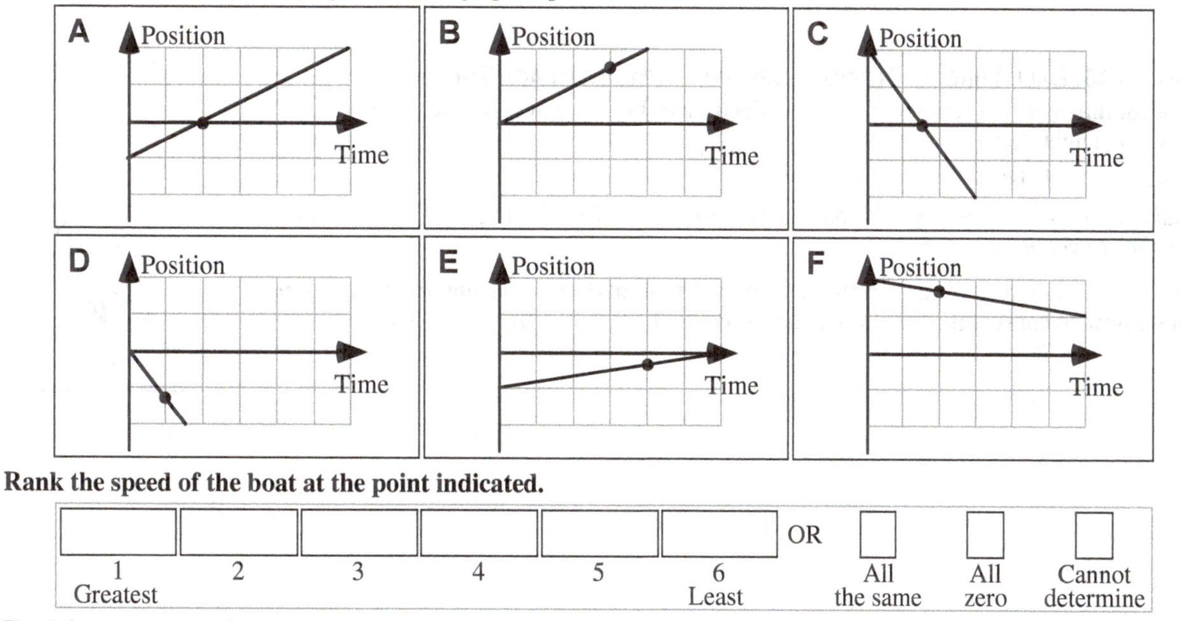

Rank the speed of the boat at the point indicated.

1 Greatest | 2 | 3 | 4 | 5 | 6 Least — OR — All the same | All zero | Cannot determine

Explain your reasoning.

B1-RT45: Position-Time Graphs—Instantaneous Velocity

The graphs below show position versus time for six boats traveling along a straight, narrow channel. The scales on both axes are the same for all of these graphs. In each graph, a point is marked with a dot.

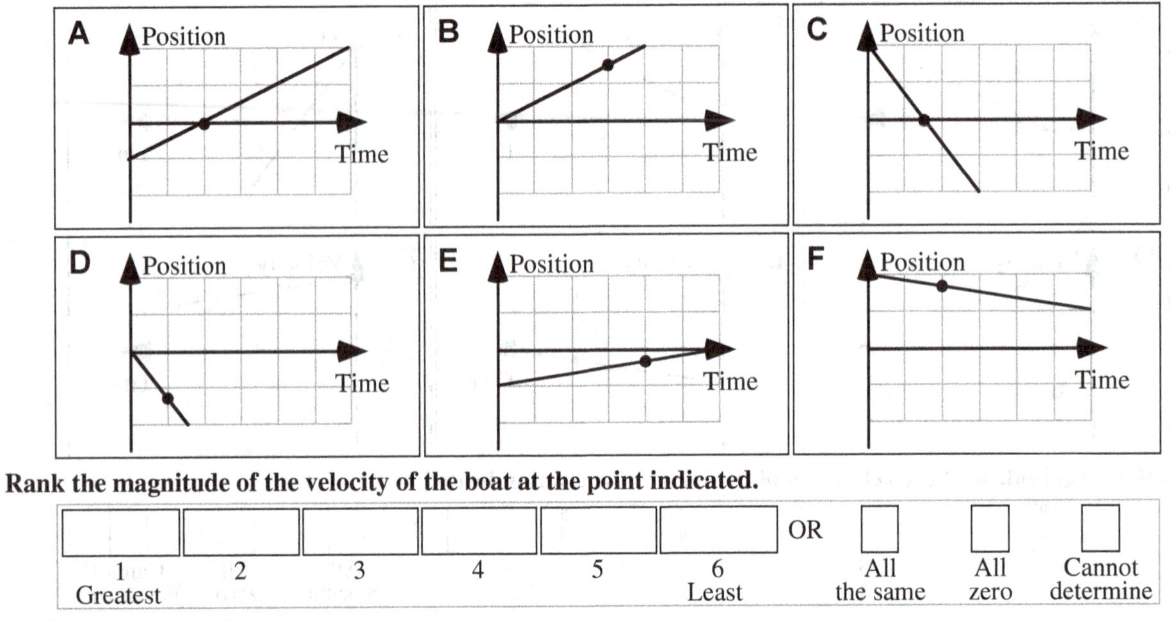

Rank the magnitude of the velocity of the boat at the point indicated.

☐	☐	☐	☐	☐	☐	OR	☐	☐	☐
1 Greatest	2	3	4	5	6 Least		All the same	All zero	Cannot determine

Explain your reasoning.

B1-WWT46: Ball Thrown Straight Upward—Time to Reach Top

A student throws a ball straight upward. A friend times how long it takes the ball to reach its maximum height.

The student predicts:

"It takes more time to go larger distances. If I throw the ball so that it goes higher, it will take more time to get there."

What, if anything, is wrong with this statement? If something is wrong, identify it and explain how to correct it. If this statement is correct, explain why.

B1-QRT47: POSITION, VELOCITY, AND ACCELERATION SIGNS—POSITION, DIRECTION, AND RATE

Eight possible combinations for the signs for the instantaneous position, velocity, and acceleration of an object moving in one dimension are given in the table. Above the table is a coordinate axis that shows the origin, marked 0, and that indicates that the positive direction is to the right. The three columns on the right-hand side of the table are to describe the location of the object (either left or right of the origin), the direction of the velocity of the object (either toward or away from the origin), and what is happening to the speed of the object (either speeding up or slowing down at the given instant). The appropriate descriptions for the first case are shown.

Complete the rest of the table for position, direction, and rate.

0

	Position	Velocity	Acceleration	Position (Left or Right)	Direction (Toward or Away from)	Rate (Speeding up or Slowing down)
A	+	+	+	Left	Away from	Speeding up
B	+	+	–			
C	+	–	+			
D	+	–	–			
E	–	+	+			
F	–	+	–			
G	–	–	+			
H	–	–	–			

Explain your reasoning.

B1-WWT48: VELOCITY-TIME GRAPHS—ACCELERATION

The graphs show the velocity versus time for two boats traveling along a straight, narrow channel. The scales on both axes are the same for the two graphs. In each graph, a point is marked with a dot.

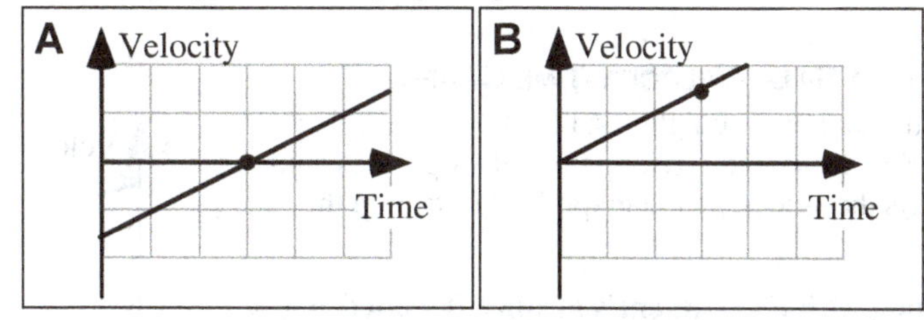

A student who is asked to compare the accelerations at the marked points on the two graphs states:

"I think that the boat in graph B has the larger acceleration because the boat in graph A is at rest at the marked point and its acceleration is zero."

What, if anything, is wrong with this student's contention? If something is wrong, identify it and explain how to correct it. If it is correct, explain why.

B1-QRT49: Position, Velocity, and Acceleration Signs—Position, Direction, and Rate

Eight possible signs of combinations for the instantaneous position, velocity, and acceleration of an object moving in one dimension are given in the table. Above the table is a coordinate axis that shows the origin, marked 0, and that indicates that the positive direction is to the left. The three columns on the right-hand side of the table are to describe the location of the object (either left or right of the origin), the direction of the motion of the object (either toward or away from the origin), and what is happening to the speed of the object (either speeding up or slowing down at the given instant). The appropriate descriptions for the first case are shown.

Complete the table for the object's location and direction of motion relative to the origin and how its speed is changing.

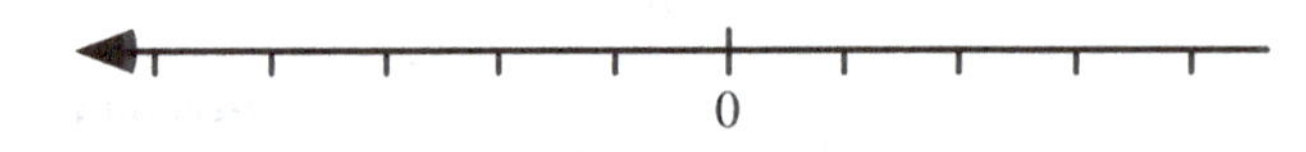

	Position	Velocity	Acceleration	Position (Left or Right)	Direction (Toward or Away from)	Rate (Speeding up or Slowing down)
A	+	+	+	Left	Away from	Speeding up
B	+	+	–			
C	+	–	+			
D	+	–	–			
E	–	+	+			
F	–	+	–			
G	–	–	+			
H	–	–	–			

Explain your reasoning.

B1-WWT50: Bicyclist on a Hill—Velocity-Time Graph

A bicyclist moving at high speed on a straight road comes to a hill that slopes upward gradually. She decides to coast up the hill. A physics student observing the bicyclist plots the velocity-time graph for her trip up the hill as shown.

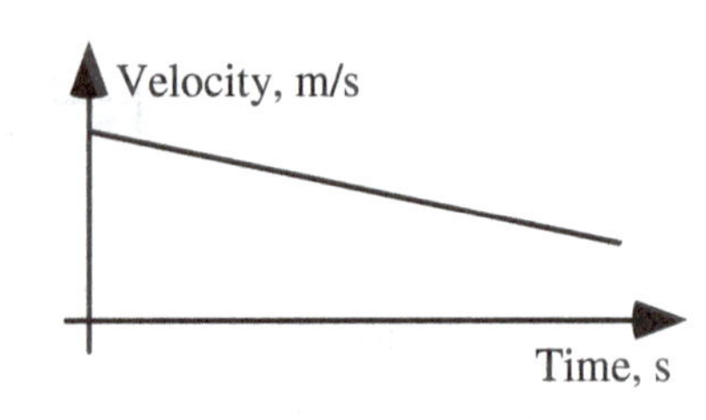

What, if anything, is wrong with this student's graph? If something is wrong, explain the error and how to correct it. If the graph is correct, explain why.

B1-QRT51: VELOCITY-TIME GRAPHS—DIRECTION

The graphs below show the velocity versus time for boats traveling along a straight, narrow channel. The scales on both axes are the same for all of these graphs, and the boats all start at the same origin. In each graph, a point is marked with a dot.

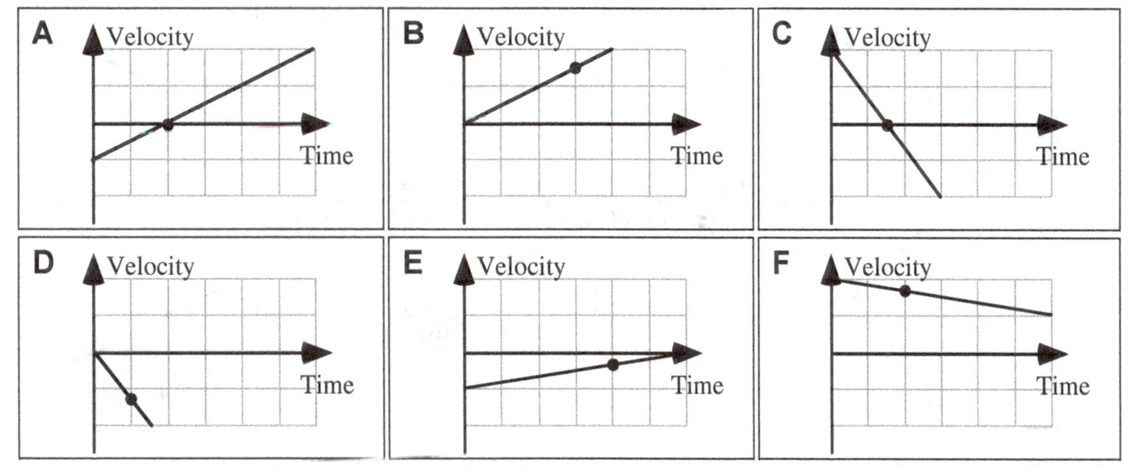

Indicate in the chart below if the position, velocity, and acceleration directions of the boat at the points indicated are in the positive (+), negative (–), or no direction (0)

	Position	Velocity	Acceleration
A			
B			
C			
D			
E			
F			

Explain your reasoning.

B1-RT52: Velocity-Time Graphs—Displacement

Graphs of velocity versus time during 4 seconds for identical objects are shown below. The objects move along a straight, horizontal surface under the action of a force exerted by an external agent.

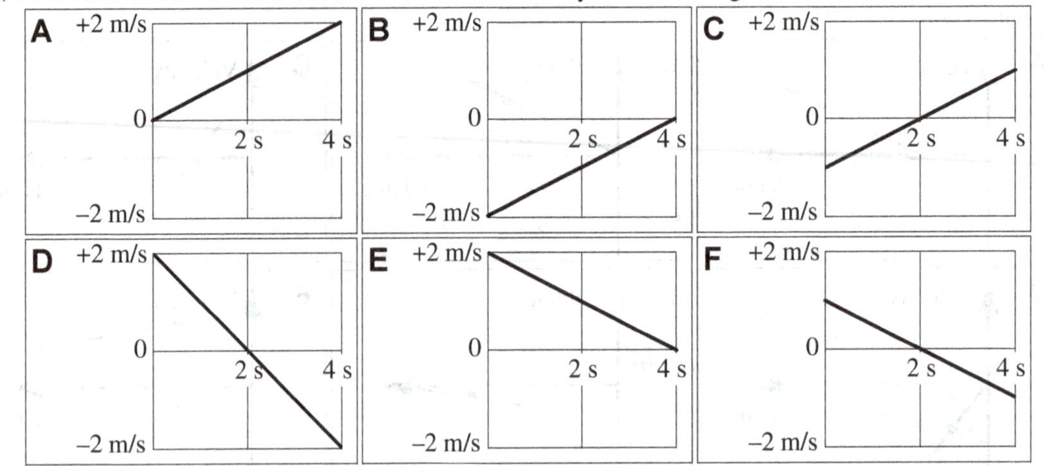

Rank the magnitudes of the displacements of the objects during each of these intervals.

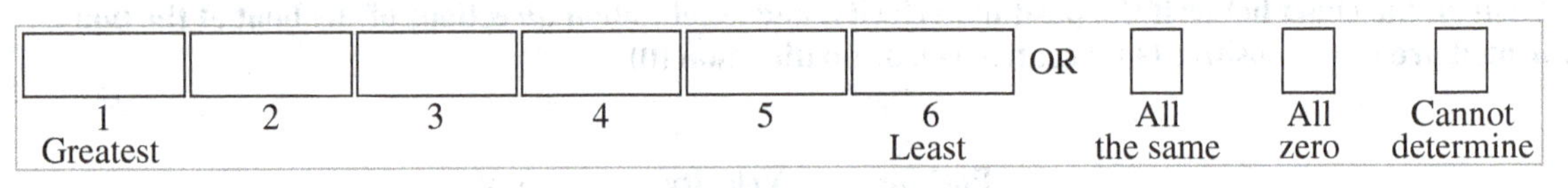

Explain your reasoning.

B1-WWT53: Walking Round Trip—Average Speed

A student walks along a trail with a constant speed of 8 km per hour for 1 hour and then turns around and walks back to her starting point at a constant speed of 6 km per hour. She contends:

"My average speed was about 6.9 km per hour since I walked a total of 16 km in two and one-third hours time."

What, if anything, is wrong with this statement? If something is wrong, identify it and explain how to correct it. If this statement is correct, explain why.

B1-RT54: VELOCITY-TIME GRAPHS—ACCELERATION

Graphs of velocity versus time during 4 seconds for identical objects are shown below. The objects move along a straight, horizontal surface under the action of a force exerted by an external agent.

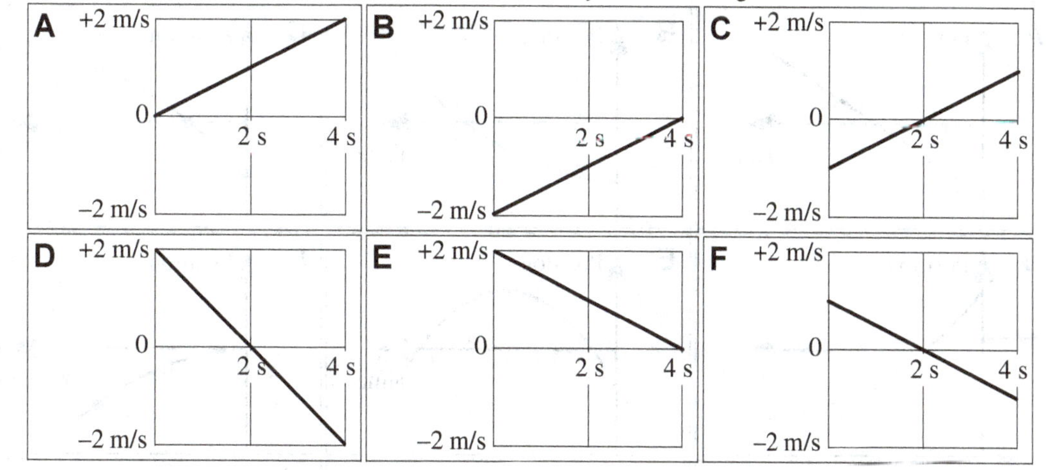

Rank the magnitudes of the accelerations of these objects during each of these intervals.

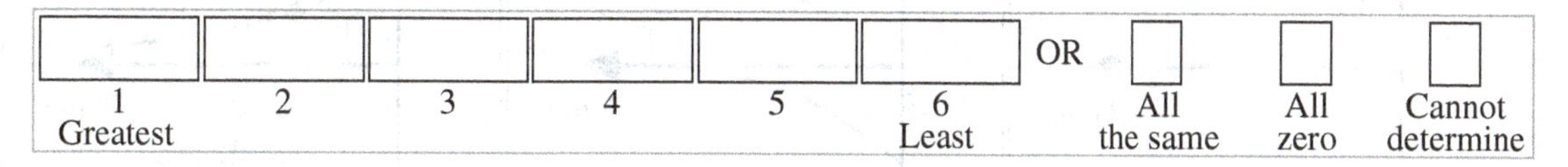

Explain your reasoning.

B1-WWT55: ROUND TRIP UP A MOUNTAIN—AVERAGE SPEED

A student contends:

"If I climb up a mountain at 1 mile per hour for 2 hours and then turn around and climb back down at 3 miles per hour, then my average speed will be 2 miles per hour.

What, if anything, is wrong with this statement? If something is wrong, identify it and explain how to correct it. If this statement is correct, explain why.

B1-QRT56: POSITION-TIME GRAPHS—ACCELERATION AND VELOCITY

Position versus time graphs for boats traveling along a narrow channel are shown below. The scales on both axes are the same for all of these graphs. In each graph, a point is marked with a dot.

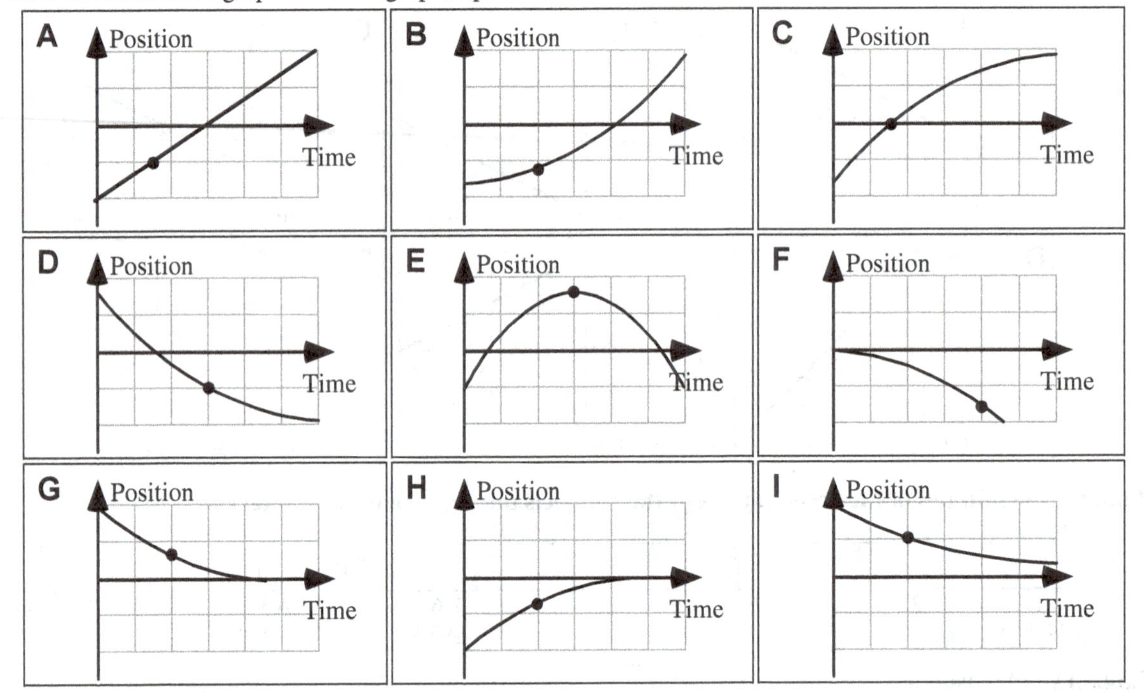

(a) For which of these cases, if any, is the position zero at the indicated point? Explain your reasoning.

(b) For which of these cases, if any, is the position negative at the indicated point? Explain your reasoning.

(c) For which of these cases, if any, is the velocity zero at the indicated point? Explain your reasoning.

(d) For which of these cases, if any, is the velocity negative at the indicated point? Explain your reasoning.

(e) For which of these cases, if any, is the acceleration zero at the indicated point? Explain your reasoning.

(f) For which of these cases, if any, is the acceleration negative at the indicated point? Explain your reasoning.

B1-RT57: PEOPLE ON TRAINS—SPEED RELATIVE TO GROUND

In each case shown, someone is running on a flatbed train car as the train moves. In cases C and D, the person is running toward the front of the train, while in cases A and B the person is running toward the rear. The speeds of the train and of each person relative to the train are given. An observer is standing beside the track watching each train go by.

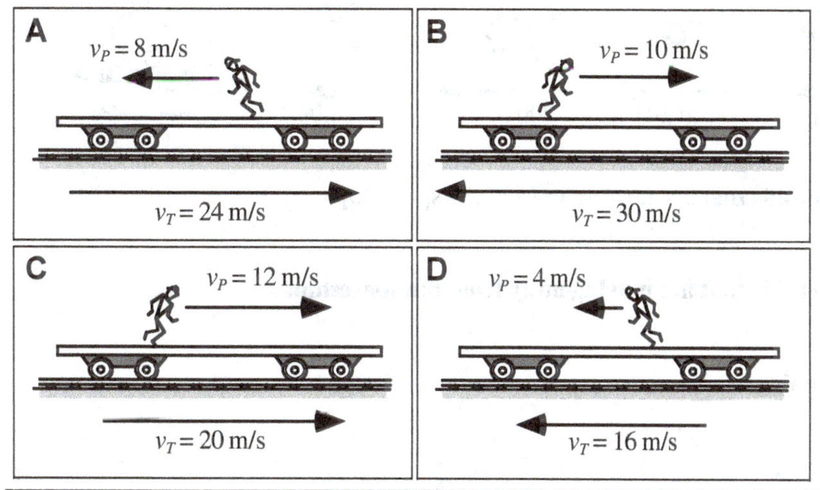

Rank the speed of the runners relative to the observer standing beside the tracks.

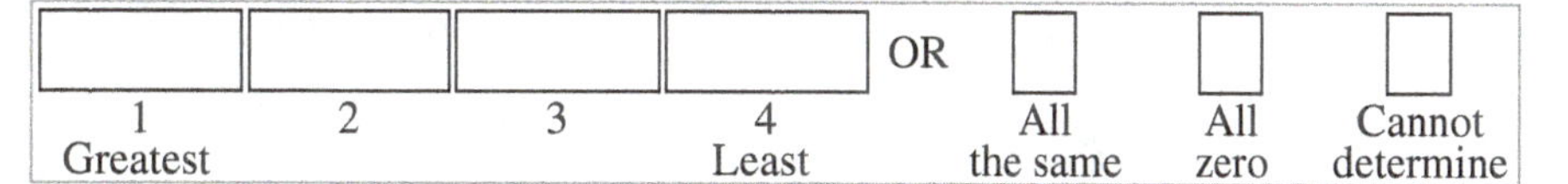

Explain your reasoning.

B1-QRT58: Moving Asteroids near Spaceship—Velocity Direction

Shown are five asteroids and a spaceship, all moving in the same direction away from Earth. The velocities of the asteroids and of the spaceship are given as measured from Earth.

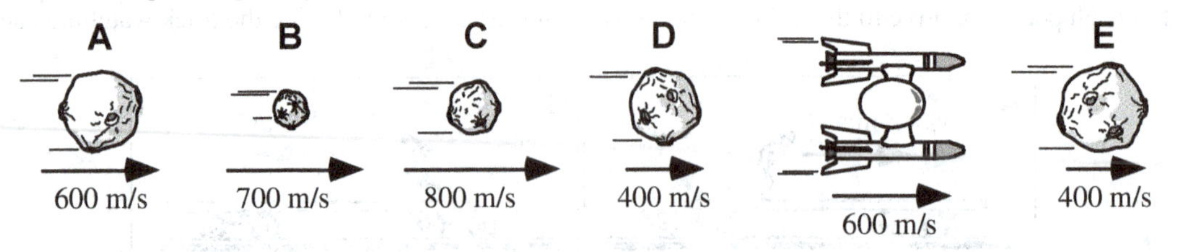

(a) List the asteroids that are moving toward the spaceship.

(b) List the asteroids that are moving away from the spaceship.

Explain your reasoning.

B2 MOTION IN TWO DIMENSIONS

B2-CT01: MOTORCYCLE TRIPS—DISPLACEMENT

Shown below are the paths two motorcyclists took on an afternoon ride. Both started at the same place, and both took the same time for the ride. Rider A traveled east for 19 km and then south for 4 km. Rider B traveled south for 7 km and then east for 16 km.

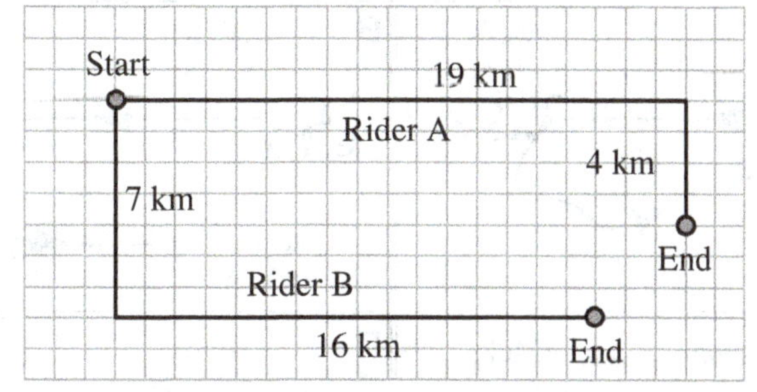

(a) Is the magnitude of the displacement of the rider (i) *greater in case A,* (ii) *greater in case B*, or (iii) *the same in both cases*? _____
Explain your reasoning.

(b) Is the magnitude of the average velocity of the rider (i) *greater in case A,* (ii) *greater in case B*, or (iii) *the same in both cases*? _____
Explain your reasoning.

B2-SCT02: MOTORCYCLE TRIPS—DISPLACEMENT

The paths three motorcyclists took on an afternoon ride are shown. Riders A and C traveled from the coffee shop to the mechanic's garage along different paths, while Rider B traveled from the garage to the coffee shop. Three physics students discussing these rides make the following contentions:

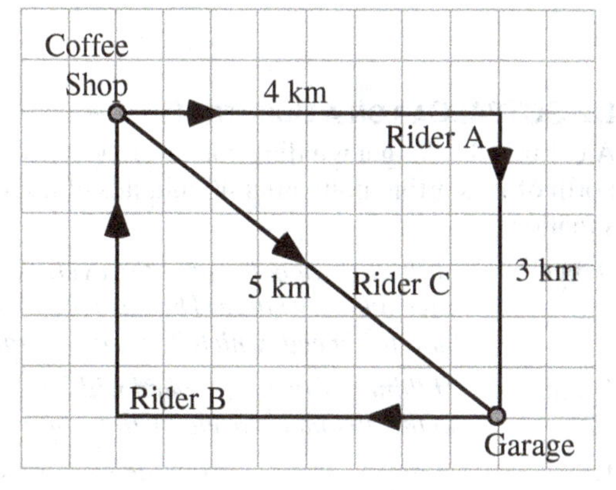

Ali: *"The lengths of the paths that Riders A and B travel are the same, so they have the same displacement. Rider C has the smallest displacement."*

Bob: *"I agree that Rider C has the smallest displacement, because the diagonal path is shortest. But the displacements of Riders A and B are actually different, because their directions are opposite each other."*

Carol: *"I think the displacements of all three riders are the same, because they go between the same two points. What path they follow doesn't matter."*

With which, if any, of these three students do you agree?

Ali _____ Bob _____ Carol _____ None of them_____

Explain your reasoning.

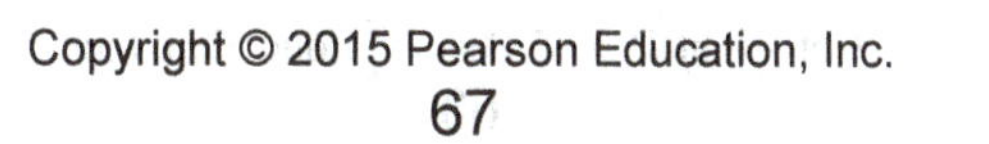

B2-RT03: Students' Journeys—Average Velocity

Four students went out for pizza to celebrate after acing their physics final. All of them went directly from their high school to the nearby pizzeria, but they returned along the paths shown, taking different times. Values for the round-trip distances they traveled and the total times they took to walk their routes are given in the figures.

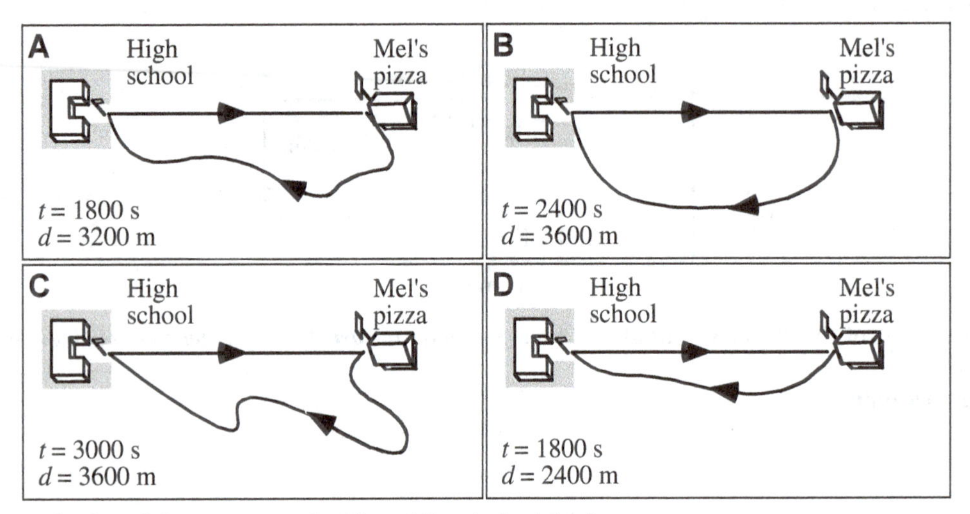

Rank the magnitudes of the average velocities of the students' trips.

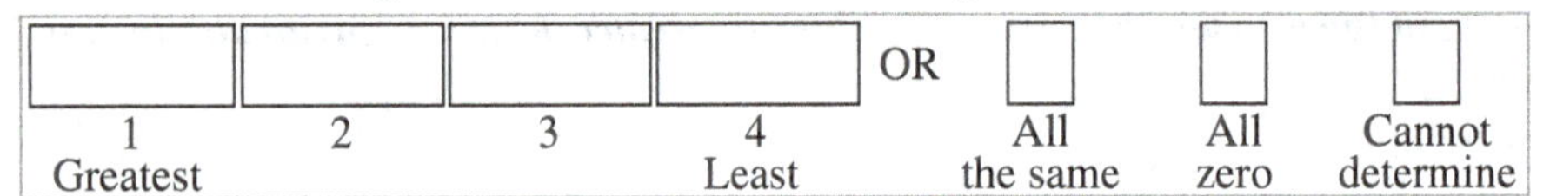

Explain your reasoning.

B2-SCT04: Car on a Country Road—Average Speed and Velocity

A car travels along a winding country road, speeding up and slowing down as it goes. The car ends up 30 km directly north of its starting point after 40 minutes of travel. Three students make the following contentions about this situation:

Adnan: *"I think if we calculated the average velocity and average speed for the car, the average velocity would have a larger value. The car moved in two dimensions, so the velocity, which is a vector, will be greater than the speed, which is a scalar quantity."*

Bunmi: *"I think the average speed will have a larger value than the average velocity. The road is not straight, so the distance traveled will be more than the displacement."*

Cici: *"I don't see any reason the average velocity and average speed would have different values. They are just different names for the same thing."*

With which, if any, of these three students do you agree?

Adnan _____ Bunmi _____ Cici _____ None of them_____

Explain your reasoning.

B2-QRT05: VELOCITY AND POSITION OF THE MOON—VELOCITY CHANGE DIRECTION

The position and velocity of the moon are shown at two times, about seven days apart.

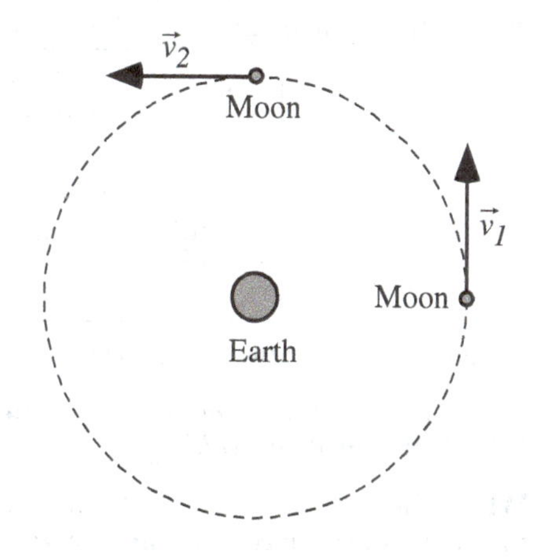

Find the direction of the change in velocity of the moon in this time interval. If the change in velocity is zero, state that explicitly.

Explain your reasoning.

B2-SCT06: MOTORCYCLE ON ROAD COURSE—ACCELERATION

A motorcycle is slowing down as it travels through a bend in a road. The path of the motorcycle is the dashed line shown in the bird's-eye view. The arrow represents the motorcycle's velocity at the instant shown. Three physics students make the following contentions about the acceleration of the motorcycle:

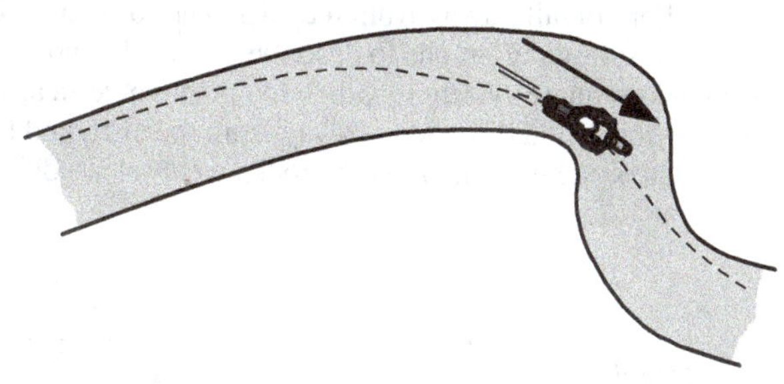

Alexi: *"The motorcycle's acceleration is in the opposite direction to the velocity since it is slowing down."*

Bindi: *"No, the acceleration will have two components, one opposite the velocity and the other toward the center of the curve."*

Carlos: *"I don't think the motorcycle has an acceleration, since it is braking."*

With which, if any, of these three students do you agree?

Alexi _____ Bindi _____ Carlos _____ None of them_____

Explain your reasoning.

B2-WWT07: Speedboats Changing Velocities—Acceleration

Two speedboats are racing on a lake. In 10 seconds, Boat A goes from traveling east at 15 m/s to traveling north at 20 m/s. In the same time interval, Boat B goes from 20 m/s east to 25 m/s east.

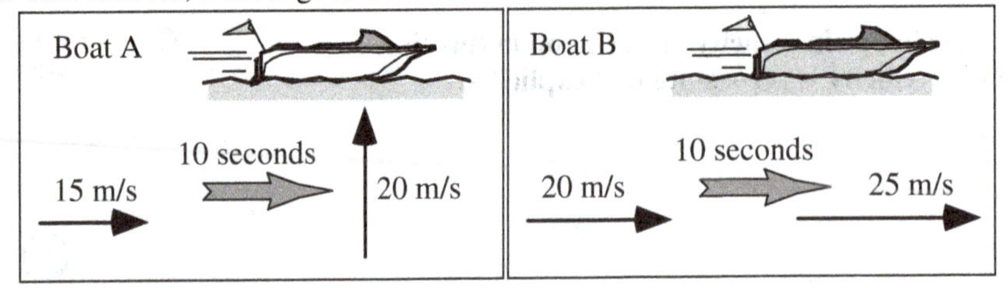

A student watching the race states:

"These two boats have the same acceleration for the 10-second interval since they both changed their velocities by 5 m/s in that time interval."

What, if anything, is wrong with this student's contention? If something is wrong, identify it and explain how to correct it. If the contention is correct, explain why.

B2-WWT08: Falling Rock and Thrown Rock—Velocity-Time Graphs

Rock A is dropped from the top of a cliff at the same instant that Rock B is thrown horizontally away from the cliff. The rocks are identical. A student draws the following graphs to describe part of the motion of the rocks, using a coordinate system in which the positive vertical direction is up, the positive horizontal direction is away from the cliff, and the origin of the coordinate system is the point the rocks were released from.

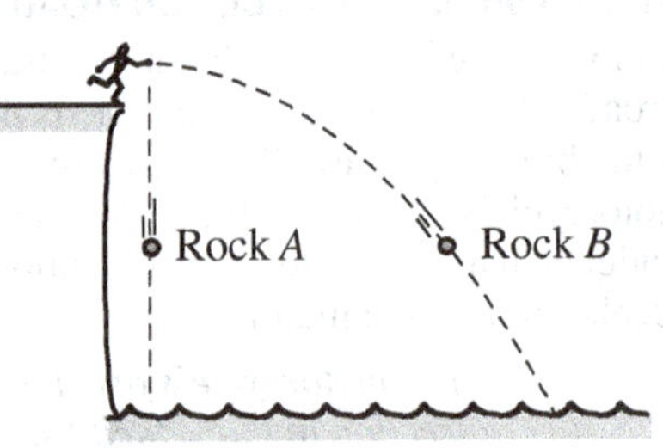

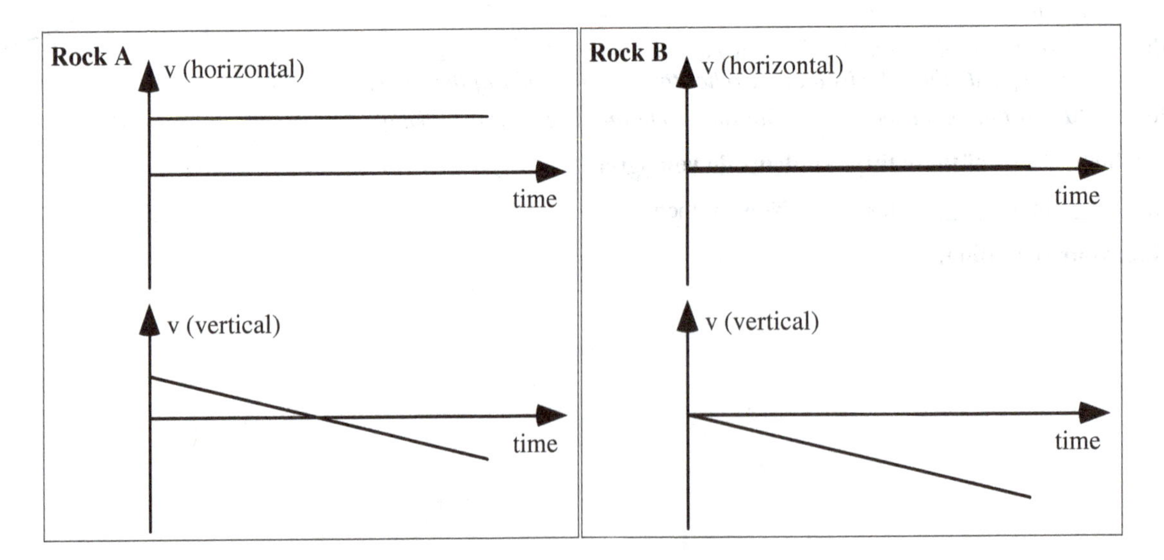

What, if anything, is wrong with these graphs for the motions of the two rocks? If something is wrong, identify it and explain how to correct it. If the graphs are correct, explain why.

B2-QRT09: PROJECTILE MOTION—VELOCITY-TIME AND ACCELERATION-TIME GRAPHS

A baseball is thrown from point *S* in right field to home plate. The dashed line in the diagram shows the path of the ball. Use a coordinate system with up as the positive vertical direction and to the right as the positive horizontal direction, with the origin at the point the ball was thrown from (point *S*).

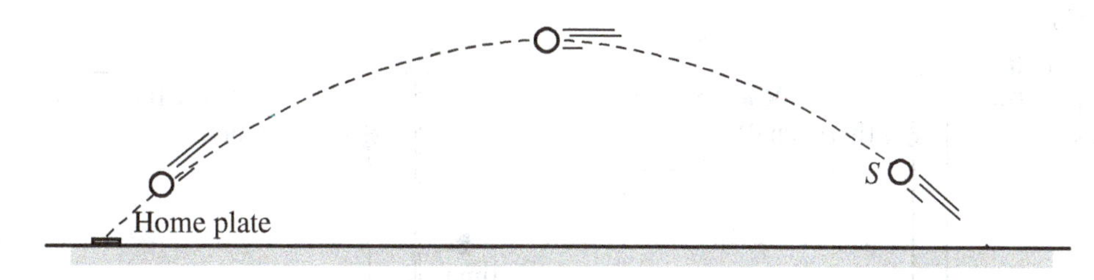

On the axes below, sketch graphs for the indicated quantities:

(a) The horizontal velocity versus time and the vertical velocity versus time.

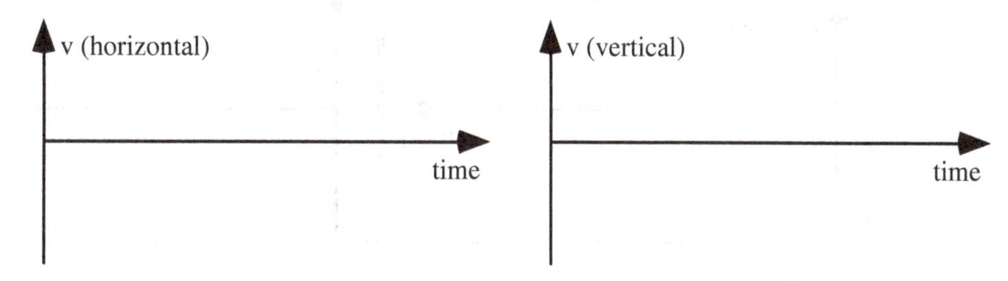

Explain your reasoning.

(b) The horizontal acceleration versus time and the vertical acceleration versus time.

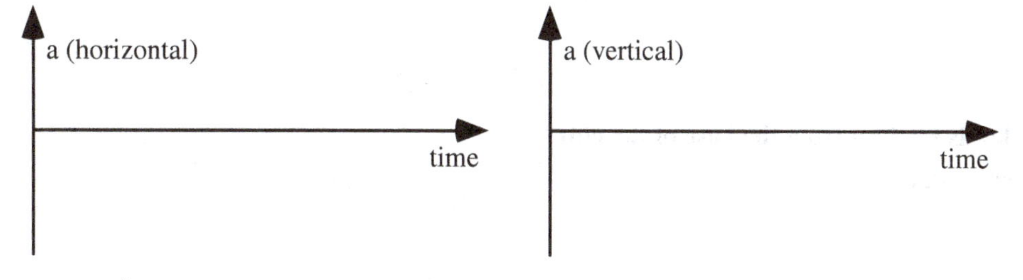

Explain your reasoning.

B2-QRT10: Projectile Motion for Two Rocks—Velocity-Time and Acceleration-Time Graphs

Two identical rocks are thrown horizontally from a cliff, with Rock A having a greater velocity at the instant it is released than Rock B. Use a coordinate system with down as the positive vertical direction, away from the cliff as the positive horizontal direction, and with the origin of the coordinate system at the bottom of the cliff directly below the release point.

(a) Sketch the velocity versus time graphs for each of the rocks.

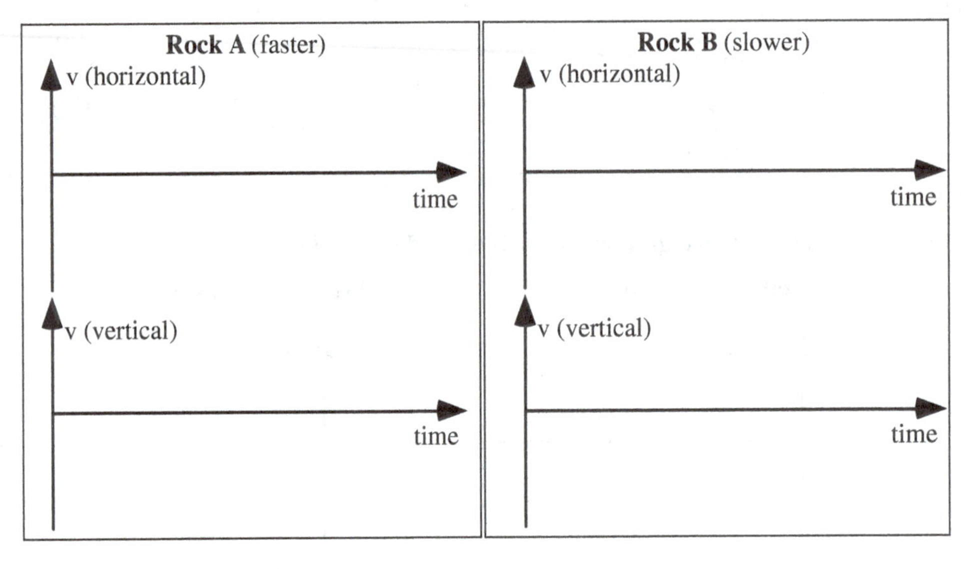

(b) Which rock hits the ground first?

Explain your reasoning.

(c) Which rock lands farthest from the base of the cliff?

Explain your reasoning.

B2-QRT11: BASEBALL PROJECTILE MOTION—VELOCITY-TIME AND ACCELERATION-TIME GRAPHS

A baseball is thrown from point S in right field to home plate. The dashed line shows the path of the ball.

Home plate

S

Use a coordinate system with up as the positive vertical direction and to the left as the positive horizontal direction, and with the origin at home plate.

Select the graph from the choices below that best represents:

(i) horizontal velocity versus time graph ____ Explain your reasoning.

(ii) horizontal acceleration versus time graph____ Explain your reasoning.

(iii) vertical velocity versus time graph____ Explain your reasoning.

(iv) vertical acceleration versus time graph____ Explain your reasoning.

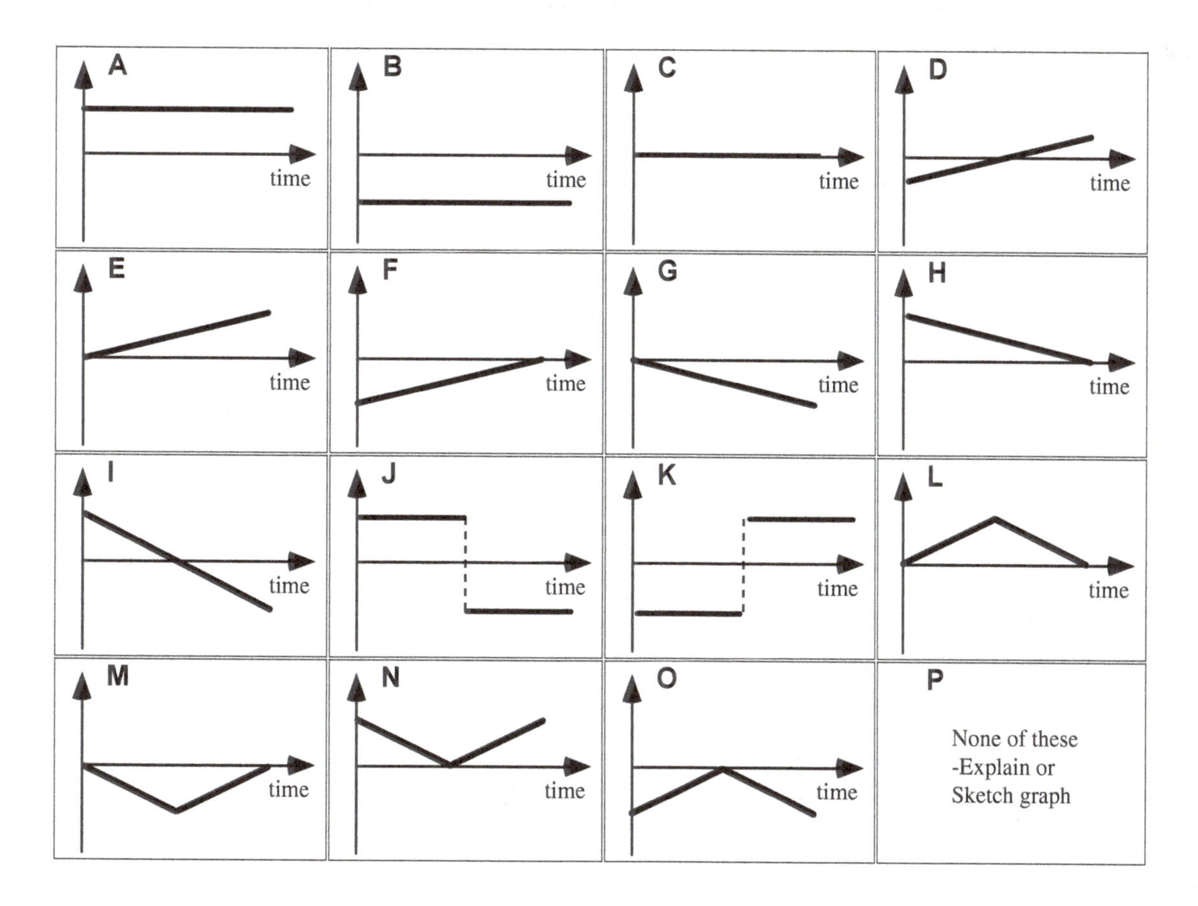

B2-CRT12: Projectile Motion for Two Rocks—Velocity-Time Graphs

Two students throw two rocks horizontally from a cliff with different velocities. Both rocks hit the water below at the same time, but Rock B hits farther from the base of the cliff. Use coordinates where up is the positive direction, away from the cliff is the positive horizontal direction, and the origin is at the top of the cliff at the point of release.

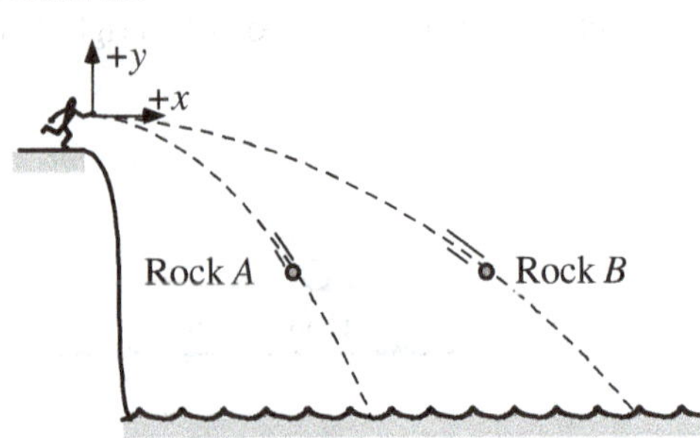

Sketch below the velocity versus time graphs for each rock.

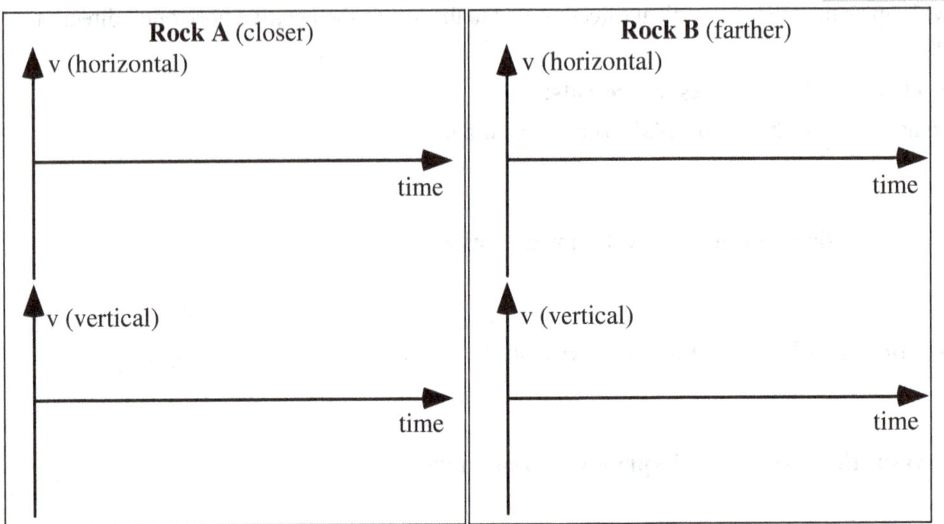

Explain your reasoning.

B2-CRT13: PROJECTILE MOTION FOR TWO ROCKS—ACCELERATION-TIME GRAPHS

Two students throw two rocks horizontally from a cliff with different velocities. Both rocks hit the water below at the same time, but Rock B hits farther from the base of the cliff. Use coordinates where up is the positive direction, away from the cliff is the positive horizontal direction, and the origin is at the top of the cliff at the point of release.

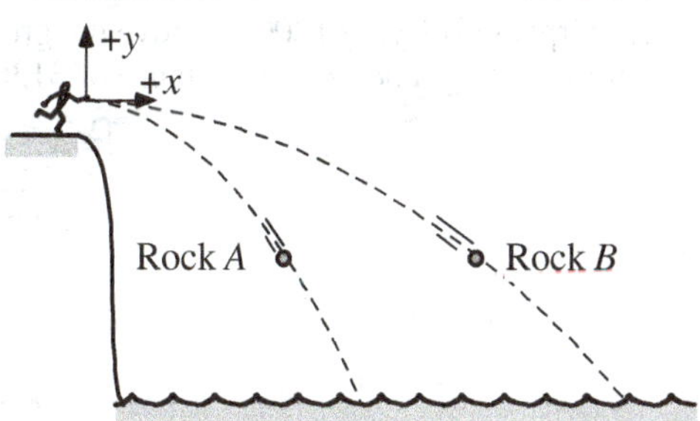

Sketch the acceleration versus time graphs for each rock.

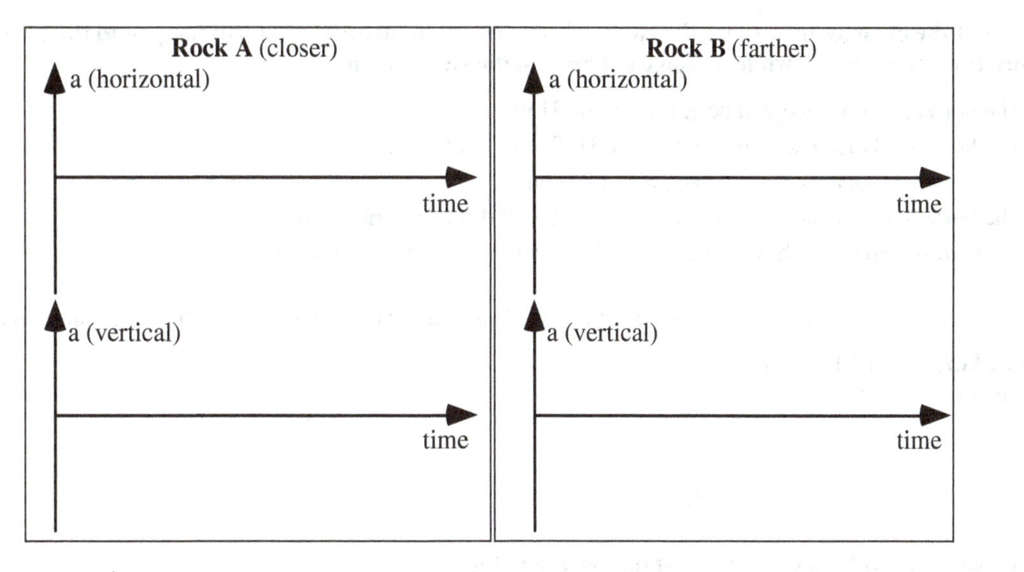

Explain your reasoning.

B2-LMCT14: Dropped Practice Bomb—Horizontal Distance Traveled

An airplane is flying 1200 m above the ground at a speed of 200 m/s. It drops a practice bomb that hits the ground after traveling a horizontal distance of 3130 m.

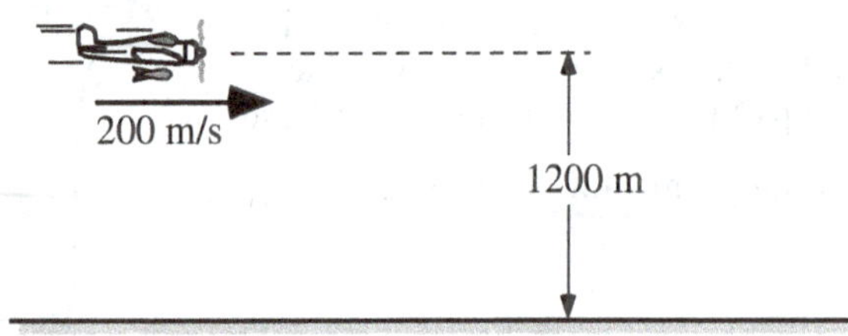

For each of the changes below, use the choices below (i)-(v) to identify what will happen to the horizontal distance the bomb travels while falling compared to the situation above.

(i) The horizontal distance will be ***greater than*** 3130 m.
(ii) The horizontal distance will be ***less than*** 3130 m but not zero.
(iii) The horizontal distance will be ***equal*** *to* 3130 m.
(iv) The horizontal distance will be ***zero*** (the bomb will drop straight down).
(v) We ***cannot determine*** how this change will affect the horizontal distance.

For each of the following changes, only the feature(s) identified is(are) modified from the given situation above.

(a) The plane's speed is tripled. _____
Explain your reasoning.

(b) The plane is climbing straight up at the release point. _____
Explain your reasoning.

(c) The plane is flying in level flight at an altitude of 1,100 m. _____
Explain your reasoning.

(d) The mass of the bomb is increased. _____
Explain your reasoning.

(e) The bomb is thrown from the plane with a vertical downward velocity of 15 m/s. _____
Explain your reasoning.

B2-RT15: RIFLE SHOTS—TIME TO HIT GROUND

The rifles in the figures are being fired horizontally (straight outward, off platforms). The bullets fired from the rifles are all identical, but the rifles propel the bullets at different speeds. The speed of each bullet and the height of each platform are given. All of the bullets miss the targets and hit the ground.

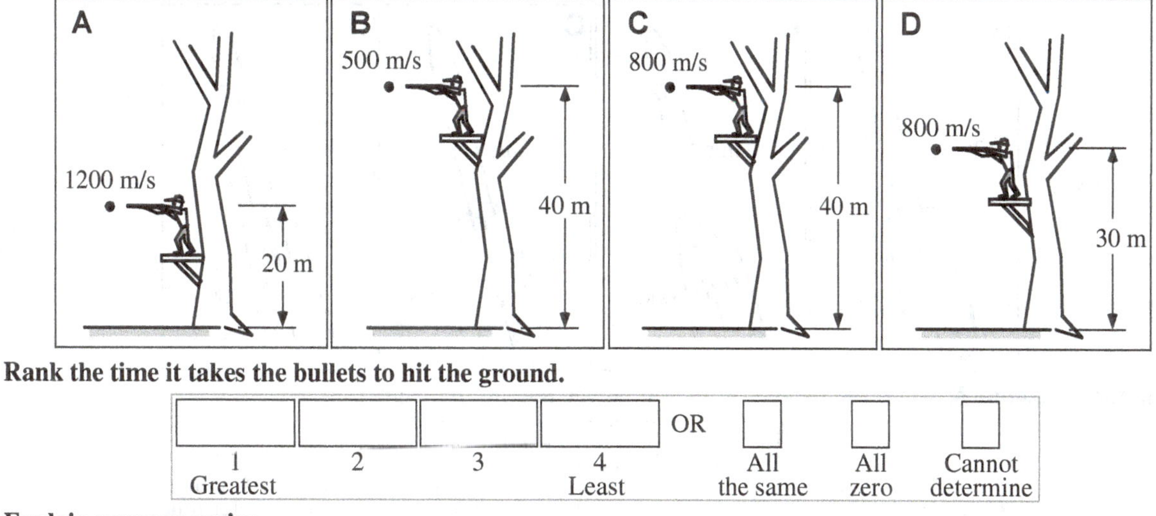

Rank the time it takes the bullets to hit the ground.

1 Greatest	2	3	4 Least	OR	All the same	All zero	Cannot determine

Explain your reasoning.

B2-RT16: ARROWS—MAXIMUM HEIGHTS

All of the arrows shown were shot from the same height and at the same angle. Though the arrows have the same size and shape, they are made of different materials, so they have different masses, and they have different speeds as they leave the bows.

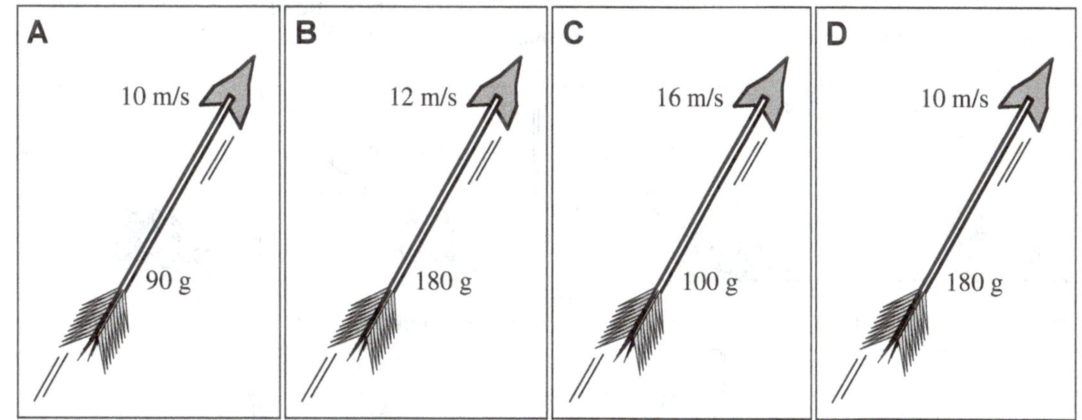

Rank the maximum heights the arrows reach.

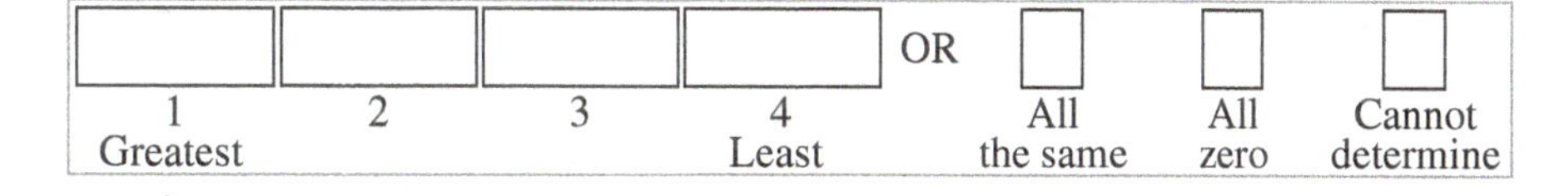

Explain your reasoning.

B2-RT17: Model Rockets Fired at an Angle—Horizontal Speed at Top

The six model rockets shown are all at the same height and have just had their engines turned off. All of the rockets are aimed upward at the same angle, but their speeds differ. Though the rockets are all the same size and shape, they carry different loads, so their masses vary.

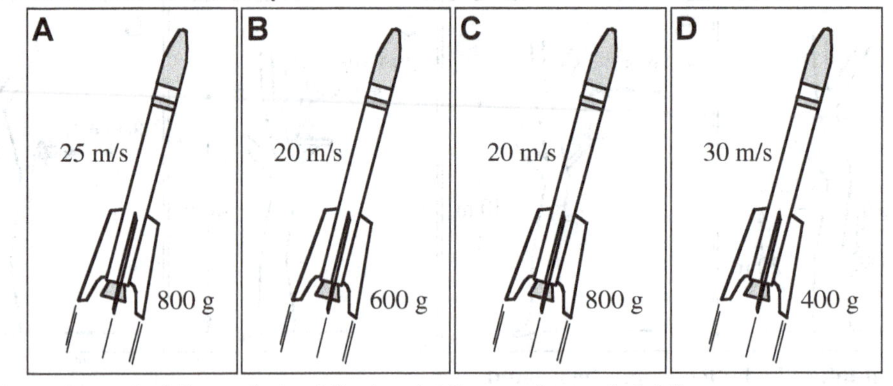

Rank the horizontal speed of the rockets at the top (at the maximum height).

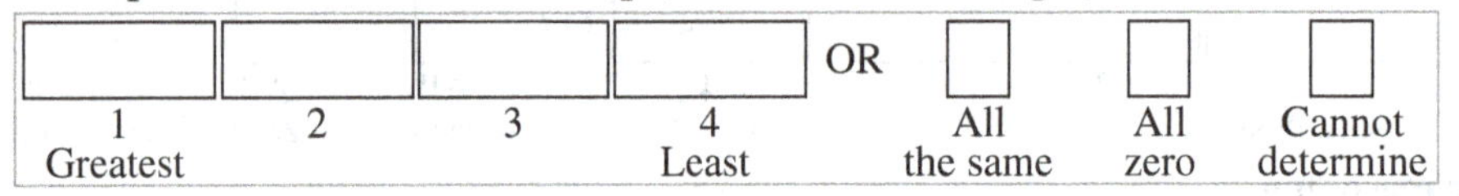

Explain your reasoning.

B2-RT18: Cannonballs—Acceleration at the Top

All of the cannons in the figures are identical, and all are aimed at the same angle of 35 degrees to the horizontal. The cannonballs are all the same size and shape, but the masses of the cannonballs, as well as their speeds as they leave the cannons, are different.

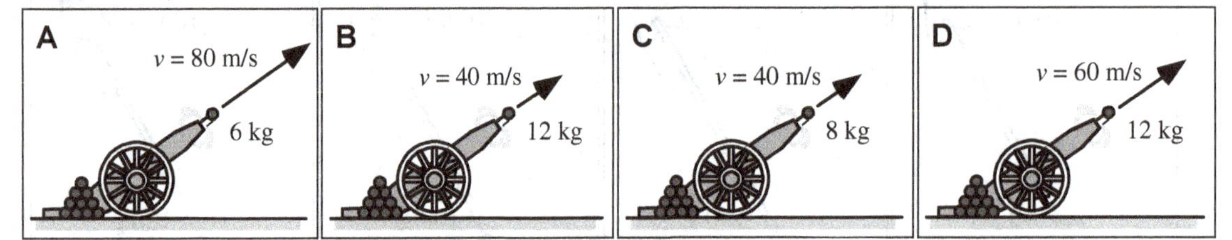

Rank the acceleration of the cannonballs when they reach their highest point.

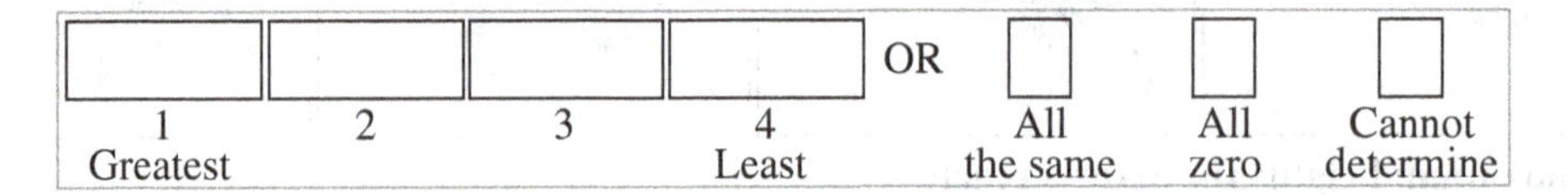

Explain your reasoning.

B2-RT19: CANNONBALLS—HORIZONTAL DISTANCE

Cannonballs of different masses are shot from cannons at various angles above the horizontal. The velocity of each cannonball as it leaves the cannon is given, along with the horizontal component of that velocity, which is the same.

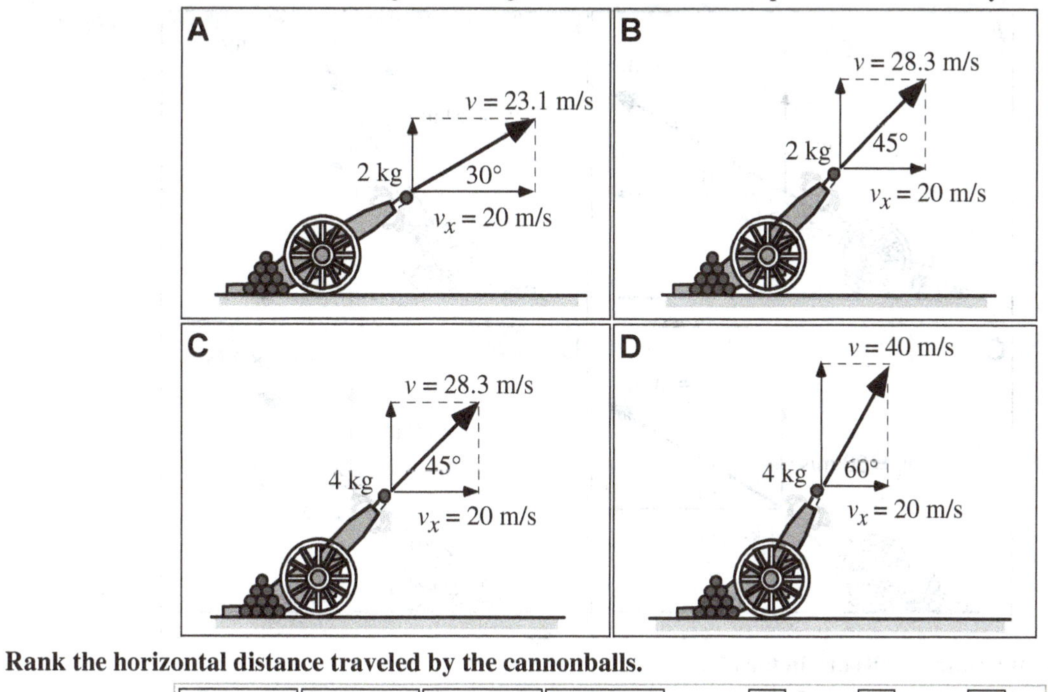

Rank the horizontal distance traveled by the cannonballs.

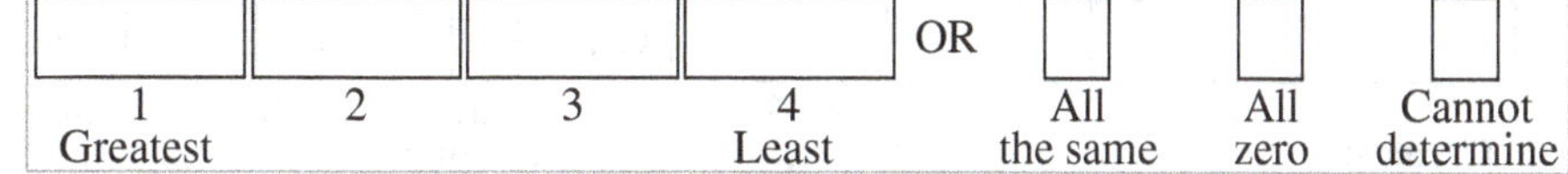

Explain your reasoning.

B2-WWT20: HORIZONTALLY LAUNCHED TOY TRUCKS—TIME IN AIR I

A toy truck is launched horizontally from a table of height H. In Case A, the toy truck leaves the table with a speed v, and in Case B the toy truck leaves the table with a speed $2v$.

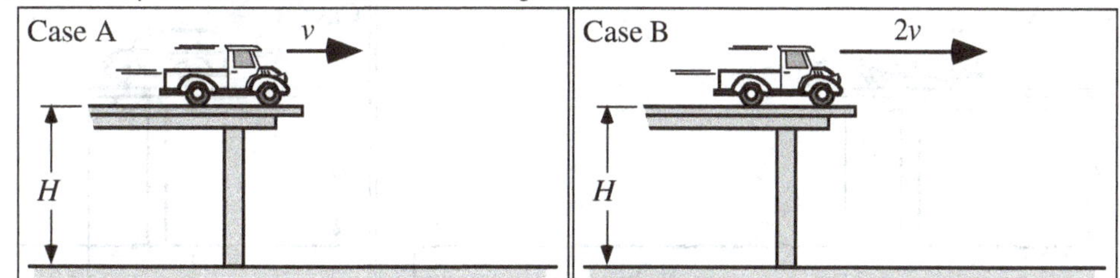

A student comparing the time the trucks are in the air in these cases states:

"The faster you go, the less time it takes to get there. The truck in Case A will be in the air longer."

What, if anything, is wrong with this statement? If something is wrong, identify it and explain how to correct it. If this statement is correct, explain why.

B2-RT21: CANNONBALLS—TIME IN AIR

Cannonballs with different masses are shot from cannons at various angles above the horizontal. The velocity of each cannonball as it leaves the cannon is given, along with the same vertical component of that velocity.

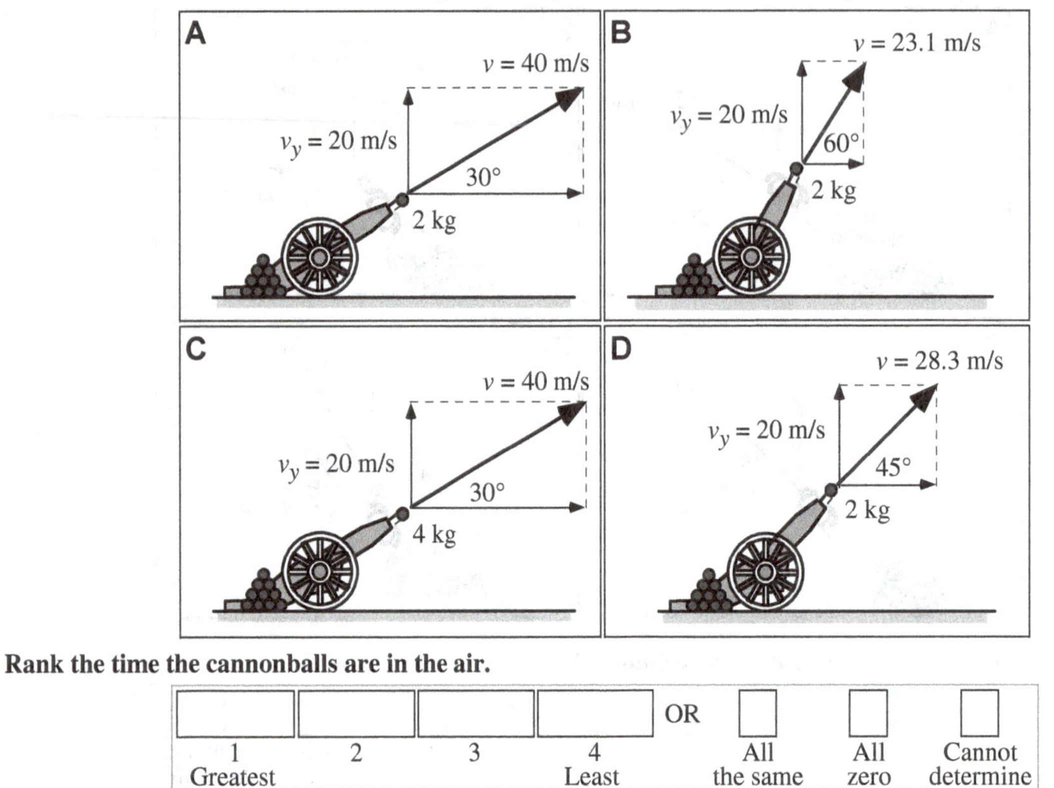

Rank the time the cannonballs are in the air.

				OR			
1 Greatest	2	3	4 Least		All the same	All zero	Cannot determine

Explain your reasoning.

B2-WWT22: HORIZONTALLY LAUNCHED TOY TRUCKS—TIME IN AIR II

A toy truck is launched horizontally from a table. In Case A the toy truck hits the floor a horizontal distance $2x$ from the edge of the table, and in Case B the toy truck hits the floor a horizontal distance x from the table.

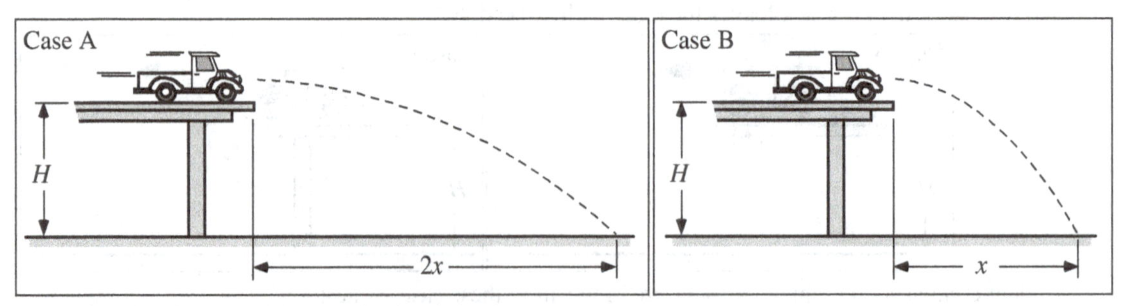

A student comparing the time the trucks are in the air in these cases states:

"The farther you go, the longer it takes to get there. The truck in Case A will be in the air longer."

What, if anything, is wrong with this statement? If something is wrong, identify it and explain how to correct it. If this statement is correct, explain why.

B2-CT23: TOY TRUCKS ROLLING FROM TABLES WITH DIFFERENT HEIGHTS—TIME

Two toy trucks roll off the ends of tables. The heights of the tables, the speeds of the trucks, and the masses of the trucks are given.

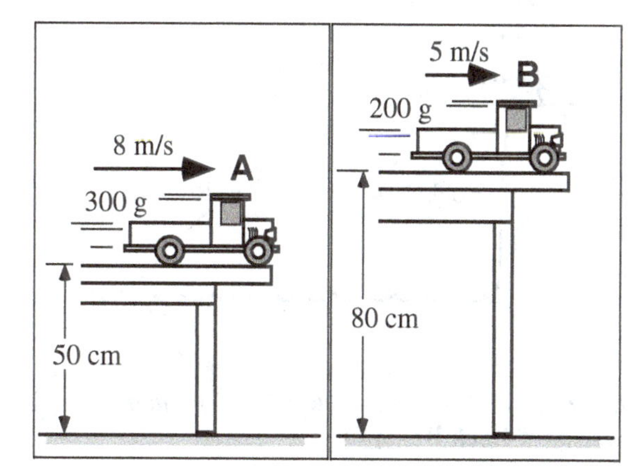

Will Truck A be in the air for (i) *a longer time,* (ii) *a shorter time*, or (iii) *the same time* as Truck B before it reaches the floor? _____

Explain your reasoning.

B2-CT24: TOY TRUCKS WITH DIFFERENT SPEEDS ROLLING FROM IDENTICAL TABLES—TIME

Two toy trucks roll off the ends of identical tables. The speeds and masses of the trucks are given.

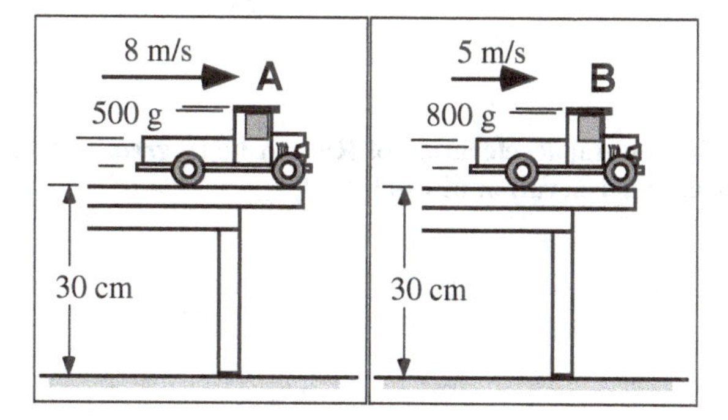

Will Truck A be in the air for (i) *a longer time,* (ii) *a shorter time*, or (iii) *the same time* as Truck B before it reaches the floor? _____

Explain your reasoning.

B2-CT25: Projectile Motion for Two Rocks—Velocity and Acceleration

Two identical rocks are thrown horizontally from a cliff with different velocities. The rocks are thrown at the same time and are shown below while they are still in the air after a few seconds.

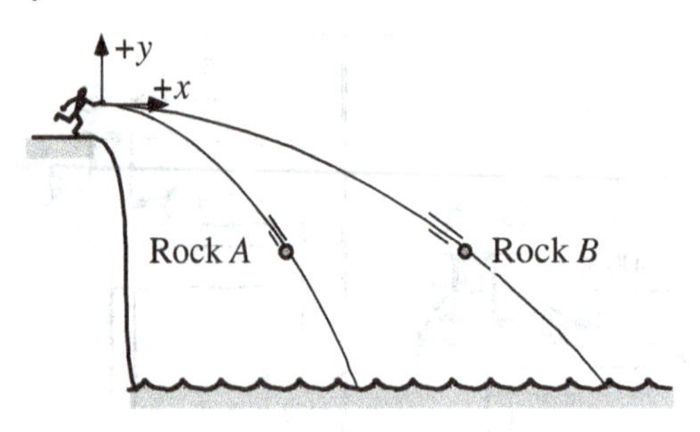

For the instant shown:

(a) Will the magnitude of the horizontal velocity of Rock A be (i) *greater than,* (ii) *less than,* or (iii) *equal to* the magnitude of the horizontal velocity of Rock B? _____

Explain your reasoning.

(b) Will the magnitude of the vertical velocity of Rock A be (i) *greater than,* (ii) *less than,* or (iii) *equal to* the magnitude of the vertical velocity of Rock B? _____

Explain your reasoning.

(c) Will the magnitude of the horizontal acceleration of Rock A be (i) *greater than,* (ii) *less than,* or (iii) *equal to* the magnitude of the horizontal acceleration of Rock B? _____

Explain your reasoning.

(d) Will the magnitude of the vertical acceleration of Rock A be (i) *greater than,* (ii) *less than,* or (iii) *equal to* the magnitude of the vertical acceleration of Rock B? _____

Explain your reasoning.

B2-QRT26: CONSTANT SPEED CAR ON OVAL TRACK—ACCELERATION AND VELOCITY DIRECTIONS

A car travels clockwise at a constant speed around an oval track.

In the table below, indicate the direction of the velocity and acceleration of the car for the labeled points. Use the direction labels in the rosette at the far right: **J** for no direction, **K** for into the page, **L** for out of the page, or **M** if none of these are correct.

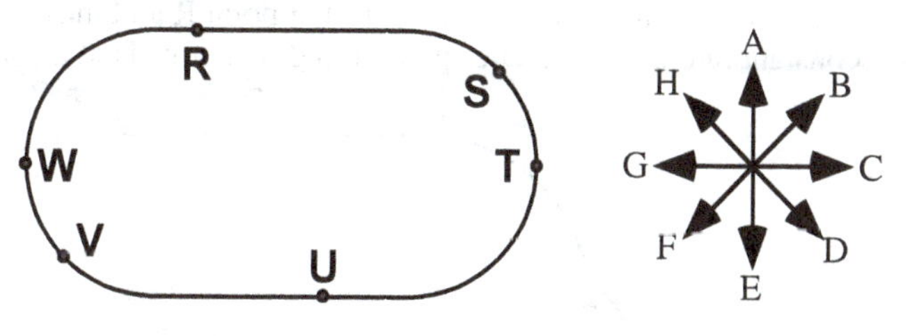

Point on track	Velocity direction	Acceleration direction
R		
S		
T		
U		
V		
W		

Explain your reasoning.

B2-QRT27: Car on Oval Track—Direction of the Acceleration and Velocity

A car on an oval track starts from rest at point R and moves clockwise around the track. It increases its speed at a constant rate until it reaches point T and then travels at a constant speed until it returns to point R.

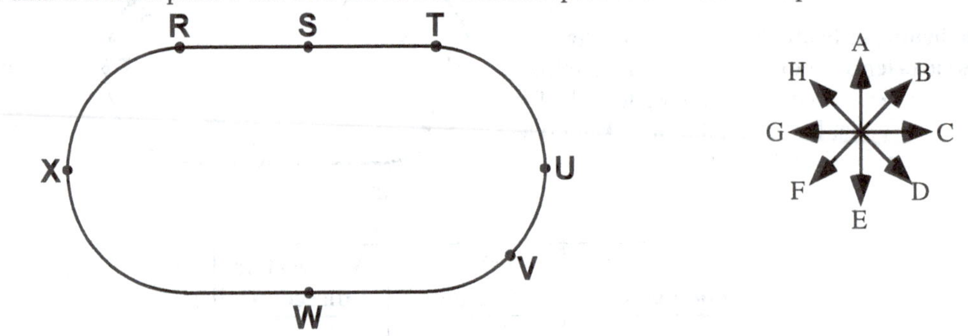

In the table below, give the direction of the velocity and acceleration of the car at the indicated points. Use the direction labels in the rosette to the right of the racetrack drawing: **J** for no direction, **K** for into page, **L** for out of page, or **M** if none of these are correct.

Point on track	Velocity direction	Acceleration direction
S		
U		
V		
W		
X		

Explain your reasoning.

B3 NEWTON'S LAWS

B3-RT01: PACKAGES MOVING ON A CONVEYOR BELT—NET FORCE

Various packages with different masses are moving on a constant-speed conveyer belt. At the instant shown below, all packages have the same constant velocity of 2 m/s directed to the right. The packages do not slip on the belt. All masses are given in the diagram in terms of M, the mass of the smallest package.

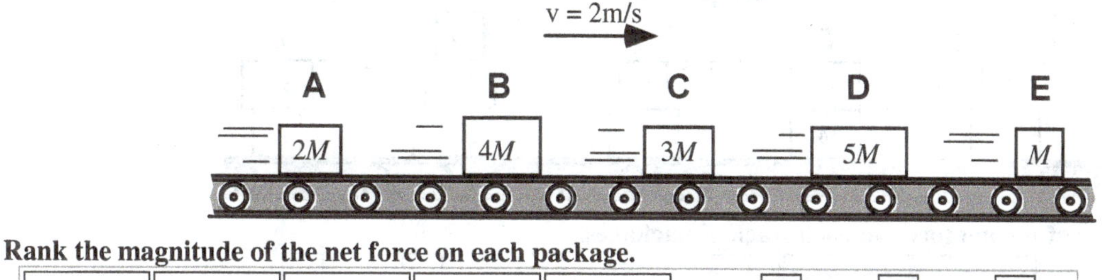

Rank the magnitude of the net force on each package.

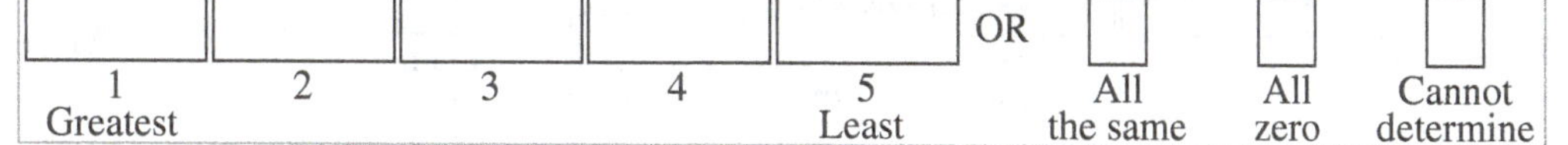

Explain your reasoning.

B3-RT02: WATER SKIERS—NET FORCE

Water skiers are pulled at a constant speed by a towrope attached to a speedboat. Because the weight of the skiers and the type of skis they are using varies, they experience different resistive forces from the water. Values for this resistive force (RF) and for the speed of the skiers are given.

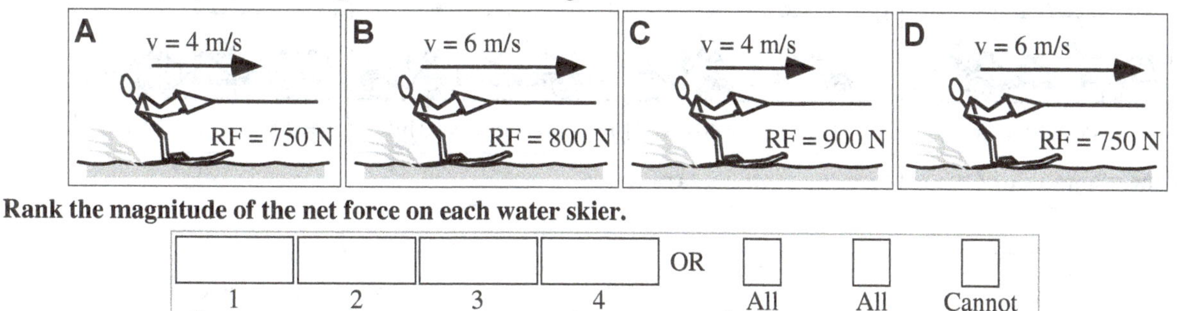

Rank the magnitude of the net force on each water skier.

1 Greatest	2	3	4 Least	OR	All the same	All zero	Cannot determine

Explain your reasoning.

B3-RT03: Stacked Packages Moving on Moving Conveyor Belt—Net Force

Various stacks of packages are traveling along a conveyer belt. At the instant shown below, all packages have the same velocity of 3 m/s to the right. The packages do not slip on the belt. All masses are given in the diagram in terms of *M*, the mass of the smallest package.

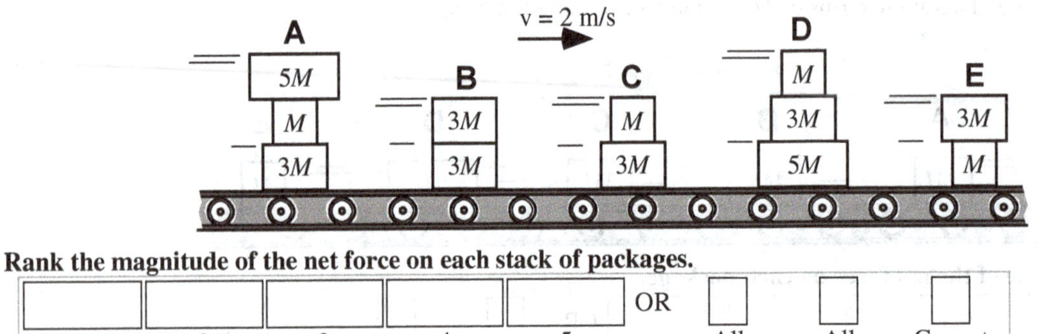

Rank the magnitude of the net force on each stack of packages.

1 Greatest	2	3	4	5 Least	OR	All the same	All zero	Cannot determine

Explain your reasoning.

B3-RT04: Moving Car with Boat Trailer—Net Force on Boat Trailer

All the trailers and cars shown are identical, but the boat trailers have different loads. In each case, the car and boat trailer are moving at the constant speed shown.

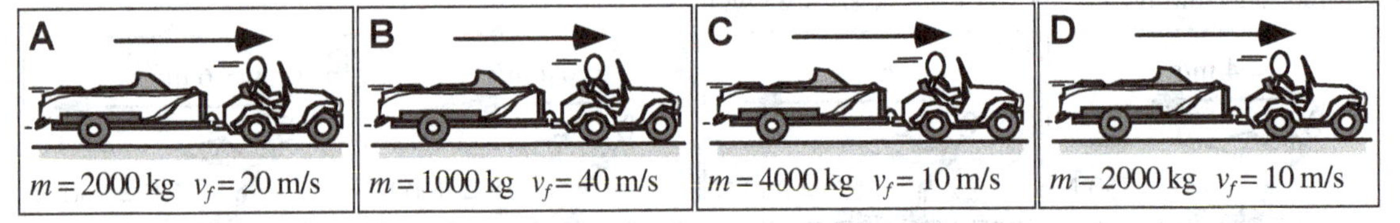

Rank the magnitude of the net force on each boat trailer.

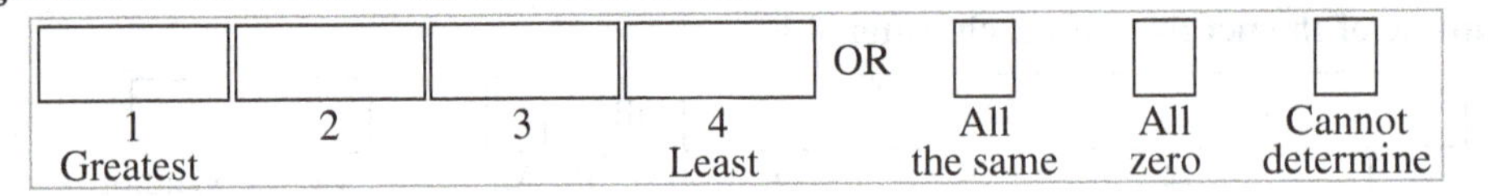

Explain your reasoning.

B3-RT05: MOVING SPACESHIP WITH FOUR CARGO PODS—TENSION IN RODS

A spaceship and four cargo pods are connected together by rods, and they are all moving at a constant velocity of 5000 m/s. All masses are given in the diagram in terms of *M*, the mass of an empty pod.

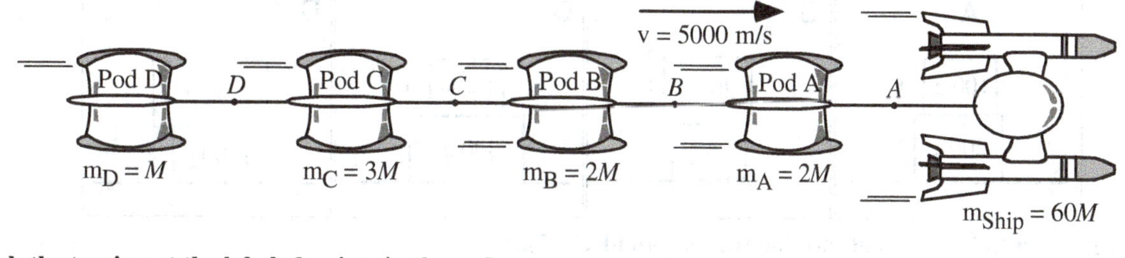

Rank the tension at the labeled points in the rods.

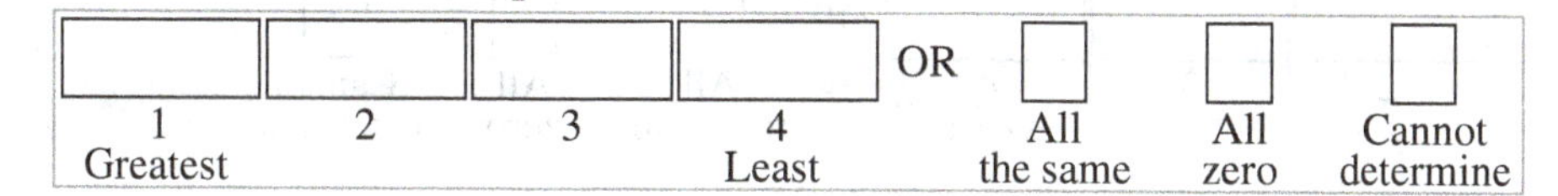

Explain your reasoning.

B3-RT06: CARTS ON INCLINES—NET FORCE

Carts that have a motor and brakes are traveling either up or down inclines at constant speeds. The carts are identical but they carry either a 2 kg or 4 kg load and are on one of two inclines. Incline angles, cart masses, and speeds are given in each figure.

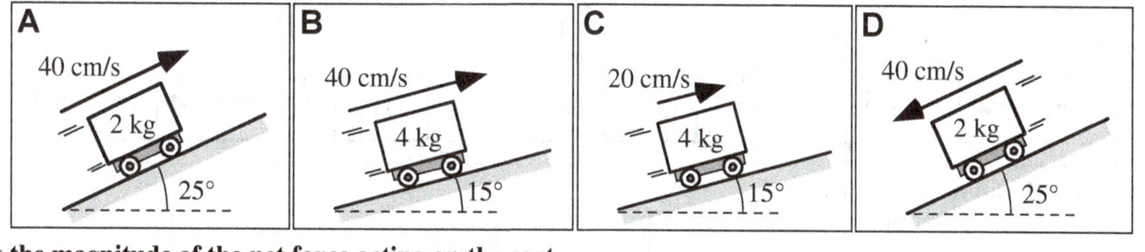

Rank the magnitude of the net force acting on the cart.

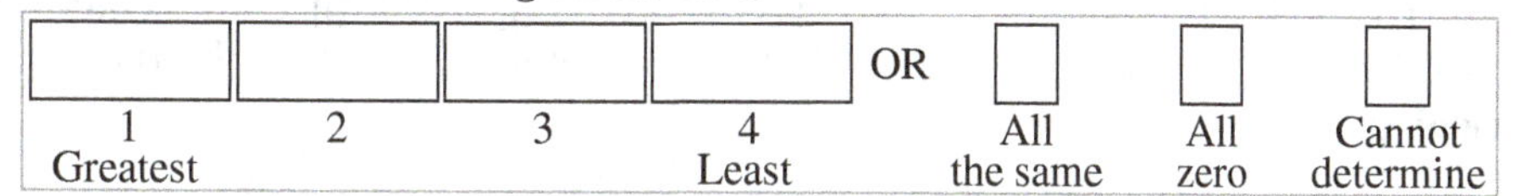

Explain your reasoning.

B3-RT07: Two Stacked Blocks at Rest—Net Force on the Bottom Block

Two wooden blocks with different masses are at rest, stacked on a table. The top block is labeled **1,** and the bottom block is labeled **2**.

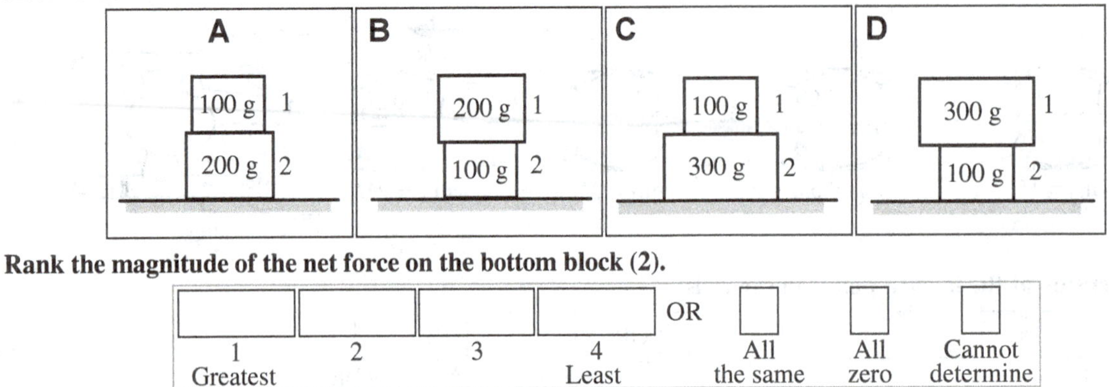

Rank the magnitude of the net force on the bottom block (2).

				OR			
1 Greatest	2	3	4 Least		All the same	All zero	Cannot determine

Explain your reasoning.

B3-RT08: Force Pushing Box—Acceleration

Various similar boxes are being pushed for 10 m across a floor by a net horizontal force as shown below. The mass of the boxes and the net horizontal force for each case are given in the indicated figures. All boxes have the same initial velocity of 10 m/s to the right.

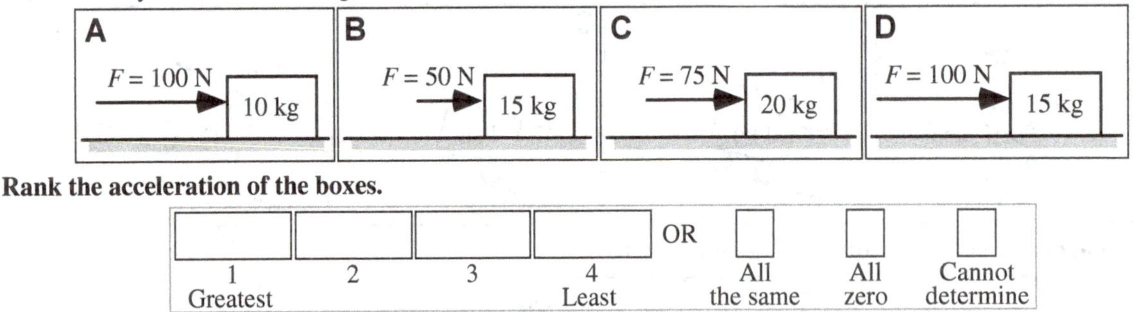

Rank the acceleration of the boxes.

				OR			
1 Greatest	2	3	4 Least		All the same	All zero	Cannot determine

Explain your reasoning.

B3-CT09: BLOCKS IN MOVING ELEVATORS—STRETCH OF SPRING

A spring is attached to the ceiling of an elevator, and a block of mass M is suspended from the spring. The cases are identical except that in Case A the elevator is moving upward with a constant speed of 7 m/s, while in Case B the elevator is moving downward with a constant speed of 9 m/s.

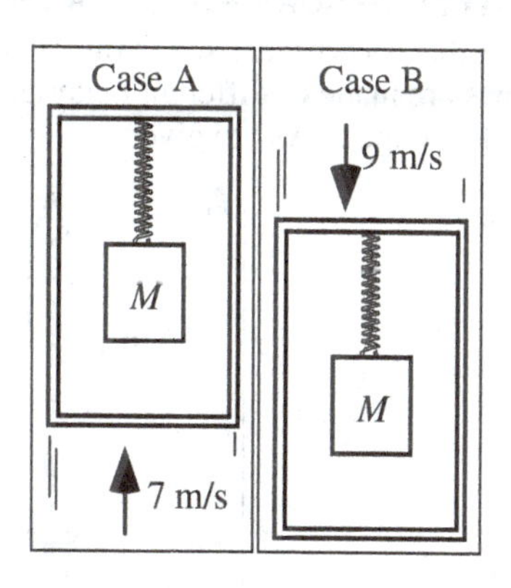

Will the spring be stretched (i) *more* in Case A, (ii) *more* in Case B, or (iii) *the same* in both cases? _____

Explain your reasoning.

B3-RT10: TWO-DIMENSIONAL FORCES ON A TREASURE CHEST—FINAL SPEED

Identical treasure chests (shown from above) each have two forces acting on them. All chests start at rest.

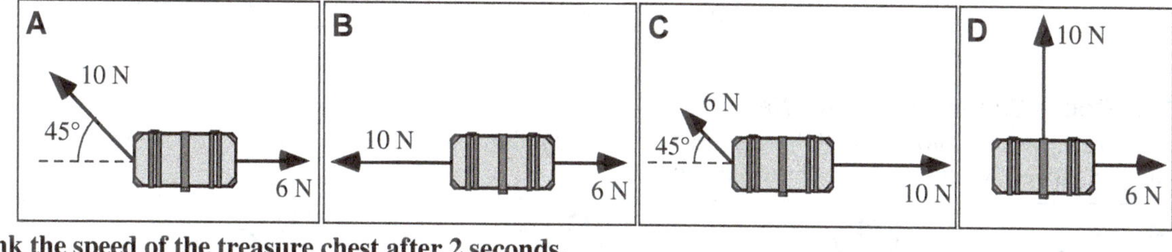

Rank the speed of the treasure chest after 2 seconds.

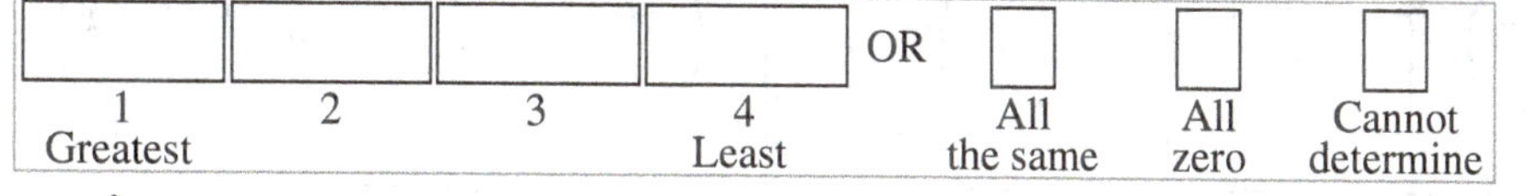

Explain your reasoning.

B3-RT11: Arrows—Acceleration

All of the arrows were shot straight up into the air from the same height, and all are the same size and shape. The arrows are made of different materials so they have different masses. The masses of the arrows and their speeds as they leave the bows are given.

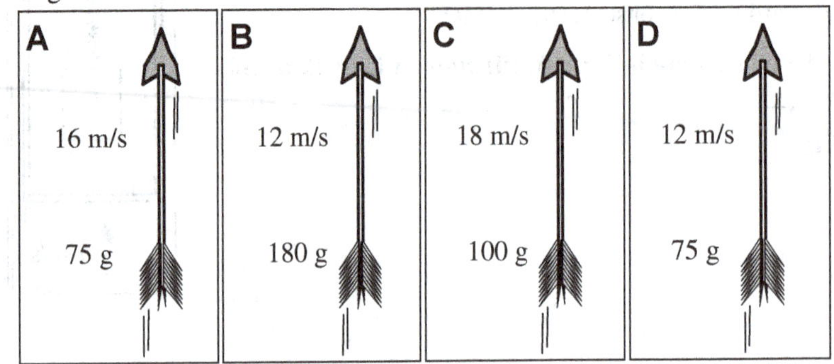

Rank the magnitude of the acceleration of the arrows at the top of their flight.

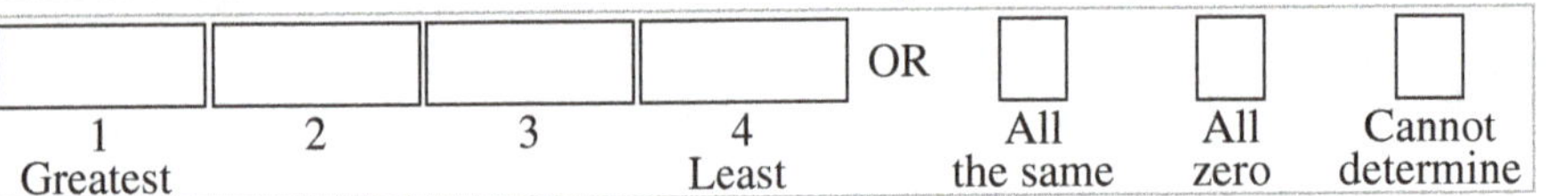

Explain your reasoning.

B3-RT12: Rocks Thrown Upward—Net Force

Rocks that are thrown up into the air all have the same shape, but they have different masses. The masses of the rocks and their speeds when they were thrown are given.

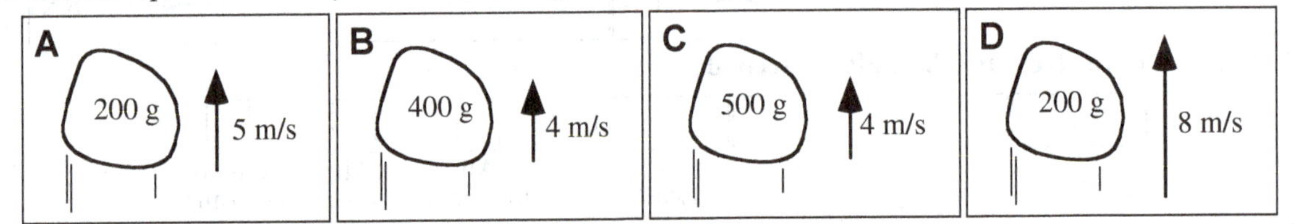

Rank the magnitude of the net force on the rocks just after they are thrown.

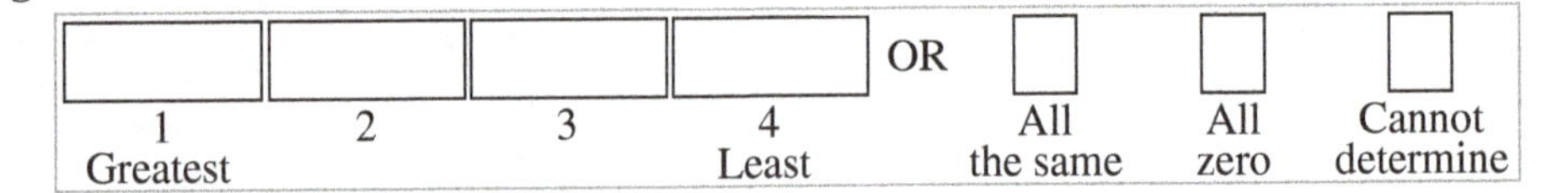

Explain your reasoning.

B3-RT13: ROCKS THROWN DOWNWARD—NET FORCE

Rocks that are thrown straight downward all have the same shape, but they have different masses. The masses of the rocks and their speeds when they were thrown are given.

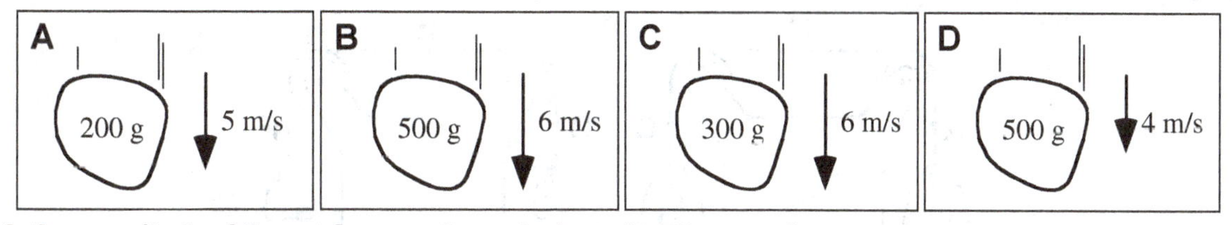

Rank the magnitude of the net force on the rocks just after they are thrown.

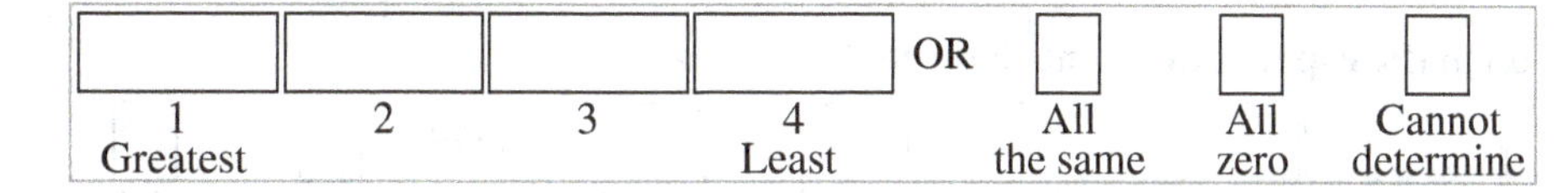

Explain your reasoning.

B3-RT14: ROCKS THROWN DOWNWARD—ACCELERATION

Rocks that are thrown straight downward all have the same shape, but they have different masses. The masses of the rocks and their speeds when they were thrown are given.

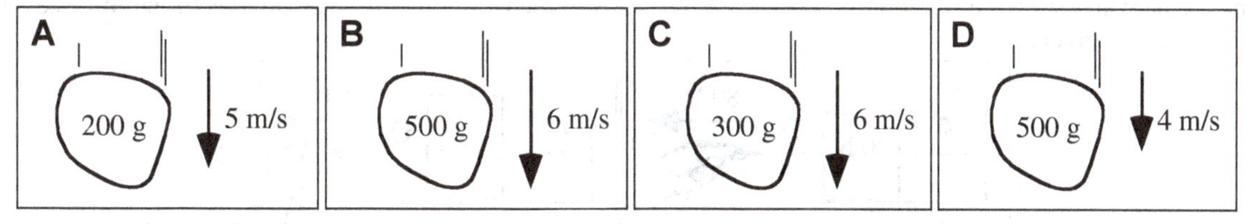

Rank the magnitude of the acceleration of the rocks just after they are thrown.

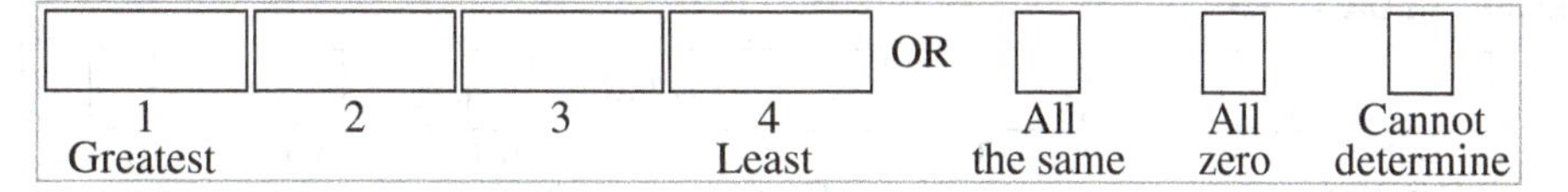

Explain your reasoning.

B3-RT15: Blocks Attached to Fixed Objects—Rope Tension

Two weights are attached by a rope and suspended from pulleys. The weights differ in the two cases, but the systems are at rest in both cases.

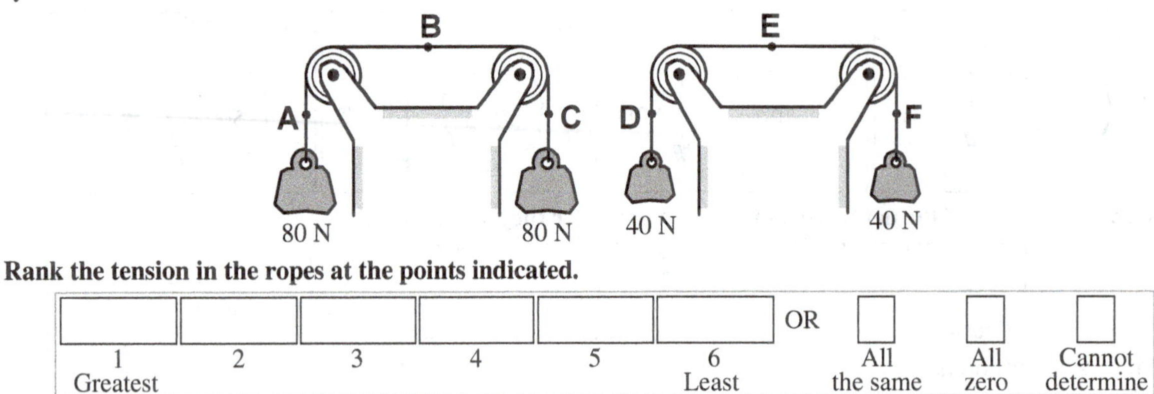

Rank the tension in the ropes at the points indicated.

1 Greatest	2	3	4	5	6 Least	OR	All the same	All zero	Cannot determine

Explain your reasoning.

B3-RT16: Blocks Attached to Wall—Rope Tension

Two blocks are attached by a rope to a wall. A child pulls horizontally on a second rope attached to each block. Both blocks remain at rest on the frictionless surface. The weights of the blocks and the magnitudes of the forces exerted by the child are given.

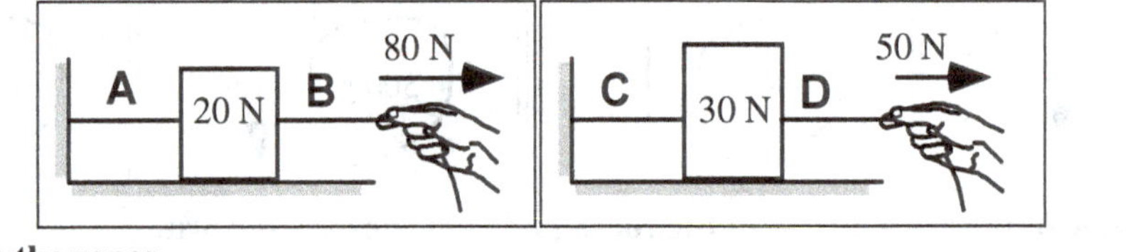

Rank the tensions in the ropes.

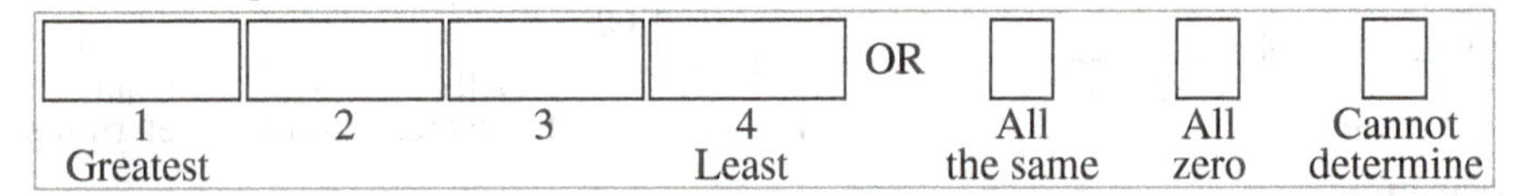

Explain your reasoning.

B3-RT17: HANGING WEIGHTS—ROPE TENSION

Two weights are hung by ropes from a ceiling as shown. All of these systems are at rest.

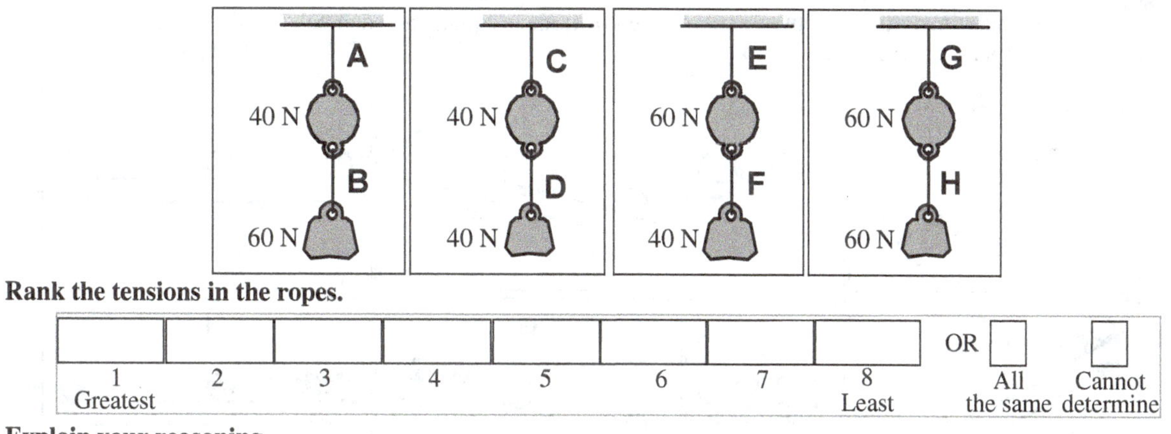

Rank the tensions in the ropes.

								OR	☐	☐
1 Greatest	2	3	4	5	6	7	8 Least		All the same	Cannot determine

Explain your reasoning.

B3-RT18: BLOCKS AND WEIGHTS AT REST—TENSION

In all of the cases shown, the systems are at rest. In Cases A and B, there is a force to the right acting on the block, which is on a frictionless surface, and in Case D there is a 40 N upward force on the weight.

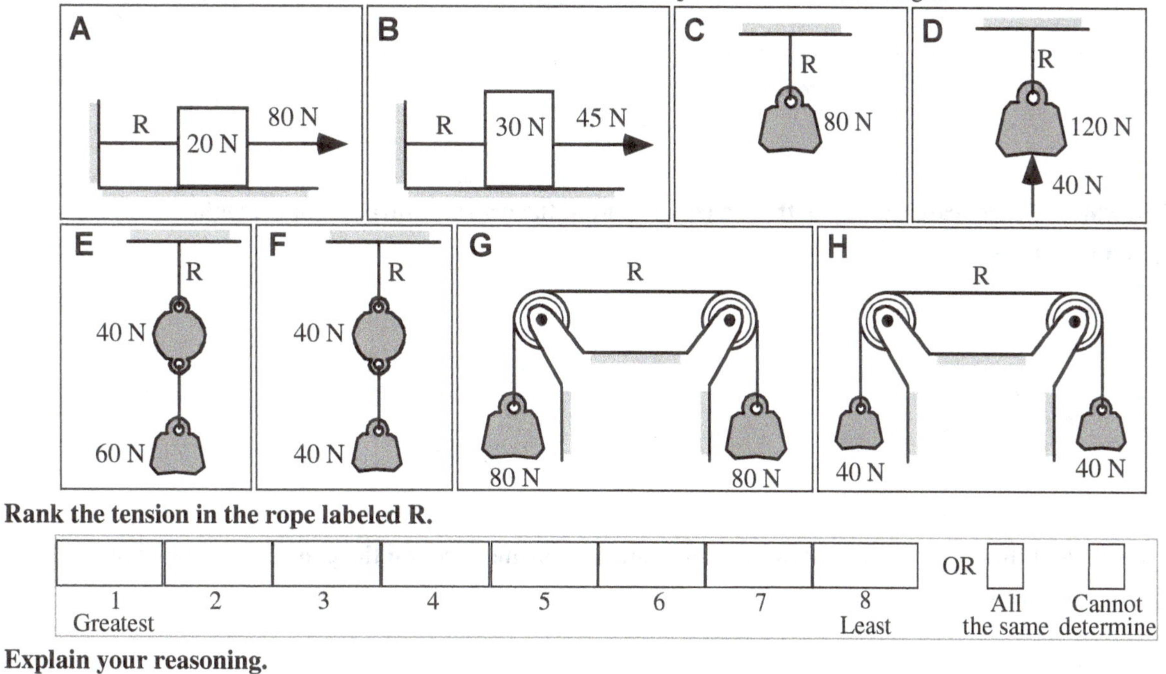

Rank the tension in the rope labeled R.

								OR	☐	☐
1 Greatest	2	3	4	5	6	7	8 Least		All the same	Cannot determine

Explain your reasoning.

B3-QRT19: Ball Strobe Motion—Net Force

The following drawings indicate the positions, using a strobe flash, of a ball moving from one side of the figure to the other as indicated by the direction of the arrow. Each circle represents the position of the ball at succeeding instants of time. Each time interval between successive positions is equal, and each ball has the same mass. Assume the acceleration, if any, for each situation to be constant.

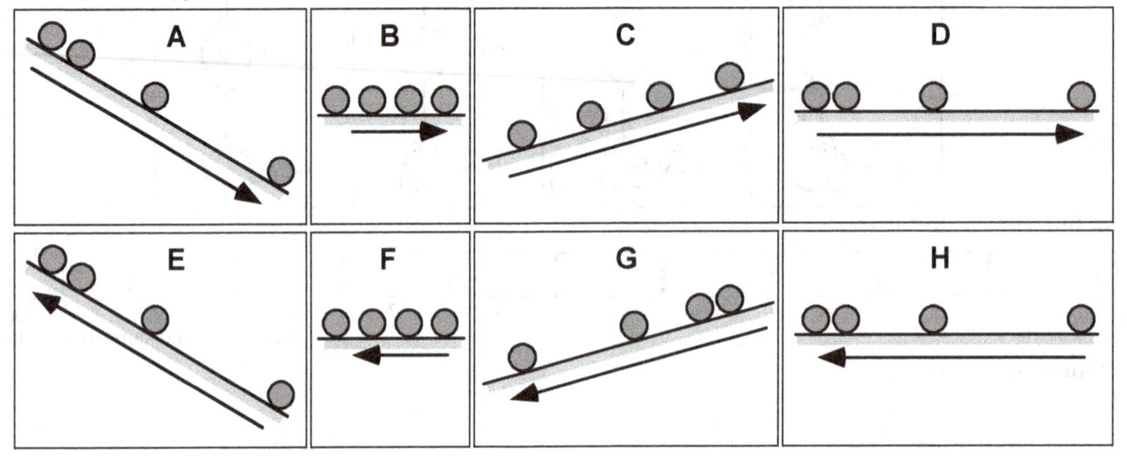

(a) In which of these cases, if any, is there a net force acting on the ball? _____
Explain your reasoning.

(b) In which of these cases, if any, is there a component of the net force directed to the right? _____
Explain your reasoning.

(c) In which of these cases, if any, is there a component of the net force on the ball directed upward? _____
Explain your reasoning.

B3-RT20: NET FORCE-ACCELERATION GRAPHS—MASS

These graphs are of net force versus acceleration for different objects. All graphs have the same scale for each respective axis.

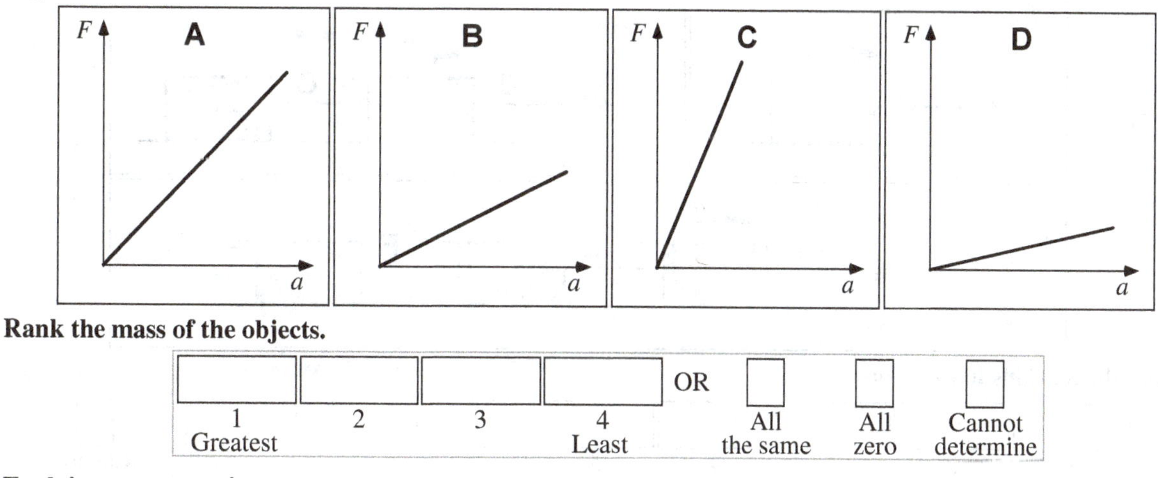

Rank the mass of the objects.

1 Greatest	2	3	4 Least	OR	All the same	All zero	Cannot determine

Explain your reasoning.

B3-RT21: ROPES PULLING BOXES—ACCELERATION

Boxes are pulled by ropes along frictionless surfaces, accelerating toward the left. All of the boxes are identical. The pulling force applied to the left-most rope is the same in each figure.

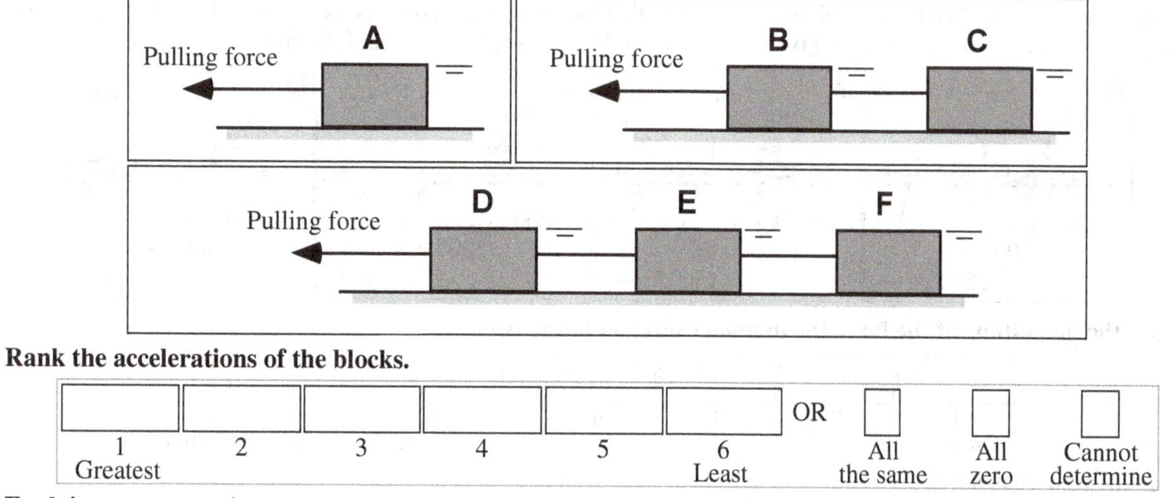

Rank the accelerations of the blocks.

1 Greatest	2	3	4	5	6 Least	OR	All the same	All zero	Cannot determine

Explain your reasoning.

B3-RT22: Ropes Pulling Boxes—Rope Tension

Boxes are pulled by ropes along frictionless surfaces, accelerating toward the left. All of the boxes are identical, and the accelerations of all three systems are the same.

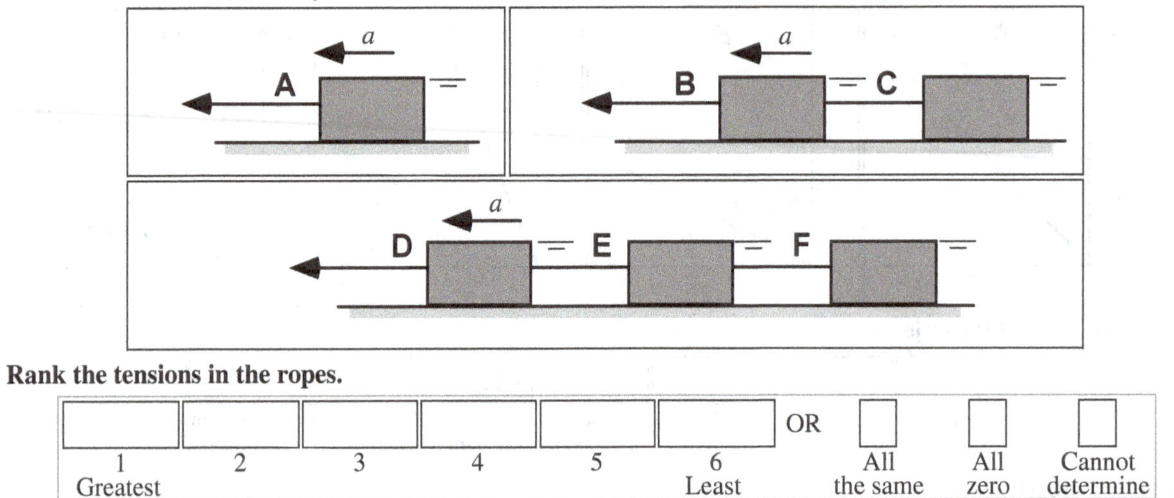

Rank the tensions in the ropes.

1 Greatest	2	3	4	5	6 Least	OR	All the same	All zero	Cannot determine

Explain your reasoning.

B3-RT23: Tugboat Pushing Barges—Force Tugboat Exerts on Lead Barge

A tugboat is pushing two barges (labeled 1 and 2) so that they speed up. The masses of the tugboats and barges and the accelerations of the systems are given for each case. Ignore the effects of fluid friction.

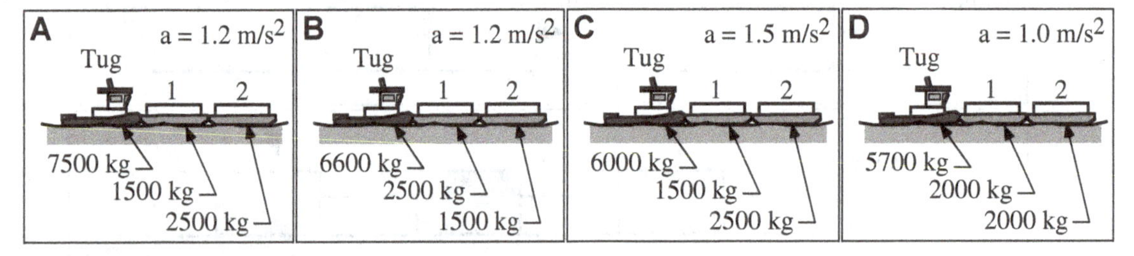

Rank the magnitude of the force the tugboat exerts on barge two.

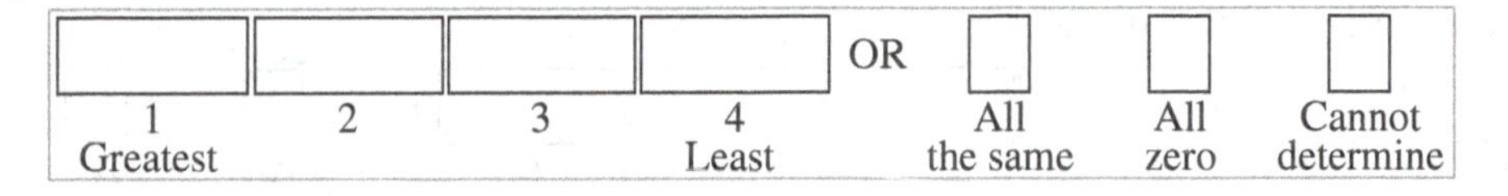

Explain your reasoning.

B3-RT24: TUGBOAT PUSHING BARGES—FORCE TUGBOAT EXERTS ON FIRST BARGE

A tugboat is pushing two barges (labeled 1 and 2) so that they speed up. The masses of the tugboats and barges and the accelerations of the systems are given for each case. Ignore the effects of fluid friction.

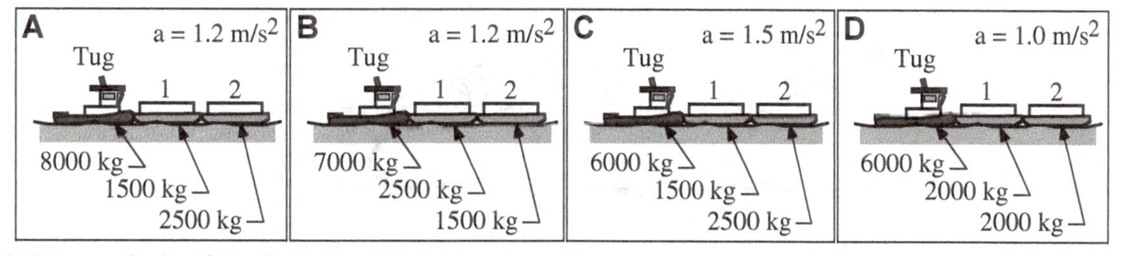

Rank the magnitude of the force the tugboat exerts on barge one.

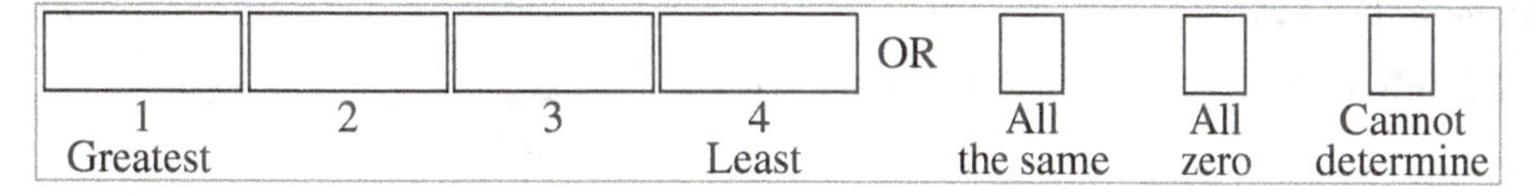

Explain your reasoning.

B3-RT25: FORCES ON BLOCKS ON SMOOTH SURFACES—SPEED

Two forces act on a block that is initially at rest on a frictionless surface.

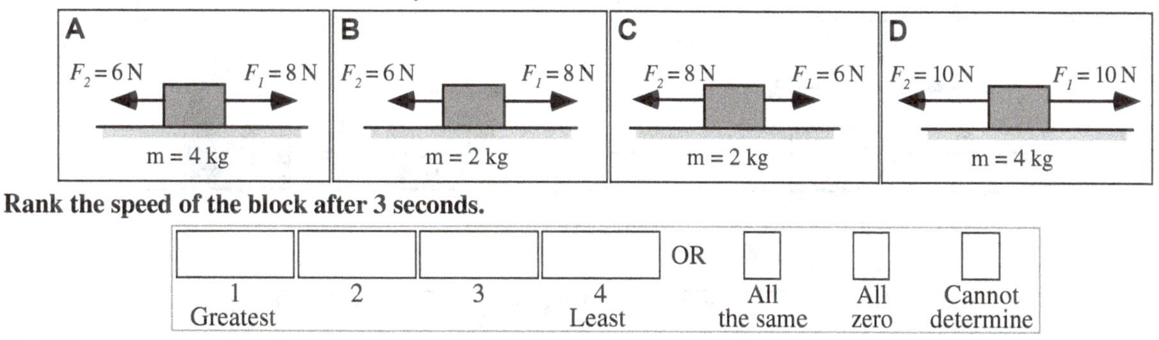

Explain your reasoning.

B3-CT26: CURLER PUSHING STONE—FORCE ON STONE

Two identical curling stones (the playing pieces in the sport of curling) are pushed horizontally on ice by curlers. The instantaneous speed and acceleration of the two stones are given. Ignore the friction between the stone and the ice.

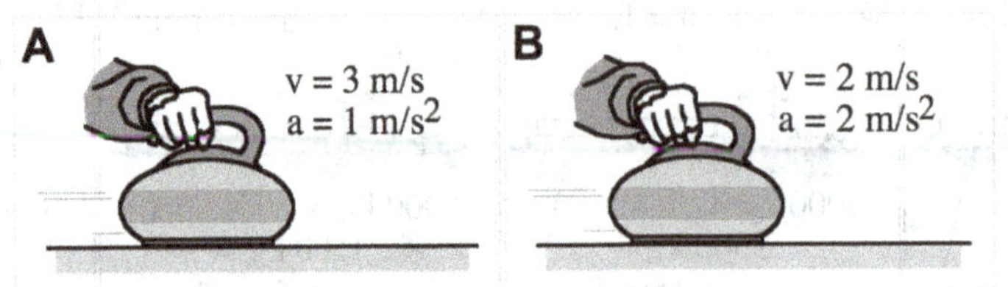

Is the magnitude of the force that the curler is exerting on the stone (i) *greater* in Case A, (ii) *greater* in Case B, or (iii) *the same* in both cases? _____

Explain your reasoning.

B3-RT27: ROPES PULLING IDENTICAL BOXES—ROPE TENSION

Boxes are pulled by ropes along frictionless surfaces, accelerating toward the left. All of the boxes are identical, and the accelerations of the boxes are indicated.

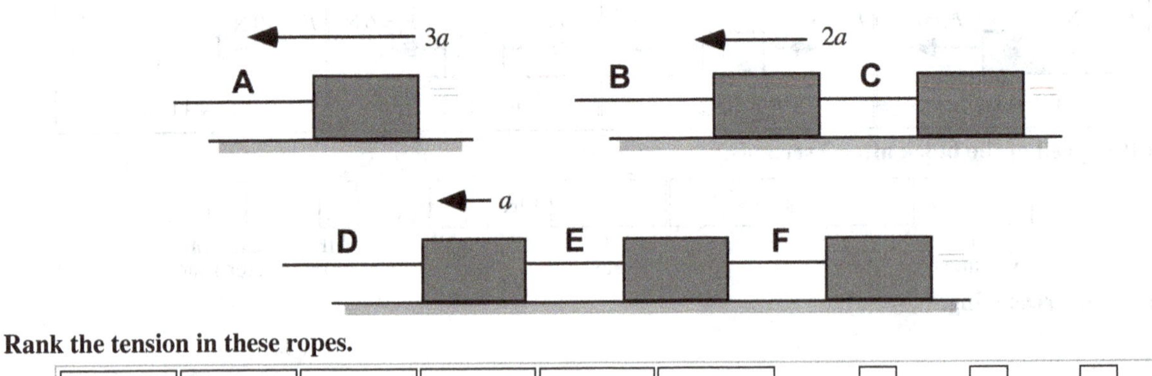

Rank the tension in these ropes.

☐	☐	☐	☐	☐	☐	OR	☐	☐	☐
1 Greatest	2	3	4	5	6 Least		All the same	All zero	Cannot determine

Explain your reasoning.

B3-QRT28: STUDENT PUSHING TWO BLOCKS—FORCE

A student pushes horizontally on two blocks, which are moving to the right. Block A has more mass than block B. There is friction between the blocks and the table.

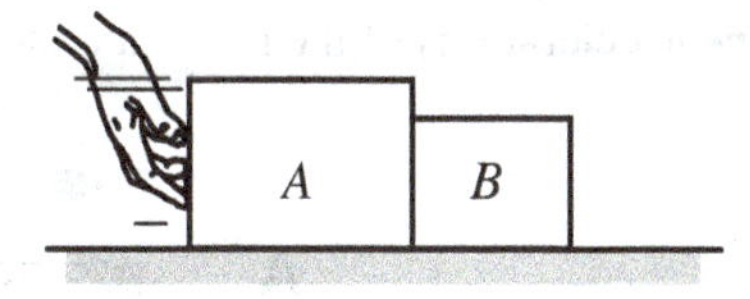

(a) For the situation where the blocks are moving at a constant speed, which of the following statements is true about the magnitude of the forces?

(i) The force that block A exerts on block B is *greater than* the force that block B exerts on block A.
(ii) The force that block A exerts on block B is *less than* the force that block B exerts on block A.
(iii) The force that block A exerts on block B is *equal to* the force that block B exerts on block A.
(iv) We cannot compare the forces unless we know how fast the blocks are slowing down.

Explain your reasoning.

(b) For the situation where the blocks are moving at a constant speed, which of the following statements is true about the net force?

(i) The net force on block A points *to the right* and is *equal to* the net force on block B.
(ii) The net force on block A points *to the left* and is *equal to* the net force on block B.
(iii) The net force on block A points *to the right* and is *greater than* the net force on block B.
(iv) The net force on block A points *to the left* and is *greater than* the net force on block B.
(v) The net force on block *A* points *to the right* and is *less than* the net force on block B.
(vi) The net force on block A points *to the left* and is *less than* the net force on block B.
(vii) None of these are correct.

Explain your reasoning.

(c) For the situation where the blocks are slowing down, which of the following statements is true about the magnitude of the forces?

(i) The force that block A exerts on block B is *greater than* the force that block B exerts on block A.
(ii) The force that block A exerts on block B is *less than* the force that block B exerts on block A.
(iii) The force that block A exerts on block B is *equal to* the force that block B exerts on block A.
(iv) We cannot compare the forces unless we know how fast the blocks are slowing down.

Explain your reasoning.

(d) For the situation where the blocks are slowing down, which of the following statements is true about the net force?

(i) The net force on block A points *to the right* and is *equal to* the net force on block B.
(ii) The net force on block A points *to the left* and is *equal to* the net force on block B.
(iii) The net force on block A points *to the right* and is *greater than* the net force on block B.
(iv) The net force on block A points *to the left* and is *greater than* the net force on block B.
(v) The net force on block A points *to the right* and is *less than* the net force on block B.
(vi) The net force on block A points *to the left* and is *less than* the net force on block B.
(vii) None of these are correct.

B3-RT29: Ropes Pulling Identical Boxes—Rope Tension

Boxes are pulled by ropes along frictionless surfaces, accelerating toward the left. All of the boxes are identical, and the accelerations of the boxes are indicated.

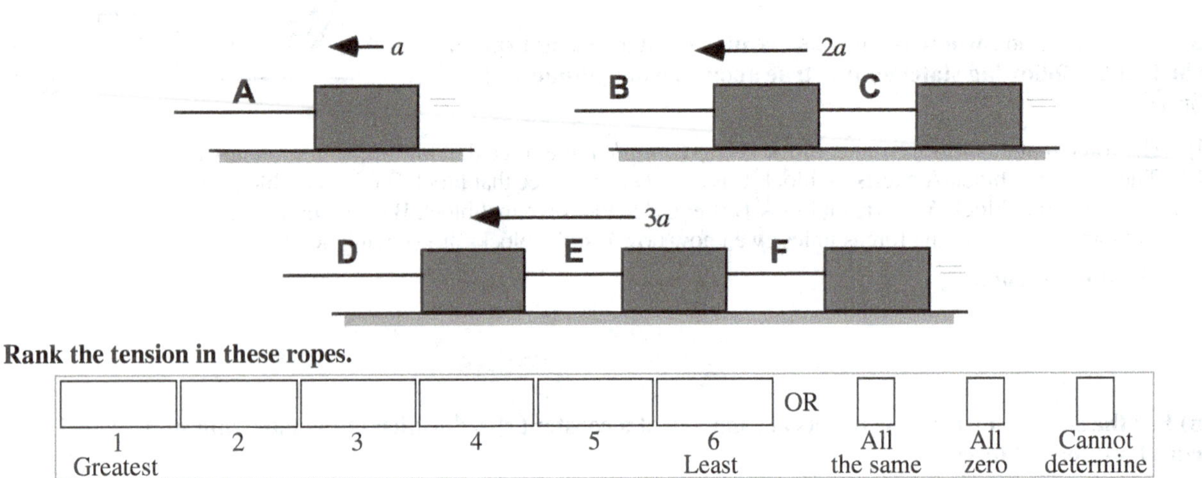

Rank the tension in these ropes.

						OR			
1 Greatest	2	3	4	5	6 Least		All the same	All zero	Cannot determine

Explain your reasoning.

B3-RT30: Forces on Objects on Rough Surfaces—Velocity Change

Two forces act on identical objects that are on rough surfaces. The forces of maximum static and kinetic friction for all cases are both 1 N. All objects start at rest.

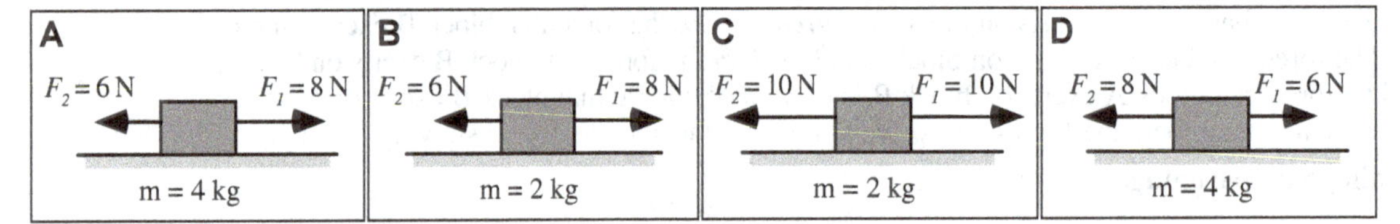

Rank the magnitude of the velocity change of the blocks in a 2-second time interval.

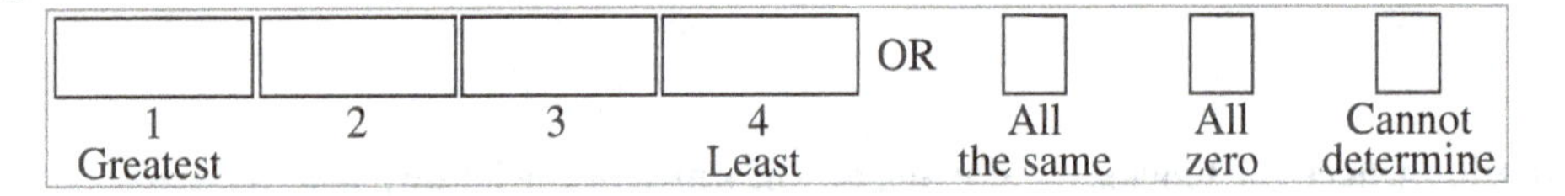

				OR			
1 Greatest	2	3	4 Least		All the same	All zero	Cannot determine

Explain your reasoning.

B3-QRT31: SKATEBOARD RIDER COASTING DOWN A HILL—ACCELERATION AND NET FORCE

At the instant shown, a skateboard rider is coasting down a hill and speeding up. Consider the skateboard and the rider as a single system and ignore friction.

(a) Use velocity vectors to find the approximate direction of the acceleration of the system.

(b) Draw a free-body diagram for the system labeling all forces, and explain how the forces in your free-body diagram add to give a net force in the direction of the acceleration.

B3-CT32: ROPES PULLING BOXES—ROPE TENSION

Two boxes are pulled by the same force F along frictionless surfaces, accelerating toward the left. The masses of the boxes are indicated in each figure.

Will the tension in rope A on the left be (i) *greater than,* (ii) *less than,* or (iii) *equal to* the tension in rope B on the right? _____

Explain your reasoning.

B3-WWT33: Lifting up a Pail—Strategy

A loaded pail is attached to a rope that passes around an overhead pulley and is tied to a ring on the floor. Linda, a construction worker, plans to untie the rope from the ring, pull on the rope to lift the pail 1 m higher, and then retie the rope. Linda weighs 800 N and is capable of lifting twice her weight, 1600 N. The loaded pail weighs 1200 N.

What, if anything, is wrong with Linda's plan? Explain how to correct it, or if the plan will work, explain why.

B3-WWT34: Velocity-Time Graphs—Net Force

Graphs are shown of the velocity versus time for two identical train engines on a straight track. A positive velocity indicates that the engine was traveling east. The scales on both axes are the same for the graphs. On each graph a point is marked with a dot.

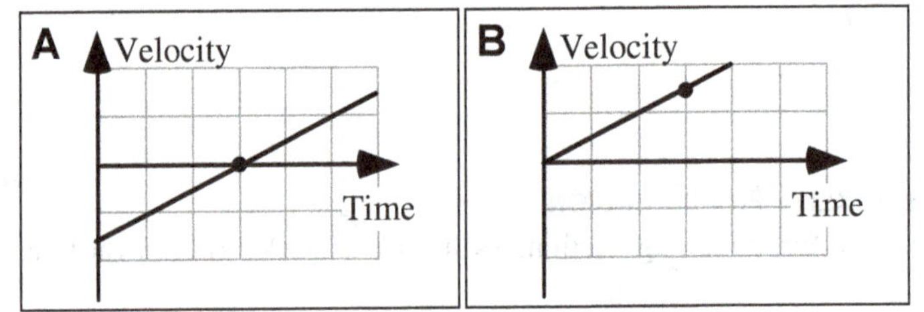

A student comparing the net force acting on the engine at the identified points in graphs A and B states:

"I think that B has the larger net force since the net force on A at the identified point is zero."

What, if anything, is wrong with this statement? If something is wrong, identify it, and explain how to correct it. If the statement is correct, explain why.

B3-CT35: BLOCKS MOVING AT CONSTANT SPEED—FORCE ON BLOCK

A block is moving to the right across a rough table at a constant speed of 2 m/s. The tables and the blocks are identical in the two cases. In Case A, the block is pushed with a stick and in Case B, the block is pulled with a string. The angle that the applied force makes with the horizontal is the same in both cases.

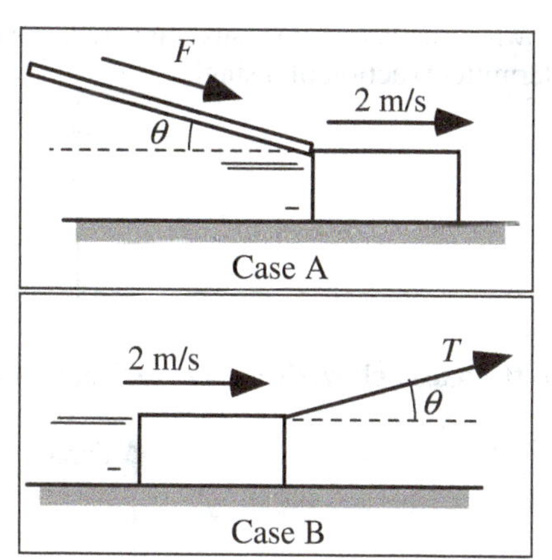

Will the magnitude of the force on the block by the stick in Case A be (i) *greater than,* (ii) *less than,* or (iii) *equal to* the tension on the block by the string in Case B? _____

Explain your reasoning.

B3-WWT36: PULLING A BLOCK ACROSS A ROUGH SURFACE—FORCE RELATIONSHIPS

A person pulls a block across a rough horizontal surface at a constant speed by applying a force ***F*** at a slight angle as shown. A free-body diagram is drawn for the block. The arrows in the diagram correctly indicate the directions but not necessarily the magnitudes of the various forces on the block. A student makes the following claim about this free-body diagram:

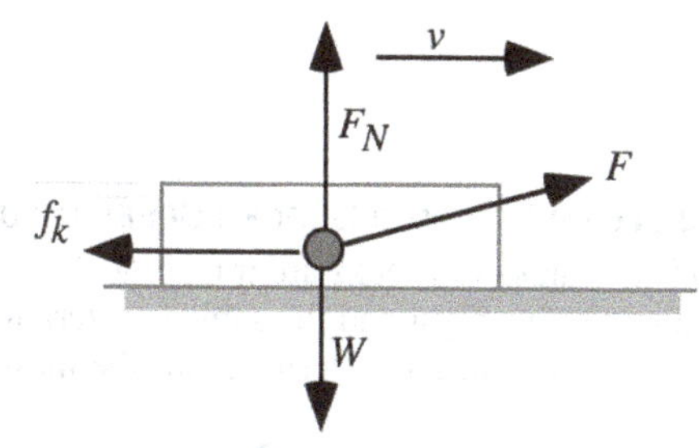

"The velocity of the block is constant, so the net force acting on the block must be zero. Thus the normal force F_N equals the weight W, and the force of friction f_k equals the applied force F."

What, if anything, is wrong with this statement? If something is wrong, identify it and explain how to correct it. If this statement is correct, explain why.

B3-CRT37: Velocity-Time Graph—Force-Time Graph

Shown is the velocity versus time graph for an object that is moving in one dimension under the (perhaps intermittent) action of a single horizontal force.

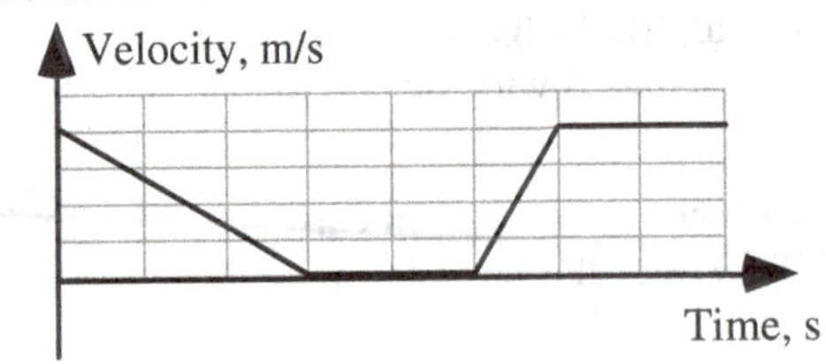

On the axes below, draw the horizontal force acting on this object as a function of time.

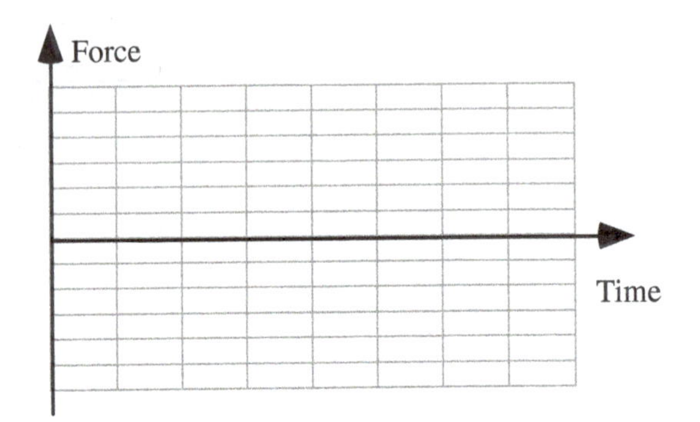

Explain your reasoning.

B3-RT38: Stacked Blocks Speeding Up on a Conveyor Belt—Net Force

Various stacks of blocks are traveling along a conveyer belt. At the instant shown, all blocks have the same velocity of 3 m/s to the right and the same acceleration of 2 m/s^2, also to the right. The blocks do not slip. All masses are given in the diagram in terms of *M*, the mass of the smallest block.

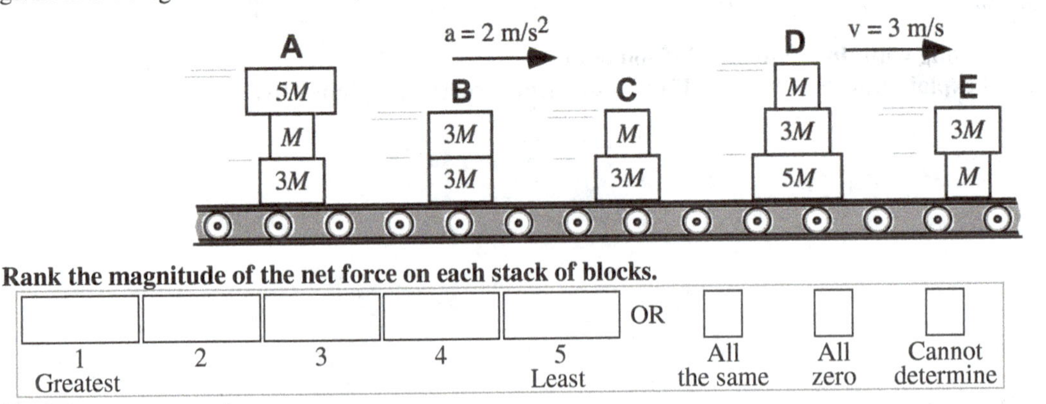

Rank the magnitude of the net force on each stack of blocks.

1 Greatest	2	3	4	5 Least	OR	All the same	All zero	Cannot determine

Explain your reasoning.

B3-CT39: SPACESHIPS PULLING TWO CARGO PODS—TENSION IN TOW RODS

In both cases a spaceship is pulling two cargo pods, one empty and one full. At the instant shown, the speed of the pods and spaceships is 300 m/s, but they have different accelerations as shown. All masses are given in terms of *M*, the mass of an empty pod.

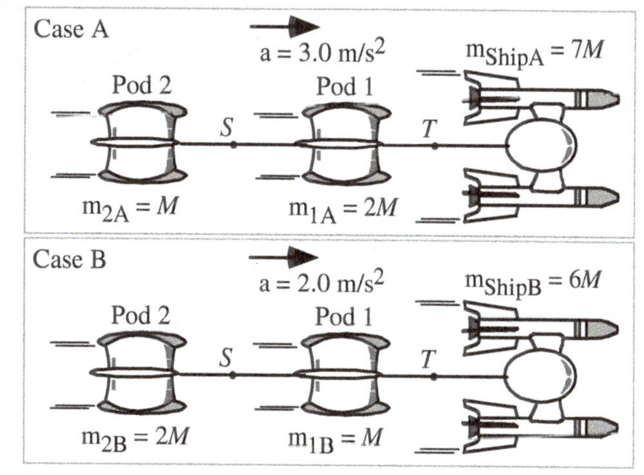

(a) Will the tension at point *S* in the tow rod be (i) *greater* in Case A, (ii) *greater* in Case B, or (iii) *the same* in both cases? _____

Explain your reasoning.

(b) Will the tension at point *T* in the tow rod be (i) *greater* in Case A, (ii) *greater* in Case B, or (iii) *the same* in both cases? _____

Explain your reasoning.

B3-QRT40: Three Vectors—Resultant

(a) In the space below, add the three vectors shown and label the resultant vector as $\vec{R}$. Be sure to clearly indicate the direction of the resultant.

Explain your reasoning.

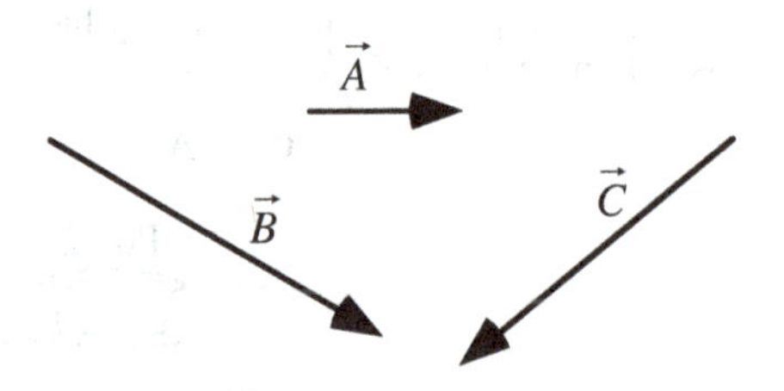

Suppose the three vectors above represent forces exerted on a slice of pepperoni pizza by three people, Abel ($\vec{A}$), Beth ($\vec{B}$), and Celia ($\vec{C}$) as shown in the top view picture to the right. A fourth person, David, also pulls on the pizza. The pizza moves to the left at a constant speed. Assume there is no friction between the pizza slice and the greasy table.

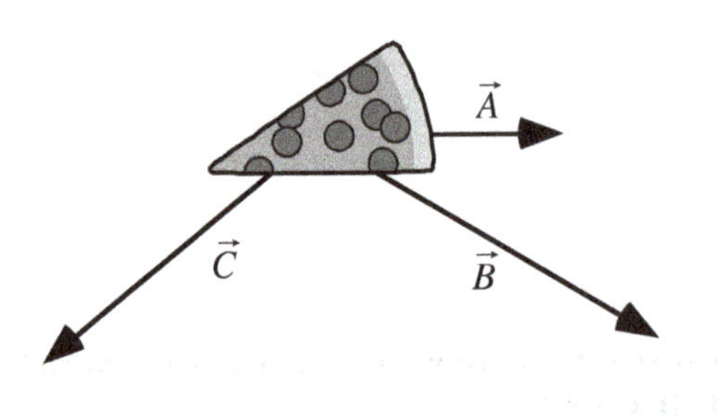

(b) In what direction is David pulling on the pizza?

Explain your reasoning.

B3-CT41: SPACESHIPS PULLING TWO CARGO PODS—TENSION OR COMPRESSION IN TOW RODS

In each case below, a spaceship is attached to two cargo pods by rods. At the instant shown, the speed of the pods and of the spaceship is 300 m/s. In Case A the acceleration of the ship and of the pods is 3 m/s^2 to the left, while in Case B it is 2 m/s^2 to the right. All masses are given in terms of *M*, the mass of an empty pod.

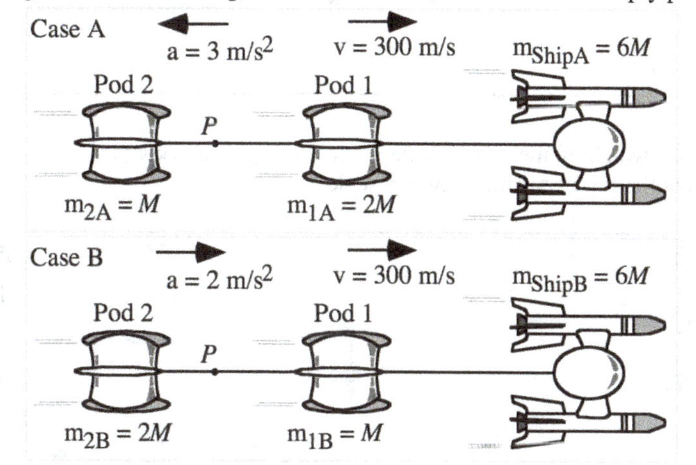

Will the tension or compression at point *P* in the tow rod be (i) *greater* in Case A, (ii) *greater* in Case B, or (iii) *the same* in both cases? _____

Explain your reasoning.

B3-QRT42: Thrown Baseball—Free-Body Diagram at the Top

A baseball is thrown from right field to home plate (HP), traveling from right to left in the diagram.

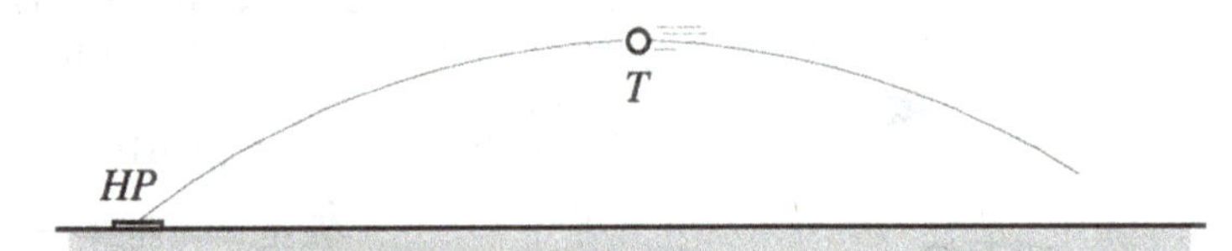

A group of physics students watching the game create the following free-body diagrams for the baseball at the top of its path (point T). Note that the forces are not drawn to scale.

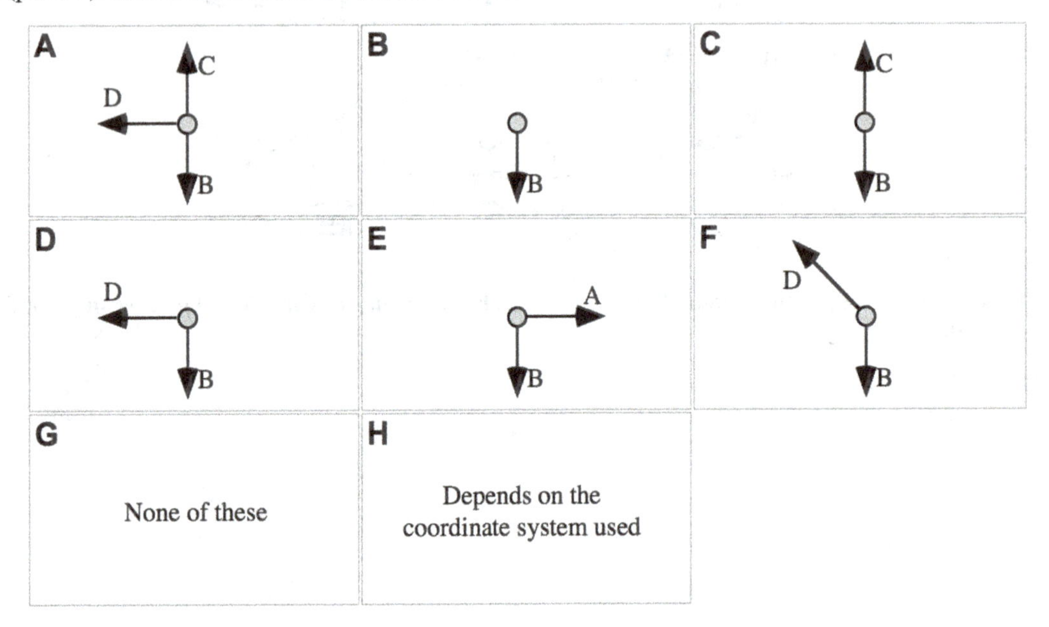

(a) If they decide to *ignore air friction,* which is the correct free-body diagram for the baseball at point T?

(b) Define all forces on the ball for this force diagram.

(c) If they decide to *include air friction,* which is the correct free-body diagram for the baseball at point T?

(d) Define all forces on the ball for this force diagram.

B3-WWT43: BOX ON INCLINE—FORCES

A heavy box is sitting at rest on an incline. There is friction between the box and the incline, and a rope is pulling on the box in a direction up and to the left, parallel to the incline. A physics student draws the free-body diagram below right for the box.

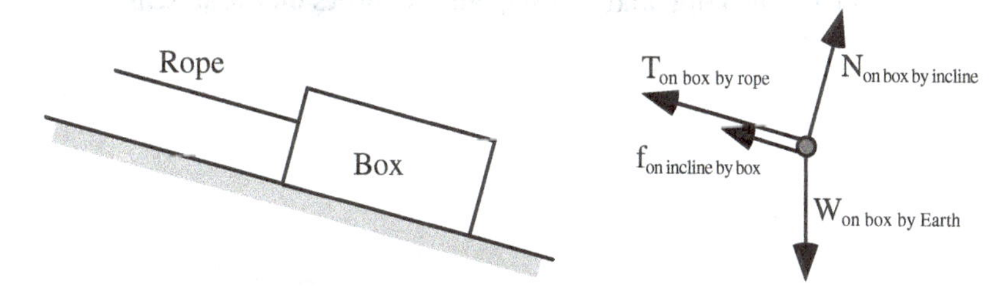

What, if anything, is wrong with this student's free-body diagram? If something is wrong, explain the error and how to correct it. If this free-body diagram is correct, explain why.

B3-WWT44: THROWN BASEBALL—FREE-BODY DIAGRAM FOR ASCENDING BASEBALL

A baseball is thrown from right field to home plate (HP), traveling from right to left in the diagram.

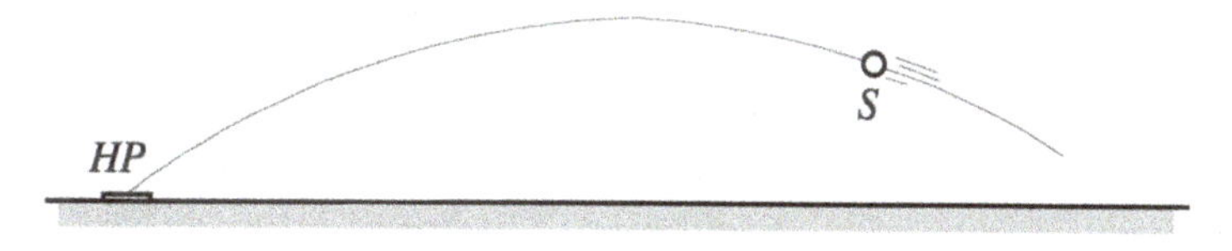

A physics student watching the game produces the free-body diagram shown below for the baseball as it moves upward at a point (*S*) along the path. She explains that in her drawing:

"I'm ignoring air resistance. ***W*** *is the weight of the baseball and* ***D*** *is the force of the throw on the baseball."*

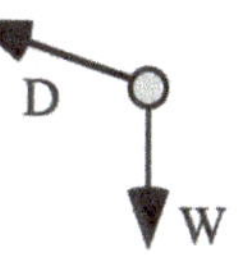

What, if anything, is wrong with this free-body diagram? If something is wrong, identify it and explain how to correct it. If this free-body diagram is correct, explain why.

B3-CRT45: Suitcase Sliding Down Ramp at Constant Speed—Forces on Suitcase

A suitcase is moving at a constant speed as it slides down a ramp angled at 45° to the horizontal.

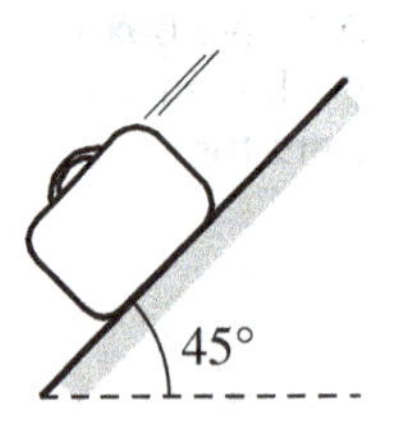

Draw a free-body diagram below, labeling and defining all the forces on the suitcase.

Rank the magnitudes of these forces on the suitcase.

Explain your ranking.

B3-RT46: TWO BLOCKS AT REST—FORCE DIFFERENCE

In each case shown, there are two blocks with different masses that are at rest and in contact with each other. One of the blocks given in each arrangement is labeled **1,** and the other is labeled **2**. The mass of each block is given in the figures.

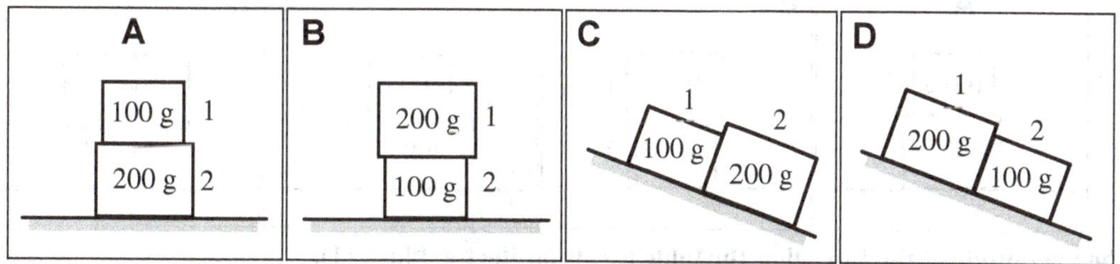

Rank the difference between the strengths (magnitudes) of the force 1 exerts on 2 and the force 2 exerts on 1.

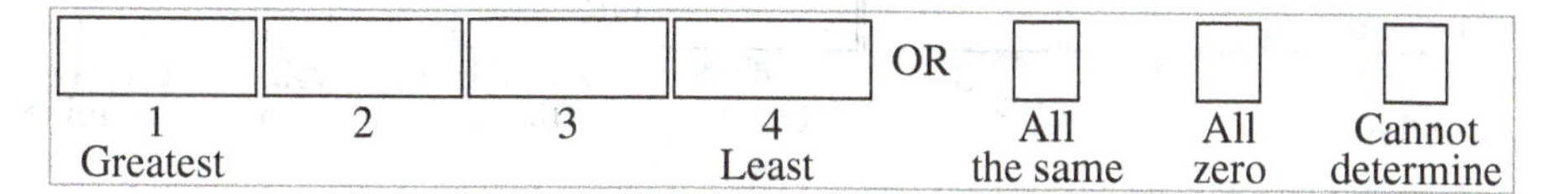

Explain your reasoning.

B3-RT47: TWO STACKED BLOCKS AT REST—FORCE ON THE TOP BLOCK BY BOTTOM BLOCK

Two wooden blocks with different masses are at rest, stacked on a table. The top block is labeled **1,** and the bottom block is labeled **2**.

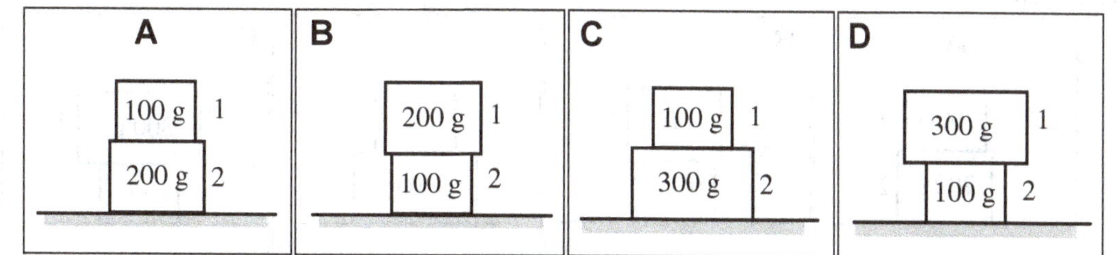

Rank the magnitude of the force that the bottom block (2) exerts on the top block (1).

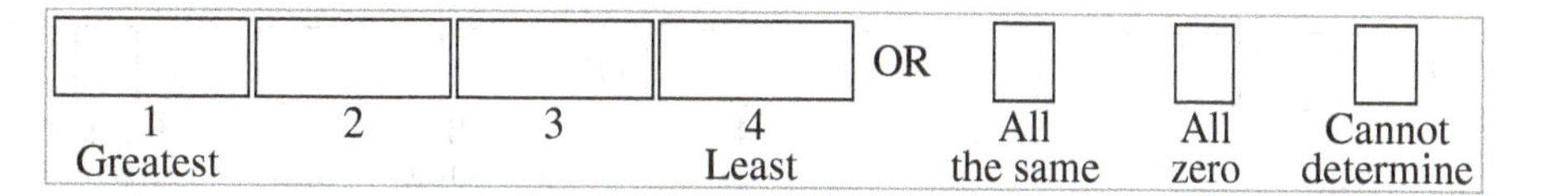

Explain your reasoning.

B3-RT48: Two Stacked Blocks at Rest—Force on the Top Block by the Table

Two wooden blocks with different masses are at rest, stacked on a table. The top block is labeled **1,** and the bottom block is labeled **2.**

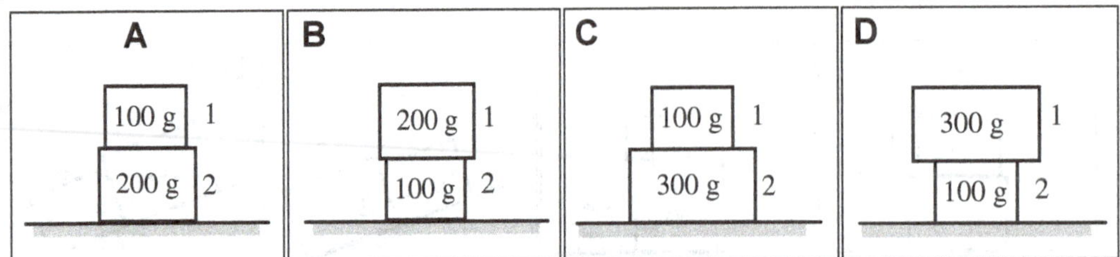

Rank the magnitude of the force that the table exerts on the top block (1).

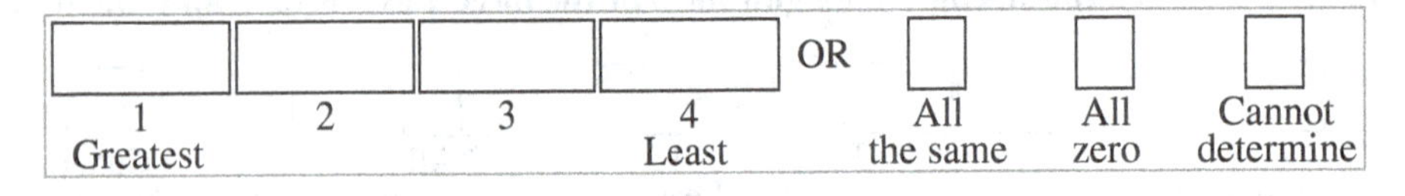

Explain your reasoning.

B3-RT49: Two Stacked Blocks at Rest—Force on the Bottom Block by Table

Two wooden blocks with different masses are at rest, stacked on a table. The top block is labeled **1,** and the bottom block is labeled **2.**

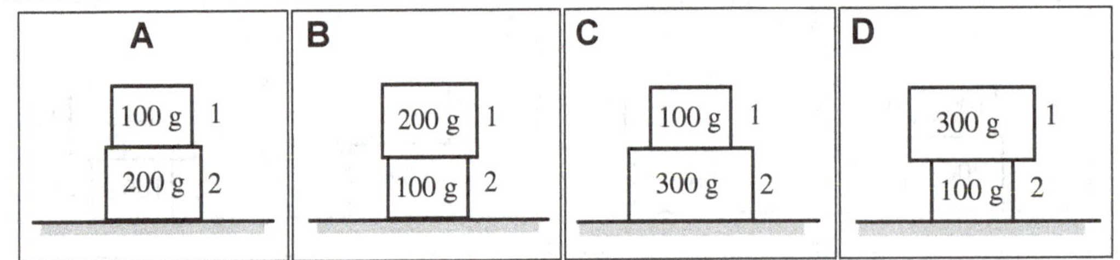

Rank the magnitude of the force that the table exerts on the bottom block (2).

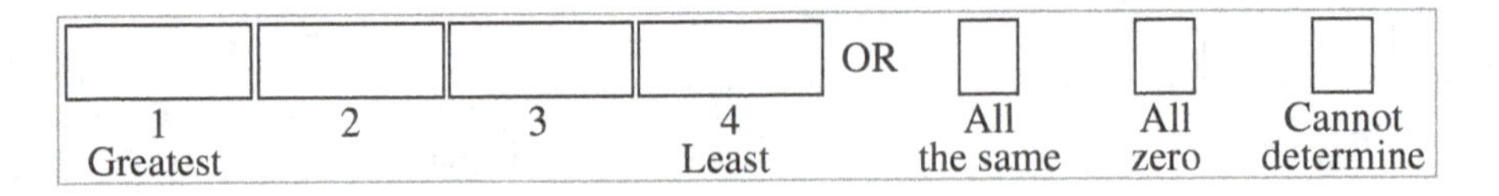

Explain your reasoning.

B3-RT50: ACCELERATING CAR AND BOAT TRAILER—FORCE DIFFERENCE

All the trailers and cars shown are identical but the boat trailers have different loads. In each case, the car and boat trailer accelerate at 1 m/s² from rest to the final speed shown.

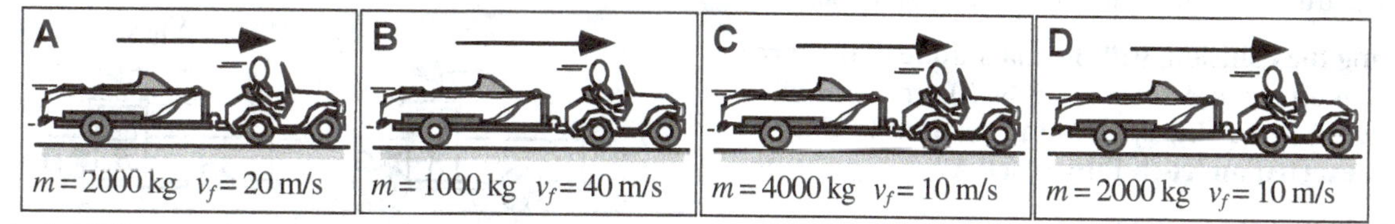

Rank the difference between the strength (magnitude) of the force the car exerts on the boat trailer and the strength of the force the trailer exerts on the car while the cars and trailers are accelerating.

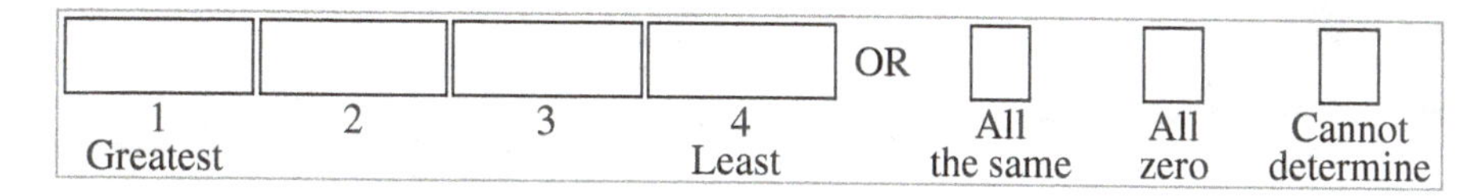

Explain your reasoning.

B3-RT51: CAR AND BOAT TRAILER ON AN INCLINE—FORCE DIFFERENCE

All the trailers and cars shown are identical, but the boat trailers have different loads. The inclines are all identical. In each case, the accelerations and velocities are given for the instant shown.

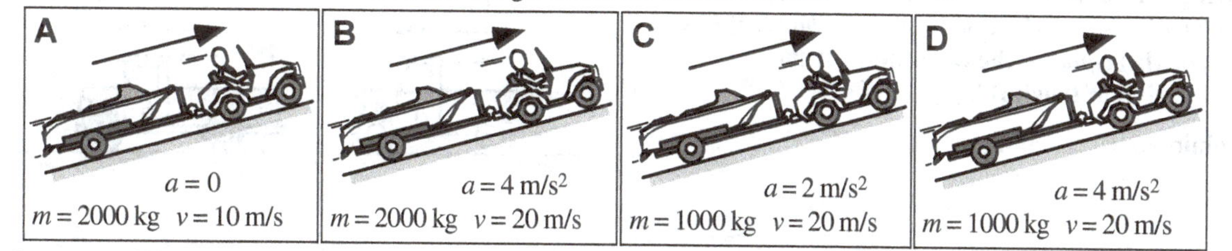

Rank on the difference between the strength (magnitude) of the force the car exerts on the boat trailer and the strength of the force the boat trailer exerts on the car.

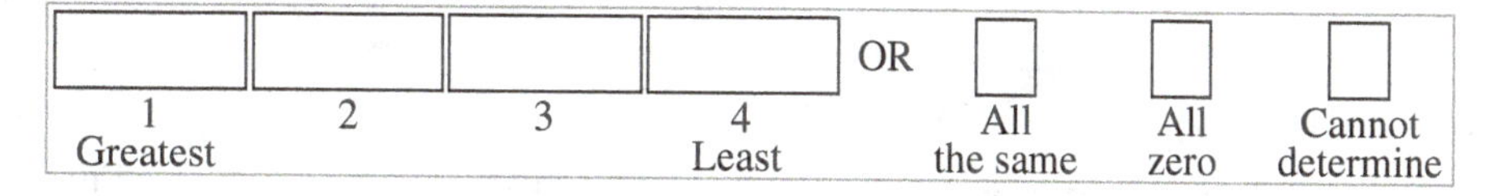

Explain your reasoning.

B3-CT52: Identical Toy Truck Collisions—Force and Acceleration

Two identical toy trucks traveling at different constant speeds are about to collide.

(a) The trucks are traveling in the same direction.

During the collision, will the magnitude of the force exerted on truck A by truck B be (i) *greater than*, (ii) *less than*, or (iii) *equal to* the magnitude of the force exerted on truck B by truck A? _____

Explain your reasoning.

(b) The trucks are traveling in opposite directions.

During the collision, will the magnitude of the force exerted on truck A by truck B be (i) *greater than*, (ii) *less than*, or (iii) *equal to* the magnitude of the force exerted on truck B by truck A? _____

Explain your reasoning.

(c) The trucks are traveling in the same direction.

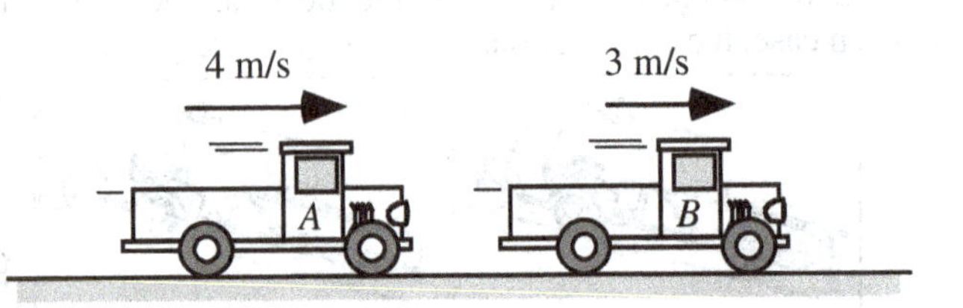

During the collision, will the magnitude of the acceleration of truck A be (i) *greater than*, (ii) *less than*, or (iii) *equal to* the magnitude of the acceleration of truck B? _____

Explain your reasoning.

(d) The trucks are traveling in opposite directions.

During the collision, will the magnitude of the acceleration of truck A be (i) *greater than*, (ii) *less than*, or (iii) *equal to* the magnitude of the acceleration of truck B? _____

Explain your reasoning.

B3-CT53: TOY TRUCK COLLISIONS—FORCE ON TRUCKS

Two toy trucks traveling at different constant speeds are about to collide.

(a) The two identical trucks are traveling in the same direction, and truck B is carrying a heavy load.

During the collision, will the magnitude of the force exerted on truck A by truck B be (i) *greater than*, (ii) *less than*, or (iii) *equal to* the magnitude of the force exerted on truck B by truck A? _____

Explain your reasoning.

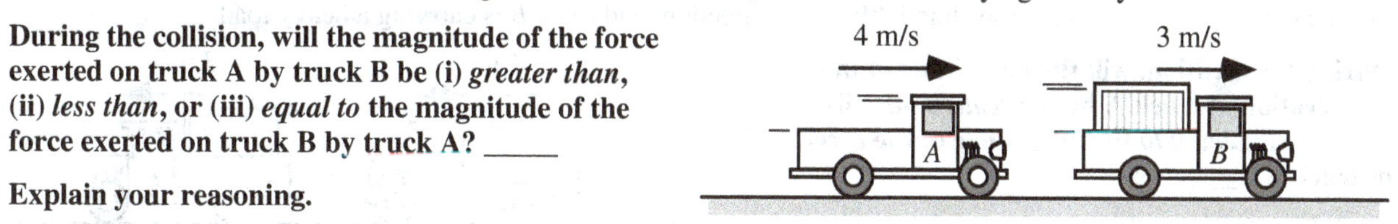

(b) The two identical trucks are traveling in opposite directions, and truck B is carrying a heavy load.

During the collision, will the magnitude of the force exerted on truck A by truck B be (i) *greater than*, (ii) *less than*, or (iii) *equal to* the magnitude of the force exerted on truck B by truck A? _____

Explain your reasoning.

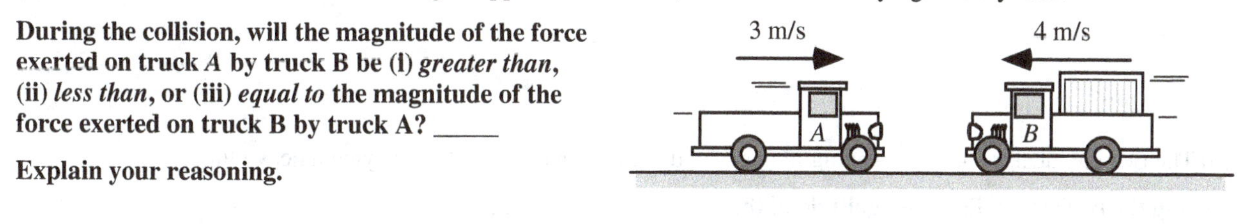

(c) The two identical trucks are traveling in the same direction, and truck A is carrying a heavy load.

During the collision, will the magnitude of the force exerted on truck A by truck B be (i) *greater than*, (ii) *less than*, or (iii) *equal to* the magnitude of the force exerted on truck B by truck A? _____

Explain your reasoning.

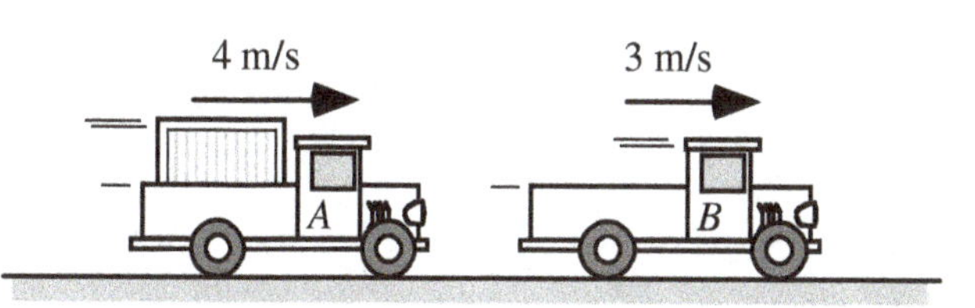

(e) The two identical trucks are traveling in opposite directions, and truck A is carrying a heavy load.

During the collision, will the magnitude of the force exerted on truck A by truck B be (i) *greater than*, (ii) *less than*, or (iii) *equal to* the magnitude of the force exerted on truck B by truck A? _____

Explain your reasoning.

B3-CT54: Toy Truck Collisions—Acceleration

Two toy trucks traveling at different constant speeds are about to collide.

(a) The two identical trucks are traveling in the same direction, and truck *B* is carrying a heavy load.

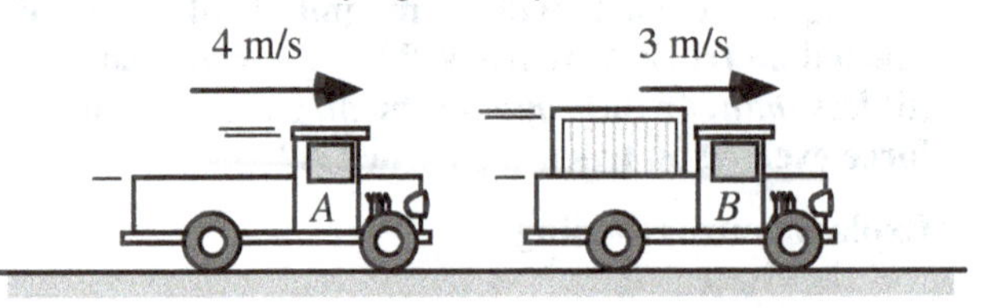

During the collision, will the magnitude of the acceleration of truck A be (i) *greater than*, (ii) *less than*, or (iii) *equal to* the magnitude of the acceleration of truck B? _____

Explain your reasoning.

(b) The two identical trucks are traveling in opposite directions, and truck B is carrying a heavy load.

During the collision, will the magnitude of the acceleration of truck *A* be (i) *greater than*, (ii) *less than*, or (iii) *equal to* the magnitude of the acceleration of truck B? _____

Explain your reasoning.

(c) The two identical trucks are traveling in the same direction, and truck A is carrying a heavy load.

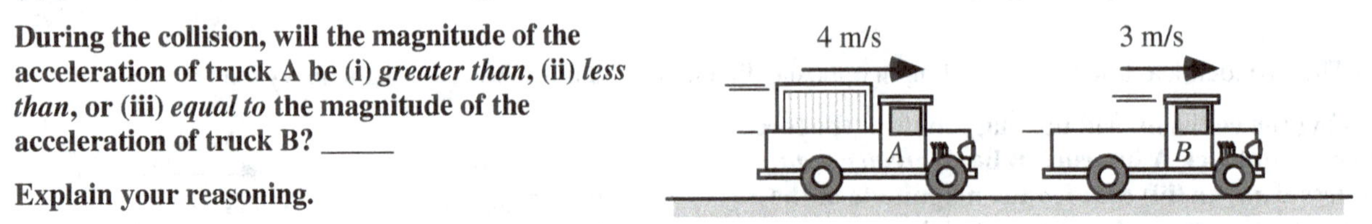

During the collision, will the magnitude of the acceleration of truck A be (i) *greater than*, (ii) *less than*, or (iii) *equal to* the magnitude of the acceleration of truck B? _____

Explain your reasoning.

B3-WWT55: TENNIS BALL AND RACQUET—FORCE

A tennis player returns a serve. A physics student watching the match makes the following contention:

> *"While the tennis ball is in contact with the tennis racquet, the racquet exerts a larger force on the tennis ball than the tennis ball does on the racquet because the racquet has to stop the tennis ball and then reverse its motion."*

What, if anything, is wrong with this contention? If something is wrong, explain the error and how to correct it. If this contention is correct, explain why.

B3-WWT56: BALL HITTING A WALL—FORCES

A student observes a rubber ball hitting a wall and rebounding. She states:

"In this situation, the wall exerts a larger force on the ball than the ball exerts on the wall, because the ball undergoes an acceleration but the wall doesn't move. That is, the ball goes from an initial speed to zero and then from zero to the rebound speed, but the wall does not accelerate since it is stationary the whole time."

What, if anything, is wrong with this contention? If something is wrong, identify it, and explain how to correct it. If this contention is correct, explain why.

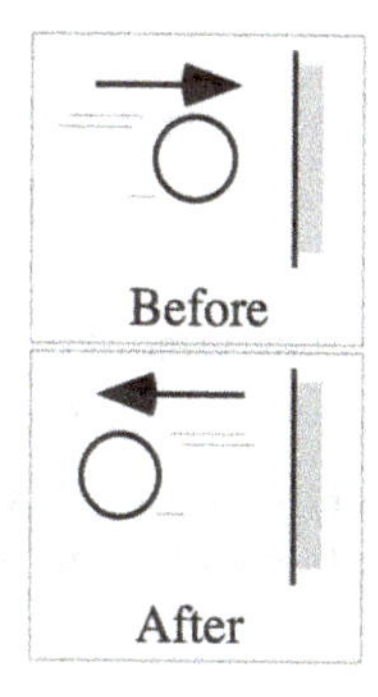

B3-WWT57: Potato on Table—Forces and Reaction Forces

A potato with a weight of 2 N is resting on a table. A student makes a number of statements about this situation.

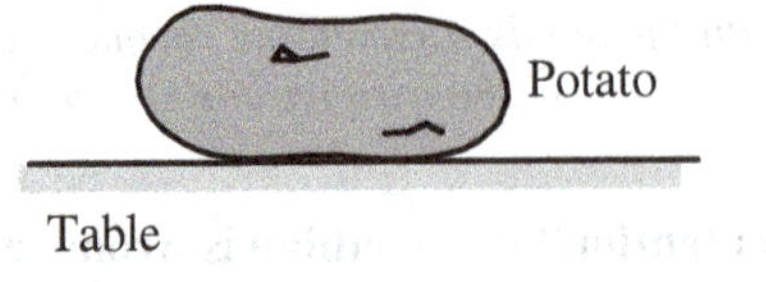

(a) *"The weight of the potato is a force of 2 N exerted by Earth in the downward direction."*

What, if anything, is wrong with this statement? If something is wrong, identify it and explain how to correct it. If this statement is correct, explain why.

(b) *"The reaction force to this weight is a force of 2 N exerted on the potato by the table in the upward direction."*

What, if anything, is wrong with this statement? If something is wrong, identify it and explain how to correct it. If this statement is correct, explain why.

(c) *"The normal force exerted on the potato by the table is a force of 2 N; the reaction force to this normal force is a force of 2 N exerted on the potato by Earth in the downward direction."*

What, if anything, is wrong with this statement? If something is wrong, identify it and explain how to correct it. If this statement is correct, explain why.

(d) *"If the 2 N potato is lifted off the table by a hand that exerts a force of 4 N upward on the potato, the reaction force to this 4 N force is a force of 4 N exerted on the hand by the potato in the downward direction."*

What, if anything, is wrong with this statement? If something is wrong, identify it and explain how to correct it. If this statement is correct, explain why.

B3-LMCT58: IDENTICAL TOY TRUCKS COLLIDING HEAD-ON—FORCE ON TRUCKS

Two identical toy trucks traveling at the same speed in opposite directions are about to collide. The magnitude of the force exerted on truck A by truck B during the collision is equal to the magnitude of the force exerted on truck B by truck A.

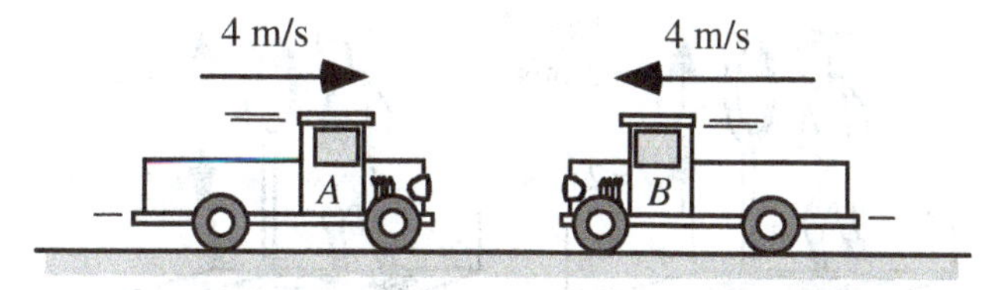

Identify from choices (i)–(iv) how each change described below will affect the magnitude of the force exerted on truck A by truck B during the collision as compared to the magnitude of the force exerted on truck B by truck A.

This change will cause the magnitude of the force exerted on truck A by truck B to be:

(i) ***greater than*** the magnitude of the force exerted on truck B by truck A.
(ii) ***less than*** the magnitude of the force exerted on truck B by truck A.
(iii) ***the*** *same* as the magnitude of the force exerted on truck B by truck A.
(iv) ***indeterminate*** *as compared to* the magnitude of the force exerted on truck B by truck A.

All of these modifications are changes to the initial situation shown in the diagram.

(a) Truck A is carrying a heavy load. _____
Explain your reasoning.

(b) Truck A is going faster than truck B. _____
Explain your reasoning.

(c) Truck A is carrying a heavy load and is going faster than truck B. _____
Explain your reasoning.

(d) Truck A is speeding up (accelerating). _____
Explain your reasoning.

(e) Truck A is speeding up (accelerating) and is carrying a heavier load. _____
Explain your reasoning.

(f) Truck A is speeding up (accelerating) while truck B is carrying a heavier load. _____
Explain your reasoning.

B3-CT59: Person in an Elevator Moving Upward—Scale Reading

A person who weighs 500 N is standing on a scale in an elevator. In both cases the elevator is identical and is moving upward, but in Case A it is accelerating upward and in Case B it is accelerating downward.

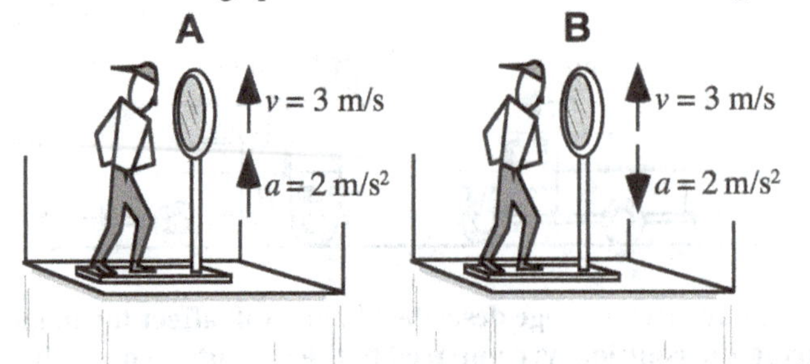

Will the scale reading be (i) *greater* in Case A, (ii) *greater* in case B, or (iii) *the same* in both cases? _____

Explain your reasoning.

B3-RT60: Person in a Moving Elevator—Scale Reading

A person who weighs 600 N is standing on a scale in an elevator. The elevator is identical in all cases. The velocity and acceleration of the elevators at the instant shown are given.

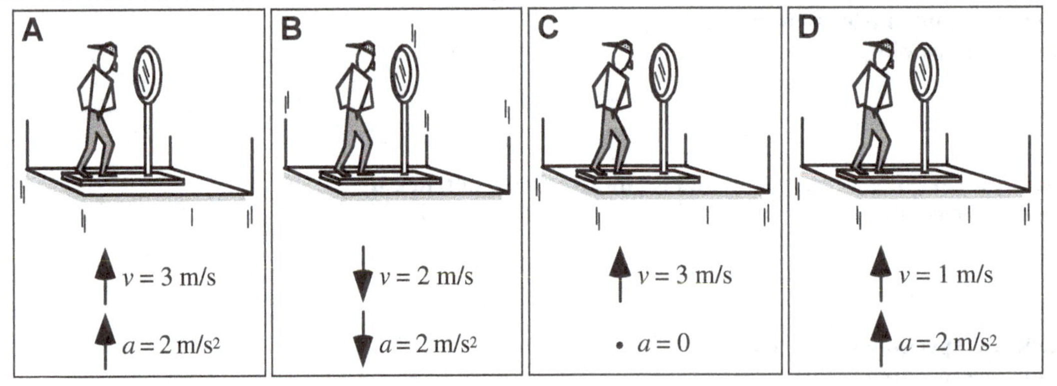

Rank the scale reading.

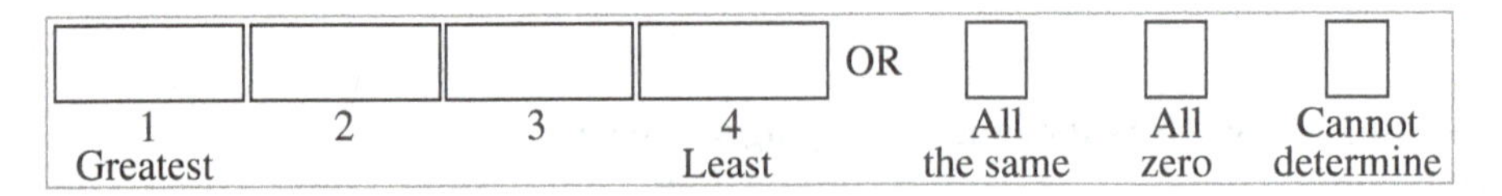

Explain your reasoning.

B3-RT61: PERSON IN AN ELEVATOR MOVING DOWNWARD—SCALE READING

A person who weighs 600 N is standing on a scale in an elevator. The elevator is identical in all cases. The velocity and acceleration of the elevators at the instant shown are given.

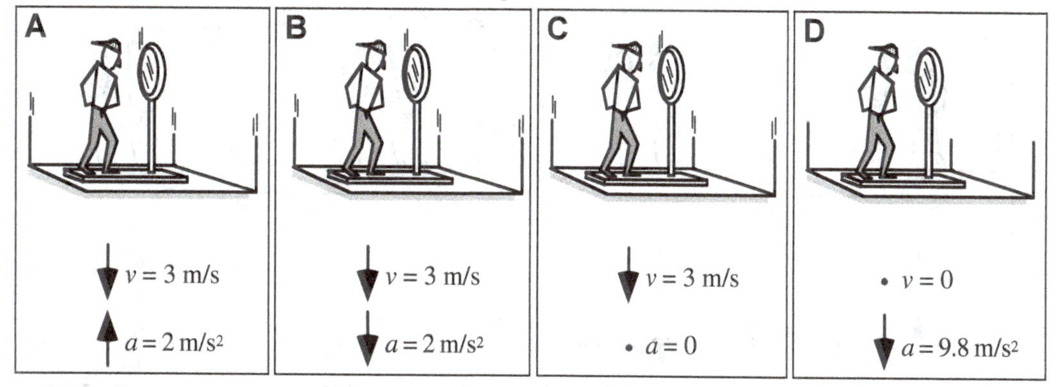

Rank the scale reading.

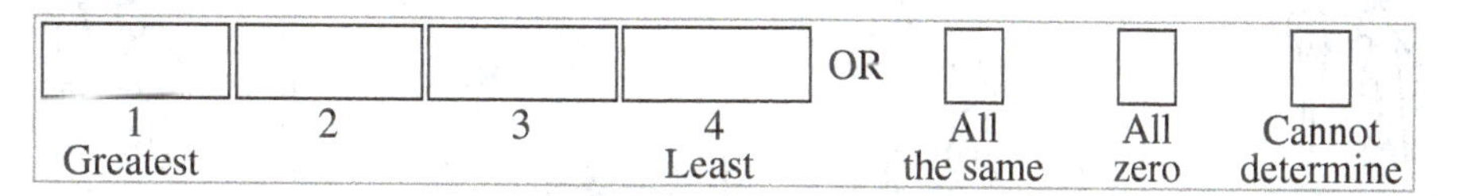

Explain your reasoning.

B3-CT62: PERSON IN AN ELEVATOR—SCALE READING

A person who weighs 500 N is standing on a scale in an elevator. The elevator is identical in both cases. In both cases the elevator is moving at a constant speed, upward in Case A and downward in Case B.

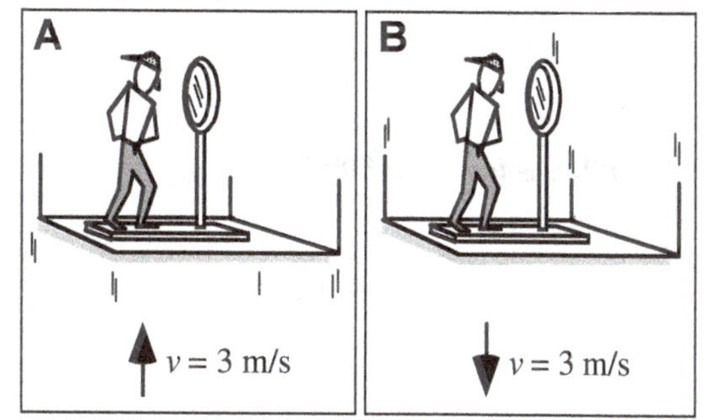

Will the scale reading be (i) *greater* in Case A, (ii) *greater* in Case B, or (iii) *the same* in both cases? _____

Explain your reasoning.

B3-QRT63: Person in an Elevator—Scale Reading

A person who weighs 500 N is standing on a scale in an elevator. The elevator is identical in all cases. The velocity and acceleration of the elevators at the instant shown are given.

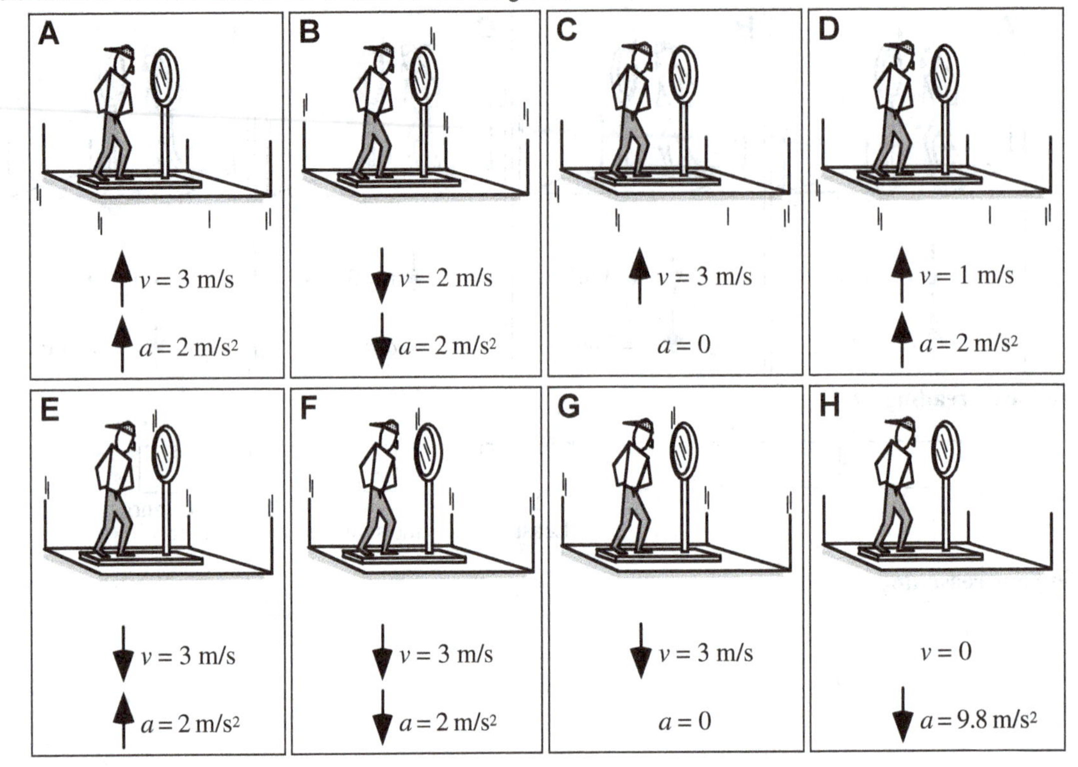

(a) List the cases where the scale reading is *greater than* 500 N. _____

Explain your reasoning.

(b) List the cases where the scale reading is *less than* 500 N. _____

Explain your reasoning.

(c) List the cases when the scale reading is *equal to* the scale reading of 500 N. _____

Explain your reasoning.

B3-CT64: PERSON IN AN ELEVATOR MOVING DOWNWARD—SCALE READING

A person who weighs 600 N is standing on a scale in an elevator. The elevator is identical in both cases. In both cases the elevator is moving downward, but in Case A it is accelerating upward and in Case B it is accelerating downward.

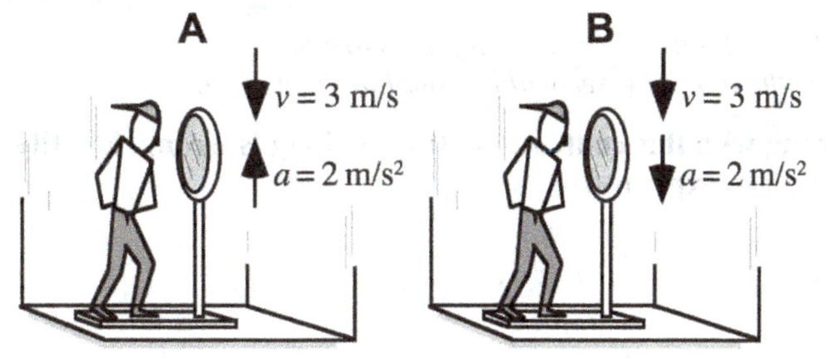

Will the scale reading be (i) *greater* in Case A, (ii) *greater* in Case B, or (iii) *the same* in both cases? _____

Explain your reasoning.

B3-CT65: BLOCK HELD ON SMOOTH RAMP—WEIGHT AND NORMAL FORCE

A block is tethered to a frictionless ramp by a horizontal string as shown. The block is at rest.

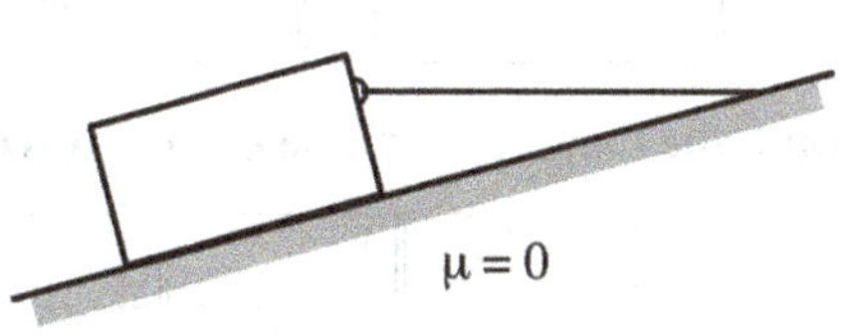

Is the normal force exerted on the block by the ramp (i) *greater than*, (ii) *less than,* or (iii) *equal to* the weight of the block? _____

Explain your reasoning.

B3-WWT66: Two Blocks at Rest—Normal Force

The two blocks are identical and both are at rest. A student comparing the normal force exerted on the block by the surface in the two cases states:

"Since both blocks are identical, I think the normal forces are the same because in each case the normal force will be equal to the weight."

What, if anything, is wrong with this contention? If something is wrong, identify it and explain how to correct it. If this contention is correct, explain why.

B3-RT67: Boxes on Rough Vertical Surface—Normal Force on Wall

Boxes are held at rest against rough, vertical walls by forces pushing horizontally on the boxes as shown.

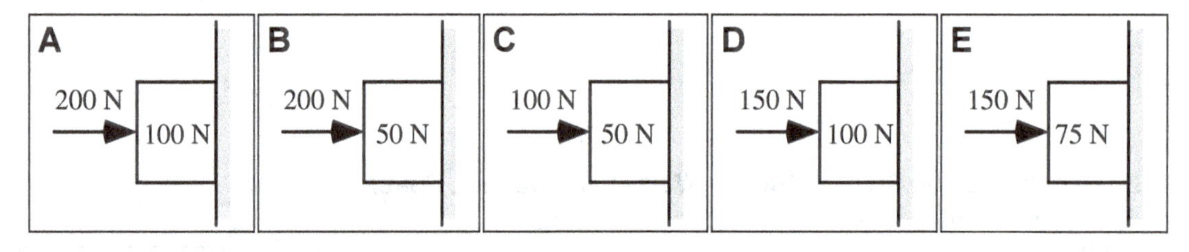

Rank the magnitude of the normal force exerted on the walls by the boxes.

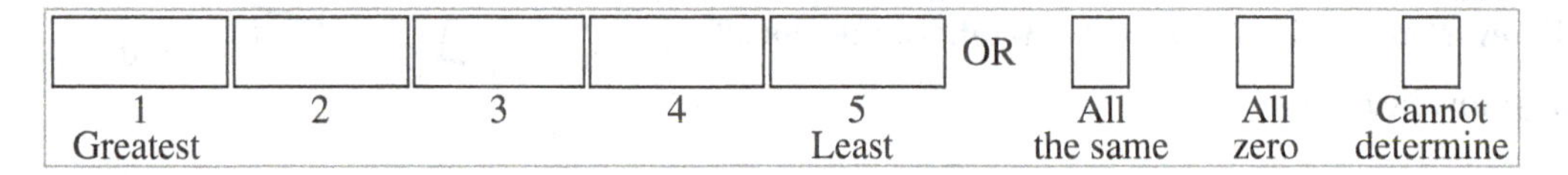

Explain your reasoning.

B3-QRT68: STACKED BLOCKS—NORMAL FORCES

A student pushes two blocks, A and B, across a desk at a constant speed. The force exerted on block A by the student is directed horizontally to the left. The mass of block A is greater than the mass of block B.

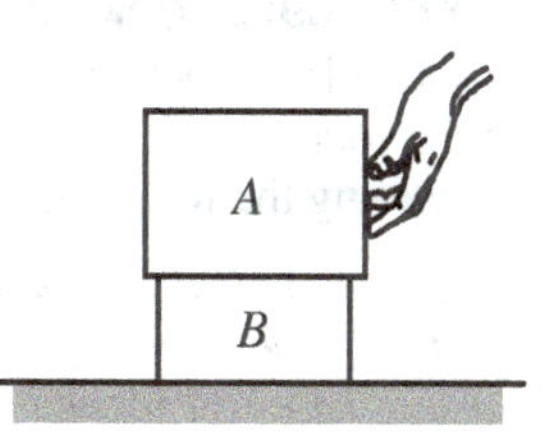

(a) The magnitude of the normal force exerted on block A by block B

(i) is *greater than* the magnitude of the normal force exerted on block B by block A.
(ii) is *less than* the magnitude of the normal force exerted on block B by block A.
(iii) is *equal to* the magnitude of the normal force exerted on block B by block A.
(iv) *cannot be compared* to the magnitude of the normal force exerted on block B by block A based on the information given.

Explain your reasoning.

(b) The magnitude of the normal force exerted on block A by block B

(i) is *greater than* the magnitude of the weight of block A.
(ii) is *less than* the magnitude of the weight of block A.
(iii) is *equal to* the magnitude of the weight of block A.
(iv) *cannot be compared* to the magnitude of the weight of block A based on the information given.

Explain your reasoning.

(c) The magnitude of the normal force exerted on block B by block A

(i) is *greater than* the magnitude of the weight of block B.
(ii) is *less than* the magnitude of the weight of block B.
(iii) is *equal to* the magnitude of the weight of block B.
(iv) *cannot be compared* to the magnitude of the weight of block B based on the information given.

Explain your reasoning.

(d) The magnitude of the normal force exerted on block B by block A

(i) is *greater than* the magnitude of the normal force exerted on block B by the desk.
(ii) is *less than* the magnitude of the normal force exerted on block B by the desk.
(iii) is *equal to* the magnitude of the normal force exerted on block B by the desk.
(iv) *cannot be compared* to the magnitude of the normal force exerted on block B by the desk based on the information given.

Explain your reasoning.

B3-WWT69: Blocks on a Rough Incline—Tension

Two identical blocks are tied together with a rope and are pulled with a second rope so that they accelerate up a rough incline. A student considering the forces acting in this situation contends:

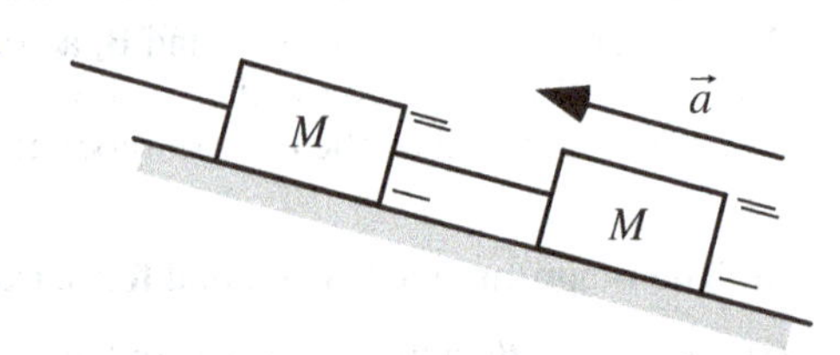

"I think the tension in the rope connecting the boxes has to be larger than the force the lower box exerts on that rope because the tension is causing the lower block to accelerate up the incline."

What, if anything, is wrong with this contention? If something is wrong, identify it and explain how to correct it. If this contention is correct, explain why.

B3-SCT70: Hanging Stone Connected to Box—Free-Body Diagrams

A massless rope connects a box on a horizontal surface and a hanging stone as shown below. The rope passes over a massless, frictionless pulley. The box is given a quick tap so that it slides to the right along the horizontal surface. The figure below shows the block after it has been pushed while it is still moving to the right. The mass of the hanging stone is larger than the mass of the box. There is friction between the box and the horizontal surface. Free-body diagrams that a student has drawn to scale for the box and for the hanging stone are shown.

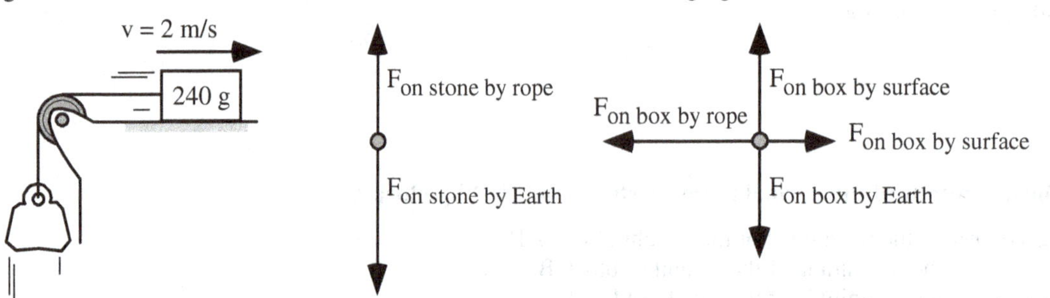

Four students discussing these free-body diagrams make the following contentions:

Ali: *"There is a problem with the free-body diagram for the hanging stone. The two forces should have the same magnitude."*

Brianna: *"But the stone is moving upward—there should be a larger force in that direction."*

Carlos: *"No, the diagram for the hanging stone is okay, but there is a problem with the diagram for the box. The frictional force is in the wrong direction."*

Dante: *"Both free-body diagrams are correct because they show the way the objects would be accelerating."*

With which, if any, of these students do you agree?

Ali _____ Brianna _____ Carlos _____ Dante _____ None of them _____

Explain your reasoning.

B3-RT71: WATER SKIERS—TENSION

Water skiers are pulled at a constant speed by a towrope attached to a speedboat. Because the weight of the skiers and the type of skis they are using varies, they experience different resistive forces from the water. Values for this resistive force (RF) and for the speed of the skiers are given.

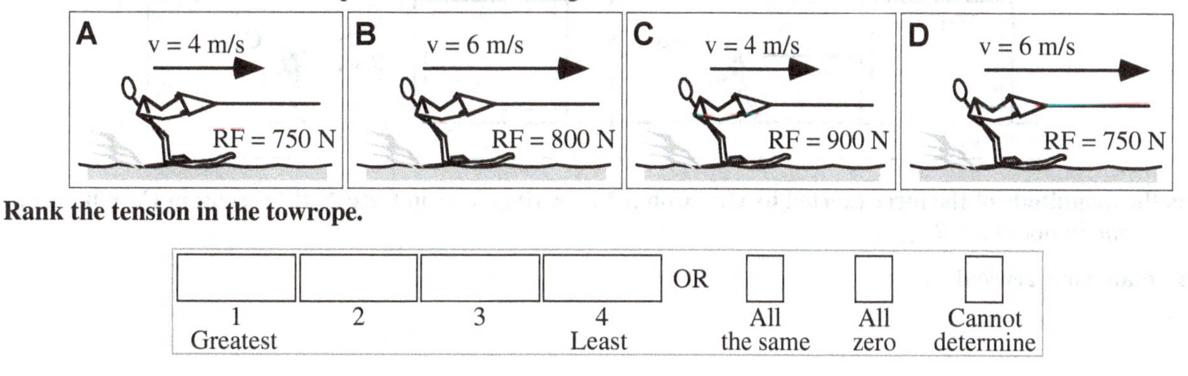

Rank the tension in the towrope.

				OR			
1 Greatest	2	3	4 Least		All the same	All zero	Cannot determine

Explain your reasoning.

B3-RT72: HANGING BLOCKS—TENSION

Two blocks are connected by strings and are pulled upward by a second string attached to the upper block. The lower block is the same in all cases, but the mass of the upper block varies. The acceleration and velocity for each system at the instant shown are given.

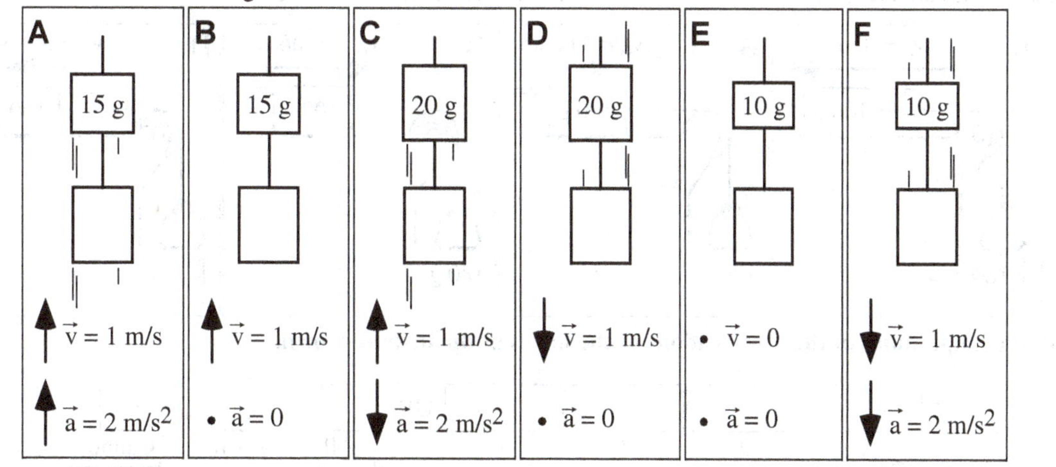

Rank the tension in the string between the blocks.

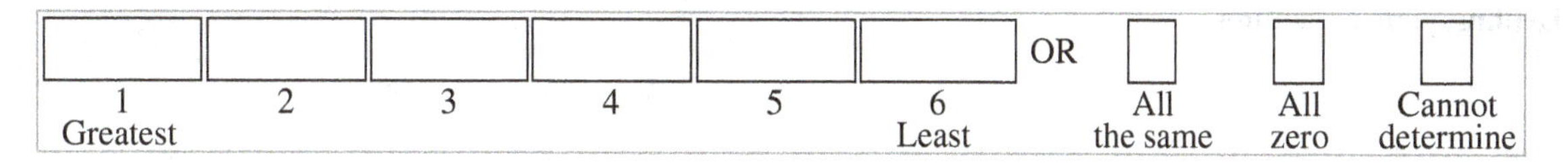

						OR			
1 Greatest	2	3	4	5	6 Least		All the same	All zero	Cannot determine

Explain your reasoning.

B3-CT73: Pulling a Crate Across Floor—Applied Force

In both cases below, Grace pulls the same large crate across a floor at a constant speed of 1.48 m per second.

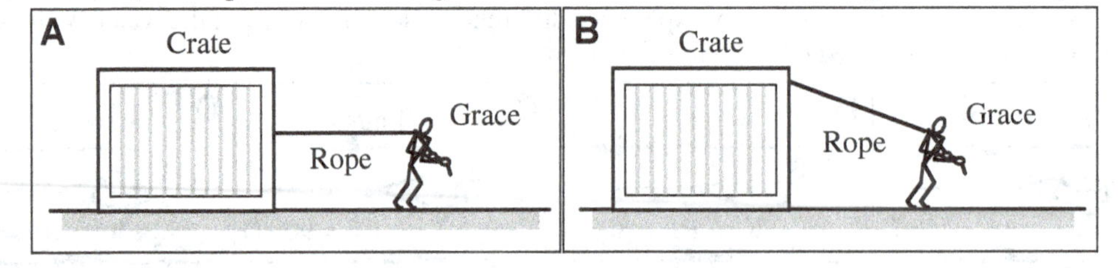

Is the magnitude of the force exerted by Grace on the rope (i) *greater* in Case A, (ii) *greater* in Case B, or (iii) *the same* in both cases? _____

Explain your reasoning.

B3-RT74: Hanging Stone Connected to Box on Rough Surface—Acceleration

In each case shown below, a box is sliding along a horizontal surface. There is friction between the box and the horizontal surface. The box is tied to a hanging stone by a massless rope running over a massless, frictionless pulley. All these cases are identical except for the different initial velocities of the boxes.

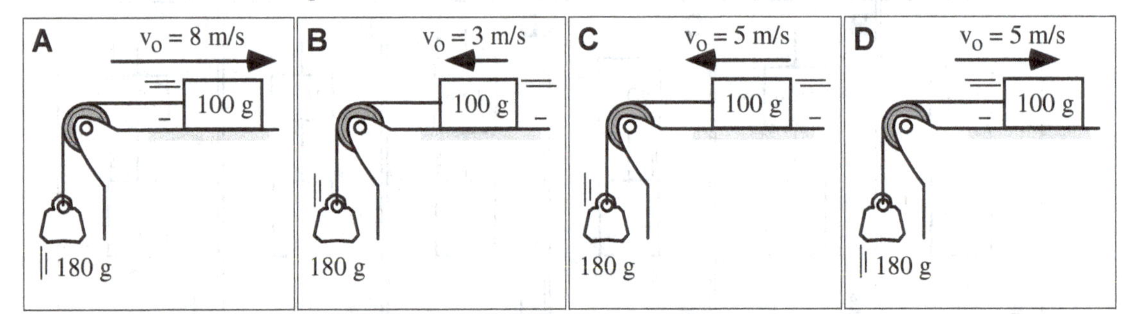

Rank the magnitudes of the accelerations of the boxes at the instant shown.

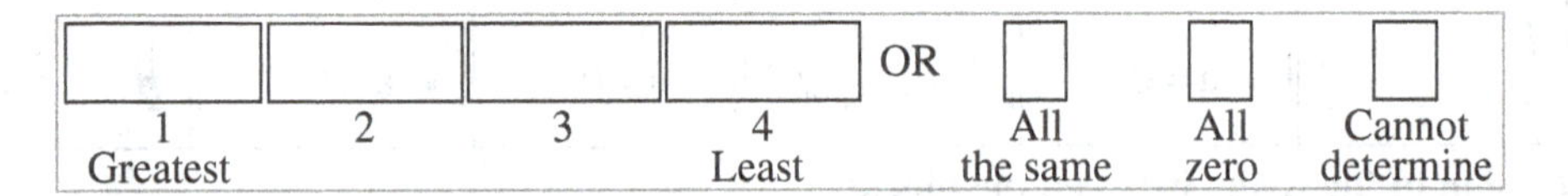

Explain your reasoning.

B3-RT75: MOVING STRING PASSING OVER A PULLEY—TENSION AT POINTS

A student pulls on a massless string that passes over a frictionless pulley and is attached to a suspended mass. He is pulling the string horizontally so that, at the instant shown, the mass is moving upward at a constant speed.

Rank the tension at the labeled points.

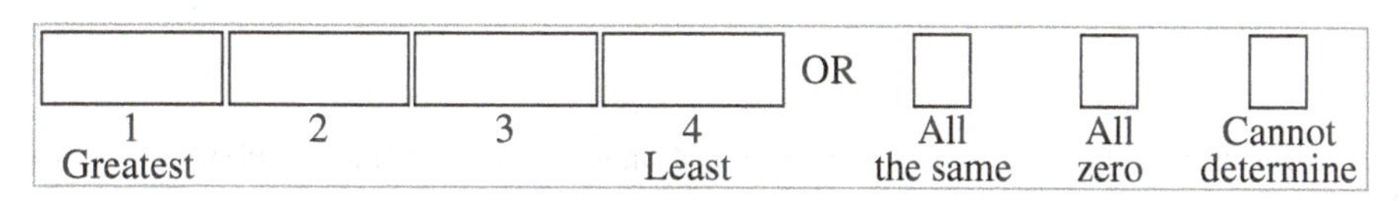

Explain your reasoning.

B3-SCT76: STRING AT ANGLE PASSING OVER A PULLEY—TENSION AT POINTS

A student holds a massless string that passes over a frictionless pulley and is attached to a suspended mass as shown. The mass is at rest. She then moves her hand so that the portion of string between her hand and the pulley moves from the horizontal to an angle as shown. Four students make the following contentions about this situation:

Aletheia: *"The tensions would all stay the same. The pulley will change the direction of the force, but the size of the force only depends on the mass, and that hasn't changed."*

Bem: *"Nothing has changed at point C, and the tension will stay the same there. But A and B are being pulled downward now as well as horizontally, so the tensions at A and B would increase."*

Charity: *"I agree with you that the tension at point C would stay the same, but I think the tensions at A and B would actually decrease. Now gravity is actually helping the hand to keep the mass in place, and so the hand won't have to pull as hard. Since A and B are close to where the hand is pulling, the tension there will go down."*

Dorothy: *I think the tension will increase at all three points. As the hand moves down so that it is more vertical, it's like it is fighting the pull of the mass more and more. So the string gets stretched tighter—the tension goes up all over."*

With which, if any, of these students do you agree?

Aletheia_____ Bem _____ Charity _____ Dorothy _____ None of them_____

Explain your reasoning.

B3-CT77: Ball Suspended from Ceiling by Two Strings—Tension

A 0.5-kg ball is suspended from a ceiling by two strings. The ball is at rest.

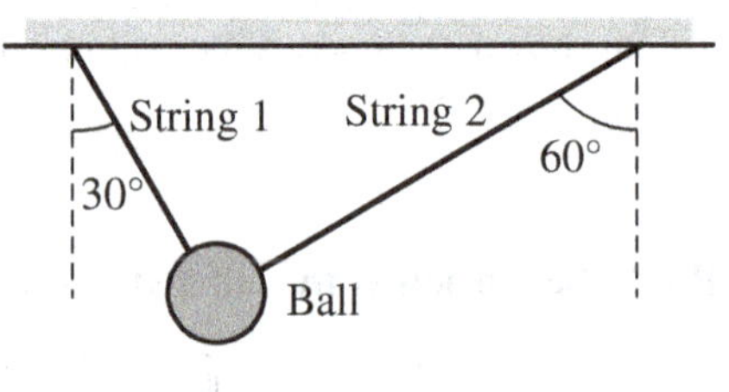

(a) Is the tension in string 1 (i) *greater than*, (ii) *less than*, or (iii) *the same as* the tension in string 2? _____

Explain your reasoning.

Suppose that the ceiling in the picture above is the ceiling of an elevator, and that the elevator is moving *down* at a constant speed of 2 m/s.

(b) Is the tension in string 1 (i) *greater than*, (ii) *less than*, or (iii) *the same as* the tension in string 1 in the previous question (a) where the ball was at rest? _____

Explain your reasoning.

B3-SCT78: Hanging Mass—Tension in Three Strings

A hanging mass is suspended midway between two walls. The string attached to the left wall is horizontal while the string attached to the right wall makes an angle with the horizontal as shown. This angle (α) in Case A is larger than the angle (β) in Case B. Four students make the following claims about the tensions in the strings:

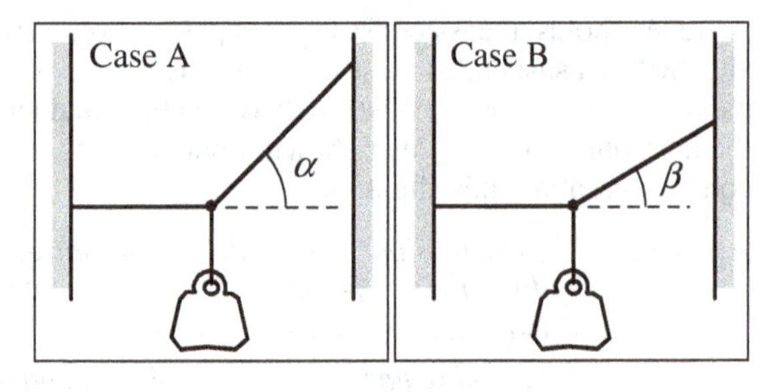

Abbie: *"I think the tensions in any string in Case A is going to be the same as the equivalent string in Case B. The weight is the same, and the weight is still going to be divided up among the three ropes."*

Bobby: *"I think the tensions in the horizontal and vertical strings are the same, because they are exactly the same in both cases. But in Case B the diagonal rope is shorter, so the tension is more concentrated there."*

Che: *"The diagonal string still has to hold the weight up by itself, because the horizontal string can't lift anything. So the diagonal string still has the same tension. But in Case B it's pulling harder against the horizontal string because of the angle, so the tension in the horizontal string has to go up."*

Damian: *"But the diagonal string is fighting harder against the weight in Case A—it is pointing more nearly opposite the weight. So it has to have a greater tension in Case A. And since the tension in the diagonal string is greater, and the tension in the vertical string is the same, the tension in the horizontal string must be less in Case A. The tensions still have to balance out so that they are the same in both cases."*

With which, if any, of these students do you agree?

Abbie _____ Bobby _____ Che _____ Damian _____ None of them_____

Explain your reasoning.

B3-SCT79: TWO CONNECTED OBJECTS ACCELERATING DOWNWARD—TENSION

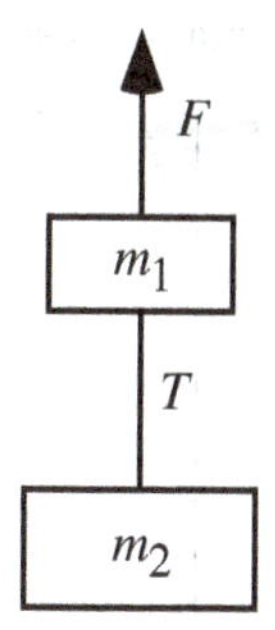

Two objects with masses of $m_1 = 6$ kg and $m_2 = 10$ kg are connected by a massless string. They are pulled upward by an applied force F. Since this force is smaller than the total weight of the objects, there is a constant downward acceleration of 3 m/s^2. The tension in the string connecting the objects is T. Four students discuss this tension:

Anh: *"The tension in the string is the net force on the lower object. Using Newton's Second law, we get $F_{net} = ma = 30$ N for the tension, since the lower object has a mass of 10 kg and it is accelerating at 3 m/s^2."*

Brandon: *"The tension in the string is more than the net force of 30 N since the lower object has a weight of about 100 N. The tension should be 130 N since the 30 N, the net force, is added to 100 N, the weight."*

Cathy: *"The tension in the string is upward and should be less than the weight since the system is accelerating downward. It should be 70 N by applying Newton's Second law and taking into account the directions of the forces."*

Deshi: *"We cannot answer it until we know which direction the system is moving. Is it moving upward or downward? Won't that make a big difference on the tension?"*

With which, if any, of these students do agree?

Anh _____ Brandon _____ Cathy _____ Deshi _____ None of them _____

Explain your reasoning.

B3-WWT80: BLOCKS ON A SMOOTH INCLINE—TENSION

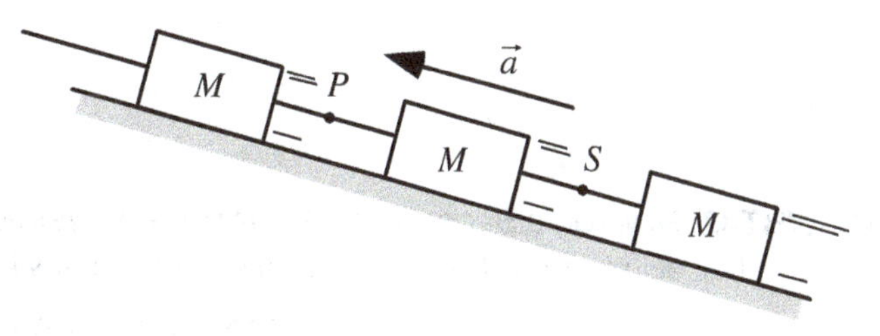

Three identical blocks are tied together with ropes and pulled up a smooth (frictionless) incline. The blocks accelerate up the incline. A student who is asked to compare the tension in the rope at point P to the tension at point S states:

"Each rope is pulling one block. All three blocks are accelerating at the same rate and they are identical. I think the tensions at points P and S will be the same."

What, if anything, is wrong with this contention? If something is wrong, identify it and explain how to correct it. If this contention is correct, explain why.

B3-RT81: Hanging Mass—String Tension

A massless string is attached to one or more identical blocks at rest. All the pulleys are frictionless and massless.

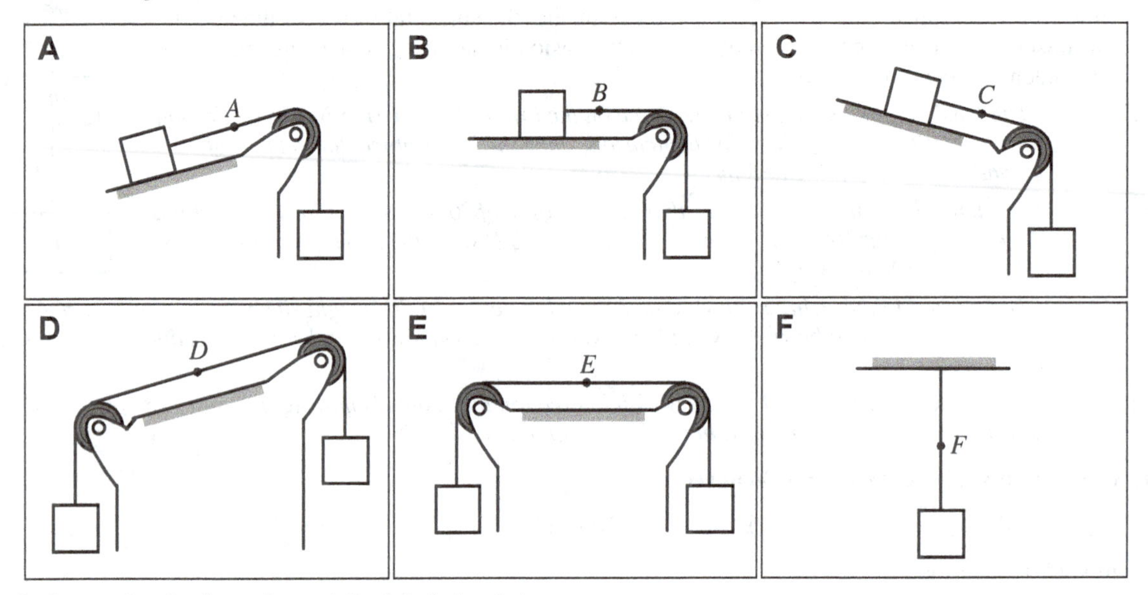

Rank the tension in the strings at the labeled points.

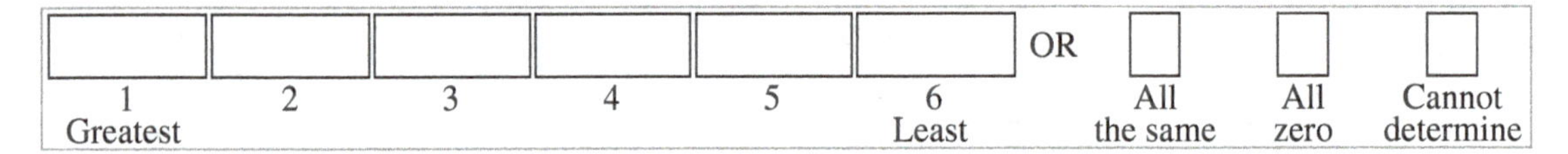

Explain your reasoning.

B3-WBT82: Newton's Second Law Equation—Physical Situation

The equation below results from the application of Newton's Laws to an object:

$$27\text{ N} - (\mu)(14\text{ kg})(9.8\text{ m/s}^2) = 0$$

Draw a physical situation that would result in this equation, and explain how your drawing is consistent with the equation.

B3-SCT83: BLOCKS ON A SMOOTH INCLINE—TENSION

Two blocks are tied together with a rope and are pulled so that they accelerate up a smooth (frictionless) incline. Three students are comparing the tension in the rope between the blocks to the magnitude of the force that the lower block exerts on that rope:

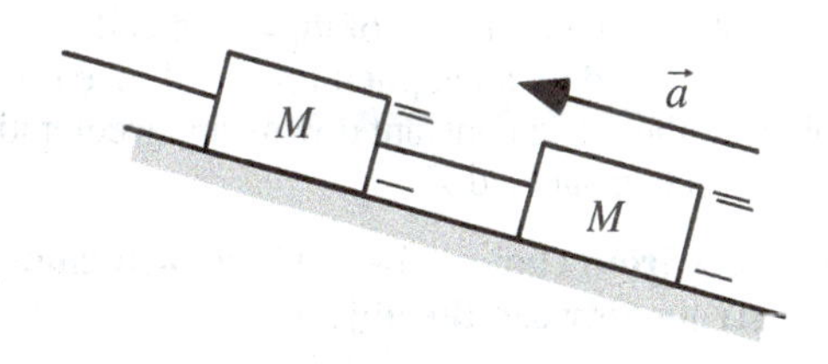

Alberto: *"I think the tension has to be larger because it is causing the lower block to accelerate up the incline. If it was the same, then the block wouldn't accelerate."*

Benifacio: *"I disagree. Force equals mass times acceleration, and the accelerations of the rope and the lower block are the same. The rope hardly weighs anything compared to the block, so it can't exert as much force. The force the block exerts has to be greater."*

Connie: *"I agree that the rope and the block have exactly the same acceleration since they are moving together. But I think that means that the force has to be the same."*

With which, if any, of these students do you agree?

Alberto____ Benifacio _____ Connie _____ None of them_____

Explain your reasoning.

B3-CT84: BLOCKS MOVING AT CONSTANT SPEED—TENSION IN CONNECTING STRING

Two identical blocks, 1 and 2, are connected by a massless string. In Case A, a student pulls on a string attached to block 2 so that the blocks travel to the right across a desk at a constant speed of 10 cm/s. In Case B, the student pulls on a string attached to block 1 so that the same blocks travel across the same desk to the left at a constant speed of 20 cm/s.

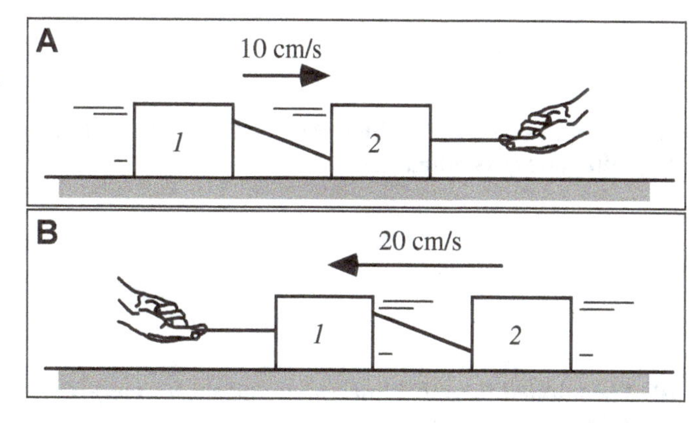

Will the tension in the diagonal string connecting the two blocks be (i) *greater* in Case A, (ii) *greater* in Case B, or (iii) *the same* in both cases? _____

Explain your reasoning.

B3-LMCT85: Two Connected Objects Accelerating Downward—Tension in String

Two objects with masses of $m_1 = 6$ kg and $m_2 = 10$ kg are connected by a massless string. They are pulled upward by an applied force F. Since this force is smaller than the total weight of the objects, there is a constant downward acceleration of 3 m/s^2. The tension in the string connecting the objects is labeled T.

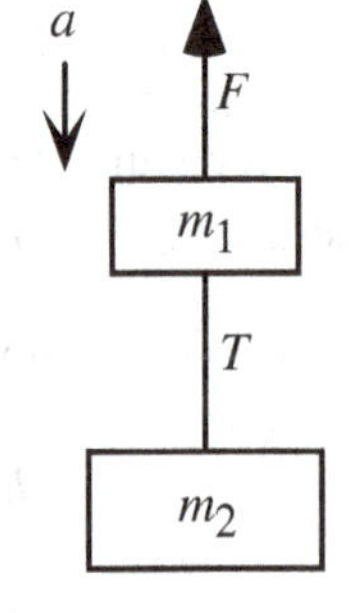

Identify from choices (i)–(iv) how each change described below will affect the tension (T) in the string between the objects.

Compared to the case above, this change will:

(i) ***increase*** the tension in the string.
(ii) ***decrease*** the tension in the string but not to zero.
(iii) ***decrease*** the tension in the string **to zero**.
(iv) *have **no effect*** on the tension in the string.
(v) *have an **indeterminate*** effect on the tension in the string.

All of these modifications are the only changes to the initial situation shown in the diagram.

(a) The mass of m_1 is decreased to 5 kg and the mass of m_2 is increased to 11 kg. _____
Explain your reasoning.

(b) The mass of m_1 is increased to 7 kg and the mass of m_2 is decreased to 9 kg. _____
Explain your reasoning.

(c) The applied force F is increased and the acceleration is 2 m/s^2 downward. _____
Explain your reasoning.

(d) The applied force F is increased and the acceleration is 4 m/s^2 upward. _____
Explain your reasoning.

(e) The applied force F is decreased and the acceleration is 4 m/s^2 downward. _____
Explain your reasoning.

B3-SCT86: BOX PULLED ON ROUGH, HORIZONTAL SURFACE—FRICTIONAL FORCE ON BOX

A 100 N box is initially at rest on a rough, horizontal surface. The coefficient of static friction is 0.6, and the coefficient of kinetic friction is 0.4. A constant 35 N horizontal force to the right is applied to the box. Four students are discussing the frictional force exerted on the box by the rough surface 1 second after the force is first applied:

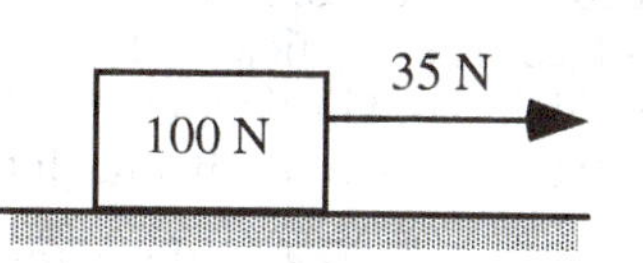

Al: *"The frictional force is 60 N since the box will not be moving and the coefficient of static friction is 0.6 with a normal force of 100 N."*

Brianna: *"The frictional force is 40 N since the coefficient of kinetic friction is 0.4 and there is a normal force of 100 N."*

Carlos: *"The frictional force is 35 N since the box will not be moving and the frictional force will cancel out the applied force of 35 N."*

David: *"It is 40 N for the kinetic frictional force and 60 N for the static frictional force. The normal force is 100 N and the coefficient of kinetic friction is 0.4, giving 40 N for the kinetic friction. Similarly, for the static frictional force it is 60 N since it has a coefficient of static friction of 0.6."*

With which, if any, of these students do you agree?

Al _____ Brianna _____ Carlos _____ David _____ None of them _____

Explain your reasoning.

B3-CT87: BOX MOVING OVER HORIZONTAL SURFACE—FRICTIONAL FORCE ON BOX

A 50 N box has an applied force on it of 40 N that makes an angle of 30° with the horizontal. The box is moving to the right at a constant speed in both cases.

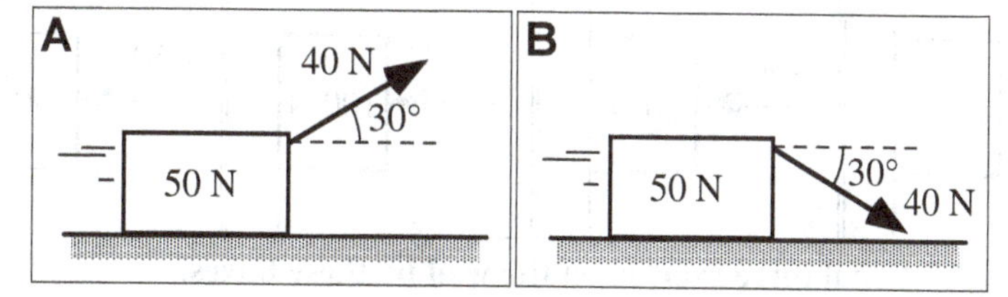

Will the frictional force exerted on the box by the rough surface be (i) *greater* in Case A, (ii) *greater* in Case B, or (iii) *the same* in both cases? _____

Explain your reasoning.

B3-SCT88: Box Held Against Vertical Surface—Frictional Force on Box

A constant horizontal force on a 200 N is applied to a box in contact with a vertical surface. The coefficient of static friction between the box and the surface is 0.6, and the coefficient of kinetic friction is 0.4. Several students are discussing the frictional force on the box 1 second after the force is first applied:

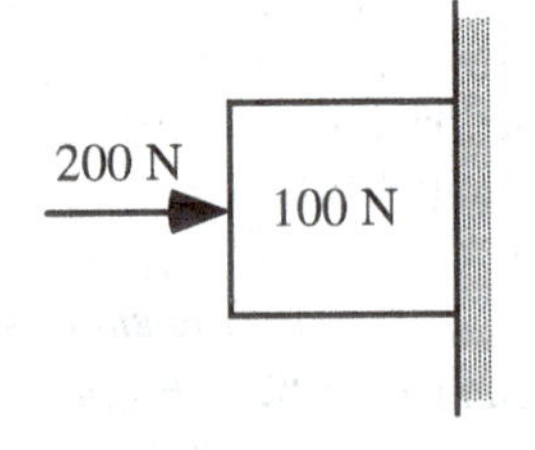

Art: *"The frictional force is 60 N since the box will not be moving and the coefficient of static friction is 0.6."*

Bratislav: *"The frictional force is 100 N upward since the box has a weight of 100 N downward."*

Celeste: *"The frictional force will be 120 N since the box will not be moving and the normal force will be 200 N."*

Dorothy: *"The frictional force will be 40 N for the kinetic frictional force and 60 N for the static frictional force. The weight is 100 N and the coefficient of kinetic friction is 0.4, giving 40 N for the kinetic friction. Likewise, for the static frictional force it has a coefficient of static friction of 0.6, giving a static frictional force of 60 N."*

With which, if any, of these students do you agree?

Art _____ Bratislav _____ Celeste _____ Dorothy _____ None of them _____

Explain your reasoning.

B3-RT89: Boxes Held Against Vertical Surfaces—Frictional Forces on the Wall

A box is held at rest against a rough, vertical surface by a force pushing horizontally as shown. Values for the applied force and the weight of the boxes are given. The boxes are all made of the same material and the walls are identical.

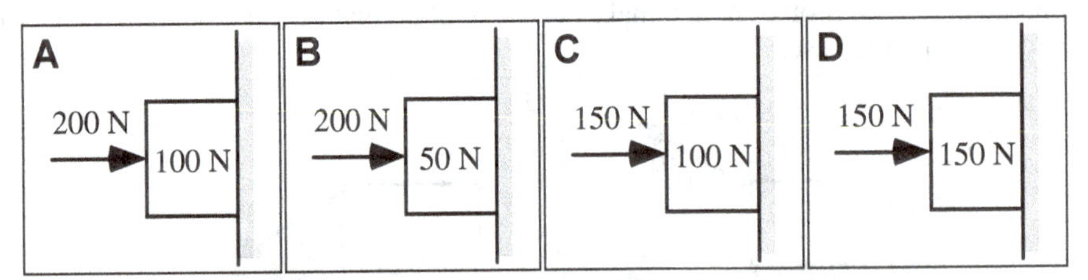

Rank the magnitude of the frictional force exerted on the wall by these boxes.

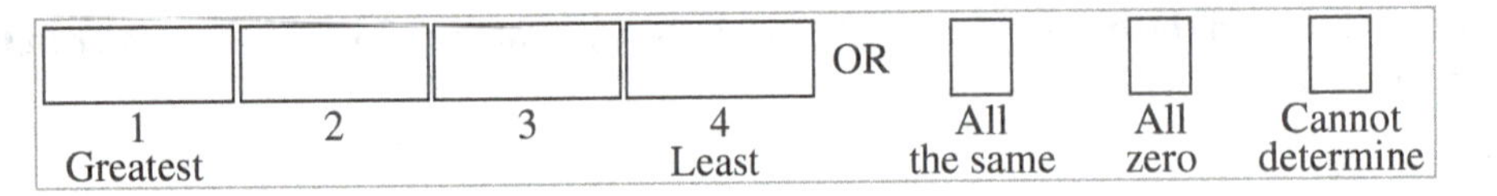

Explain your reasoning.

B3-QRT90: MOVING STACKED BLOCKS—FRICTION FORCES

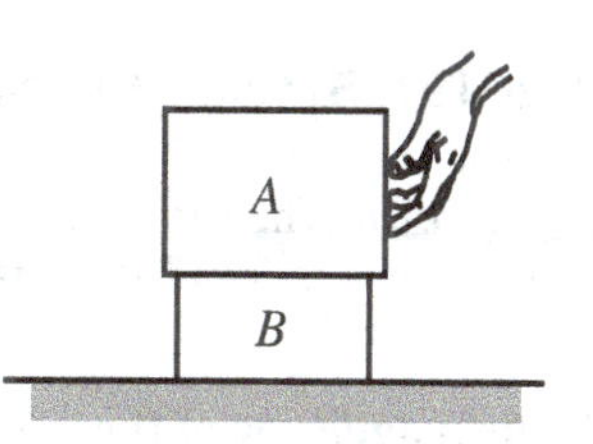

A student pushes two blocks across a desk at a constant speed. The force exerted on block A by the student is directed horizontally to the left. The mass of block A is greater than the mass of block B.

(a) The magnitude of the friction force exerted on block A by block B

(i) is *greater than* the magnitude of the friction force exerted on block B by block A.
(ii) is *less than* the magnitude of the friction force exerted on block B by block A.
(iii) is *equal to* the magnitude of the friction force exerted on block B by block A.
(iv) cannot be compared to the magnitude of the friction force exerted on block B by block A based on the information given.

Explain your reasoning.

(b) The magnitude of the friction force exerted on block B by the desk

(i) is *greater than* the magnitude of the friction force exerted on block B by block A.
(ii) is *less than* the magnitude of the friction force exerted on block B by block A.
(iii) is *equal to* the magnitude of the friction force exerted on block B by block A.
(iv) cannot be compared to the magnitude of the friction force exerted on block B by block A based on the information given.

Explain your reasoning.

(c) The magnitude of the friction force exerted on block A by block B

(i) is *greater than* the magnitude of the force exerted on block A by the hand.
(ii) is *less than* the magnitude of the force exerted on block A by the hand.
(iii) is *equal to* the magnitude of the force exerted on block A by the hand.
(iv) cannot be compared to the magnitude of the force exerted on block A by the hand based on the information given.

Explain your reasoning.

B3-QRT91: Stacked Blocks Slowing Down—Friction Forces

A student pushes two blocks across a desk. At the instant shown, the blocks are *slowing down.* The force exerted on block A by the student is directed horizontally to the left. The mass of block A is greater than the mass of block B.

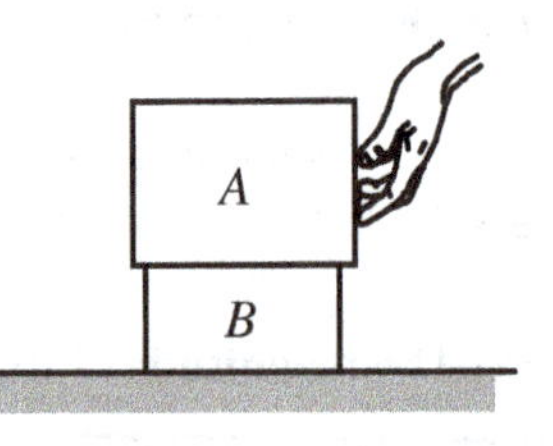

(a) The magnitude of the friction force exerted on block A by block B

(i) is *greater than* the magnitude of the friction force exerted on block B by block A.
(ii) is *less than* the magnitude of the friction force exerted on block B by block A.
(iii) is *equal to* the magnitude of the friction force exerted on block B by block A.
(iv) cannot be compared to the magnitude of the friction force exerted on block B by block A based on the information given.

Explain your reasoning.

(b) The magnitude of the friction force exerted on block B by the desk

(i) is *greater than* the magnitude of the friction force exerted on block B by block A.
(ii) is *less than* the magnitude of the friction force exerted on block B by block A.
(iii) is *equal to* the magnitude of the friction force exerted on block B by block A.
(iv) cannot be compared to the magnitude of the friction force exerted on block B by block A based on the information given.

Explain your reasoning.

(c) The magnitude of the friction force exerted on block A by block B

(i) is *greater than* the magnitude of the force exerted on block A by the hand.
(ii) is *less than* the magnitude of the force exerted on block A by the hand.
(iii) is *equal to* the magnitude of the force exerted on block A by the hand.
(iv) cannot be compared to the magnitude of the force exerted on block A by the hand based on the information given.

Explain your reasoning.

B3-WWT92: TWO ASTEROIDS—GRAVITATIONAL FORCE ON EACH

Two asteroids with masses of m and $3m$ exert gravitational forces on each other. A student contends that the forces will be in different directions and of different magnitudes as shown below.

What's wrong, if anything, with this student's contention? If something is wrong, identify it and explain how to correct it. If this student's contention is correct, explain why.

B3-SCT93: THREE ASTEROIDS IN A LINE—CALCULATION OF MASS

At the instant shown, three asteroids are in a line, and the distance between A and B is twice the distance between B and C. Asteroid C has mass M. There is no net force on asteroid B due to the other asteroids.

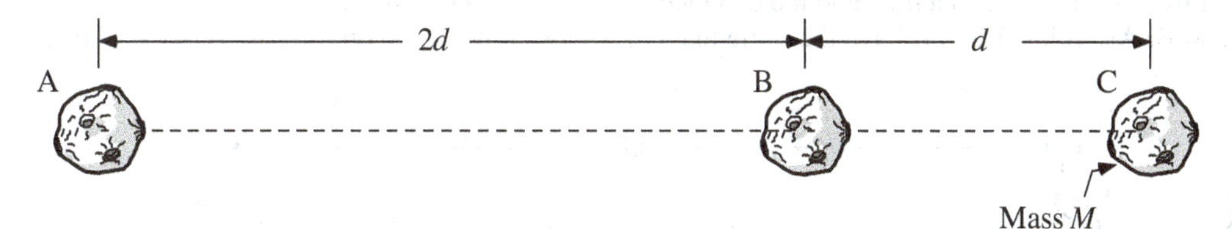

Three students are discussing how they might find the mass of asteroid A:

Ari: *"We don't really have enough information to find the mass of A. Since there's no net force on B, the force from A has to cancel the force from C. To find the force on B from C, we'd use Newton's law of universal gravitation. But since the force is proportional to the product of the masses, we'd need to know both masses."*

Bira: *"I don't think we really need the mass of B. Asteroid A is twice as far away as C, so if it also has a mass M it will exert half as much force as C does. Since it has to exert the same force for the net force on B to be zero, it has to have twice the mass."*

Cole: *"It's true that A pulls on B to the left, and C pulls on B to the right. But you can't just use Newton's law of universal gravitation, because that only allows you to calculate the force between two masses. Here there are three masses, and asteroid A is exerting some of its force on B and some on C."*

With which, if any, of these students do you agree?

Ari _____ **Bira** _____ **Cole** _____ None of them_____

Explain your reasoning.

B3-QRT94: Three Objects Exerting Gravitational Forces—Net Force

Three objects each with a mass of M exert gravitational forces on each other. **Which of the arrows below shows the direction of the net force on mass B?**

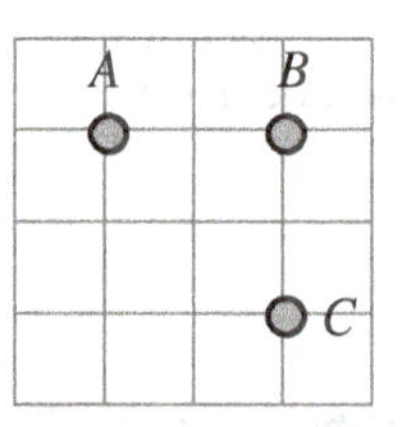

Explain your reasoning.

B3-WWT95: Three Asteroids in a Line—Mass of Asteroid

At the instant shown, three asteroids are in a line, and the distance between B and C is 4 times as large as the distance between A and B. Asteroids A and B have the same mass. There is no net force on asteroid B due to the other two asteroids.

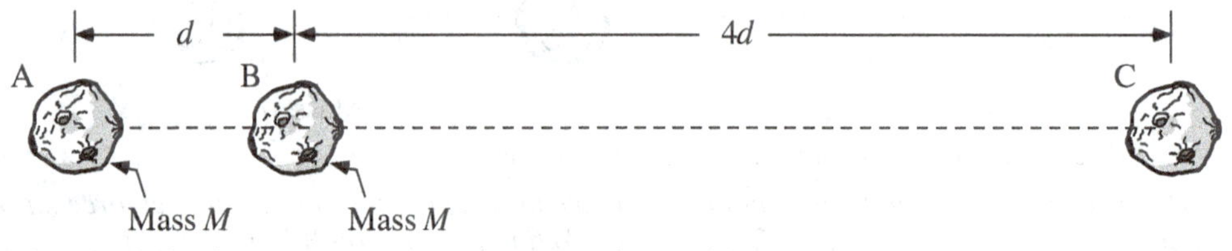

A student makes the following comment about the mass of asteroid C:

"Since C is four times as far from B as A is, it is only going to have one-quarter the effect on B. To get the forces on B to balance, you'd need the mass of C to be four times as large."

What, if anything, is wrong with the student's statement? If nothing is wrong, state that explicitly, and explain why it is correct. If the statement is incorrect, state what is wrong and how you would correct it.

B3-QRT96: TWO OBJECTS—GRAVITATIONAL FORCE ON EACH

Object A has twice the mass of Object B.

$2m$ A m B

Identify the pair of force vectors (the arrows) that correctly compare the gravitational force exerted on A by B with the gravitational force exerted on B by A.

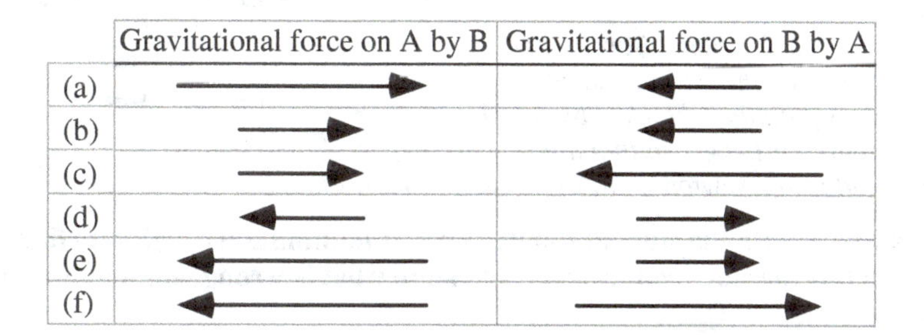

Explain your reasoning.

B3-RT97: THREE ASTEROIDS IN A LINE—NET FORCE

At the instant shown, three asteroids are in a line, and the distance between A and B is the same as the distance between B and C. Asteroids B and C have the same mass, while asteroid A has twice the mass.

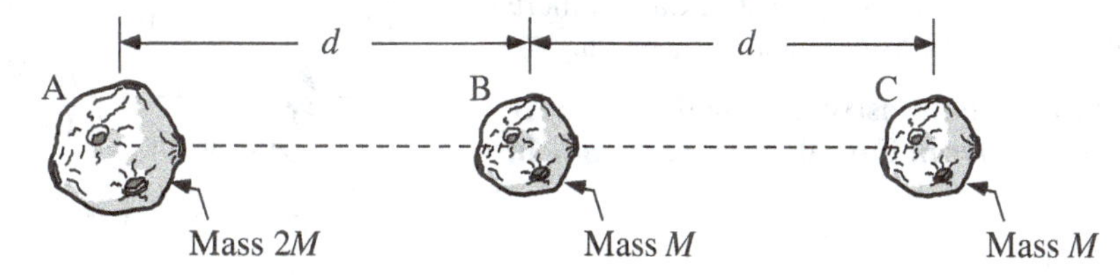

Rank the magnitude of the net force on each asteroid due to the other two asteroids.

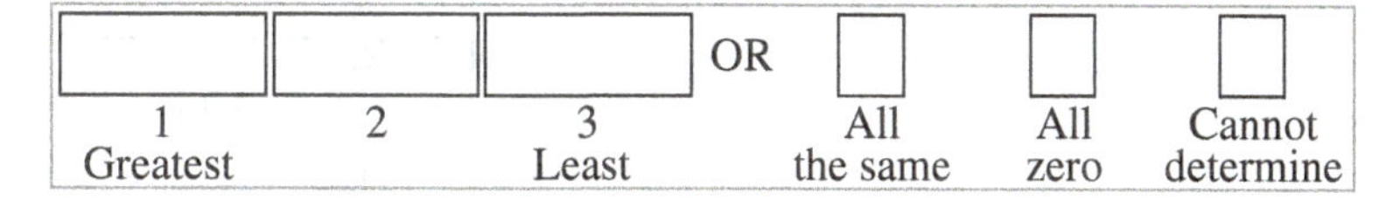

Explain your reasoning.

B3-WWT98: Astronaut near a Planet and a Moon—Gravitational Force

At the instant shown, Catalina, an astronaut, is in space a distance $2d$ from the planet Barolo. A moon of Barolo, Aquitania, is a distance d on the other side of Barolo. A physics student trying to calculate the gravitational force on Catalina states:

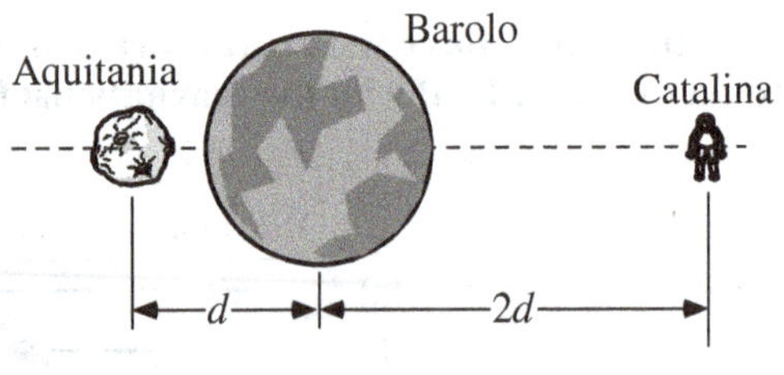

"We only need to worry about the planet Barolo, not its moon. Since Catalina is on the other side of Barolo, the gravitational field due to Aquitania is blocked by Barolo. The only force acting on Catalina is the gravitational force due to Barolo, which we can find using Newton's law of universal gravitation."

What, if anything, is wrong with the student's statement? If nothing is wrong, state that explicitly, and then explain why it is correct. If the statement is incorrect, state what is wrong and how you would correct it.

B3-CT99: Astronaut near a Moon—Gravitational Force

In both cases shown at right, an astronaut is a distance x from a moon. The two cases are identical except that in Case A there is a large planet directly between the astronaut and the moon.

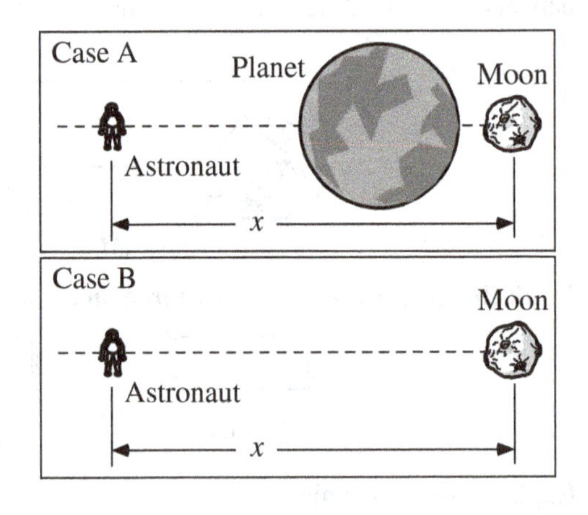

Is the gravitational force on the astronaut by the moon (i) *greater* in Case A, (ii) *greater* in Case B, or (iii) *the same* in both cases? _____

Explain your reasoning.

B3-SCT100: BALL WHIRLED IN VERTICAL CIRCLE—NET FORCE ON BALL

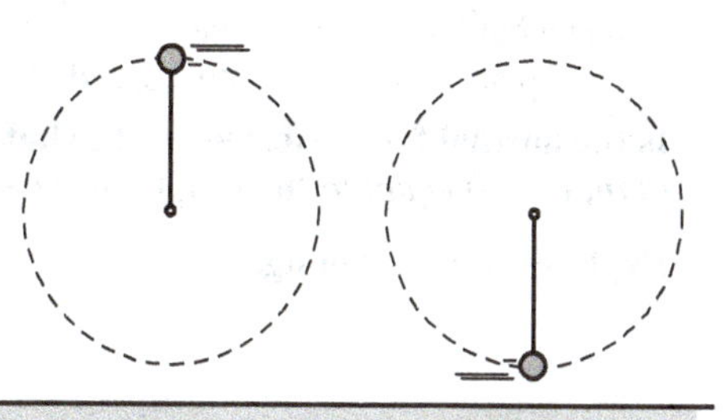

A ball with a weight of 2 N is attached to the end of a cord of length 2 m. The ball is whirled in a vertical circle counterclockwise. The tension in the cord at the top of the circle is 7 N, and at the bottom it is 15 N. (The speed of the ball is not the same at these points.)

(a) Three students discuss the net force on the ball at the top.

Angelica: *"The net force on the ball at the top position is 7 N since the net force is the same as the tension."*

Bo: *"The net force on the ball at the top position is 9 N. Both the tension and the weight are acting downward so you have to add them."*

Charles: *"No, you are both wrong. You need to figure out the centripetal force (mv^2/r) and include it in the net force."*

With which, if any, of these students do you agree?

Angelica _____ Bo _____ Charles _____ None of them _____

Explain your reasoning.

(b) Now the students discuss the net force on the ball at the bottom.

Angelica: *"The net force on the ball at the bottom position is 15 N since the net force is the same as the tension."*

Bo: *"The net force on the ball at the bottom position is 17 N, since you need to add the weight of 2 N to the tension of 15 N."*

Charles: *"The net force on the ball at the bottom position is 13 N. I agree that you need to take into account both the weight and the tension, but they are in different directions so they will subtract."*

With which, if any, of these students do you agree?

Angelica _____ Bo _____ Charles _____ None of them _____

Explain your reasoning.

B3-CT101: Skateboarder on Circular Bump—Weight and Normal Force

A skateboarder is skating over a circular bump. At the instant shown, she is at the top of the bump and is moving with a speed of 5 m/s.

Is the normal force exerted on the skateboarder by the bump (i) *greater than,* (ii) *less than,* or (iii) *equal to* the weight of the skateboarder? _____

Explain your reasoning.

B3-SCT102: Child on a Swing—Tension

A child is swinging back and forth on a tire swing that is attached to a tree branch by a single rope. Shown are two positions during a swing from right to left. Three students are discussing the tension in the rope at the bottom of the swing.

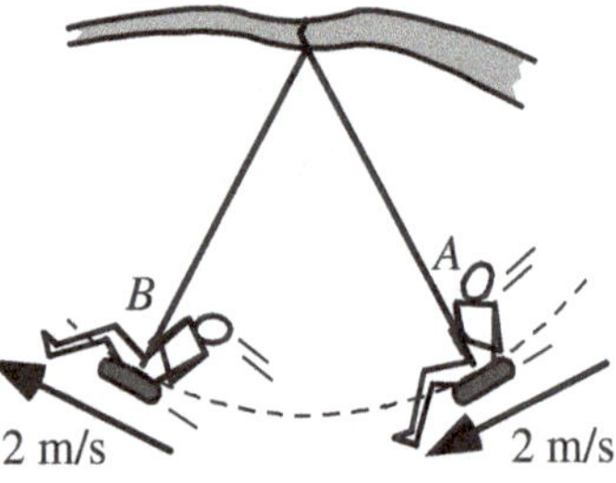

Alia: *"At the bottom of the swing, she will be moving exactly horizontally. Since she is not moving vertically at that instant, the vertical forces cancel. The tension in the rope at that instant equals the weight."*

Brian: *"Just looking at the velocity vectors, the change in velocity points upward between A and B. So that is the direction of the acceleration, and also of the net force. To get a net force pointing upward, the tension would have to be greater than the weight."*

Clara: *"But there aren't just two forces acting on her at the bottom of the swing. Since she's moving in a circle, there's also the centripetal force, which acts toward the center of the circle. Since both the tension and the centripetal force point upward, and the weight points downward, to get zero net force the tension actually has to be less than the weight. The tension plus the centripetal force equals the weight."*

With which, if any, of these students do you agree?

Alia _____ Brian _____ Clara _____ None of them_____

Explain your reasoning.

B4 WORK AND ENERGY

B4-RT01: MOVING BALLS I—KINETIC ENERGY

In the figures below, balls are traveling in different directions. The balls have the same size and shape, but they have different masses and are traveling at different velocities as shown.

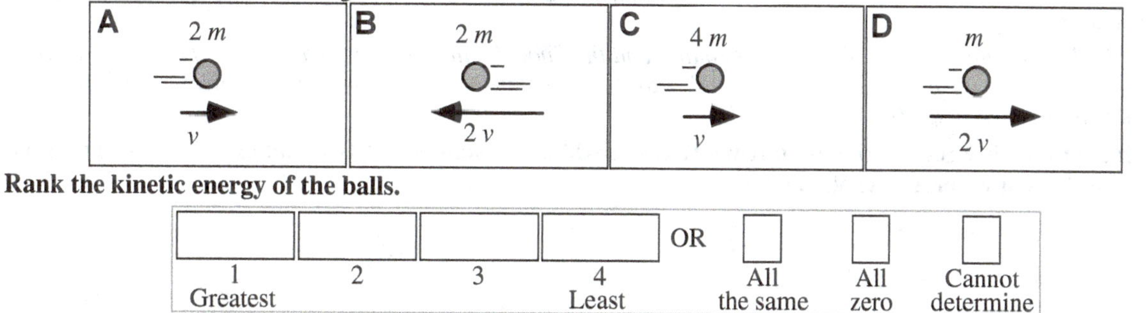

Rank the kinetic energy of the balls.

				OR			
1 Greatest	2	3	4 Least		All the same	All zero	Cannot determine

Explain your reasoning.

B4-RT02: MOVING BALLS II—KINETIC ENERGY

In the figures below, balls are traveling in different directions. The balls have the same size, mass, and shape, but they are traveling with different velocities as shown.

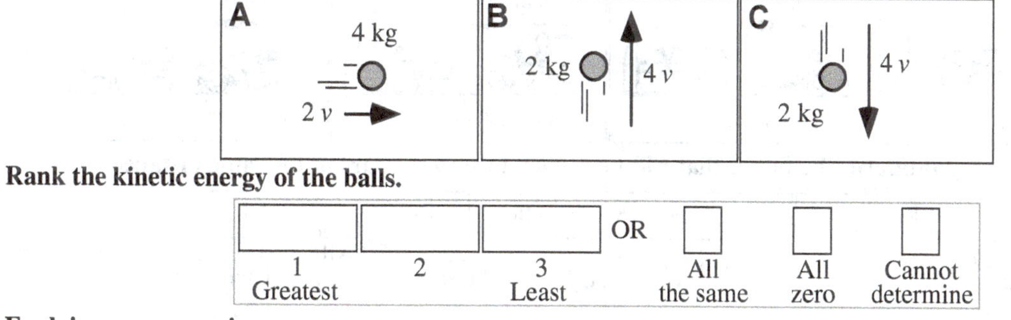

Rank the kinetic energy of the balls.

			OR			
1 Greatest	2	3 Least		All the same	All zero	Cannot determine

Explain your reasoning.

B4-WWT03: Object Changing Velocity—Work

A 2-kg object accelerates as a net force acts on it. During the 5 seconds this force acts, the object changes its velocity from 3 m/s east to 7 m/s west.

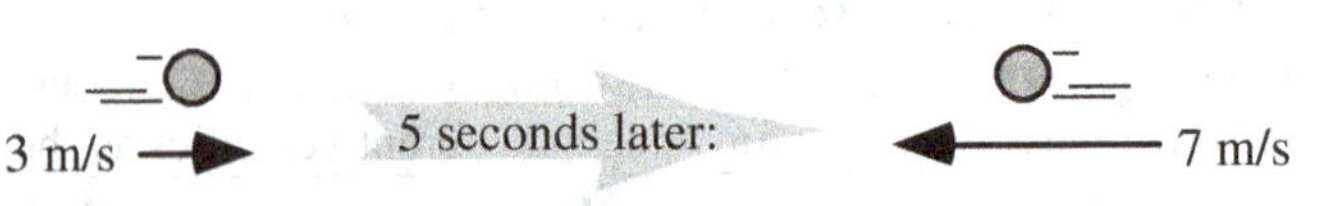

A student states:

"The initial kinetic energy of the object was 9 Joules, and the final kinetic energy was 49 Joules. Thus the change in kinetic energy of this object during these 5 seconds was 40 J, and thus the work done on this object by the net force during this period was also 40 J."

What, if anything, is wrong with this statement? If something is wrong, identify it and explain how to correct it. If this statement is correct, explain why.

B4-RT04: Cars and Barriers—Stopping Force in Same Distance

Cars that are moving along horizontal roads are going to be stopped by plowing into barrel barriers. All of the cars are the same size and shape, but they are carrying loads with different masses. All of the cars are going to be stopped in the same distance.

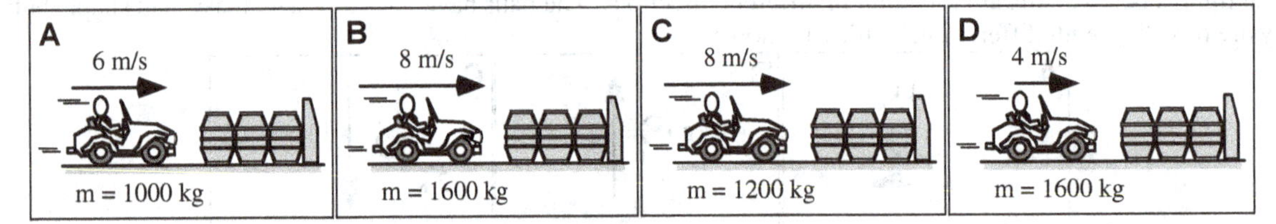

Rank the strength (magnitude) of the forces that will be needed to stop the cars in the same distance.

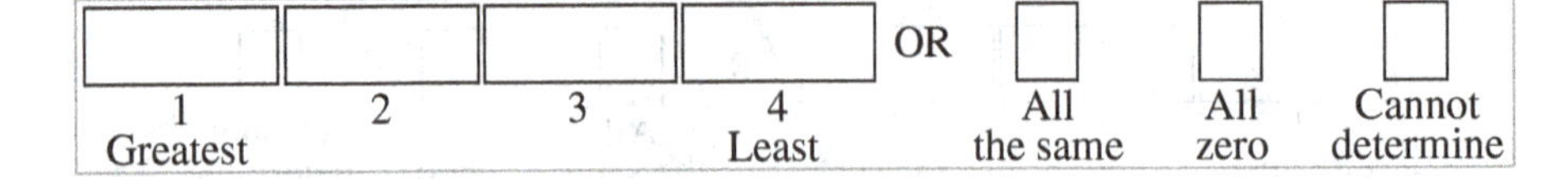

Explain your reasoning.

B4-RT05: CARS AND BARRIERS—STOPPING DISTANCE WITH THE SAME FORCE

Cars that are moving along horizontal roads are going to be stopped by plowing into identical barriers. All of the cars are the same size and shape, but they are carrying loads with different masses. All of the cars are going to be stopped by the same constant force by the barrier.

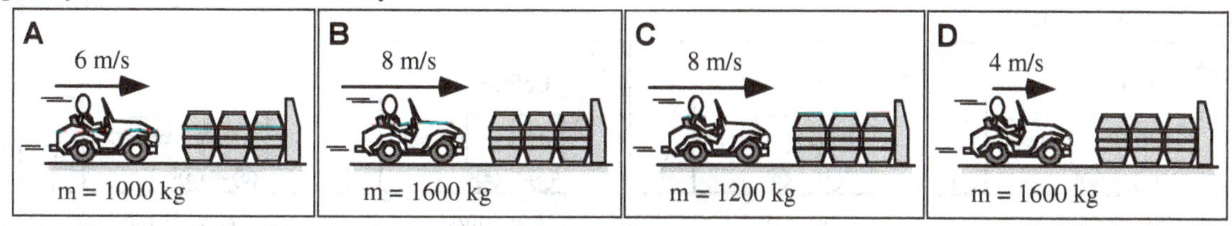

Rank the distance that will be needed to stop the cars with the same force.

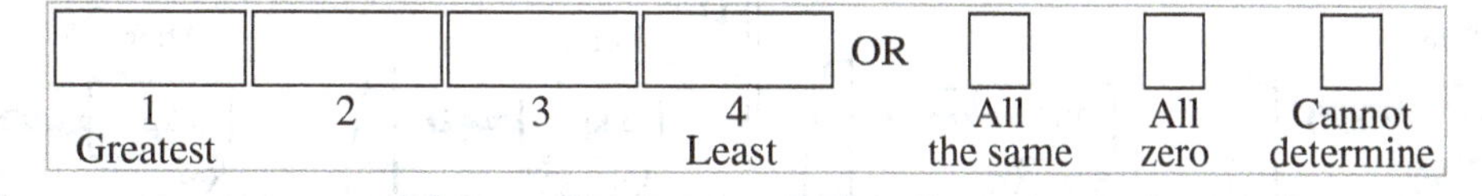

Explain your reasoning.

B4-RT06: CARS CHANGING VELOCITY—WORK DONE

The situations below show before and after "snapshots" of a car's velocity. All cars have the same mass.

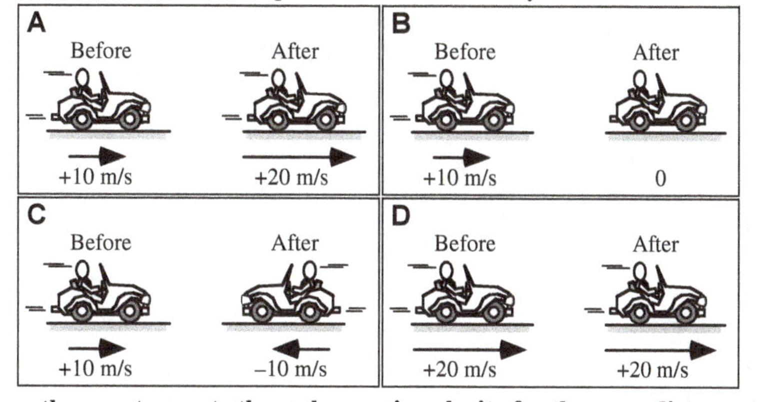

Rank the work done on the cars to create these changes in velocity for the same distance traveled.

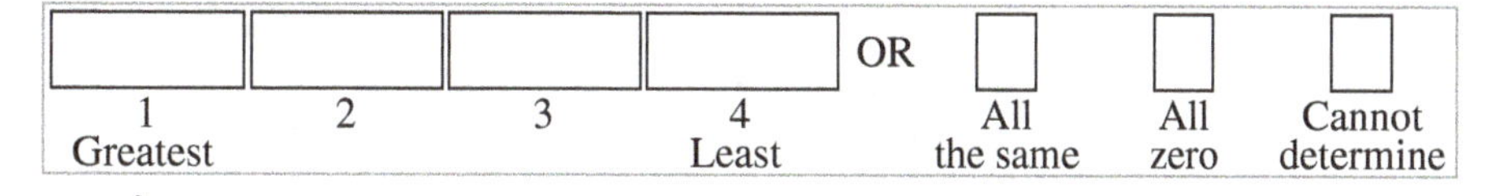

Explain your reasoning.

B4-RT07: Bouncing Cart—Change in Kinetic Energy

A cart with a spring plunger runs into a fixed barrier. The mass of the cart, its velocity just before impact with the barrier, and its velocity right after collision are given in each figure. (All velocities before the collision are given as positive since the cart is moving to the right. After the collision the cart is either moving to the left, indicated with a negative velocity, or is at rest.)

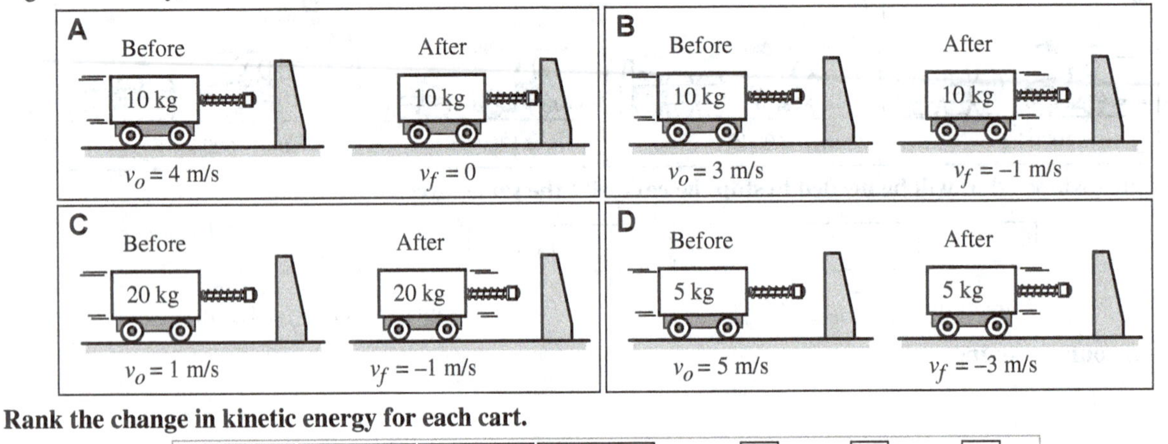

Rank the change in kinetic energy for each cart.

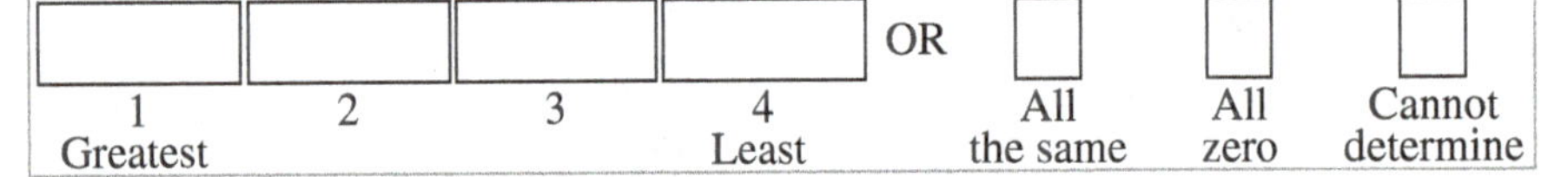

Explain your reasoning.

B4-RT08: EQUAL FORCES ON BOXES—WORK DONE ON BOX

In the figures below, identical boxes of mass 10 kg are moving at the same initial velocity to the right on a flat surface. The same magnitude force, *F*, is applied to each box for the distance, *d*, indicated in the figures.

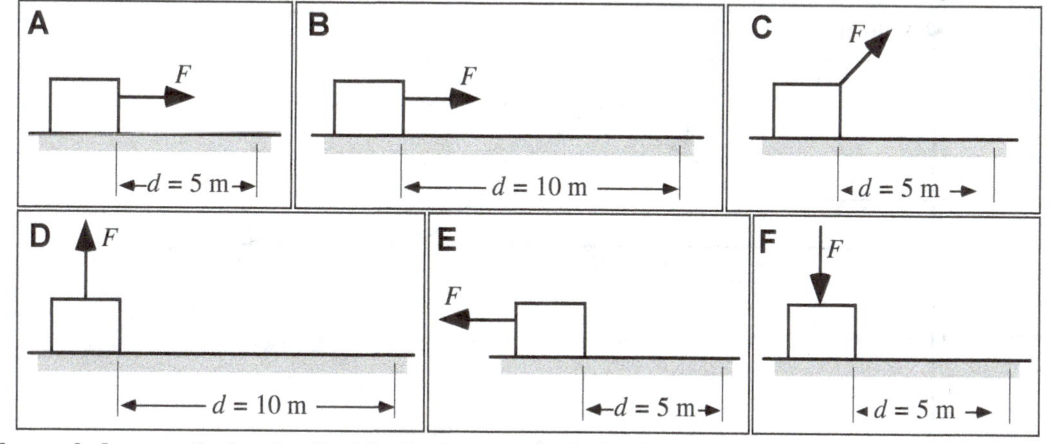

Rank the work done on the box by *F* while the box moves the indicated distance.

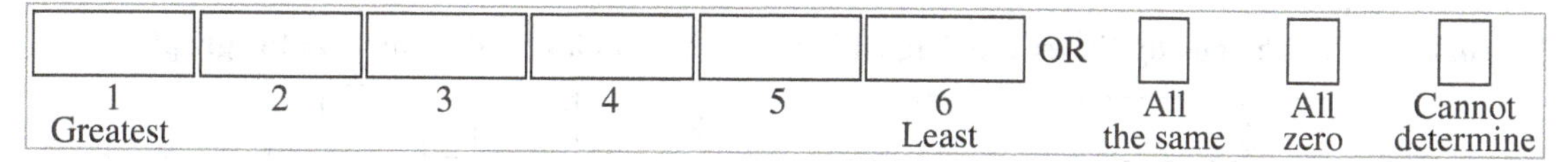

Explain your reasoning.

B4-RT09: VELOCITY-TIME GRAPH I—WORK DONE ON BOX

Shown below is a graph of velocity versus time for an object that moves along a straight, horizontal line under the perhaps intermittent action of a single force exerted by an external agent.

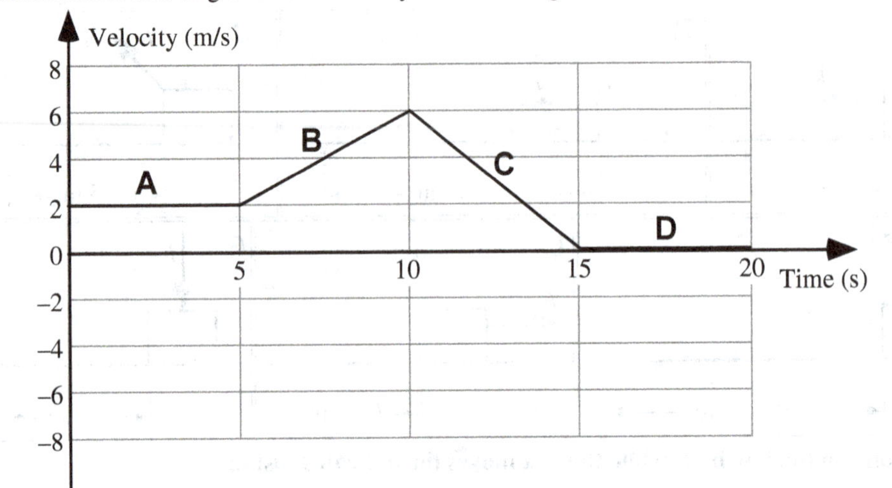

Rank the work done on the box by the external agent for the 5-second intervals shown on the graph.

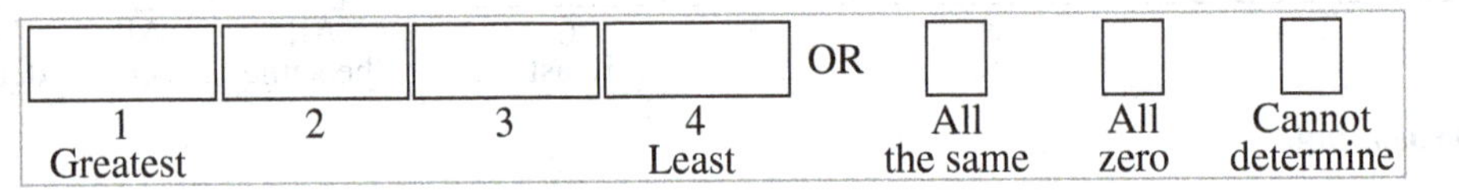

Explain your reasoning.

B4-RT10: VELOCITY-TIME GRAPH II—WORK DONE ON BOX

Shown below is a graph of velocity versus time for an object that moves along a straight, horizontal line under the perhaps intermittent action of a single force exerted by an external agent.

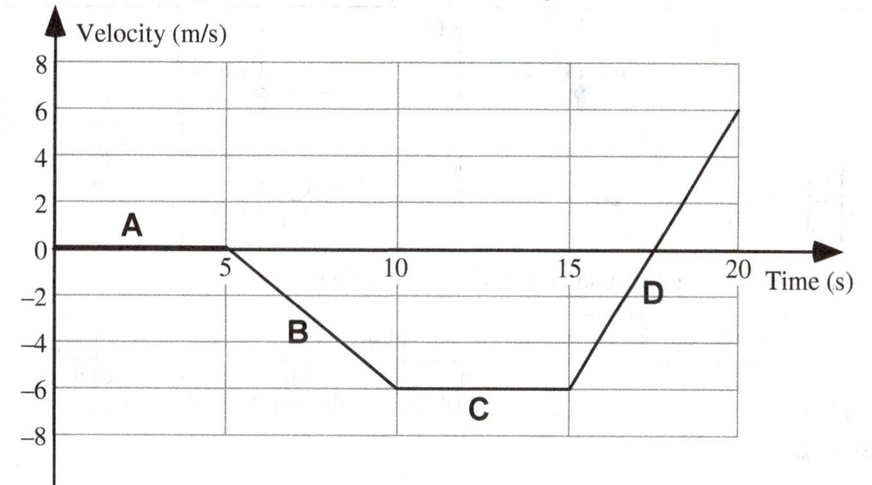

Rank the work done on the box by the external agent for the 5-second intervals shown on the graph.

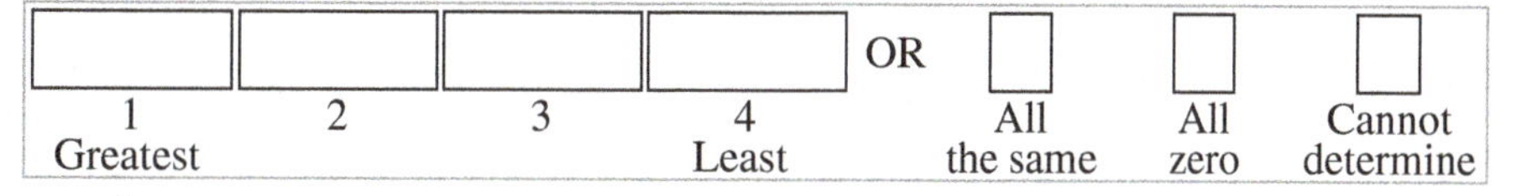

Explain your reasoning.

B4-RT11: FORCE PUSHING BOX—CHANGE IN KINETIC ENERGY

A box is pushed 10 m across a floor in each case shown. All boxes have an initial velocity of 10 m/s to the right. The mass of the box and the net horizontal force for each case are given.

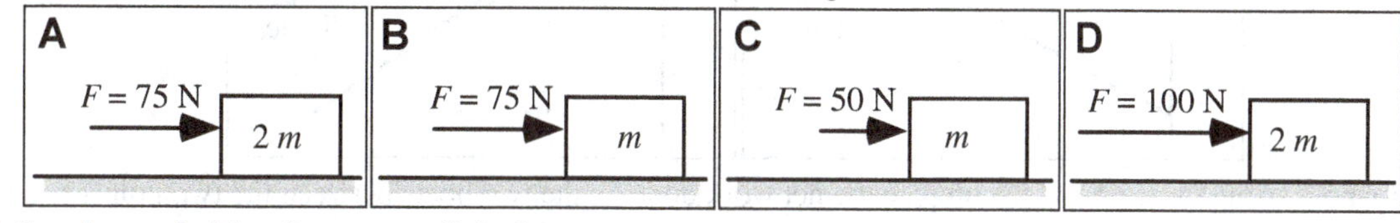

Rank the change in kinetic energy of the boxes.

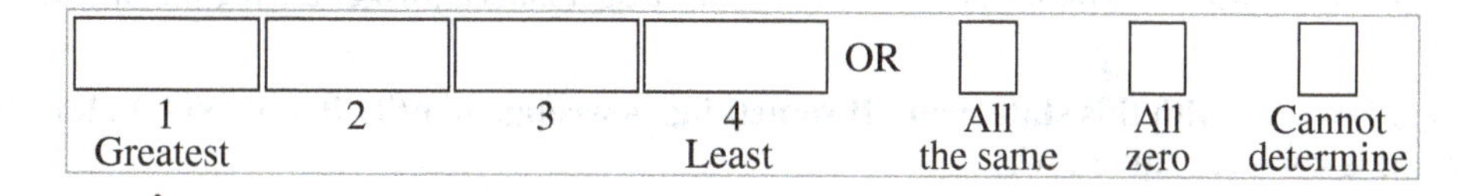

Explain your reasoning.

B4-QRT12: Two Ball Systems—Kinetic Energy of System

In the figures below, systems of two balls are traveling in different directions. The balls are identical in size and shape, but they have different masses and are traveling at different velocities as shown.

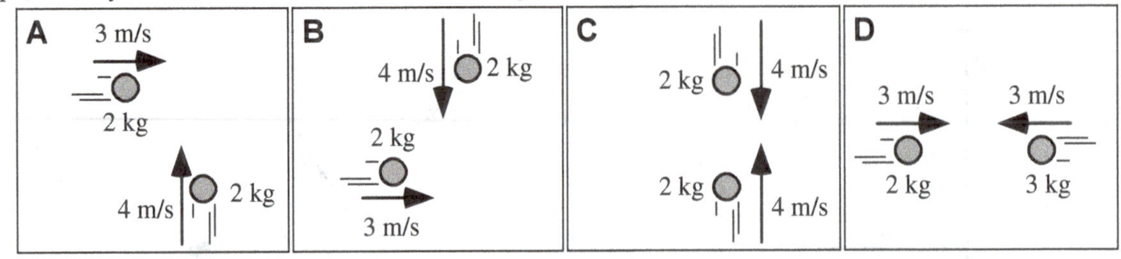

Rank the total kinetic energy of the two-ball systems before they interact.

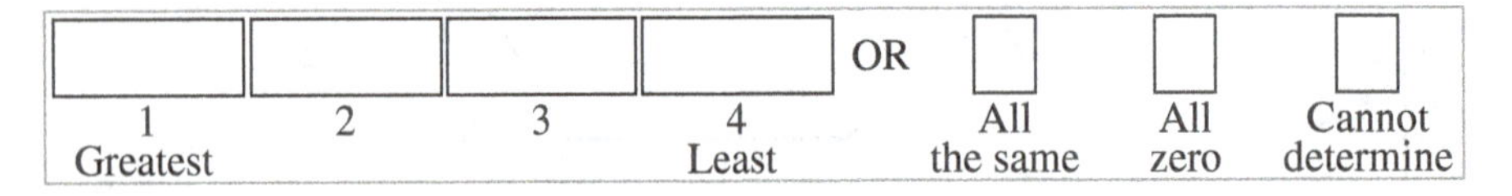

Explain your reasoning.

B4-WWT13: Boat Position-Time Graphs—Work

Shown are graphs of the position versus time for two boats traveling along a narrow channel. The scales on both axes are the same for the graphs. In each graph, two points are marked with dots.

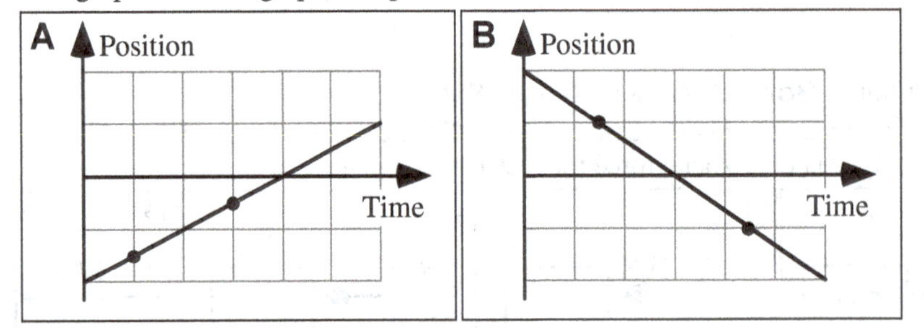

A student who is using these graphs to compare the net work done on the two boats between the two points says:

"I think that more net work was done on the boat in graph B because it moved farther during the interval between the points."

What, if anything, is wrong with this statement? If something is wrong, identify it and explain how to correct it. If this statement is correct, explain why.

B4-BCT14: TUGBOAT CHANGING VELOCITY I—WORK & KINETIC ENERGY BAR CHART

(a) The velocity of a tugboat increases from 2 m/s to 4 m/s in the same direction while a force is applied to the tugboat for 20 seconds.

Complete the work and kinetic energy bar chart for this process. The bar heights should be in correct proportion to one another.

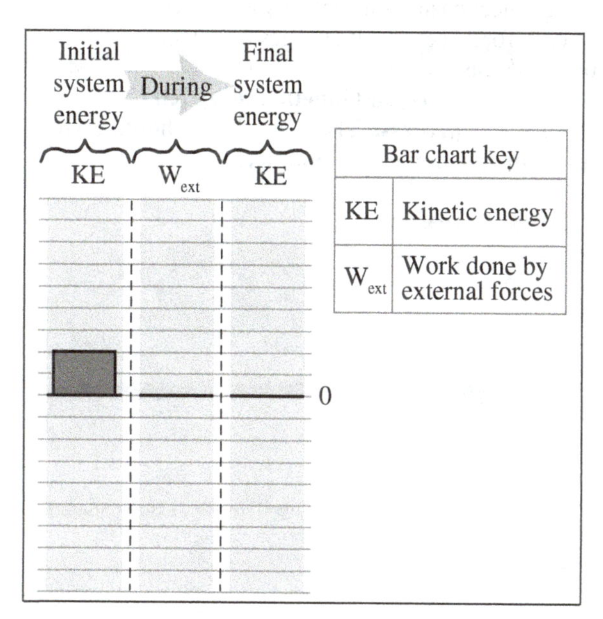

Explain.

(b) The velocity of a tugboat changes from 2 m/s to 4 m/s in the opposite direction while a force is applied to the tugboat for 20 seconds.

Complete the work and kinetic energy bar chart for this process. The bar heights should be in correct proportion to one another.

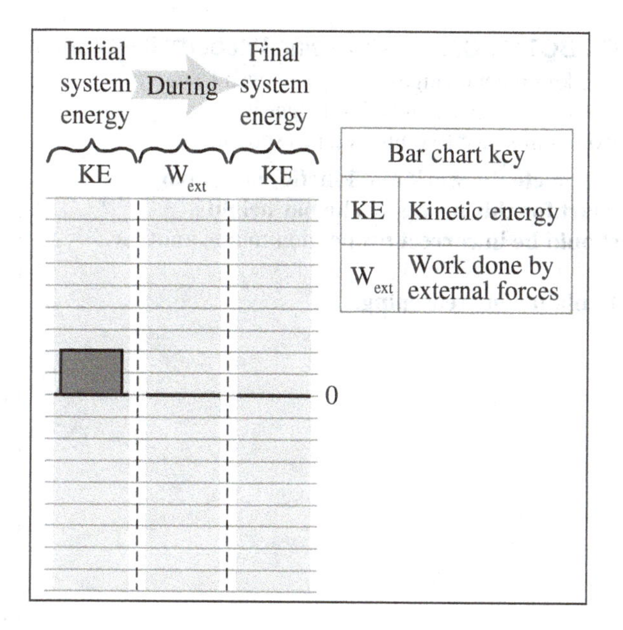

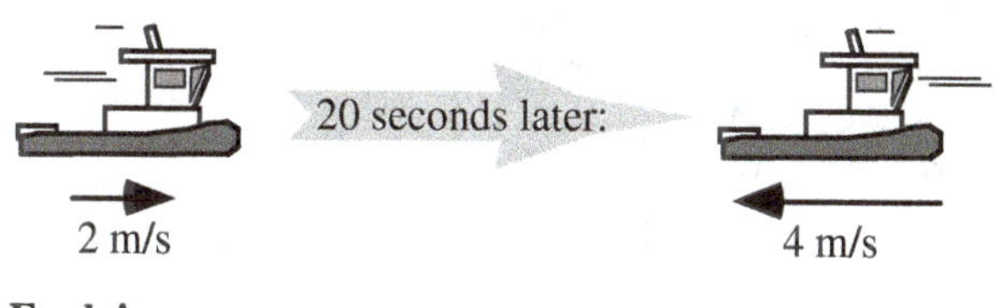

Explain.

B4-BCT15: Object Changing Velocity I—Work and Kinetic Bar Chart

A 2-kg object changes its velocity as a force acts on it for 5 seconds. It changes its velocity from 4 m/s east to 6 m/s east as shown.

Complete the work and kinetic energy bar chart for this process. The bar heights should be in correct proportion to one another.

Explain your reasoning.

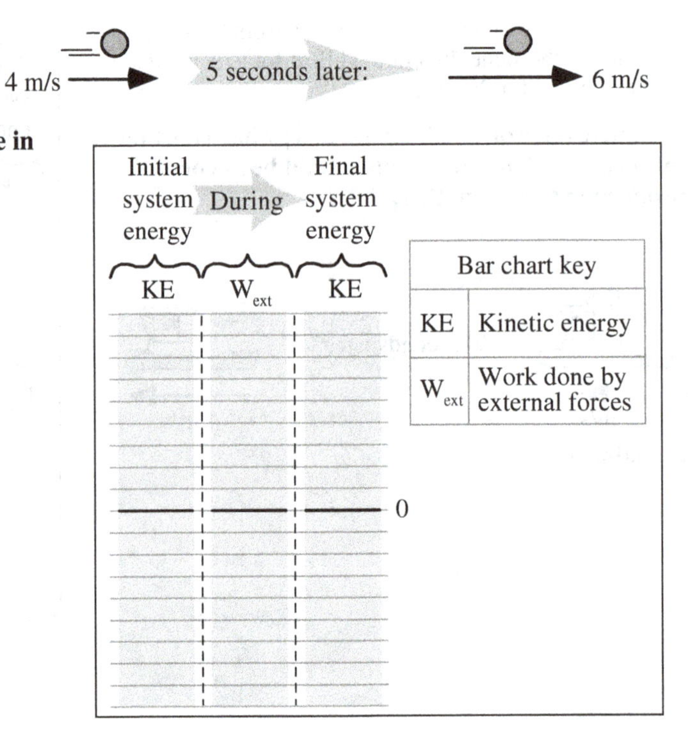

B4-BCT16: Object Changing Velocity II—Work and Kinetic Bar Chart

A 2-kg object changes its velocity as a force acts on it for 5 seconds. It changes its velocity from 4 m/s east to 6 m/s west as shown.

Complete the work and kinetic energy bar chart for this process. The bar heights should be in correct proportion to one another.

Explain your reasoning.

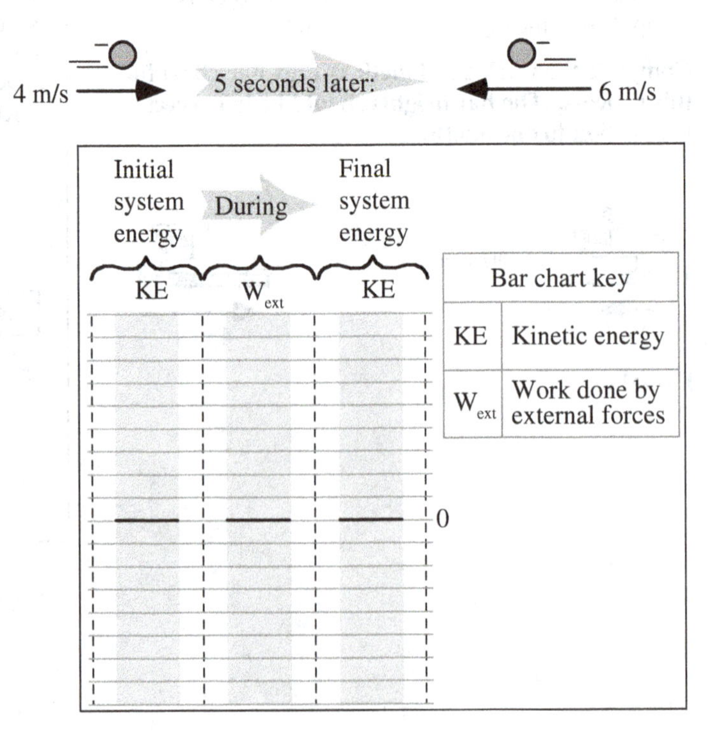

B4-BCT17: TUGBOAT CHANGING VELOCITY II—WORK AND KINETIC ENERGY BAR CHART

(a) The velocity of a tugboat changes from 2 m/s west to 4 m/s west while a force is applied to the tugboat for 20 seconds.

Complete the work and kinetic energy bar chart for this process. The bar heights should be in correct proportion to one another.

Initial system energy | During | Final system energy

KE | W_{ext} | KE

0

Bar chart key	
KE	Kinetic energy
W_{ext}	Work done by external forces

Explain.

(b) The velocity of a tugboat changes from 4 m/s west to 2 m/s west while a force is applied to the tugboat for 20 seconds.

Complete the work and kinetic energy bar chart for this process. The bar heights should be in correct proportion to one another.

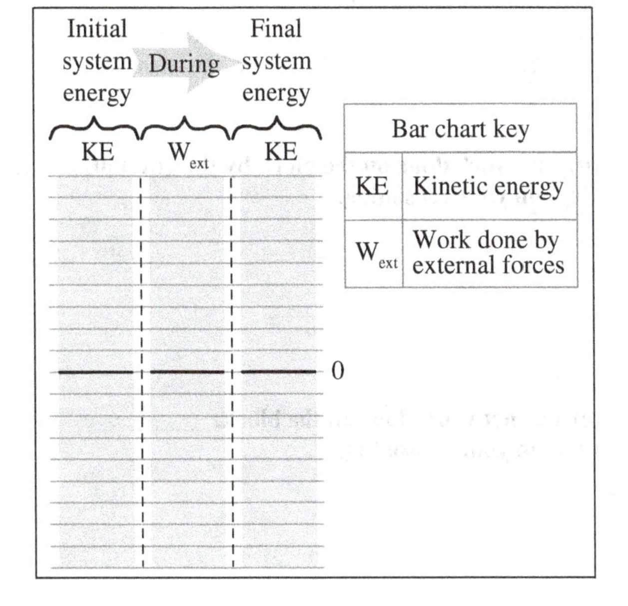

Explain.

B4-LMCT18: Block Pushed on Incline—Work Done

A block is pushed so that it moves up a ramp at constant speed.

B

A

Identify from choices (i)–(iv) below the appropriate description for the work done by the specified force while the block moves from point *A* to point *B*.

(i) is ***zero***.

(ii) is ***less than*** zero.

(iii) is ***greater than*** zero.

(iv) could be ***positive or negative*** depending on the choice of coordinate systems.

(v) ***cannot*** *be determined*.

(a) The work done on the block by the hand. _____
Explain your reasoning.

(b) The work done on the block by the normal force from the ramp. _____
Explain your reasoning.

(c) The work done on the block by friction. _____
Explain your reasoning.

(d) The work done on the block by the gravitational force. _____
Explain your reasoning.

(e) The net work done on the block. ______
Explain your reasoning.

B4-SCT19: BLOCKS SLIDING DOWN FRICTIONLESS RAMPS—WORK BY THE NORMAL FORCE

Two identical blocks are released from rest at the same height. Block A slides down a steeper ramp than Block B. Both ramps are frictionless. The blocks reach the same final height indicated by the lower dashed line. Three students are comparing the work done on the two blocks by the normal force:

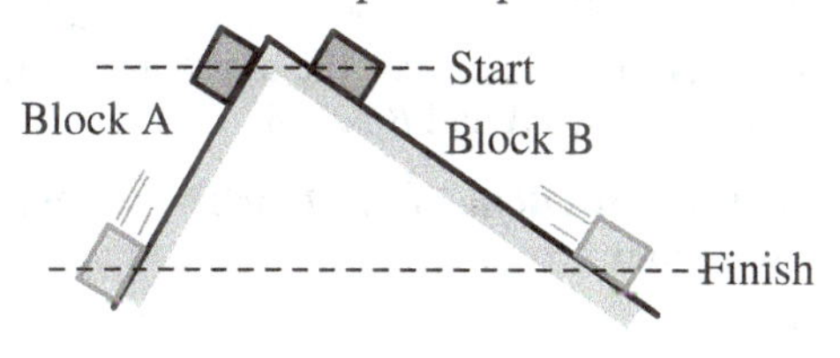

Annika: *"I think the normal force doesn't do any work on either block. The force on the block by the ramp is perpendicular to the ramp, and the displacement is parallel to the ramp. So the dot product is zero."*

BoBae: *"Work is force times displacement. The work done on Block A is negative, while the work done on Block B is positive, because the displacement for B is in the positive direction, while the displacement for A is in the negative direction."*

Craig: *"Since work is force times distance, and the distance the block travels is greater for Block B, the work done is greater for Block B."*

With which, if any, of these students do you agree?

Annika _____ BoBae _____ Craig _____ None of them_____

Explain your reasoning.

B4-SCT20: BLOCKS SLIDING DOWN FRICTIONLESS RAMPS—WORK BY THE EARTH

Two identical blocks are released from rest at the same height. Block A slides down a steeper ramp than Block B. Both ramps are frictionless. The blocks reach the same final height indicated by the lower dashed line. Three students are comparing the work done on the two blocks by the gravitational force (the weight of the blocks):

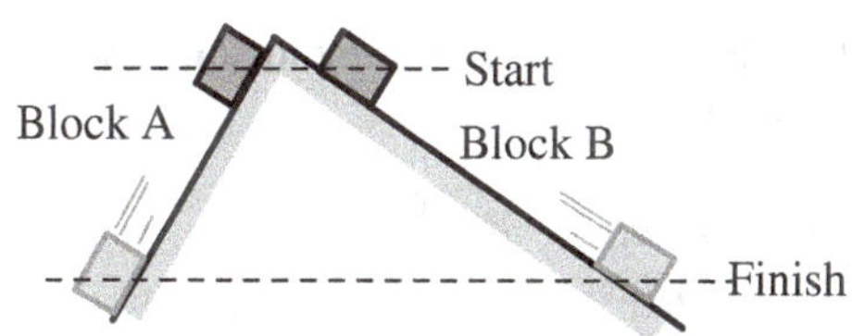

Asmita: *"Work is related to the product of force and displacement, and the weight is the same since the blocks are identical. But Block B travels farther, so more work is done on Block B by the gravitational force than on Block A."*

Ben: *"Both blocks fall the same vertical distance, so the work done is the same."*

Cocheta: *"By Newton's third law, the force exerted on the block by Earth is exactly cancelled by the force exerted on Earth by the block. The work done is zero."*

Danae: *"The work depends on the angle that the force makes with the displacement. If we put the displacement and force vectors tail-to-tail, the angle is smaller for Block B than for Block A, and so the work done is greater."*

With which, if any, of these students do you agree?

Asmita _____ Ben _____ Cocheta _____ Danae _____ None of them_____

Explain your reasoning.

B4-QRT21: Block on Ramp with Friction—Work

A block is pushed at constant speed up a ramp from point *A* to point *B*. The direction of the force on the block by the hand is horizontal. There is friction between the block and the ramp. The distance between points *A* and *B* is 1 m.

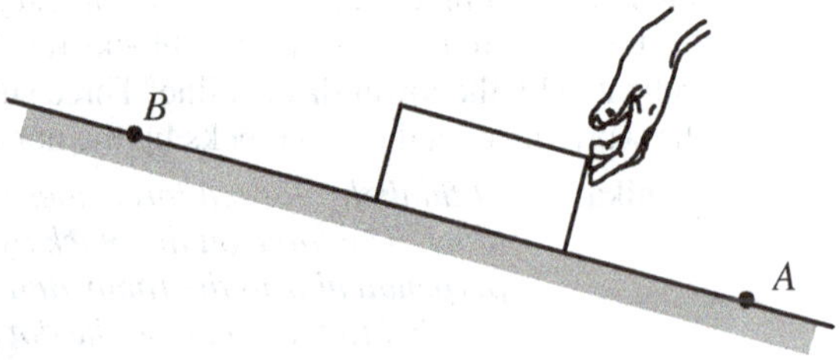

(a) The work done on the block by the hand as the block travels from point *A* to point *B*

(i) is *zero*.
(ii) is *negative*.
(iii) is *positive*.
(iv) *could be positive or negative* depending on the choice of coordinate systems.

Explain your reasoning.

(b) The work done on the block by the normal force from the ramp as the block travels from point *A* to point *B*

(i) is *zero*.
(ii) is *negative*.
(iii) is *positive*.
(iv) *could be positive or negative* depending on the choice of coordinate systems.

Explain your reasoning.

(c) The work done on the block by the friction force from the ramp as the block travels from point *A* to point *B*

(i) is *zero*.
(ii) is *negative*.
(iii) is *positive*.
(iv) *could be positive or negative* depending on the choice of coordinate systems.

Explain your reasoning.

(d) The work done on the block by the gravitational force of the earth as the block travels from point *A* to point *B*

(i) is *zero*.
(ii) is *negative*.
(iii) is *positive*.
(iv) *could be positive or negative* depending on the choice of coordinate systems.

Explain your reasoning.

B4-QRT22: BLOCK ON RAMP WITH FRICTION—WORK AND ENERGY

A block is pushed at constant speed up a ramp from point *A* to point *B*. The direction of the force on the block by the hand is horizontal. There is friction between the block and the ramp. The distance between points *A* and *B* is 1 m.

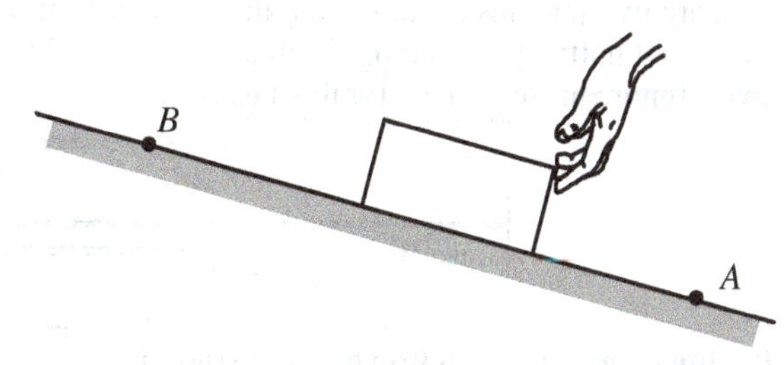

(a) The kinetic energy of the block at point *B*

(i) is *greater than* the kinetic energy of the block at point *A*.
(ii) is *less than* the kinetic energy of the block at point *A*.
(iii) is *equal to* the kinetic energy of the block at point *A*.
(iv) *cannot be compared* to the kinetic energy of the block at point *A* unless we know the height difference between *A* and *B*.

Explain your reasoning.

(b) The net work done on the block as it travels from point *A* to point *B*

(i) is *zero*.
(ii) is *negative*.
(iii) is *positive*.
(iv) *could be positive or negative* depending on the choice of coordinate systems.

Explain your reasoning.

(c) The work done on the block by the hand as the block travels from point *A* to point *B*

(i) is *equal* to 1 m times the magnitude of the force exerted on the block by the hand.
(ii) is *greater than* 1 m times the magnitude of the force exerted on the block by the hand.
(iii) is *less than* 1 m times the magnitude of the force exerted on the block by the hand but not zero.
(iv) is *zero*.
(v) *cannot be compared* to the magnitude of the force exerted on the block by the hand based on the information given.

Explain your reasoning.

B4-CT23: Thrown Javelins—Horizontal Force

Shown are two javelins (light spears) that have been thrown at targets. We are viewing the javelins when they are in the air about halfway to landing. Both javelins have the same mass, but they have different kinetic energies as shown. (Ignore air resistance for this task.)

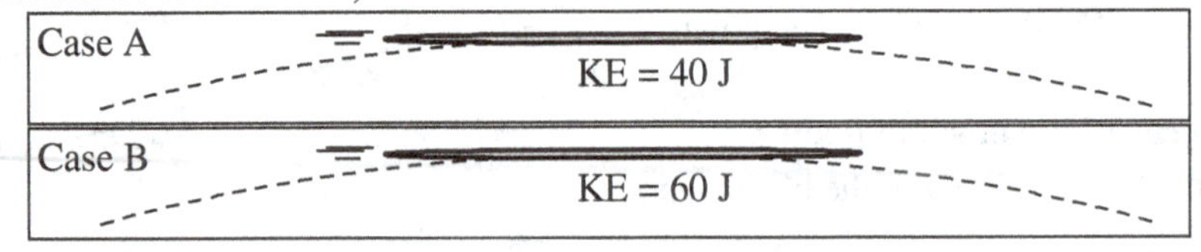

Is the horizontal force acting on the javelin in Case A (i) *greater than*, (ii) *less than*, or (iii) *equal to* the horizontal force acting on the javelin in Case B? _____

Explain your reasoning.

B4-SCT24: Skaters Pushing off Each Other—Force

Two skaters—a small girl and a large boy—are initially standing face-to-face but then push off each other. After they are no longer touching, the girl has more kinetic energy than the boy. Three physics students make the following contentions about the forces the boy and girl exerted on each other:

Arianna: *"I think the boy pushed harder on the girl because he is bigger, so she ended up with more kinetic energy than he did."*

Boris: *"I disagree. They pushed equally hard on each other, but the girl moved farther while they were pushing on each other, so she ended up with more kinetic energy."*

Carmen: *"I think the girl had to push harder to get the boy moving since he is bigger, but that caused her to accelerate more as she recoiled."*

With which, if any, of these students do you agree?

Arianna _____ Boris _____ Carmen _____ None of them_____

Explain your reasoning.

B4-RT25: ARROWS SHOT FROM BUILDINGS—FINAL SPEED

In each case below, an arrow has been shot from the top of a building either up at a 45° angle, straight out horizontally, or down at a 45° angle. All arrows are identical and are shot at the same speed, and the heights of the buildings and the direction the arrows are shot are given. Ignore air resistance.

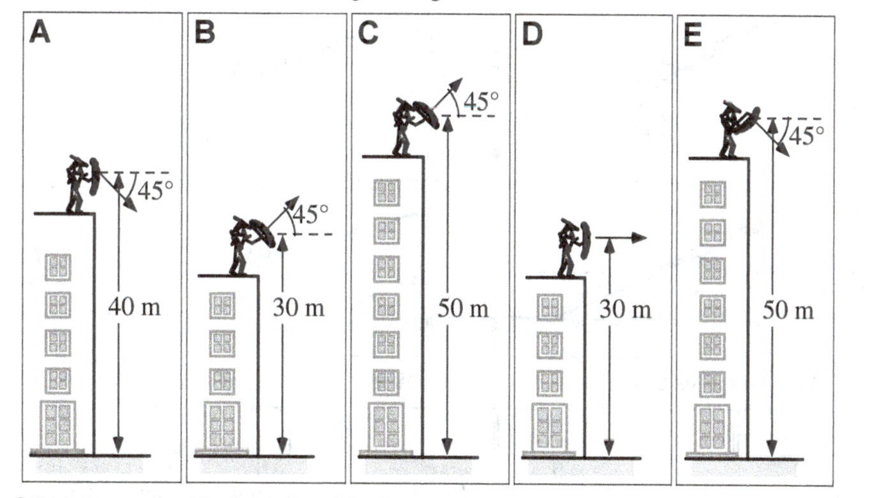

Rank the speed of the arrows just before they hit the ground below.

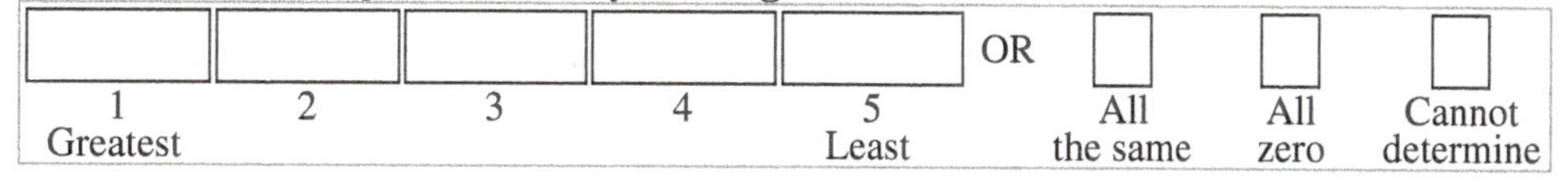

Explain your reasoning.

B4-RT26: Toboggans Going Down Slippery Hills—Speed at Bottom

In each case below, a toboggan starts from rest and slides without friction down a snowy hill. The toboggans are all identical, and the starting heights (vertical distance above the flat bottom of the incline) and angles of the hills are given.

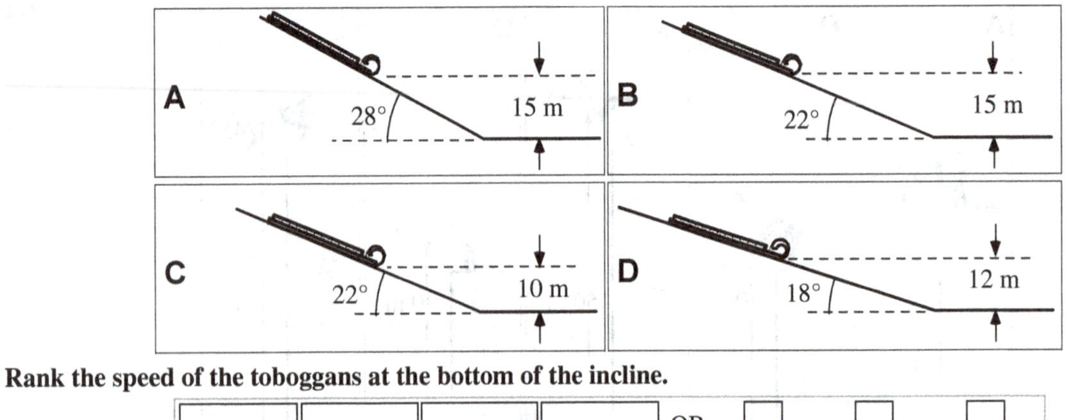

Rank the speed of the toboggans at the bottom of the incline.

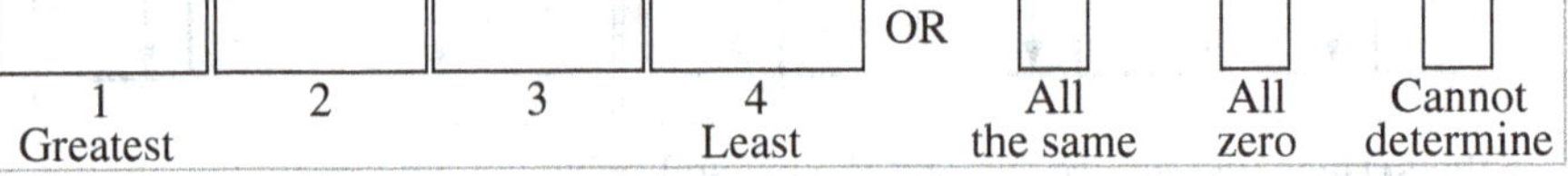

Explain your reasoning.

B4-CT27: Roller Coaster Ride over Lagoon—Maximum Height

For extra excitement, a new roller coaster ride is designed to launch the riders over an alligator-infested lagoon. The frictionless coaster starts at rest at point *A*. The coaster lands on a ramp on the other side of the lagoon.

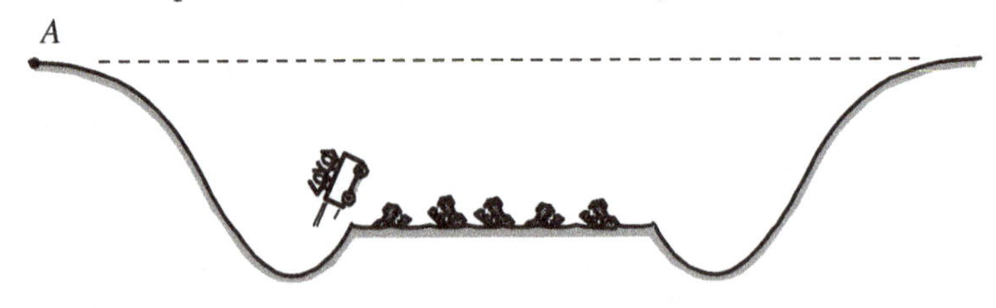

After it is airborne, will the maximum height of the coaster be (i) *greater than*, (ii) *less than*, or (iii) *equal to* the height at point *A*? _____

Explain your reasoning.

B4-CT28: SKATEBOARDERS ON A HILL—TIME, SPEED, KINETIC ENERGY, AND WORK

Starting from rest, Angel and Britney skateboard down a hill as shown. Angel rides down the steep side while Britney rides down the shallow side. Angel has more mass than Britney. Assume that friction and air resistance are negligible.

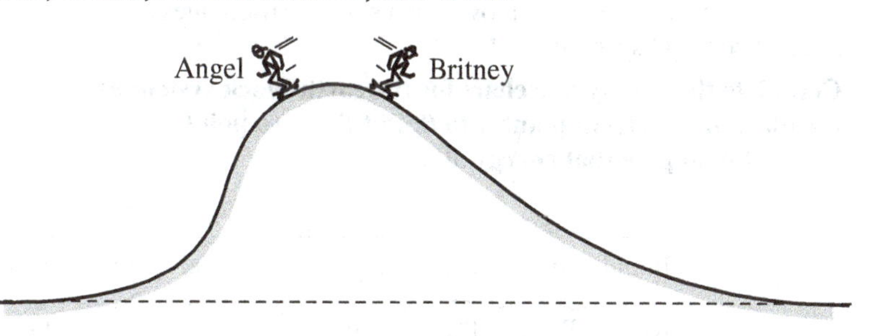

(a) Is the speed at the bottom of the hill (i) *greater* for Angel, (ii) *greater* for Britney, or (iii) *the same* for both skateboarders? _____

Explain your reasoning.

(b) Is the time it takes to get to the bottom of the hill (i) *greater* for Angel, (ii) *greater* for Britney, or (iii) *the same* for both skateboarders? _____

Explain your reasoning.

(c) Is the work done by the gravitational force on the skateboarder (i) *greater* for Angel, (ii) *greater* for Britney, or (iii) *the same* for both skateboarders? _____

Explain your reasoning.

(d) Is the work done by the normal force on the skateboarder (i) *greater* for Angel, (ii) greater for Britney, or (iii) *the same* for both skateboarders? _____

Explain your reasoning.

(e) Is the kinetic energy at the bottom of the hill (i) *greater* for Angel, (ii) *greater* for Britney, or (iii) *the same* for both skateboarders? _____

Explain your reasoning.

B4-BCT29: Block Pushed on Smooth Ramp—Energy Bar Chart

A block is pushed so that it moves up a smooth (frictionless) ramp at constant speed from *A* to *B*.

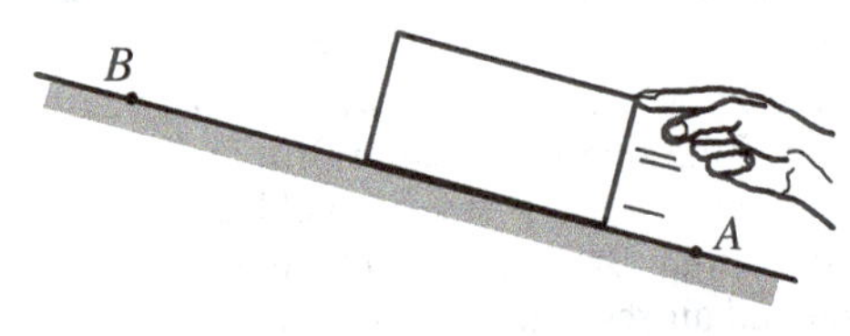

Complete the energy bar chart for the earth-block system as the block moves from point *A* to *B*. Put the zero point for the gravitational potential energy at *A*.

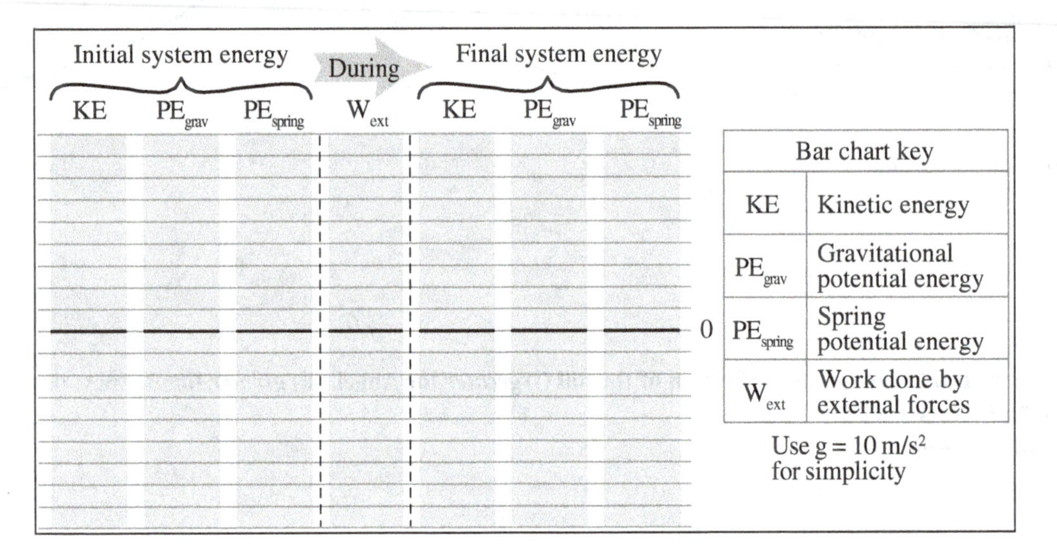

Explain your reasoning.

B4-BCT30: BOX PULLED ON SMOOTH SURFACE—ENERGY BAR CHART

A 100-N box is initially at rest at point *A* on a smooth (frictionless) horizontal surface. A student applies a horizontal force of 80 N to the right on the box as shown.

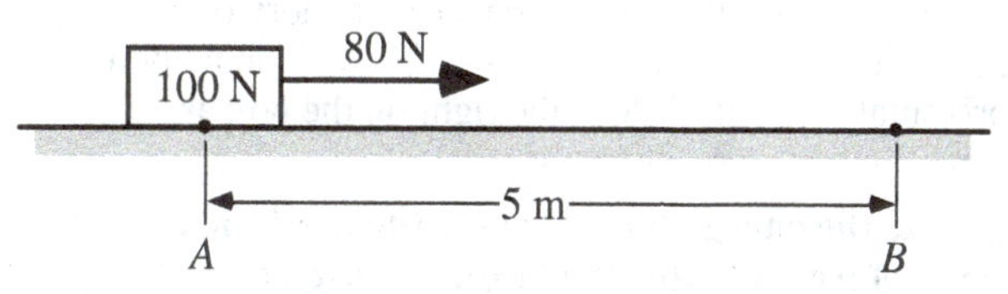

Complete the energy bar chart for the earth-box system before and after the box has moved a horizontal distance of 5.0 m. Put the zero point for the gravitational potential energy at the surface.

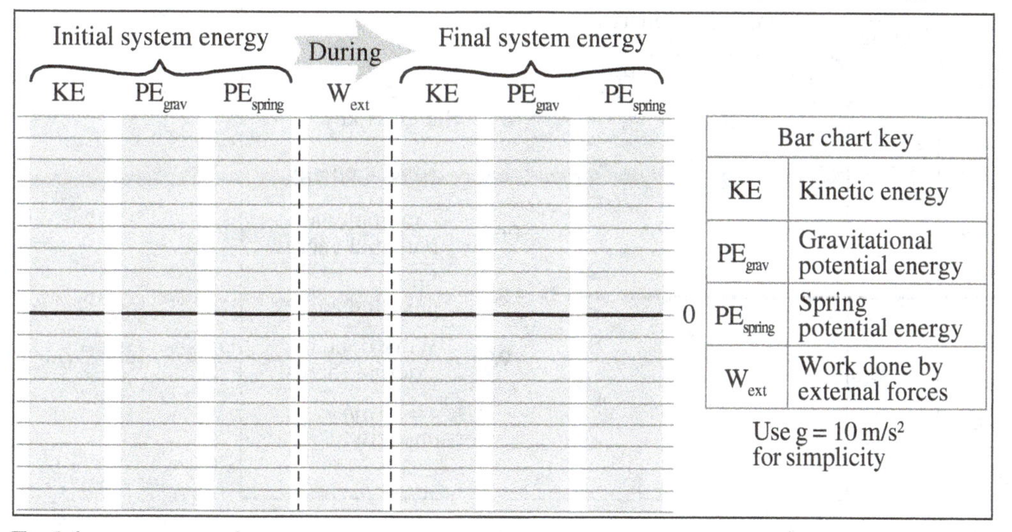

Explain your reasoning.

B4-BCT31: Box Pulled on Rough Surface—Energy Bar Chart

A 100-N box is initially at rest on a rough, horizontal surface where the friction force is 40 N. A student applies a horizontal force of 80 N to the right on the box as shown. The box starts at rest at point *A*.

100 N
80 N
5 m
A
B

Complete the energy bar chart for the earth-box system before and after the box has moved a horizontal distance of 5.0 m. Put the zero point for the gravitational potential energy at the surface.

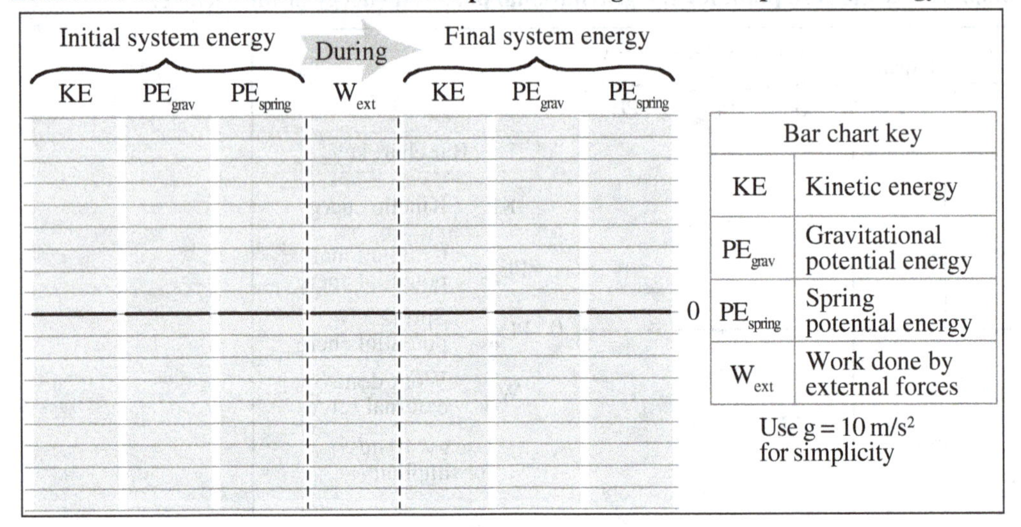

Explain your reasoning.

B4-BCT32: LIFTED BOX MOVING UPWARD I—ENERGY BAR CHART

A 100-N box is initially 0.40 m above the surface of a table and is moving upward with a kinetic energy of 80 J. A man is applying a constant upward force of 80 N with his hand to the box.

Complete the energy bar chart for the earth-box system before and after the box has moved upward a distance of 1.0 m. Put the zero point for the gravitational potential energy at the surface of the table.

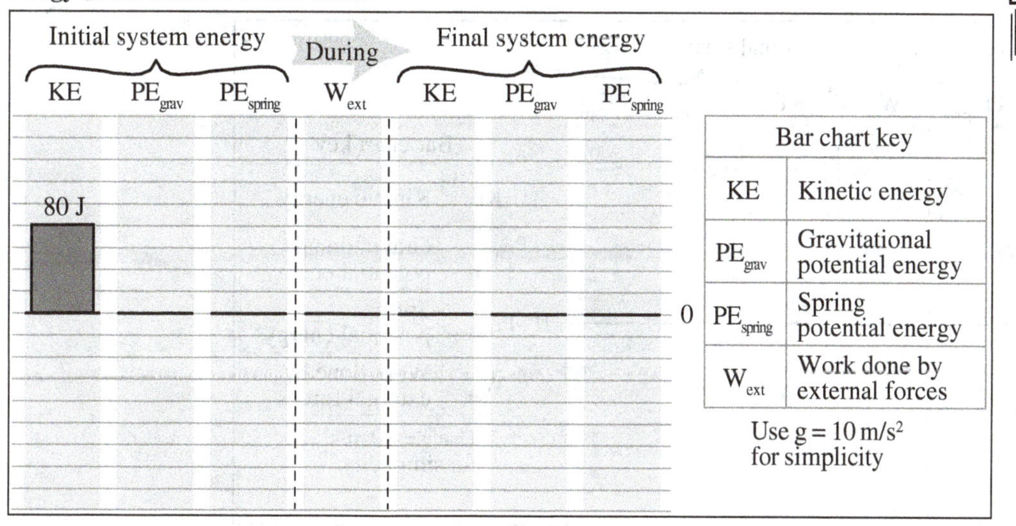

Explain your reasoning.

B4-BCT33: Lifted Box Moving Upward II—Energy Bar Chart I

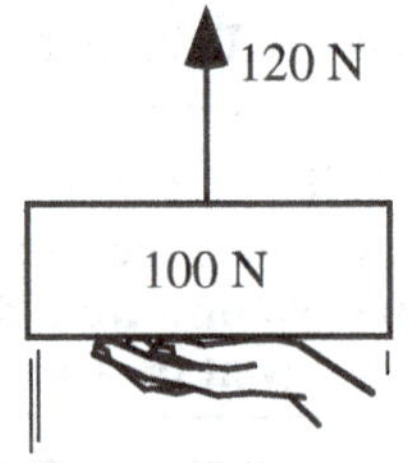

A 100-N box is initially 1.0 m above the ground while moving upward at 10 m/s. A student starts applying a vertical force of 120 N upward with her hand at this point.

Complete the energy bar chart for the earth-box system before and after the box has moved upward a distance of 1.0 m. Put the zero point for the gravitational potential energy at the surface of the ground.

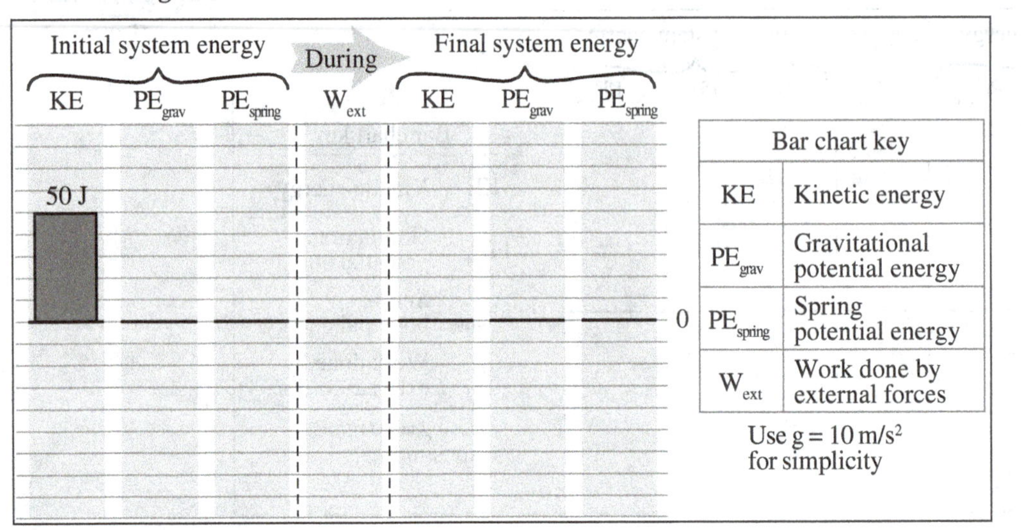

Explain your reasoning.

B4-BCT34: LIFTED BOX MOVING UPWARD II—ENERGY BAR CHART II

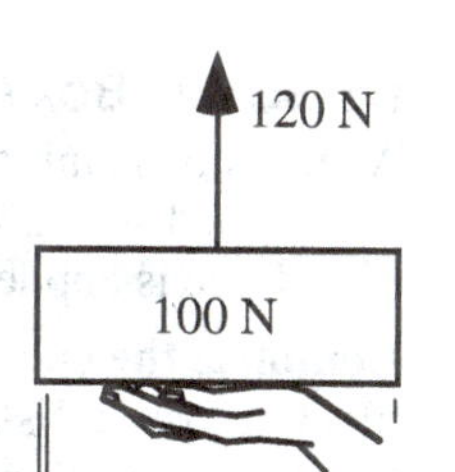

A 100-N box is initially 1.0 m above the ground while moving upward at 10 m/s. A student starts applying a vertical force of 120 N upward with her hand at this point.

Complete the energy bar chart for the earth-box system before and after the box has moved upward a distance of 1.0 m. Put the zero point for the gravitational potential energy at the final location at 2.0 m above of ground.

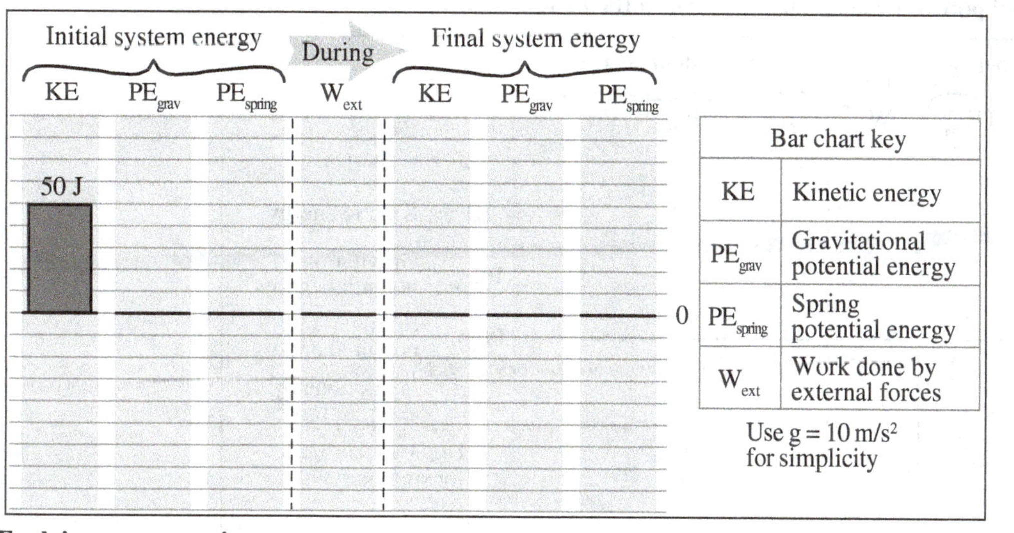

Bar chart key	
KE	Kinetic energy
PE_{grav}	Gravitational potential energy
PE_{spring}	Spring potential energy
W_{ext}	Work done by external forces

Explain your reasoning.

B4-BCT35: BOX ATTACHED TO SPRING—ENERGY BAR CHART

A 40-N box is initially at rest on a smooth (frictionless) horizontal surface. An unstretched spring with spring constant 10 N/m connects the box to the wall. A 60 N force is applied horizontally to the right.

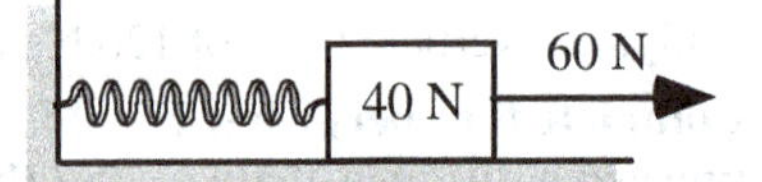

Complete the energy bar chart for the spring-block-earth system as the block moves a distance of 2 m. Label the column heights. Set the zero point for the gravitational potential energy at the center of the block.

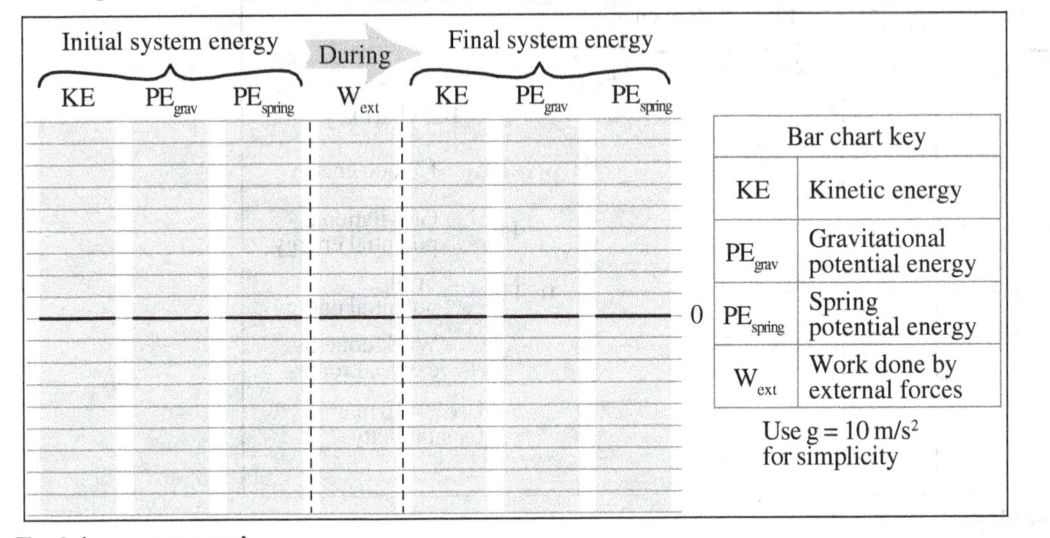

Explain your reasoning.

B4-BCT36: MOVING BLOCK PUSHED UP A SMOOTH RAMP—ENERGY BAR CHART

A moving block is pushed so that it moves up a smooth (frictionless) ramp at increasing speed from *A* to *B*.

Complete the bar charts for the earth-block system as the block moves from point *A* to *B*. Label the column heights. Set the zero point for the gravitational potential energy of the system at *A*.

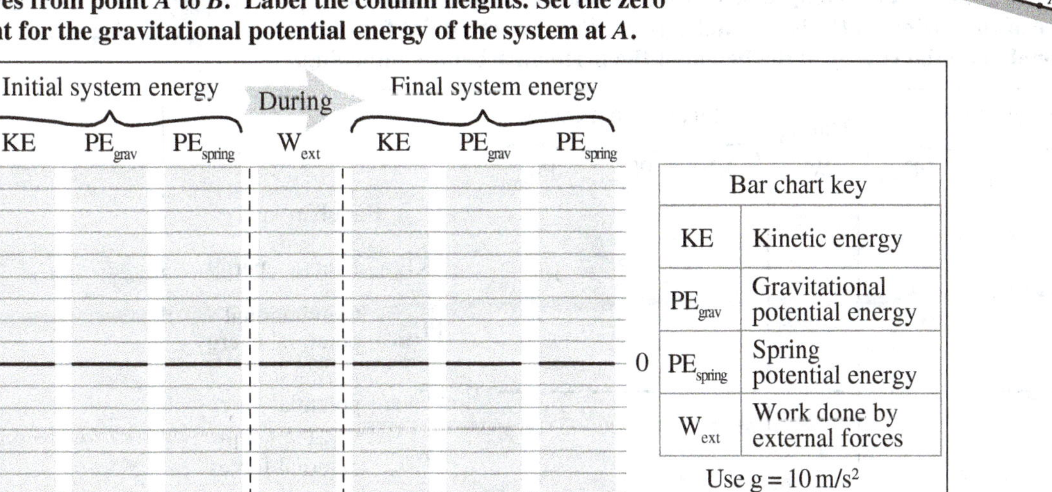

Explain your reasoning.

B4-BCT37: Skateboarder Launched by a Spring I—Energy Bar Chart

A performer on a skateboard is launched by a spring initially compressed a distance Δx. His speed on the horizontal portion of the ramp is v. Ignore friction effects.

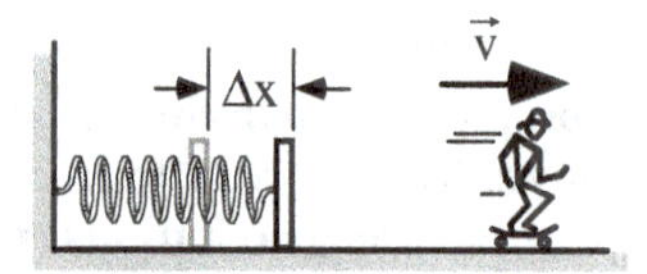

Draw an energy bar chart for the earth-skateboarder-spring system as the skateboarder goes from the compressed spring position at rest to where he moves free of the spring on the horizontal surface. Put the zero point for the gravitational potential energy at the height of the performer before launching.

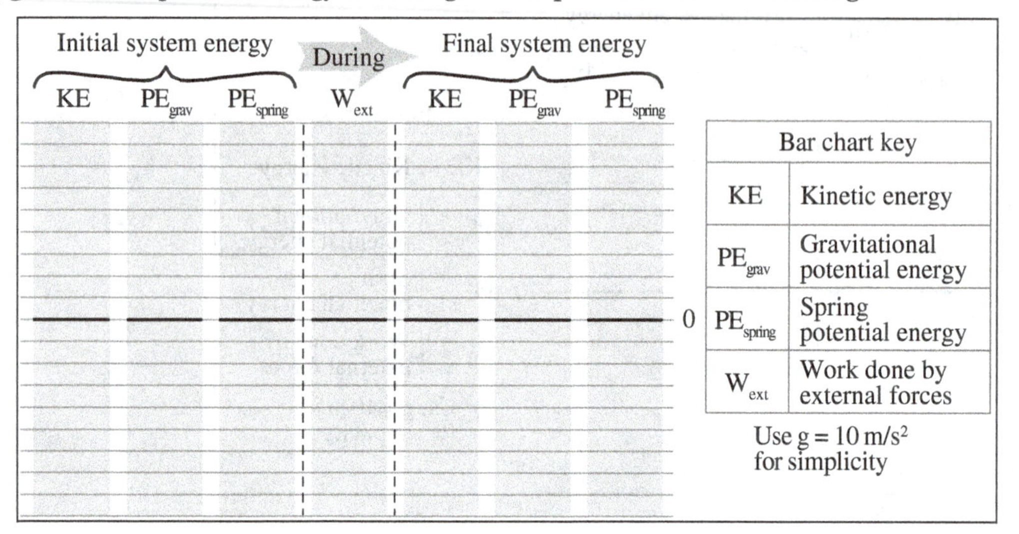

Explain your reasoning.

B4-BCT38: SKATEBOARDER LAUNCHED BY A SPRING II—ENERGY BAR CHART

A performer on a skateboard is launched by a spring initially compressed a distance Δx as shown. His speed on the horizontal portion of the ramp is v, and he rises to a height H after he leaves the ramp. Ignore friction effects.

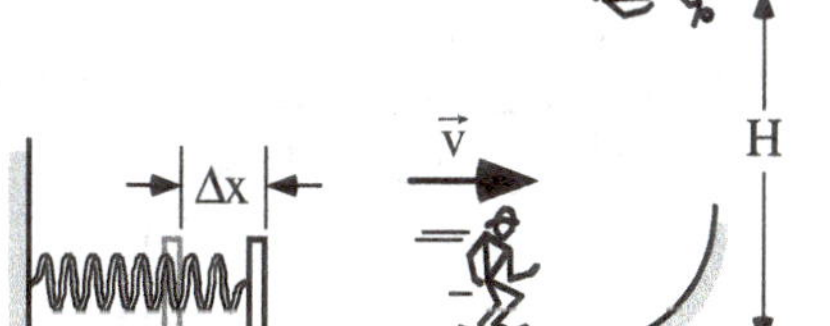

Draw an energy bar chart for the earth- skateboarder-spring system as he goes from the compressed spring position at rest to when he reaches the height H. Put the zero point for the gravitational potential energy at the initial height of the performer before launching.

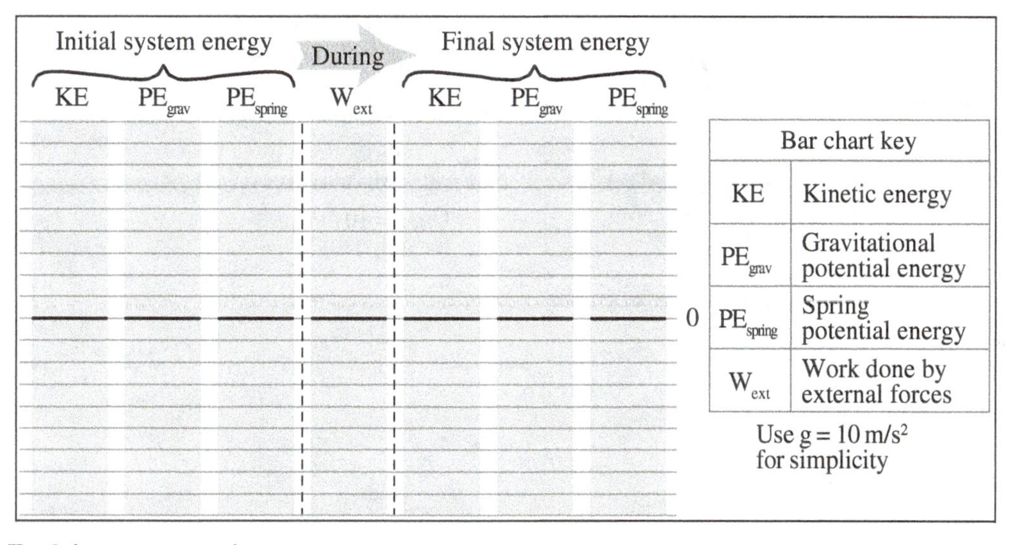

Explain your reasoning.

B4-WBT39: Energy Bar Chart I—Physical Situation

Describe a physical situation and a system to which this energy bar chart could apply.

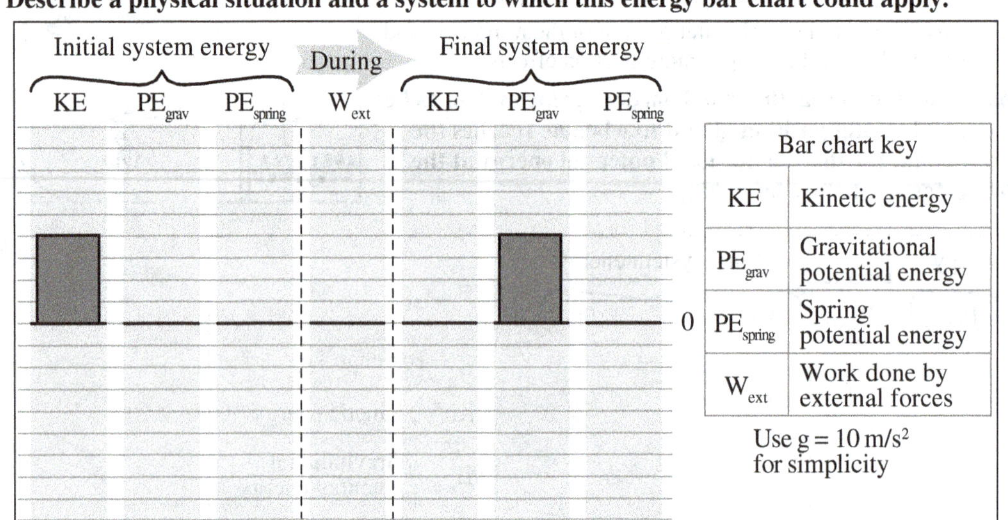

Explain your reasoning.

B4-WBT40: Energy Bar Chart II—Physical Situation

Describe a physical situation and a system to which this energy bar chart could apply.

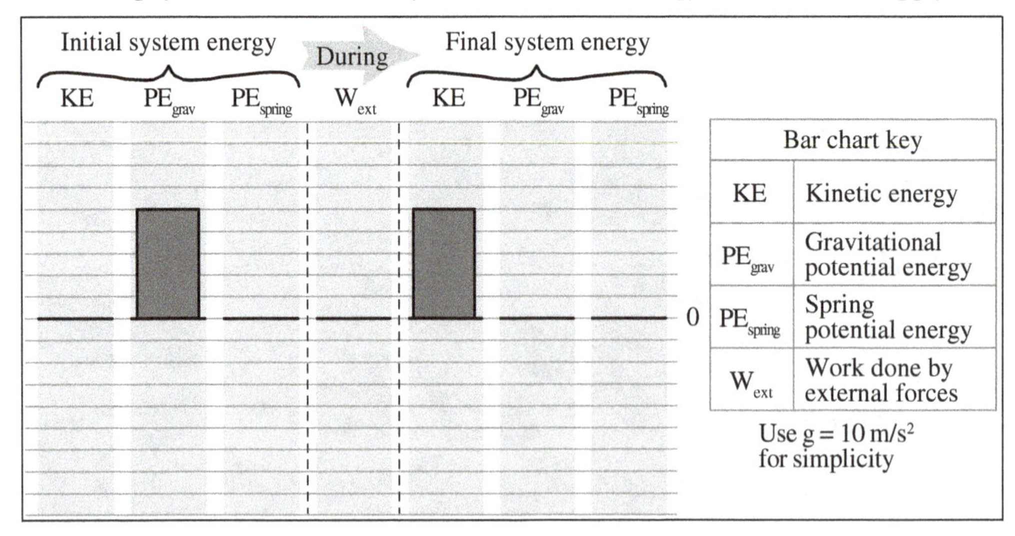

Explain your reasoning.

B4-WWT41: SLIDING BOX—ENERGY BAR CHART

Shown is an energy bar chart drawn by a student.

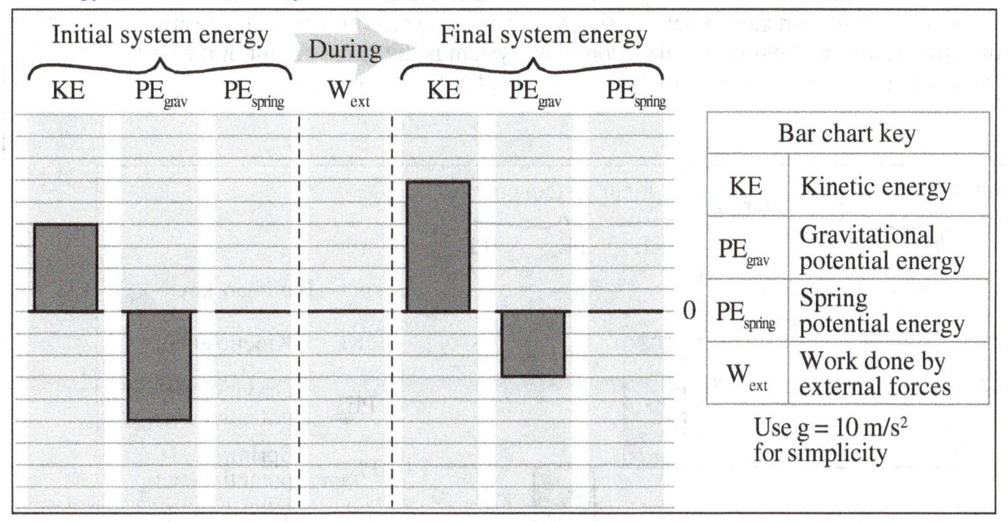

A student who drew this chart says:

"This chart is for a moving box sliding up a smooth slope from a lower point to a higher one. The zero point for the gravitational potential energy is set at the ground level."

What, if anything, is wrong with this chart? If something is wrong, identify it and explain how to correct it. If this statement is correct, explain why.

B4-WWT42: BOX ON SLOPING HILL—ENERGY BAR CHART

Shown is an energy bar chart drawn by a student about a box on a sloping hill.

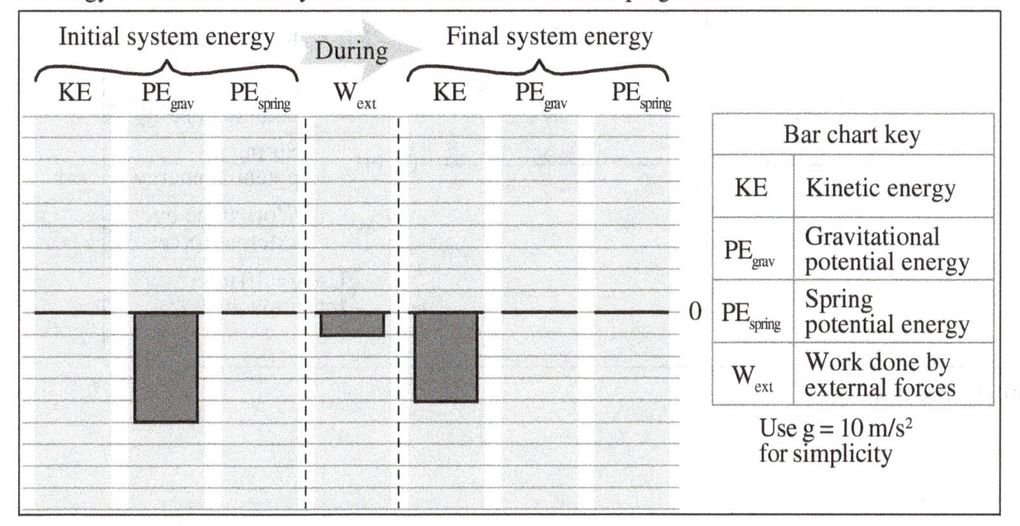

A second student says:

"No, this is not correct since the work done must be positive."

What, if anything, is wrong with this chart? If something is wrong, identify it and explain how to correct it. If this statement is correct, explain why.

B4-QRT43: Dropped Rock—Energy Bar Chart

A rock is dropped by a student from the top of a cliff and falls straight to the ground below. He constructs an energy bar chart shown below using a coordinate system in which the positive vertical direction is up and the origin of the coordinate system is the release point of the rock which is also selected as the zero point for the gravitational potentials energy.

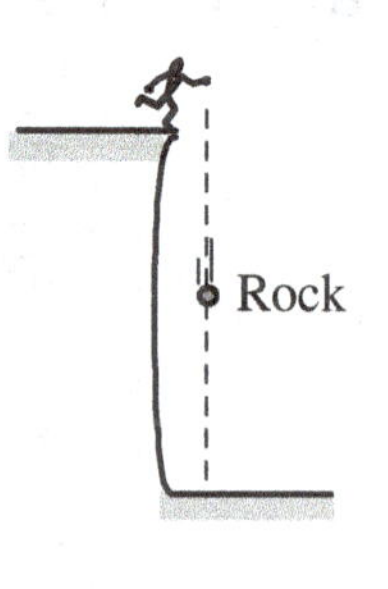

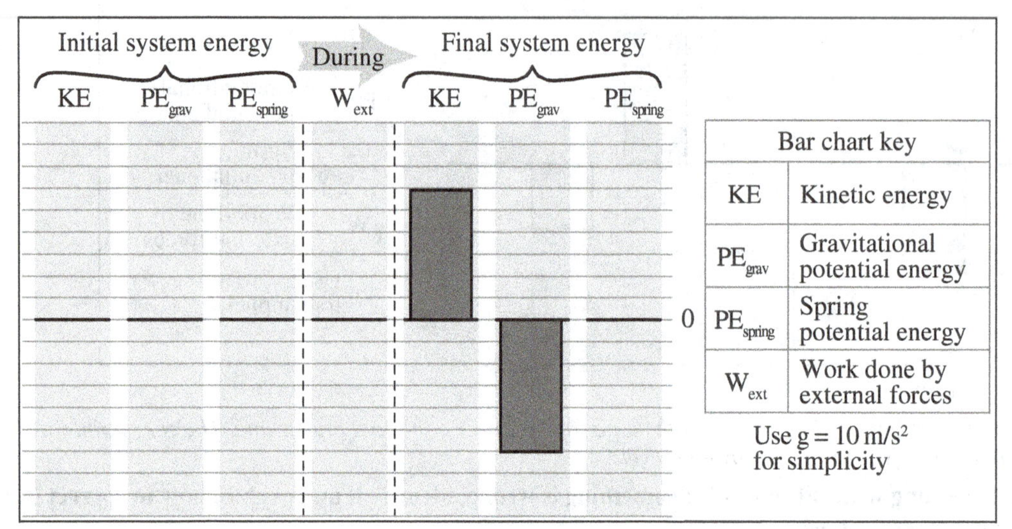

Bar chart key	
KE	Kinetic energy
PE_{grav}	Gravitational potential energy
PE_{spring}	Spring potential energy
W_{ext}	Work done by external forces

Use $g = 10 \text{ m/s}^2$ for simplicity

Draw a new energy bar chart for this event but use the ground as the zero point for the potential gravitational energy.

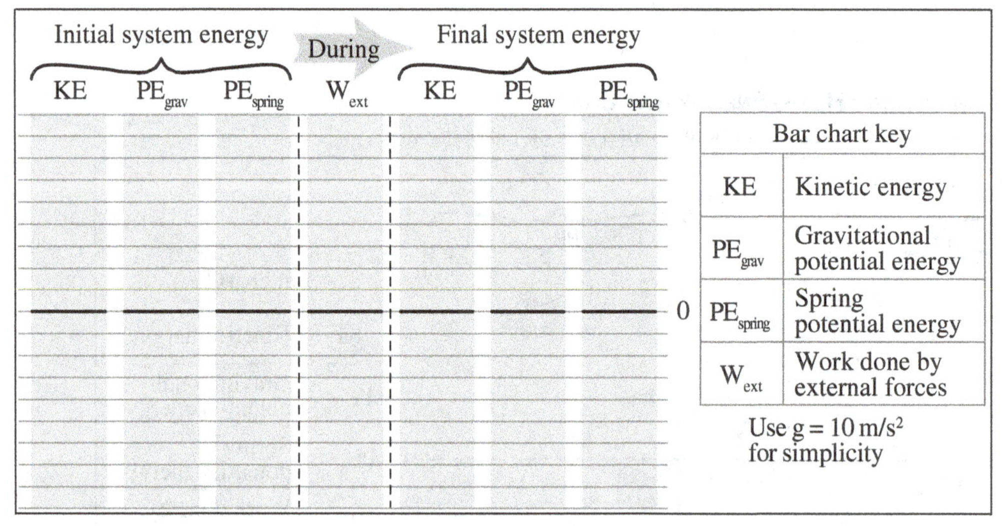

Bar chart key	
KE	Kinetic energy
PE_{grav}	Gravitational potential energy
PE_{spring}	Spring potential energy
W_{ext}	Work done by external forces

Use $g = 10 \text{ m/s}^2$ for simplicity

Explain your reasoning.

B4-WBT44: ENERGY BAR CHART—PHYSICAL SITUATION

Describe a physical situation and a system to which this energy bar chart could apply.

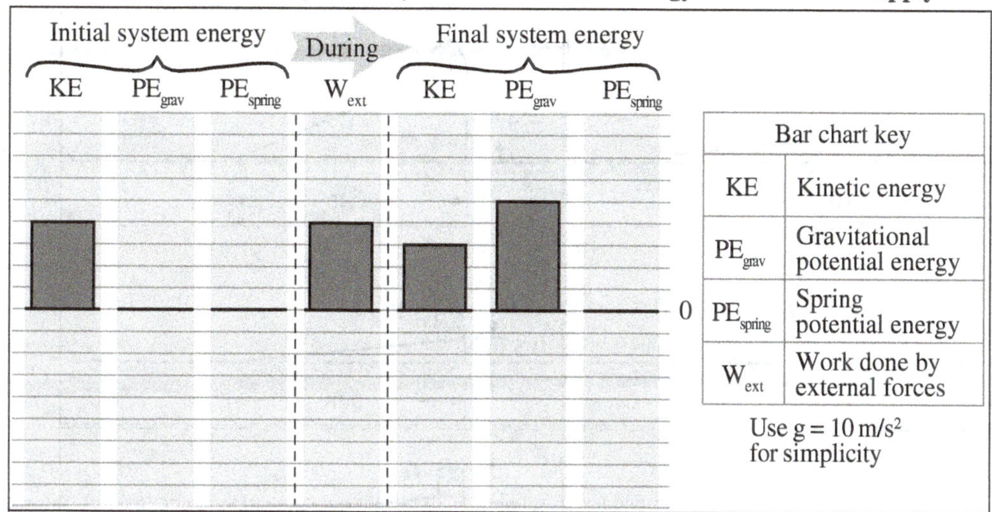

Explain your reasoning.

B4-RT45: Sliding Masses on Incline—Kinetic Energy

Shown are blocks that slide down frictionless inclines. All masses start from rest at the top of the incline.

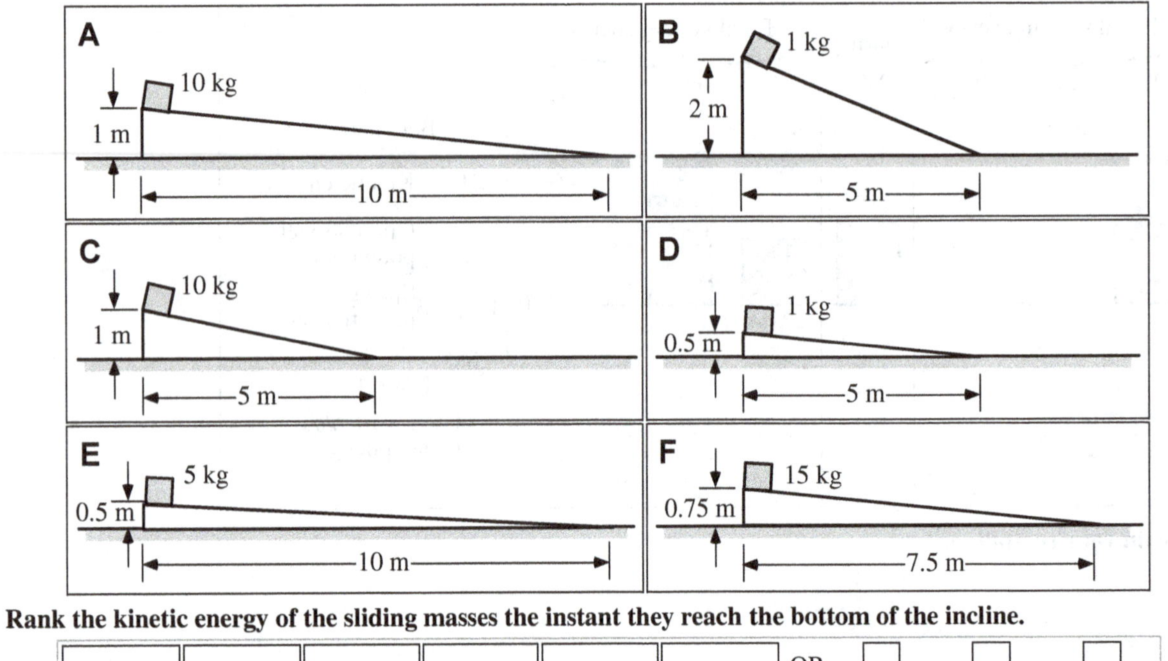

Rank the kinetic energy of the sliding masses the instant they reach the bottom of the incline.

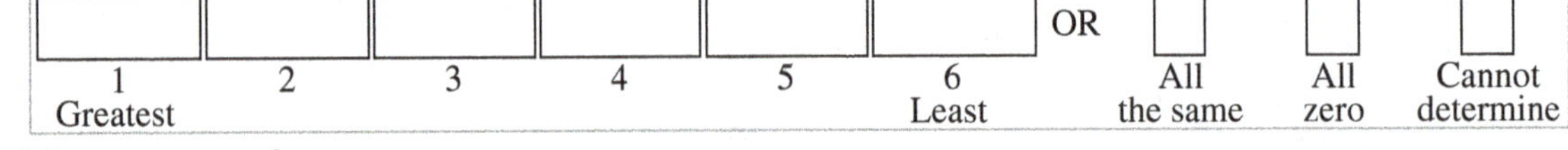

Explain your reasoning.

B4-RT46: SLIDING MASSES ON INCLINE—CHANGE IN POTENTIAL ENERGY

Shown are blocks that slide down frictionless inclines. All masses start from rest at the top of the incline.

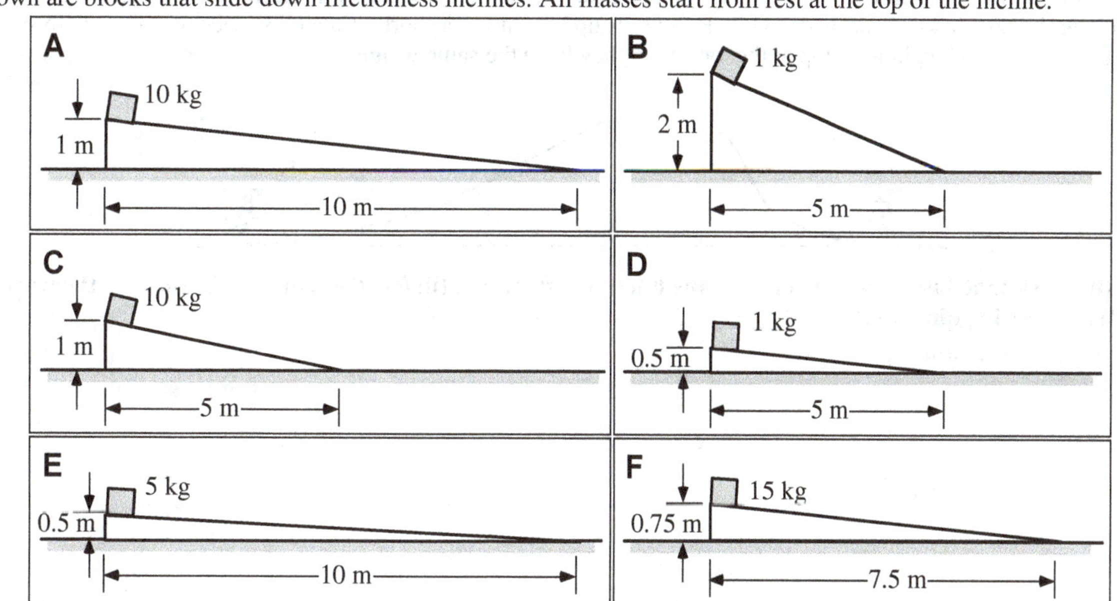

Rank the change in gravitational potential energy of the sliding masses from the top of the incline to the bottom of the incline.

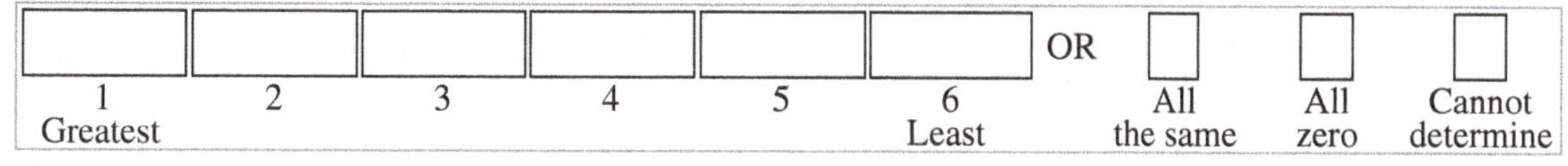

Explain your reasoning.

B4-CT47: Race up a Hill I—Work and Power

Jason and Brent race up a hill that is 30 m high. Jason takes a path that is 60 m while Brent uses a longer path that is 100 m long. It takes Jason 40 seconds, while Brent runs up his path in a shorter time of 30 seconds. They both start from rest at the same height and stop at the top. Also, they have the same weight.

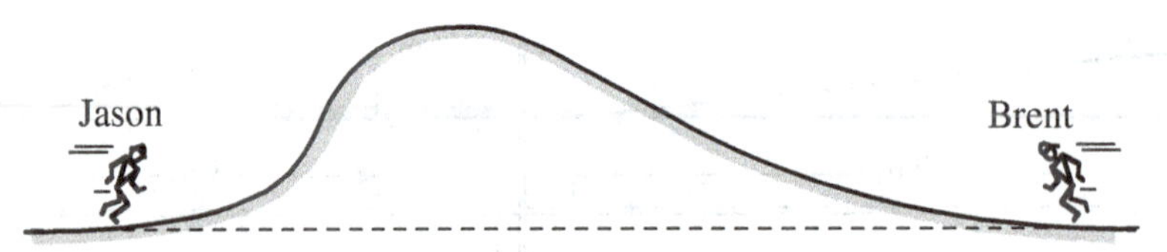

(a) Is the work that Jason does in going up the hill (i) *greater than*, (ii) *less than*, or (iii) *the same as* the work that Brent does in going up the hill? _____

Explain your reasoning.

(b) Is the power generated by Jason in going up the hill (i) *greater than*, (ii) *less than*, or (iii) *the same as* the power generated by Brent in going up the hill? _____

Explain your reasoning.

B4-CT48: RACE UP A HILL II—WORK AND POWER

Abbie and Bonita decide to race up a hill that is 30 m high. Abbie takes a path that is 60 m long while Bonita uses a path that is 100 m long. It takes Abbie 40 seconds because her route is steep, while Bonita runs up her path in 30 seconds. They both start from rest at the same height and stop at the top. Abbie has a weight of 700 N, and Bonita has a weight of 500 N.

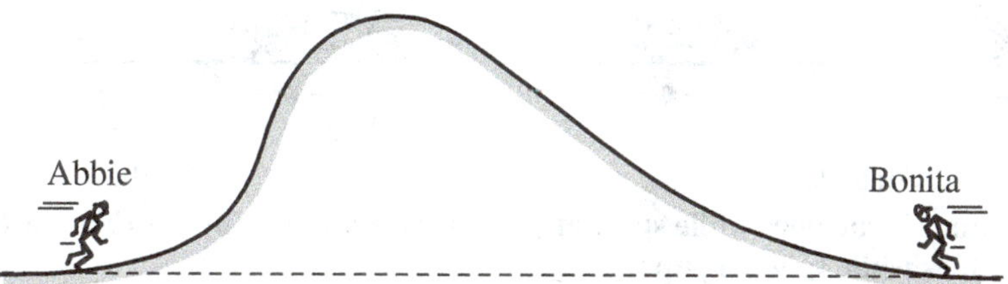

(a) Is the work that Abbie does in going up the hill (i) *greater than*, (ii) *less than*, or (iii) *the same as* the work that Bonita does in going up the hill? _____

Explain your reasoning.

(b) Is the power generated by Abbie in going up the hill *greater than*, *less than*, or *the same as* the power generated by Bonita in going up the hill? _____

Explain your reasoning.

B4-CT49: Car Race—Work and Power

Amanda and Bertha are in a car race. Their cars have the same mass. At one point in the race, they both change their speeds by 10 m/s in 2 seconds. Ignore air friction.

(a) Is the work that Amanda's car does while speeding up (i) *greater than*, (ii) *less than*, or (iii) *the same as* the work that Bertha's car does while speeding up? _____

Explain your reasoning.

(b) Is the power generated by Amanda's car while speeding up (i) *greater than*, (ii) *less than*, or (iii) *the same as* the power generated by Bertha's car while speeding up? _____

Explain your reasoning.

B5 MOMENTUM AND IMPULSE

B5-RT01: MOVING BALL I—MOMENTUM AND KINETIC ENERGY

In the figures below, balls are traveling in different directions. The balls have the same size and shape, but they have different masses and are traveling at different velocities as shown.

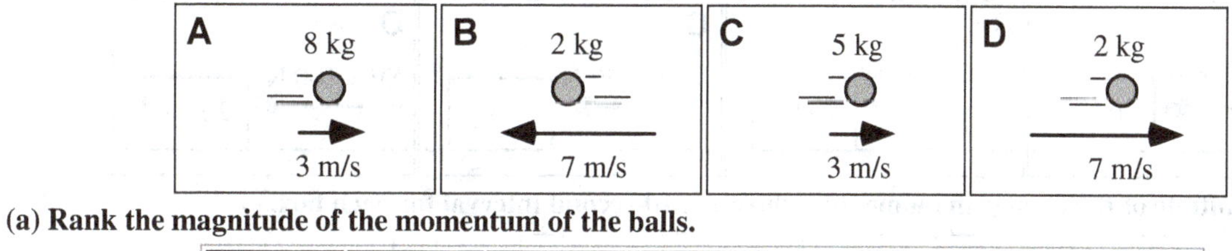

(a) Rank the magnitude of the momentum of the balls.

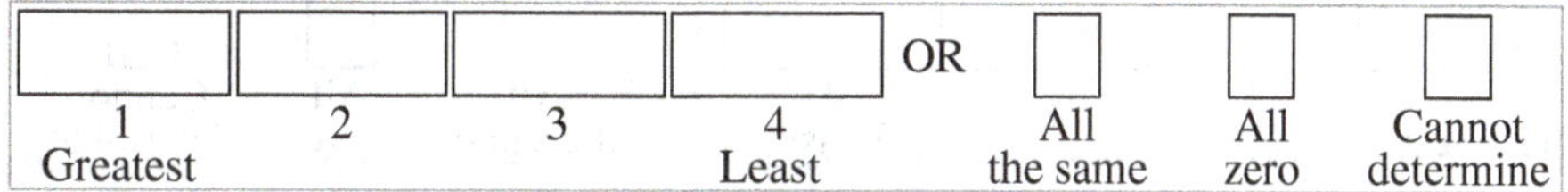

Explain your reasoning.

(b) Rank the kinetic energy of the balls.

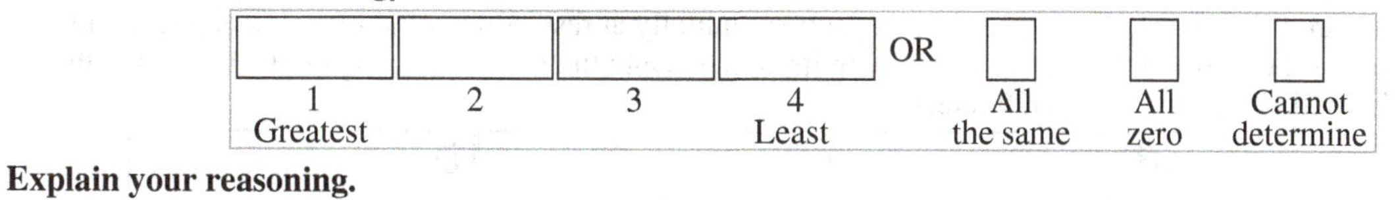

Explain your reasoning.

B5-RT02: Force Pushing Box I—Change in Momentum

Identical boxes that are filled with different objects are initially at rest. A horizontal force is applied for 10 seconds, and the boxes move across the floor. The mass of the box with its contents and the *net* force acting on the box while the horizontal force are applied is given in each figure.

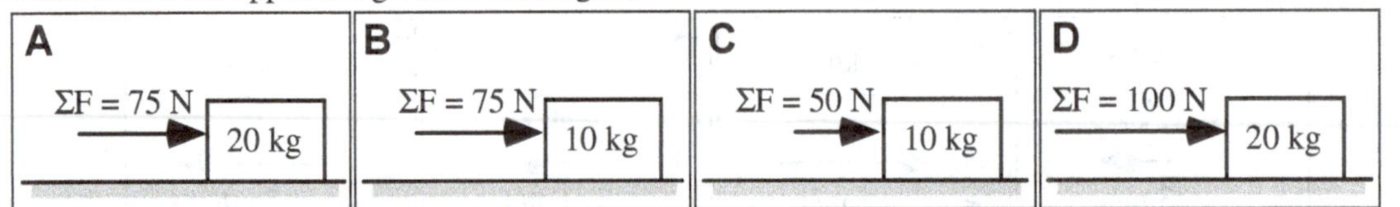

Rank the magnitude of the change in momentum during a 10-second interval for each box.

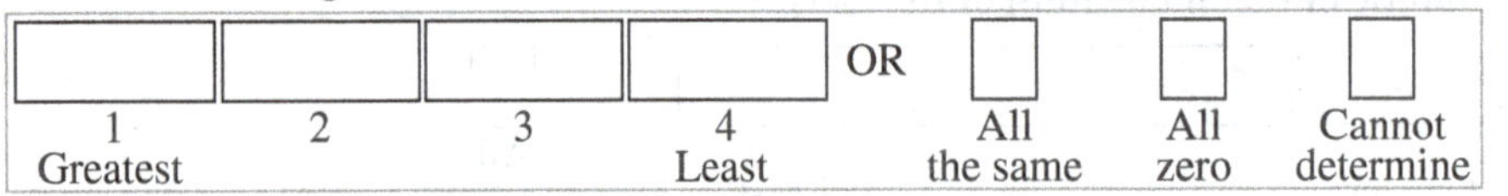

Explain your reasoning.

B5-RT03: Force Pushing Box II—Impulse

Identical boxes that are filled with different amounts of sand are initially at rest. A horizontal force is applied, and the boxes move across the floor. The mass of the box with its contents and the *net* force acting on the box while the horizontal force is applied are given in each figure.

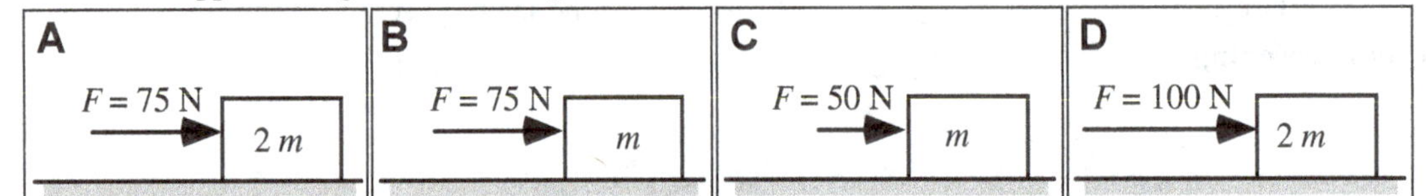

Rank the magnitude of the impulse on each box for a 2-second time interval.

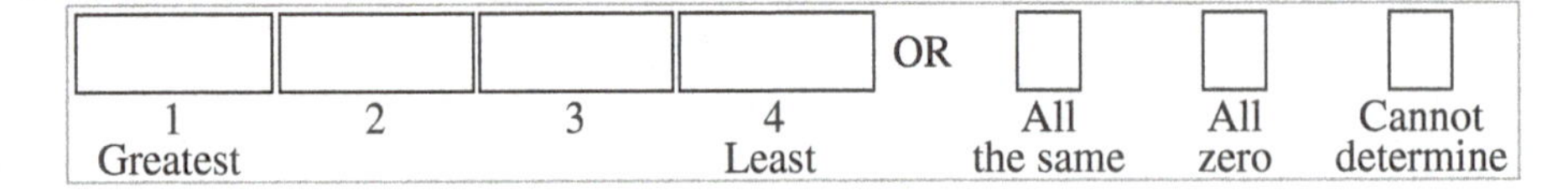

Explain your reasoning.

B5-RT04: FORCE PUSHING BOX III—CHANGE IN MOMENTUM

Identical boxes that are filled with different amounts of sand are initially at rest. A horizontal force is applied, and the boxes move across the floor. The mass of the box with its contents and the *net* force acting on the box while the horizontal force is applied are given in each figure.

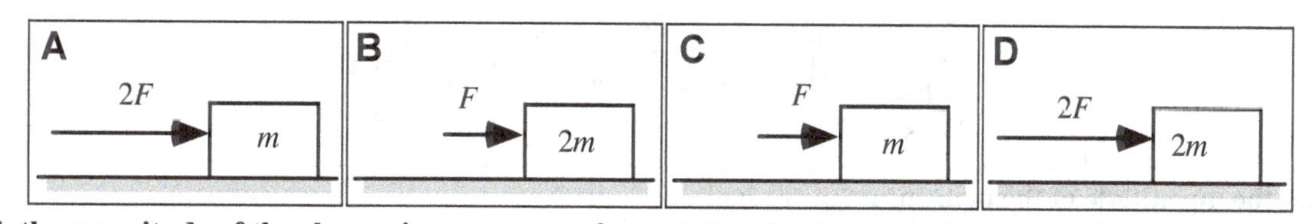

Rank the magnitude of the change in momentum for each box for the same time interval.

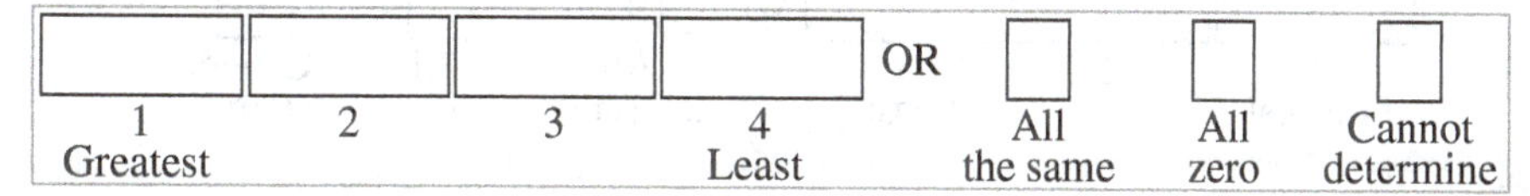

Explain your reasoning.

B5-QRT05: AMANDA AND BERTHA'S CAR RACE—WORK AND IMPULSE

Amanda and Bertha are driving cars in a race. Their two cars, including Amanda and Bertha, have the same mass. At one point in the race, they both change their speeds by 10 m/s in 2 seconds. Ignore air friction.

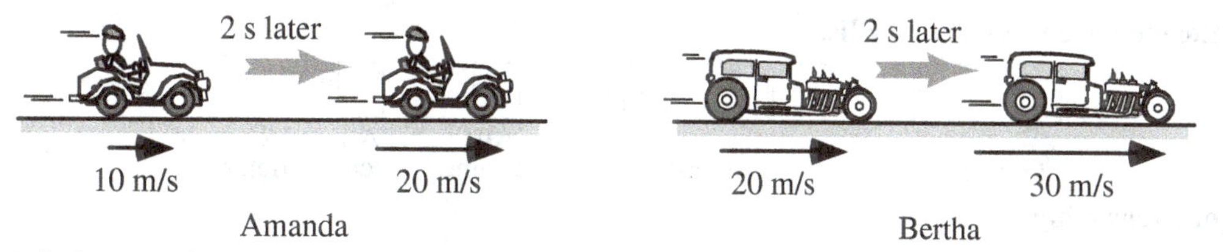

(a) Is the work done on Amanda's car while speeding up (i) *greater than*, (ii) *less than*, or (iii) *the same as* the work done on Bertha's car while speeding up? _____

Explain your reasoning.

(b) Is the impulse on Amanda's car while speeding up (i) *greater than*, (ii) *less than*, or (iii) *the same as* the work done on Bertha's car does while speeding up? _____

Explain your reasoning.

B5-RT06: Moving Ball II—Momentum and Kinetic Energy

In the figures below, balls are traveling in different directions. The balls have the same size, mass and shape, but they are traveling with different velocities as shown.

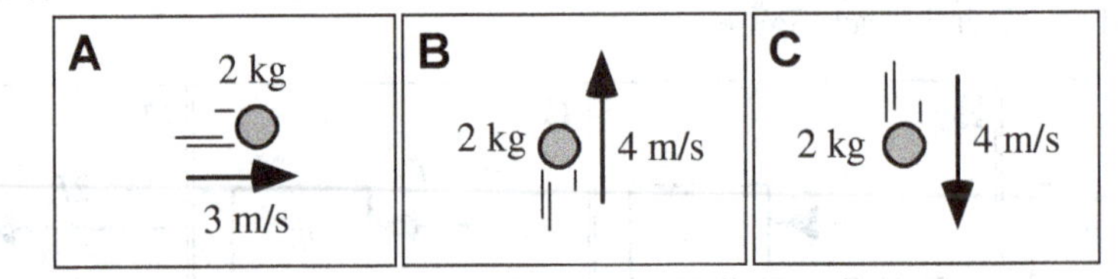

(a) Rank the magnitude of the momentum of the balls.

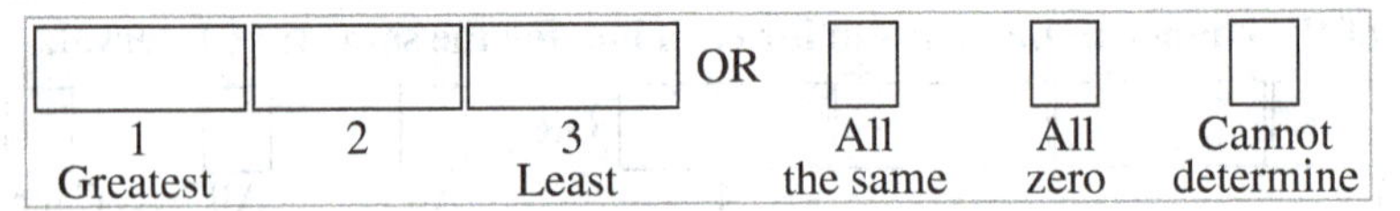

Explain your reasoning.

(b) Rank the kinetic energy of the balls.

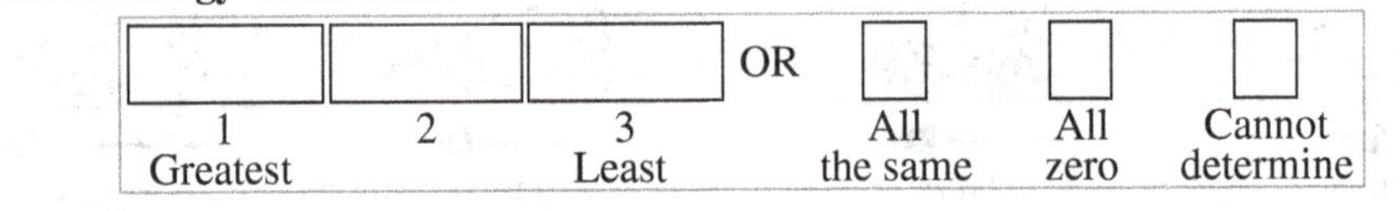

Explain your reasoning.

B5-QRT07: BLOCKS SLIDING DOWN FRICTIONLESS RAMPS—WORK AND MOMENTUM

Two blocks are released from rest at the same height. Block A slides down a steeper ramp than Block B. Both ramps are frictionless. The blocks reach the same final height indicated by the lower dashed line. Block B weighs more than Block A.

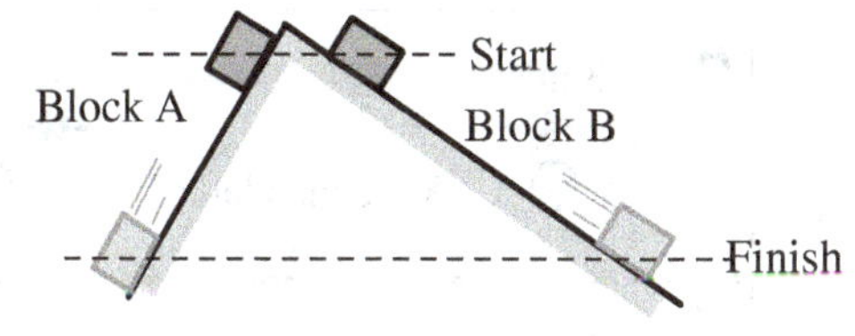

(a) Is the work done by the gravitational force on Block A (i) *greater than*, (ii) *less than*, or (iii) *the same as* the work done by the gravitational force on Block B? _____
Explain your reasoning.

(b) Is the speed of Block A (i) *greater than*, (ii) *less than*, or (iii) *the same as* the speed of Block B? _____
Explain your reasoning.

(c) Is the momentum of Block A (i) *greater than*, (ii) *less than*, or (iii) *the same as* the momentum of Block B? _____
Explain your reasoning.

B5-RT08: CARS STOPPED BY CONSTANT FORCE BARRIERS—STOPPING TIME

Cars moving along horizontal roads are about to be stopped when they hit a protective barrier. All of the cars are the same size and shape, but they are moving at different speeds and have different masses. The barriers are all identical and exert the same constant force.

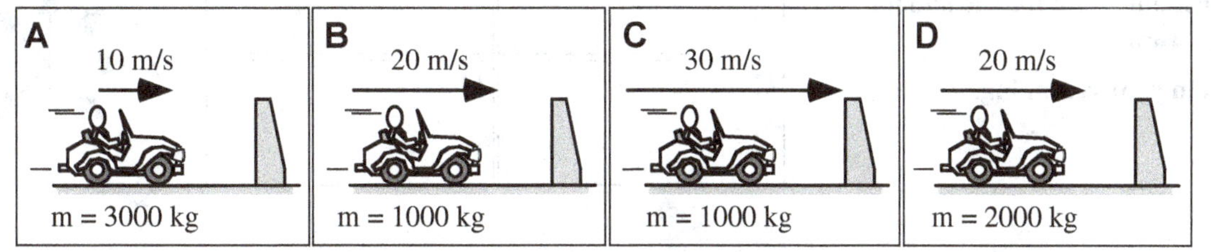

Rank the time that it takes to stop the cars as the barriers apply the same constant force.

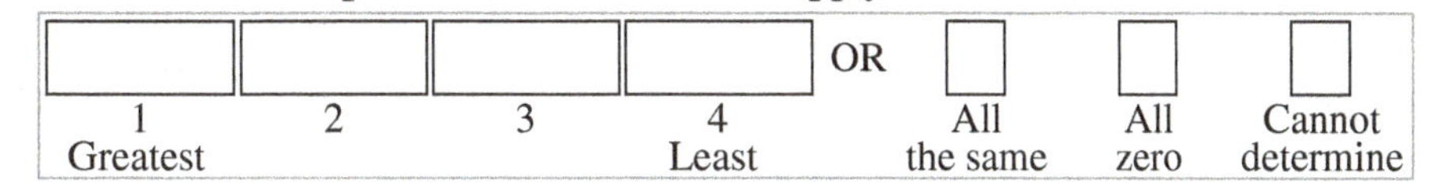

Explain your reasoning.

B5-QRT09: Object Changing Velocity—Direction of the Impulse

An object changes its velocity as forces act on it for 5 seconds in various ways shown below.

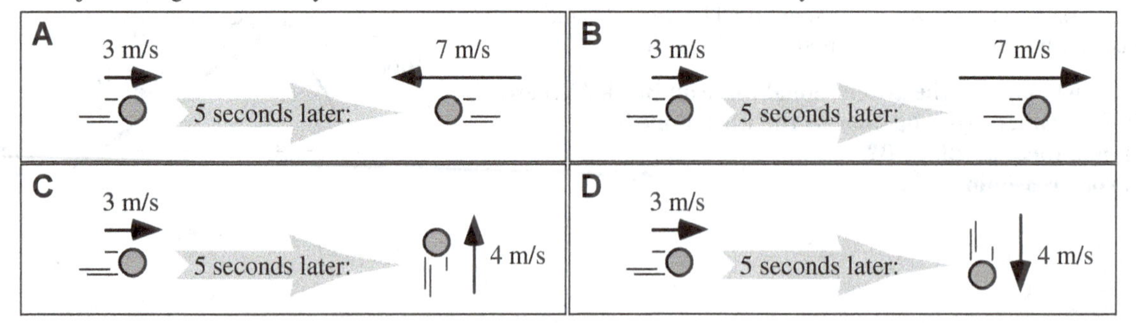

For the question below, use the directions indicated by the arrows in the direction rosette, or use **J** for no direction, **K** for into the page, or **L** for out of the page.

(a) Identify the closest directional match for the direction of the impulse on the ball for these cases.

Explain your reasoning.

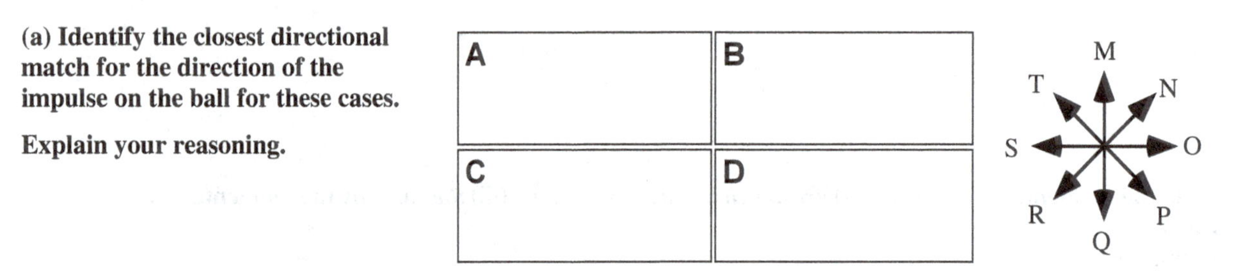

(b) Identify the closest directional match for the direction of the change in the momentum for the ball for these cases.

Explain your reasoning.

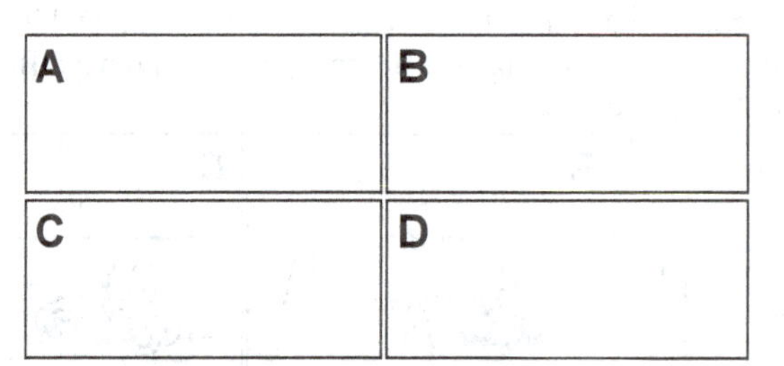

B5-RT10: BOUNCING CART I—CHANGE IN MOMENTUM

Carts with spring plungers run into fixed barriers. The carts are identical but are carrying different loads and so have different masses. The velocity of the cart just before and just after impact is given.

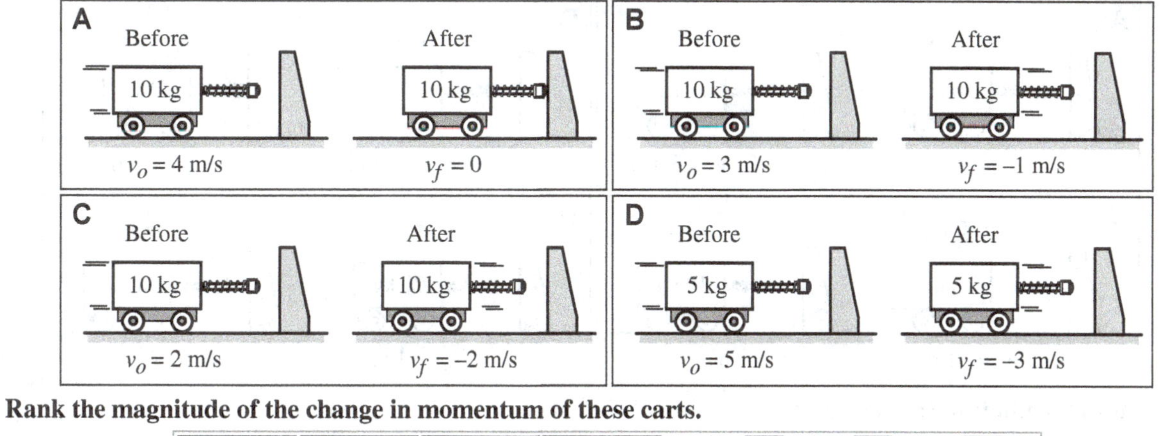

Rank the magnitude of the change in momentum of these carts.

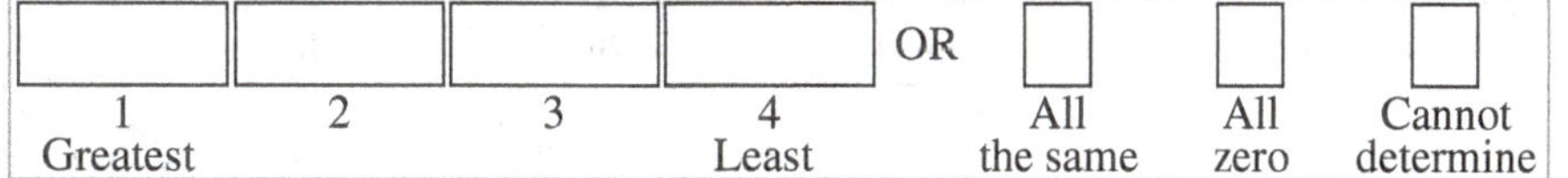

Explain your reasoning.

B5-RT11: Bouncing Cart II—Change in Momentum

Carts with spring plungers run into fixed barriers. The carts are identical but are carrying different loads and so have different masses. The velocity of each cart just before and just after impact is given.

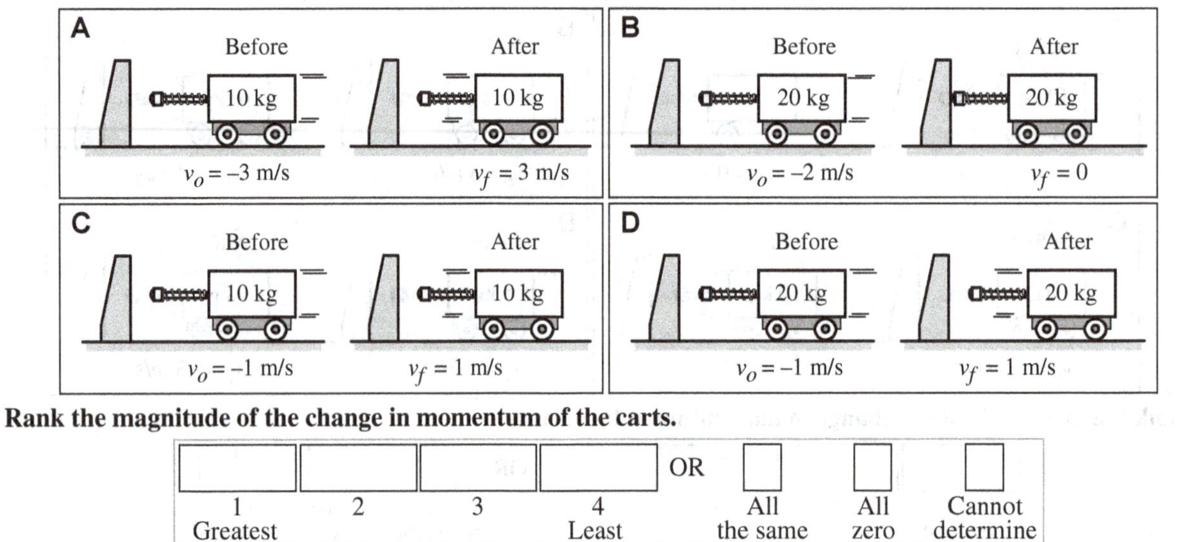

Rank the magnitude of the change in momentum of the carts.

Explain your reasoning.

B5-QRT12: BOUNCING CART—DIRECTION OF THE CHANGE IN MOMENTUM

Carts with spring plungers run into fixed barriers. The carts are identical but are carrying different loads and so have different masses. The velocity of the cart just before and just after impact is given.

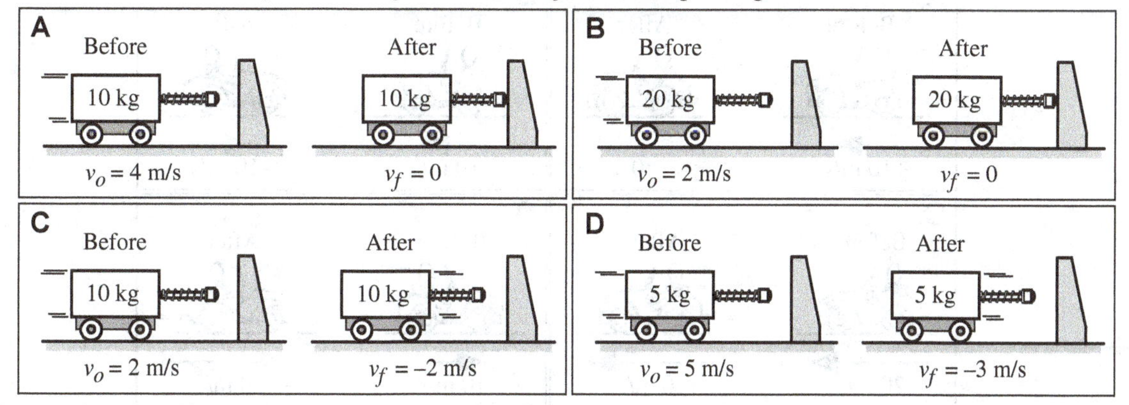

(a) Is the direction of the change in momentum in Case A *to the left* or *to the right*? If the change in momentum *cannot be determined,* state that explicitly. _____
Explain your reasoning.

(b) Is the direction of the change in momentum in Case B *to the left* or *to the right*? If the change in momentum cannot be determined, state that explicitly. _____
Explain your reasoning.

(c) Is the direction of the change in momentum in Case C *to the left* or *to the right*? If the change in momentum *cannot be determined,* state that explicitly. _____
Explain your reasoning.

(d) Is the direction of the change in momentum in Case D *to the left* or *to the right?* If the change in momentum *cannot be determined,* state that explicitly. _____
Explain your reasoning.

B5-RT13: Cars—Change in Momentum During a Change of Velocity

Before and after "snapshots" of a car's velocity are shown. All cars have the same mass.

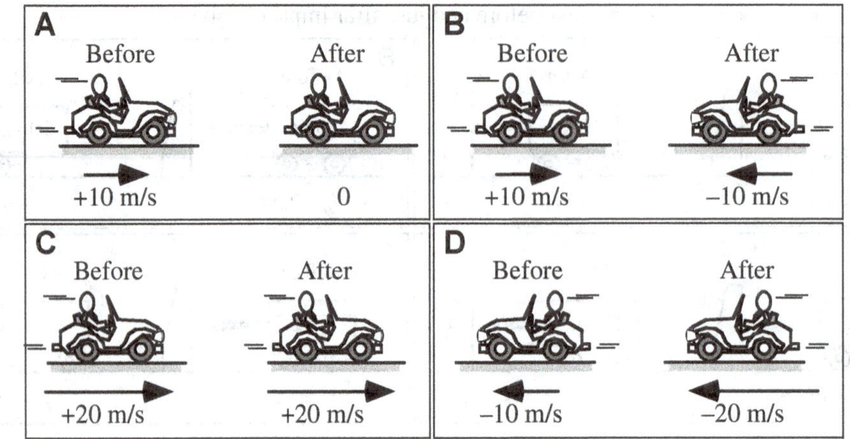

Rank the magnitude of the change in momentum of the cars.

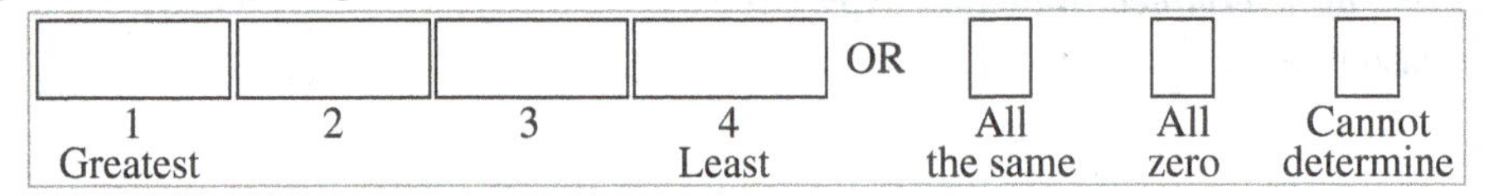

Explain your reasoning.

B5-SCT14: Object Changing Velocity I—Impulse

A 2-kg object accelerates as a net external force is applied to it. During the 5-second interval that the force is applied, the object's velocity changes from 3 m/s to the right to 7 m/s to the left.

Several students discussing the impulse on this object state the following:

Andre: *"The impulse is equal to the change in momentum, which is (2 kg)(3 m/s + 7 m/s) = 20 kg·m/s."*

Bela: *"But the change in velocity is 4 m/s. We multiply by the mass to get the change in momentum, and also the impulse, which is 8 kg·m/s."*

Carleton: *"The change in momentum of this object during these 5 seconds was 8 kg·m/s so the impulse applied to this object during these 5 seconds was 8/5 kg·m/s."*

Dylan: *"The impulse is the force F times the time t, and since we don't know the force, we can't find the impulse for this situation."*

With which, if any, of these students do you agree?

Andre _____ Bela _____ Carleton _____ Dylan _____ None of them_____

Explain your reasoning.

B5-WWT15: OBJECT CHANGING VELOCITY II—IMPULSE

A 2-kg object accelerates as a net external force is applied to it. During the 5-second interval that the force is applied, the object's velocity changes from 3 m/s to the right to 7 m/s to the left.

A student states:

"The change in momentum of this object during these 5 seconds was 8 kg·m/s, so the impulse applied to this object during these 5 seconds was 8/5 kg·m/s."

What, if anything, is wrong with this statement? If something is wrong, identify it and explain how to correct all errors. If this statement is correct, explain why.

B5-WWT16: OBJECT CHANGING VELOCITY III—IMPULSE

A 2-kg object accelerates as a net external force is applied to it. During the 5-second interval that the force is applied, the object's velocity changes from 3 m/s to the right to 7 m/s to the left.

A student states:

"The change in velocity for this 2 kg object was 4 m/s, so the change in momentum, and also the impulse, was 8 kg·m/s."

What, if anything, is wrong with this statement? If something is wrong, identify it and explain how to correct it. If this statement is correct, explain why.

B5-WWT17: Object Changing Velocity IV—Impulse

A student proposes the following description for the impulse on a 2-kg object that changes direction and speed as shown:

"The object goes from moving at 3 m/s in the positive x-direction to 7 m/s in the positive y-direction in 5 seconds. So the impulse given to it is 8 kg·m/s, since the impulse equals the change in momentum. The 5 seconds does not enter into the calculation of this impulse."

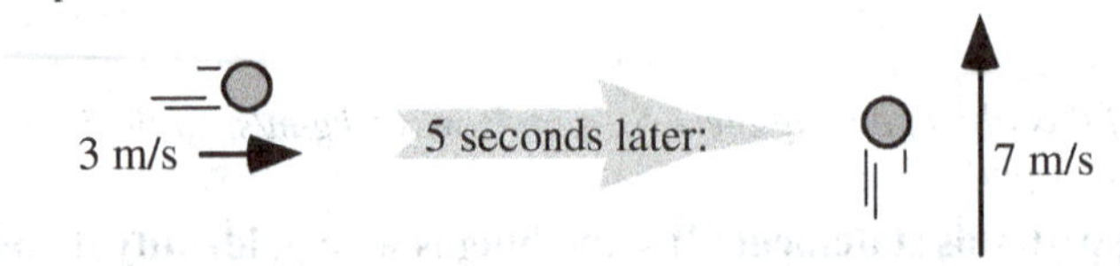

What, if anything, is wrong with this statement? If anything is wrong, identify it and explain how to correct it. If this statement is correct, explain why.

B5-RT18: Force-Time Graph I—Impulse Applied to Box

A 10-kg box, initially at rest, moves along a frictionless horizontal surface. A horizontal force to the right is applied to the box. The magnitude of the force changes as a function of time as shown.

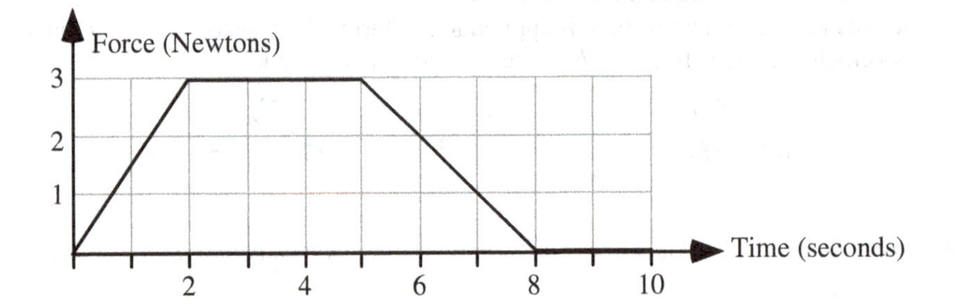

Rank the impulse applied to the box by this force during each 2-second interval indicated below.

A. 0 to 2 s **B. 2 to 4 s** **C. 4 to 6 s** **D. 6 to 8 s** **E. 8 to 10 s**

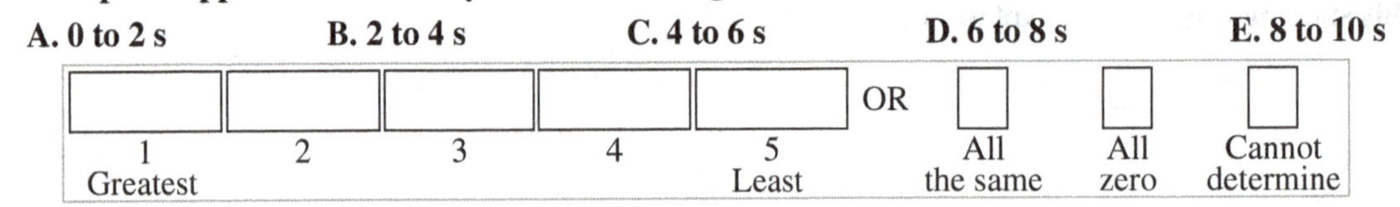

Explain your reasoning.

B5-WWT19: FORCE-TIME GRAPH II—IMPULSE APPLIED TO BOX

A 10-kg box, initially at rest, moves along a frictionless horizontal surface. A horizontal force to the right is applied to the box. The magnitude of the force changes as a function of time as shown.

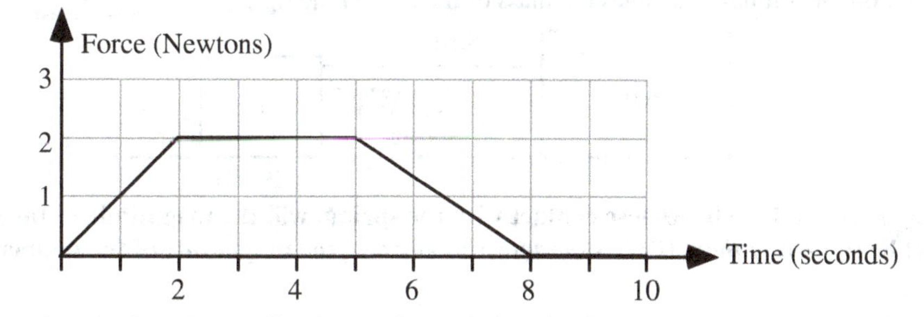

A student calculates that the impulse applied by the force during the first 2 seconds is 4 N·s and that the impulse applied during the following 3 seconds is 6 N·s.

What, if anything, is wrong with these calculations? If something is wrong, identify it and explain how to correct it. If these calculations are correct, explain why.

B5-WWT20: TWO SKATERS PUSHING OFF EACH OTHER—FORCE

Two skaters, a large girl and a small boy, are initially standing face-to-face but then push off each other. After they are no longer touching, the boy has more kinetic energy than the girl. A physics student who is watching makes the following contention about the forces that the boy and girl exerted on each other:

"Since the boy has more kinetic energy, he also has more momentum, so the girl had to have pushed harder on him than he pushed on her."

What, if anything, is wrong with this contention? If something is wrong, identify all problems and explain how to correct them. If this contention is correct, explain why.

B5-CT21: Two Boxes on a Frictionless Surface—Momentum and Speed

Two boxes are tied together by a string and are sitting at rest on a frictionless surface. Between the two boxes is a massless compressed spring. The string tying the two boxes together is cut and the spring expands, pushing the boxes apart. The box on the left has four times the mass of the box on the right.

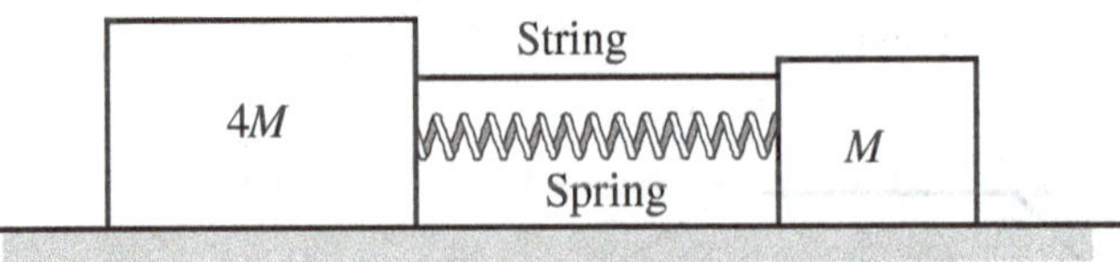

(a) After the string is cut and the boxes lose contact with the spring, will the magnitude of the momentum of the box on the left be (i) *greater than*, (ii) *less than*, or (iii) *equal to* the magnitude of the momentum of the box on the right? _____
Explain your reasoning.

(b) At the instant (after the string is cut) that the boxes lose contact with the spring, will the speed of the box on the left be (i) *greater than*, (ii) *less than*, or (iii) *equal to* the speed of the box on the right? _____
Explain your reasoning.

B5-WWT22: Ball Hitting a Wall—Momentum

A student observing a rubber ball hitting a wall and rebounding states:

"The change in momentum for the ball is equal and opposite to the change in momentum for the wall, because in this situation momentum has to be conserved."

What, if anything, is wrong with this statement? If something is wrong, identify it and explain how to correct it. If this statement is correct, explain why.

B5-RT23: COLLIDING CARTS STICKING TOGETHER—FINAL SPEED

Two carts traveling in opposite directions are about to collide. The carts are all identical in size and shape, but they carry different loads and are traveling at different speeds. The carts stick together after the collision. There is no friction between the carts and the ground.

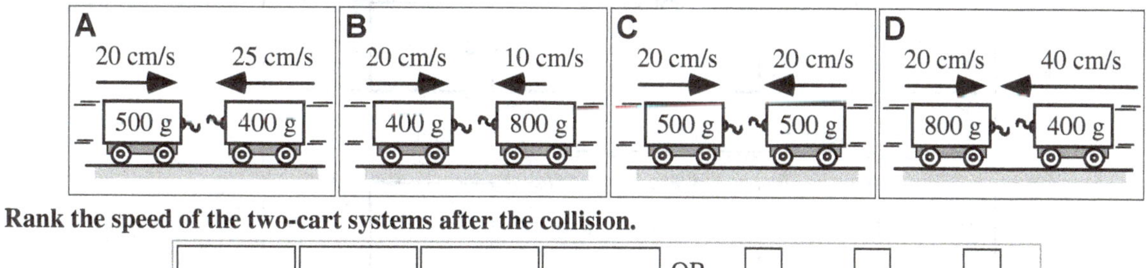

Rank the speed of the two-cart systems after the collision.

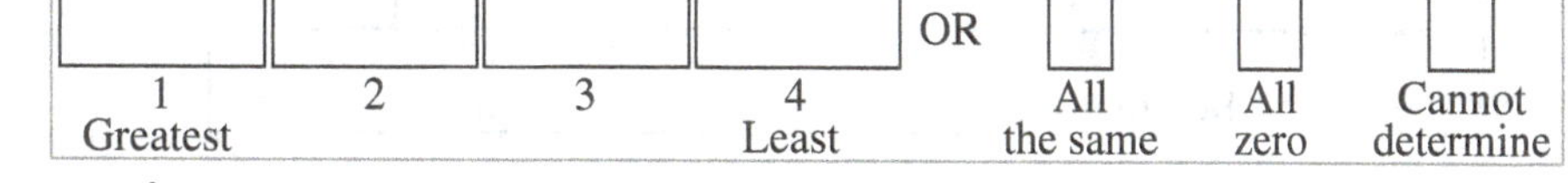

Explain your reasoning.

B5-SCT24: TWO MOVING CARTS—RESULT OF COLLISION

Carts A and B are shown just before they collide. Four students discussing this situation make the following contentions:

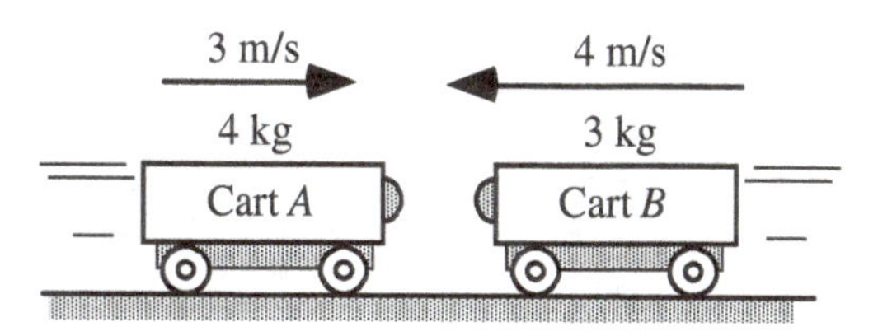

Alma: *"After the collision, the carts will stick together and move off to the left. Cart B has more speed, and its speed is going to determine which cart dominates in the collision."*

Baxter: *"I think they'll stick together and move off to the right because Cart A is heavier. It's like when a heavy truck hits a car: The truck is going to win no matter which one's going fastest, just because it's heavier."*

Callie: *"I think the speed and the mass compensate, and the carts are going to be at rest after the collision."*

Dante: *"The carts must have the same momentum after the collision as before the collision, and the only way this is going to happen is if they keep the same speeds. All the collision does is change their directions, so that Cart A will be moving to the left at 3 m/s and Cart B will be moving to the right at 4 m/s."*

With which, if any, of these students do you agree?

Alma _____ Baxter _____ Callie _____ Dante _____ None of them_____

Explain your reasoning.

B5-CT25: Bullet Strikes a Wooden Block—Block and Bullet Speed After Impact

In Case A, a metal bullet penetrates a wooden block. In Case B, a rubber bullet with the same initial speed and mass bounces off of an identical wooden block.

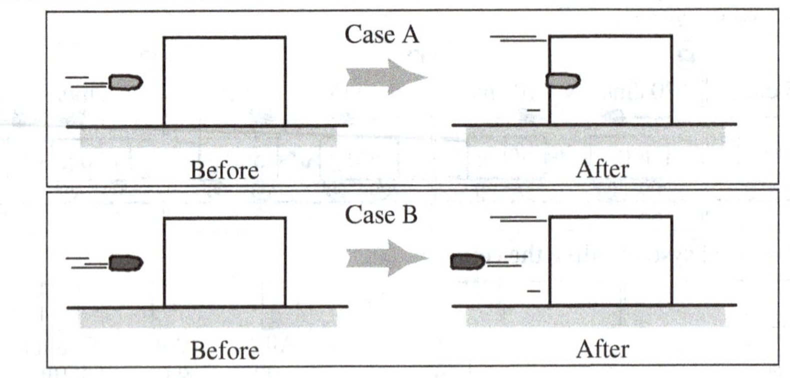

(a) Will the speed of the wooden block after the collision be (i) *greater* in Case A, (ii) *greater* in Case B, or (iii) *the same* in both cases? _____

Explain your reasoning.

(b) In Case B, will the speed of the bullet after the collision be (i) *greater than*, (ii) *less than*, or (iii) *the same as* the speed of the bullet just before the collision? _____

Explain your reasoning.

B5-SCT26: COLLIDING CARTS THAT STICK TOGETHER—FINAL KINETIC ENERGY

Two identical carts traveling in opposite directions are shown just before they collide. The carts carry different loads and are initially traveling at different speeds. The carts stick together after the collision.

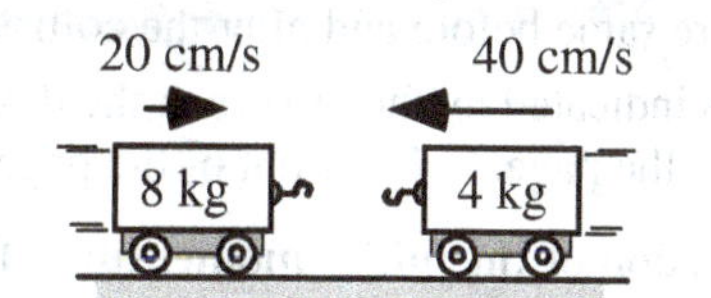

Three physics students discussing this situation make the following contentions:

Alex: *"These carts will both be at rest after the collision since the initial momentum of the system is zero, and the final momentum has to be zero also."*

Belinda: *"If that were true it would mean that they would have zero kinetic energy after the collision, and that would violate conservation of energy. Since the right-hand cart has more kinetic energy, the combined carts will be moving slowly to the left after the collision."*

Chano: *"I think that after the collision the pair of carts will be traveling left at 20 cm/s. That way conservation of momentum and conservation of energy are both satisfied."*

With which, if any, of these students do you agree?

Alex _____ Belinda _____ Chano _____ None of them_____

Explain your reasoning.

B5-QRT27: Colliding Steel Balls—Momentum and Impulse Direction

Two identical steel balls, *S* and *T*, are shown at the instant that they collide. The paths and velocities of the two balls before and after the collision are indicated by the dashed lines and arrows. The speeds of the balls are same before and after the collision.

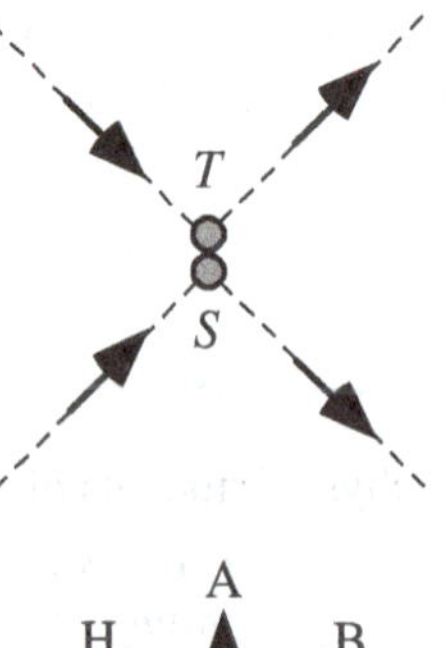

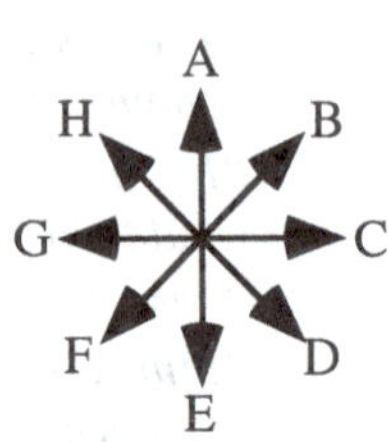

For the questions below, use the directions indicated by the arrows in the direction rosette, or use **J** for no direction, **K** for into the page, or **L** for out of the page.

(a) Which letter best represents the direction of the initial momentum of ball *T*?

_____

Explain your reasoning.

(b) Which letter best represents the direction of the final momentum of ball *T*? _____

Explain your reasoning.

(c) Which letter best represents the direction of the change in momentum for ball *T*? _____

Explain your reasoning.

(d) Which letter best represents the direction of the change in momentum for ball *S*? _____

Explain your reasoning.

(e) Which letter best represents the direction of the impulse on ball *T*? _____

Explain your reasoning.

(f) Which letter best represents the direction of the impulse on ball *S*? _____

Explain your reasoning.

B5-RT28: COLLIDING BALL SYSTEMS—MOMENTUM BEFORE AND AFTER COLLIDING

In the figures below, two balls traveling in different directions are about to collide. The balls are identical in size and shape, but they have different masses and are traveling at different velocities as shown.

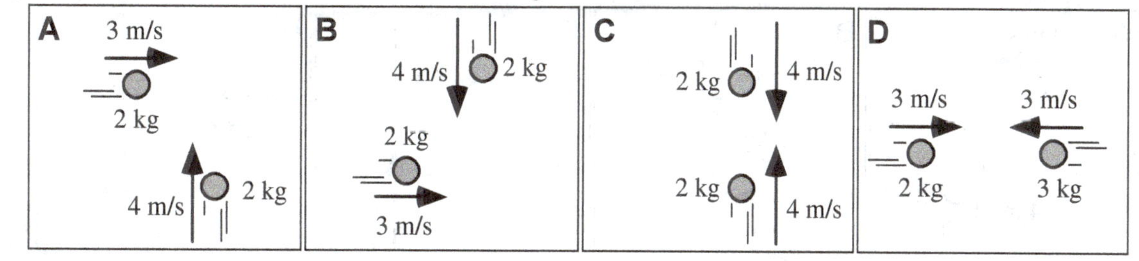

(a) Rank the magnitude of the momentum of the two-ball systems before they collide.

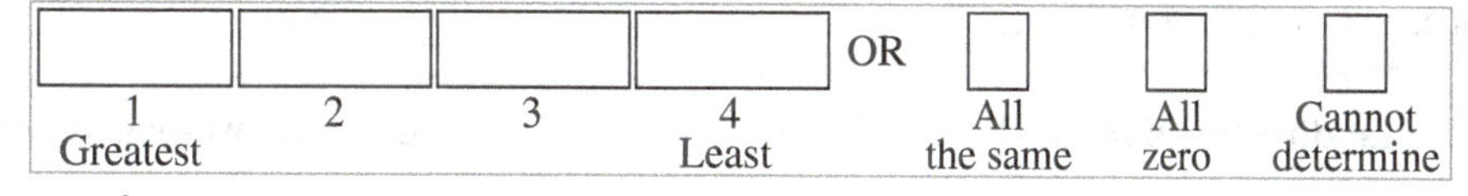

Explain your reasoning.

(b) Rank the magnitude of the momentum of the two-ball systems after they collide if the balls stick together.

OR

1 Greatest 2 3 4 Least All the same All zero Cannot determine

Explain your reasoning.

(c) Rank the magnitude of the momentum of the two-ball systems after they collide elastically (energy conserved).

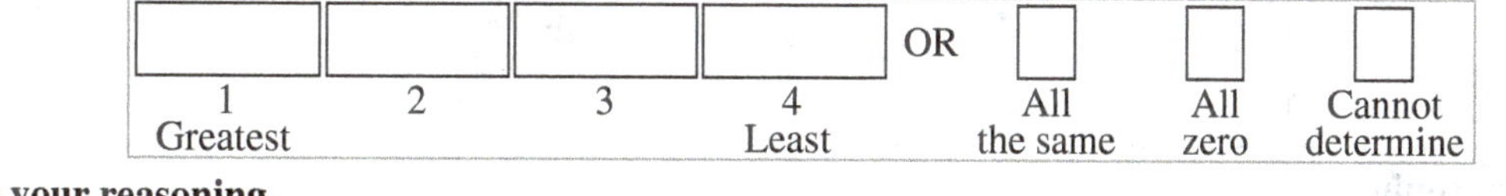

Explain your reasoning.

B5-QRT29: Colliding Ball Systems—Momentum Direction before and after Colliding

In the figures below, two balls traveling in different directions are about to collide. The balls have the same size and shape, but they have different masses and are traveling at different velocities as shown.

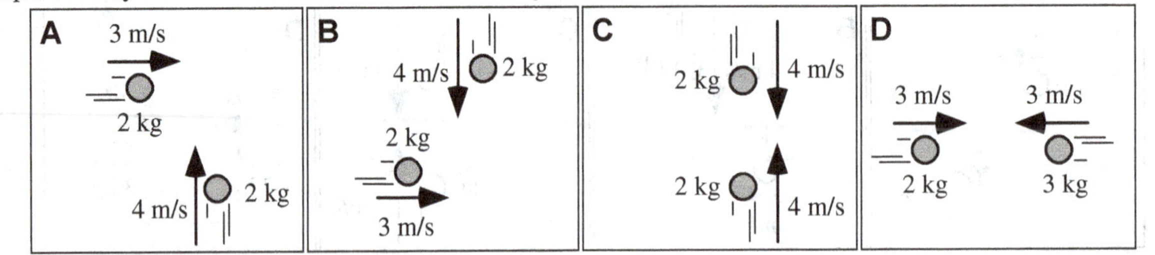

For the questions below, use the directions indicated by the arrows in the direction rosette, or use **J** for no direction, **K** for into the page, or **L** for out of the page.

(a) Identify the closest directional match for the direction of the momentum of the two-ball systems before they collide.

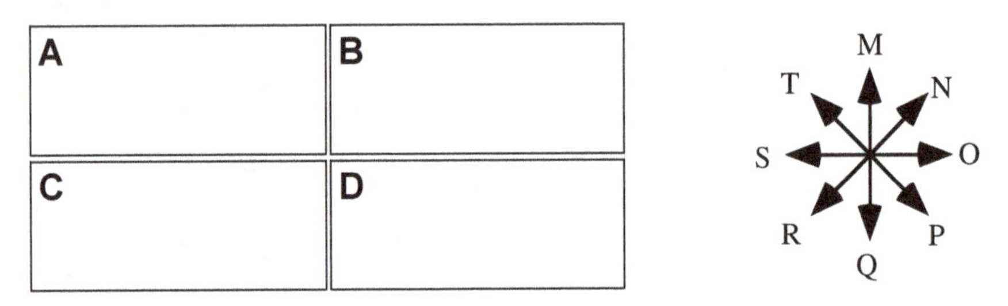

Explain your reasoning.

(b) Identify the closest directional match for the direction of the momentum of the two-ball systems after they collide if the balls stick together.

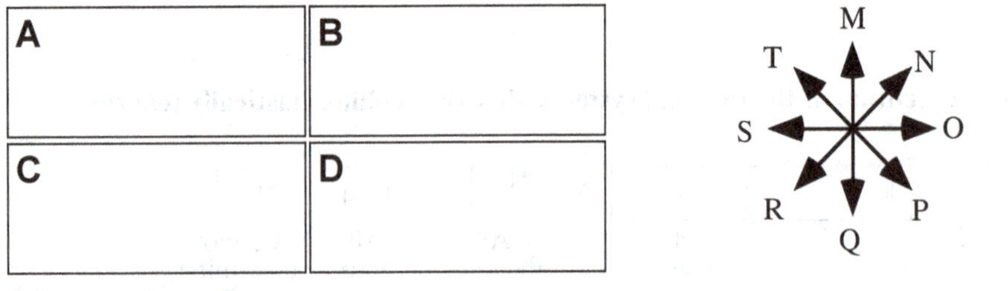

Explain your reasoning.

B6 ROTATION

B6-CRT01: PULLEY AND WEIGHT—ANGULAR VELOCITY-TIME AND ACCELERATION-TIME GRAPHS

A weight is tied to a rope that is wrapped around a pulley. The pulley is initially rotating counterclockwise and is pulling the weight up. The tension in the rope creates a torque on the pulley that opposes this rotation. The weight slows down, stops momentarily, and then moves back downward.

ω

v

Axis of rotation

(a) **Graph of the angular velocity (ω) versus time for the period from the initial instant shown until the weight comes back down to the same height.** Take the initial angular velocity as positive.

(b) **Graph the angular acceleration (α) versus time for the same time period.**

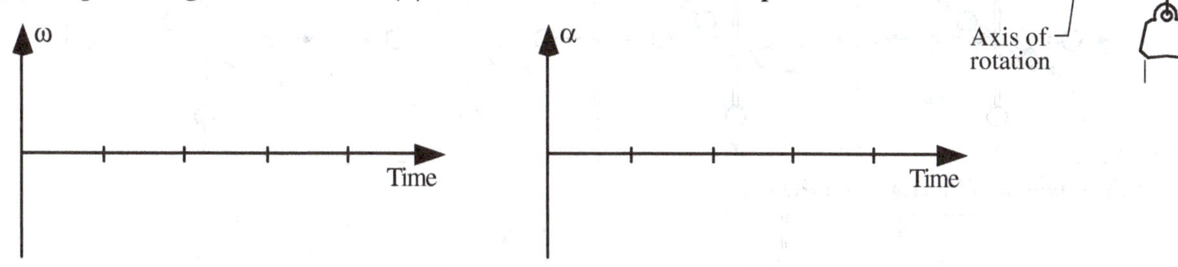

Explain your reasoning.

B6-CRT02: ANGULAR VELOCITY-TIME GRAPH—ANGULAR ACCELERATION-TIME GRAPH

Sketch an angular acceleration versus time graph given the angular velocity versus time graph shown for the same time interval.

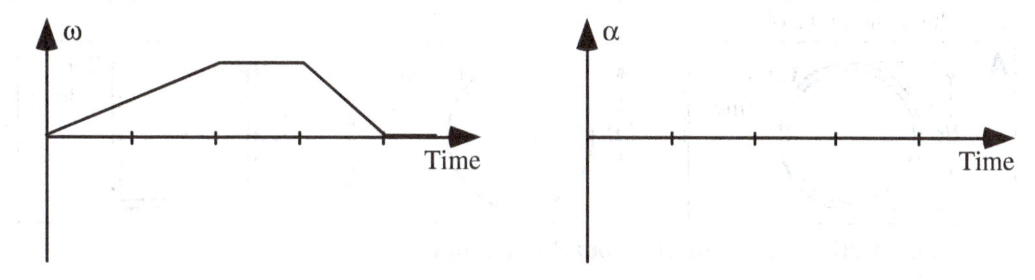

Explain your reasoning.

B6-RT03: Three-Dimensional Point Objects—Moment of Inertia about the x-Axis

Six small brass and aluminum spheres are connected by three stiff, lightweight rods to form a rigid object shaped like a jack. The rods are joined at their centers, are mutually perpendicular, and lie along the axes of the coordinate system shown. All spheres are the same distance from the connection point of the three rods at the origin of the coordinate axis. The brass spheres are shaded in the diagram and are identical. The aluminum spheres are identical, have less mass than the brass spheres, and are unshaded in the diagram. For this problem, ignore the mass of the connecting rods.

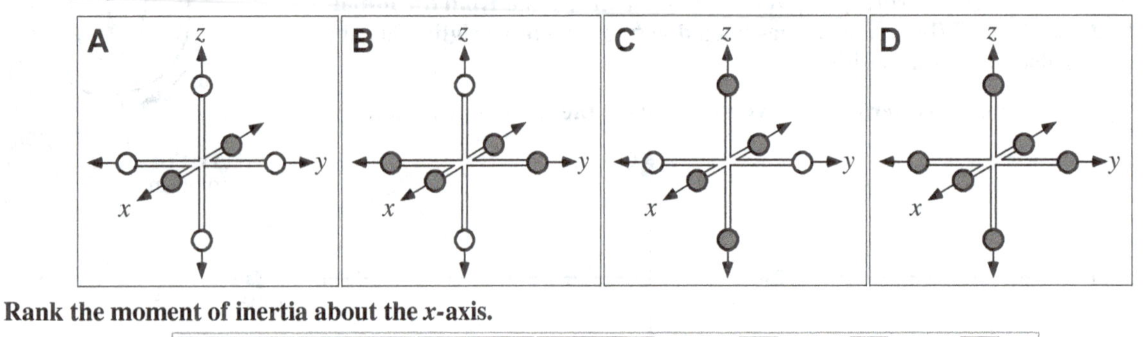

Rank the moment of inertia about the x-axis.

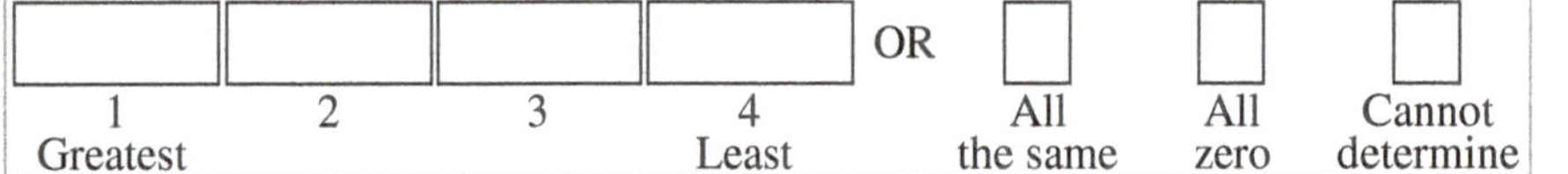

Explain your reasoning.

B6-RT04: Flat Objects—Moment of Inertia Perpendicular to Surface

Three flat objects (circular ring, circular disc, and square loop) have the same mass M and the same outer dimension (circular objects have diameters of $2R$ and the square loop has sides of $2R$). The small circle at the center of each figure represents the axis of rotation for these objects. This axis of rotation passes through the center of mass and is perpendicular to the plane of the objects.

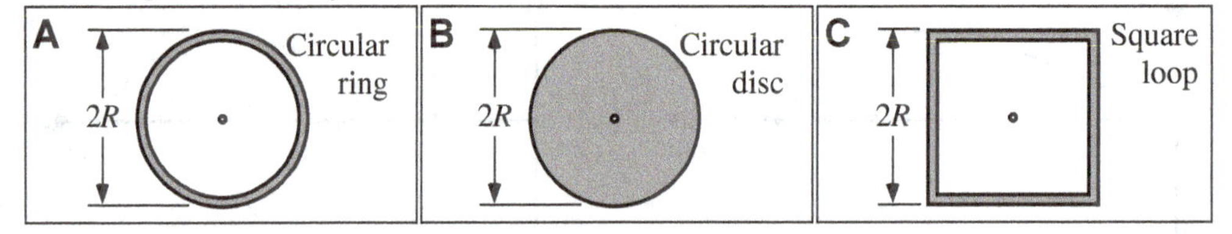

Rank the moment of inertia of these objects about this axis of rotation.

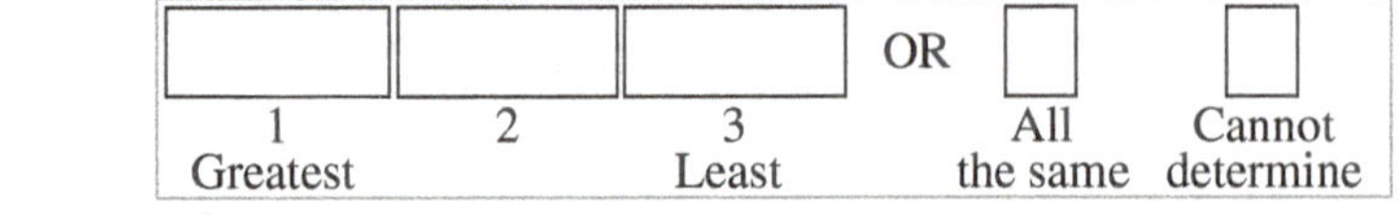

Explain your reasoning.

B6-QRT05: PULLEYS WITH DIFFERENT RADII—ROTATION AND TORQUE

A wheel is composed of two pulleys with different radii (labeled *a* and *b*) that are attached to one another so that they rotate together. Each pulley has a string wrapped around it with a weight hanging from it as shown. The pulleys rotate about a horizontal axis at the center. When the wheel is released it is found to have an angular acceleration that is directed out of the page or counterclockwise.

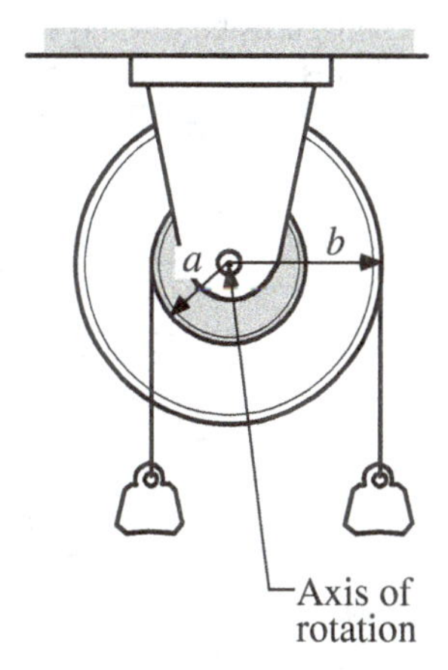

(a) Is the wheel going to rotate (i) *clockwise,* (ii) *counterclockwise,* or (iii) *none*?

______

Explain your reasoning.

(b) Is the direction of the net torque on the pulley wheels (i) *clockwise,* (ii) *counterclockwise,* or (iii) *none*? _____

How do you know?

(c) How do the masses of the two weights compare?

Explain your reasoning.

B6-RT06: Spheres Rolling—Radius

The figures below show hollow spheres (not drawn to scale) that are rolling at a constant rate without slipping. The spheres all have the same mass, but their radii as well as their linear and angular speeds vary.

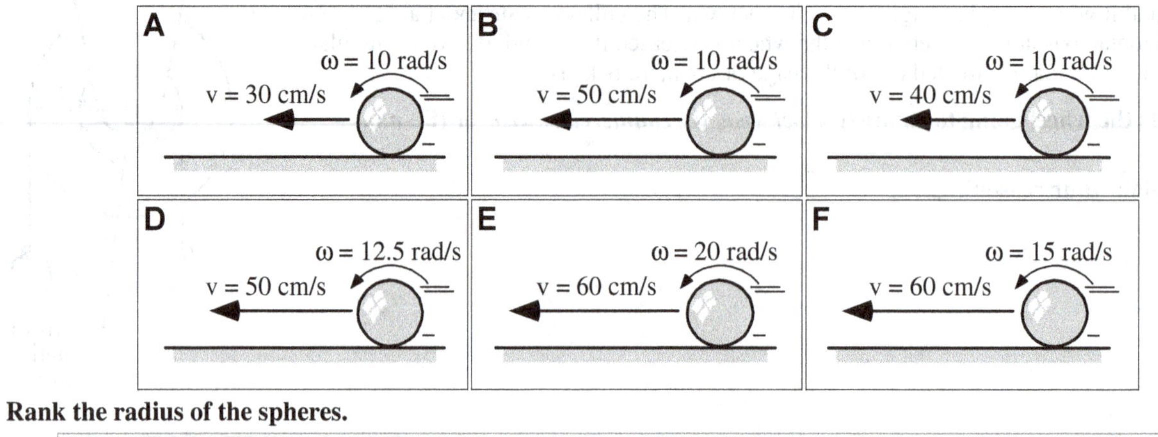

Rank the radius of the spheres.

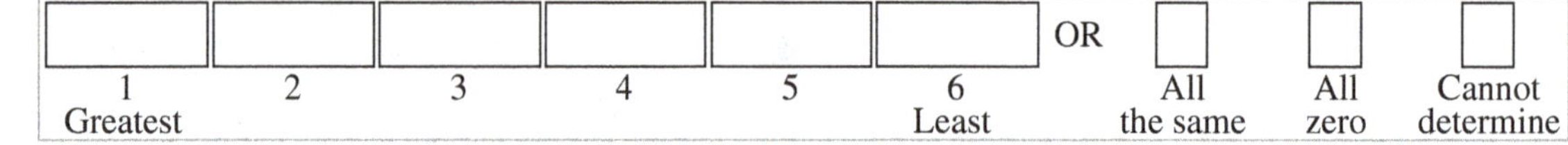

Explain your reasoning.

B6-QRT07: Three Equal Forces Applied to a Rectangle—Net Torque Direction

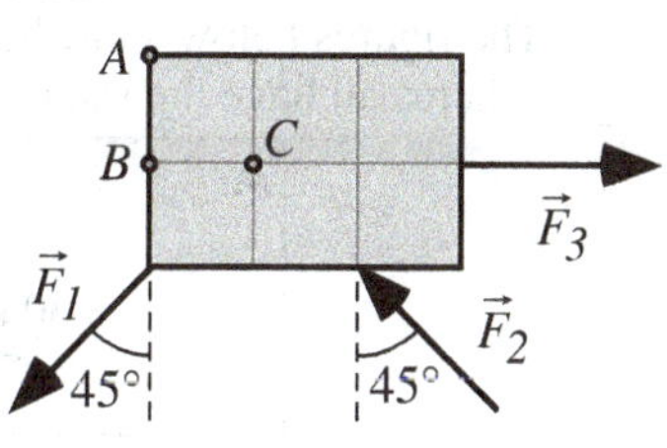

Three forces of equal magnitude are applied to a 3-m by 2-m rectangle. Forces $\vec{F}_1$ and $\vec{F}_2$ act at 45° angles to the vertical as shown, while $\vec{F}_3$ acts horizontally.

(a) Is the net torque about point *A* (i) *clockwise*, (ii) *counterclockwise*, or (iii) *zero*? _____

Explain your reasoning.

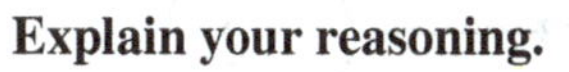

(b) Is the net torque about point *B* (i) *clockwise*, (ii) *counterclockwise*, or (iii) *zero*? _____

Explain your reasoning.

(c) Is the net torque about point *C* (i) *clockwise*, (ii) *counterclockwise*, or (iii) *zero*? _____

Explain how you determined your answer.

B6-RT08: Spheres Rolling—Rotational Kinetic Energy

The figures below show hollow spheres (not drawn to scale) that are rolling at a constant rate without slipping. The spheres all have the same mass, but their radii as well as their linear and angular speeds vary.

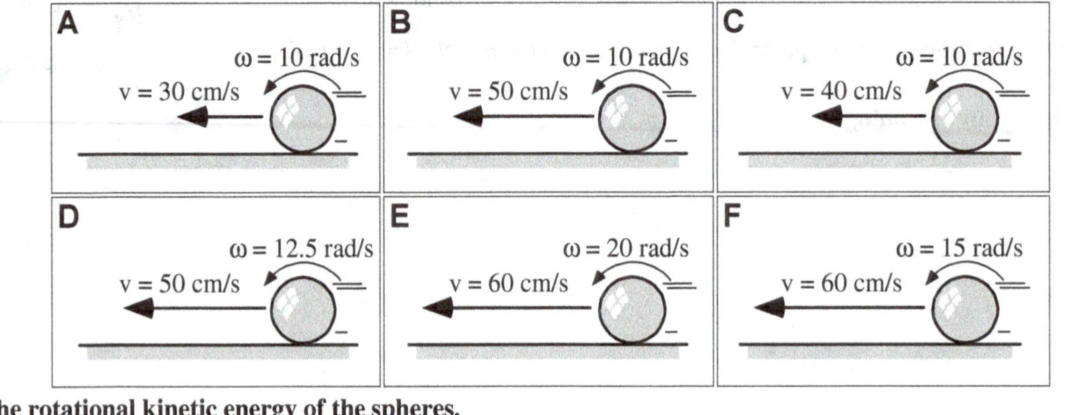

Rank the rotational kinetic energy of the spheres.

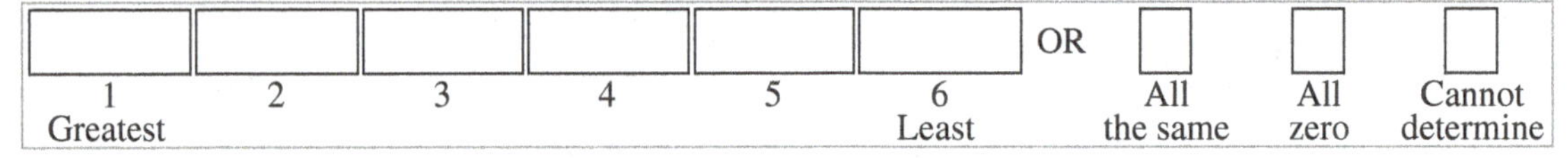

Explain your reasoning.

B6-QRT09: THREE FORCES APPLIED TO A RECTANGLE—TORQUE DIRECTION

Three forces of equal magnitude are applied to a 3-m by 2-m rectangle. Forces $\vec{F}_1$ and $\vec{F}_2$ act at 45° angles to the vertical as shown, while $\vec{F}_3$ acts horizontally.

(a) Is the torque by $\vec{F}_1$ about point *A* (i) *clockwise*, (ii) *counterclockwise*, or (iii) *zero*? _____

Explain your reasoning.

(b) Is the torque by $\vec{F}_1$ about point *B* (i) *clockwise*, (ii) *counterclockwise*, or (iii) *zero*? _____

Explain your reasoning.

(c) Is the torque by $\vec{F}_1$ about point *C* (i) *clockwise*, (ii) *counterclockwise*, or (iii) *zero*? _____

Explain your reasoning.

(d) Is the torque by $\vec{F}_2$ about point *A* (i) *clockwise*, (ii) *counterclockwise*, or (iii) *zero*? _____

Explain your reasoning.

(e) Is the torque by $\vec{F}_2$ about point *B* (i) *clockwise*, (ii) *counterclockwise*, or (iii) *zero*? _____

Explain your reasoning.

(f) Is the torque by $\vec{F}_2$ about point *C* (i) *clockwise*, (ii) *counterclockwise*, or (iii) *zero*? _____

Explain your reasoning.

(g) Is the torque by $\vec{F}_3$ about point *A* (i) *clockwise*, (ii) *counterclockwise*, or (iii) *zero*? _____

Explain your reasoning.

(h) Is the torque by $\vec{F}_3$ about point *B* (i) *clockwise*, (ii) *counterclockwise*, or (iii) *zero*? _____

Explain your reasoning.

(i) Is the torque by $\vec{F}_3$ about point *C* (i) *clockwise*, (ii) *counterclockwise*, or (iii) *zero*? _____

Explain your reasoning.

B6-CT10: Fishing Rod—Weight of Two Pieces

An angler balances a fishing rod on her finger as shown.

If she were to cut the rod along the dashed line, would the weight of the piece on the left-hand side be (i) *greater than*, (ii) *less than*, or (iii) *equal to* the weight of the piece on the right-hand side? _____

Explain your reasoning.

B6-RT11: Suspended Signs—Torque

Signs are suspended from equal-length rods on the side of a building. For each case, the mass of the sign compared to the mass of the rod is small and can be ignored. The mass of the sign is given in each figure. In Cases B and D, the rod is horizontal; in the other cases, the angle that the rod makes with the vertical is given.

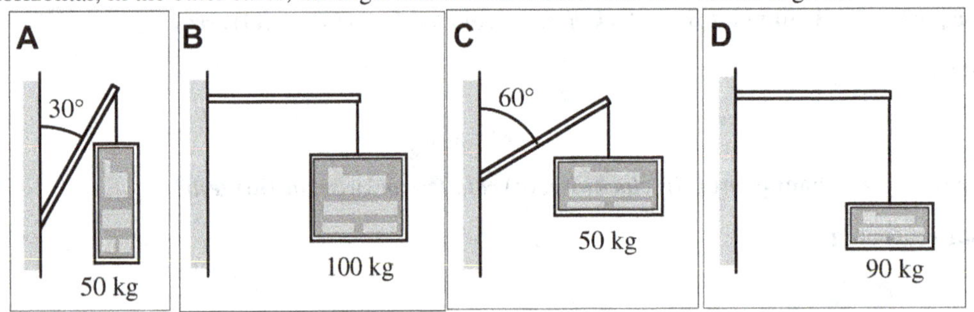

Rank the magnitude of the torque the signs exert about the point at which the rod is attached to the side of the building.

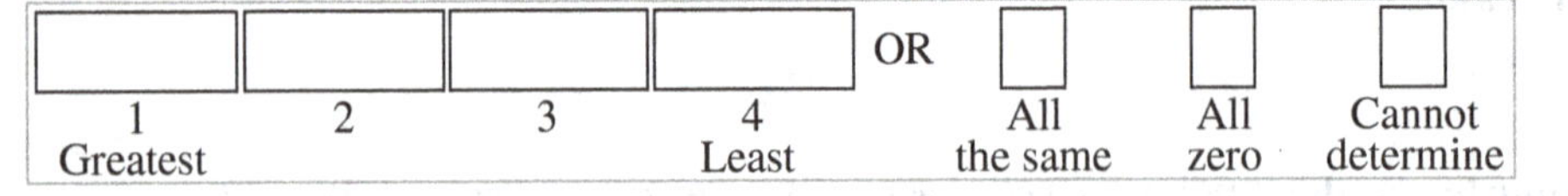

Explain your reasoning.

B6-RT12: FOUR FORCES ACTING ON A HEXAGON—TORQUE ABOUT CENTER

Four forces act on a plywood hexagon as shown in the diagram. The sides of the hexagon each have a length of 1 m.

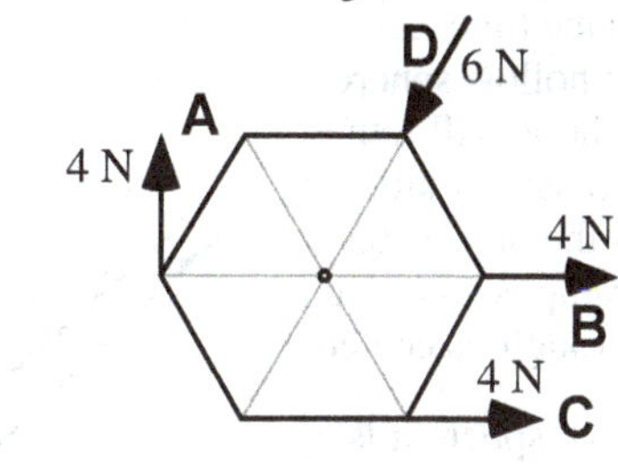

Rank the magnitude of the torque applied about the center of the hexagon by each force.

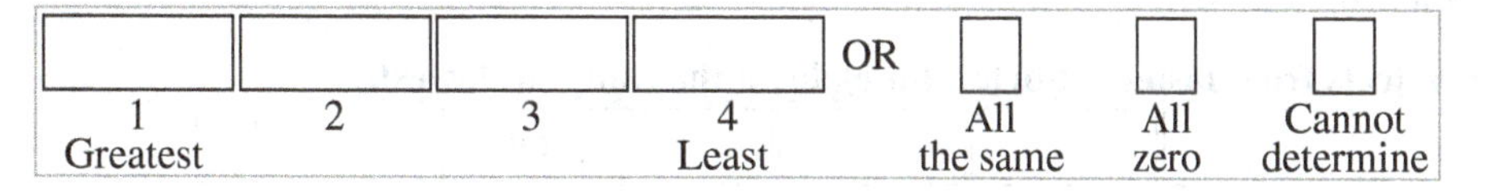

Explain your reasoning.

B6-QRT13: BALANCE BEAM—MOTION AFTER RELEASE

Five identical keys are suspended from a balance, which is held horizontally as shown. The two keys on the left are attached to the balance 6 cm from the pivot and the three keys on the right are attached 5 cm from the pivot.

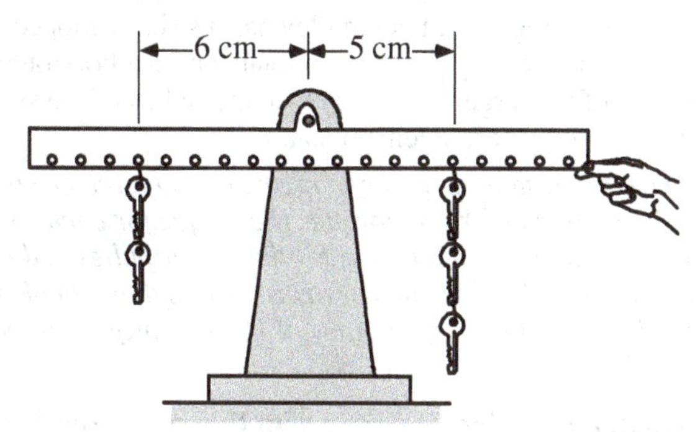

What will happen when the person lets go of the balance beam?

Explain.

B6-RT14: Rolling Objects Released from Rest—Time Down Ramp

Four objects are placed in a row at the same height near the top of a ramp and are released from rest at the same time. The objects are (i) a 1-kg solid sphere; (ii) a 1-kg hollow sphere; (iii) a 2-kg solid sphere; and (iv) a 1-kg thin hoop. All four objects have the same diameter, and the hoop has a width that is one-quarter its diameter. The time it takes the objects to reach the finish line near the bottom of the ramp is recorded. The moment of inertia for an axis passing through its center of mass for a solid sphere is $\frac{2}{5}MR^2$; for a hollow sphere it is $\frac{2}{3}MR^2$; and for a hoop it is MR^2.

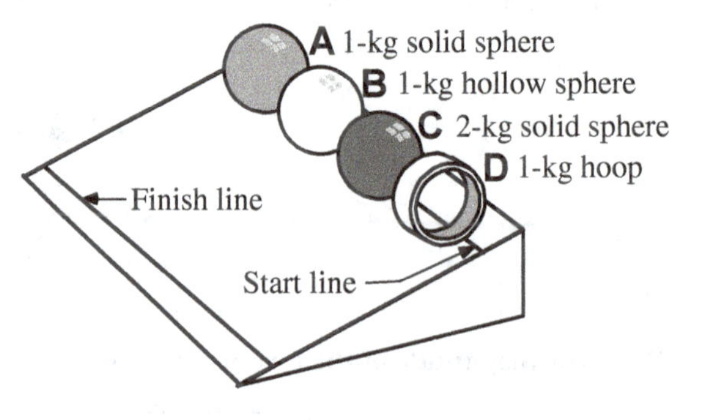

Rank the four objects from fastest (shortest time) down the ramp to slowest.

Explain your reasoning.

B6-WWT15: Pulley with Hanging Weights—Angular Acceleration

Two pulleys with different radii (labeled a and b) are attached to one another so that they rotate together. Each pulley has a string wrapped around it with a weight hanging from it. The pulleys are free to rotate about a horizontal axis through the center. The radius of the larger pulley is twice the radius of the smaller one ($b = 2a$). A student describing this arrangement states:

"The larger mass is going to create a counterclockwise torque and the smaller mass will create a clockwise torque. The torque for each will be the weight times the radius, and since the radius for the larger pulley is double the radius of the smaller, and the weight of the heavier mass is less than double the weight of the smaller one, the larger pulley is going to win. The net torque will be clockwise, and so the angular acceleration will be clockwise."

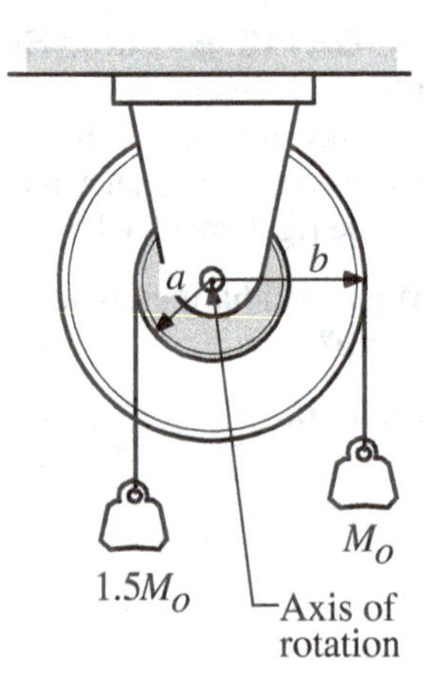

What, if anything, is wrong with this contention? If something is wrong, explain how to correct it. If this contention is correct, explain why.

B6-RT16: TILTED PIVOTED RODS WITH VARIOUS LOADS—FORCE TO HOLD RODS

Six identical massless rods are supported by a fulcrum and are tilted at the same angle to the horizontal. A mass is suspended from the left end of the rod, and the rods are held motionless by a downward force on the right end. Each rod is marked at 1-m intervals.

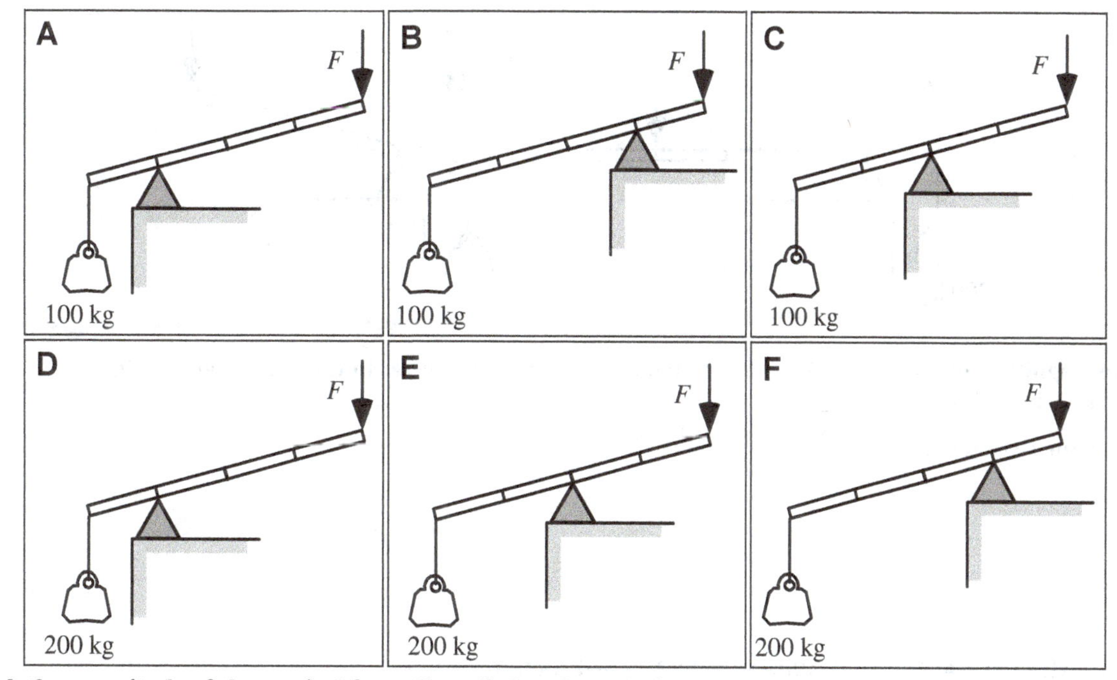

Rank the magnitude of the vertical force *F* applied to the end of the rod.

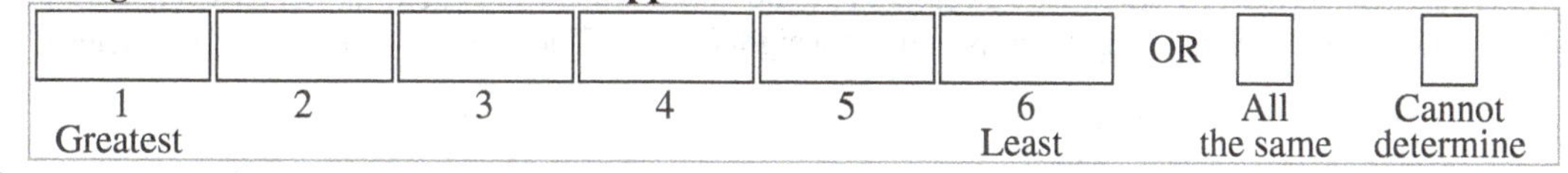

Explain your reasoning.

B6-CT17: SPECIAL ROD—MOMENT OF INERTIA

A rod is made of three segments of equal length with different masses. The total mass of the rod is $6m$.

m | $2m$ | $3m$

Will the moment of inertia of the rod be (i) *greater* about the left end, (ii) *greater* about the right end, or (iii) *the same* about both ends? _____

Explain your reasoning.

B6-CT18: Tilted Pivoted Rods with Various Loads—Force to Hold Rods

In both cases, a massless rod is supported by a fulcrum, and a 200-kg hanging mass is suspended from the left end of the rod by a cable. A downward force *F* keeps the rod at rest. The rod in Case A is 50 cm long, and the rod in Case B is 40 cm long. (Each rod is marked at 10-cm intervals.)

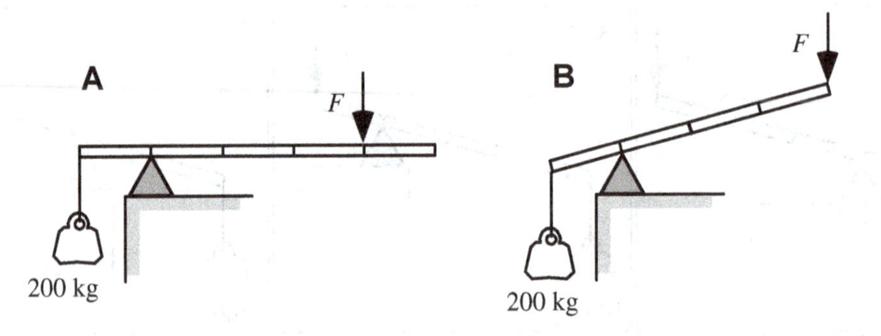

Will the magnitude of the vertical force *F* exerted on the rod be (i) *greater* in Case A, (ii) *greater* in Case B, or (iii) *the same* in both cases? _____

Explain your reasoning.

B6-RT19: Horizontal Pivoted Rods with Loads I—Force to Hold

A 2-m long massless rod supports a 12-Newton weight. The left end of each rod is held in place by a frictionless pin. In each case, a vertical force *F* is holding the rods and the weights at rest. The rods are marked at half-meter intervals.

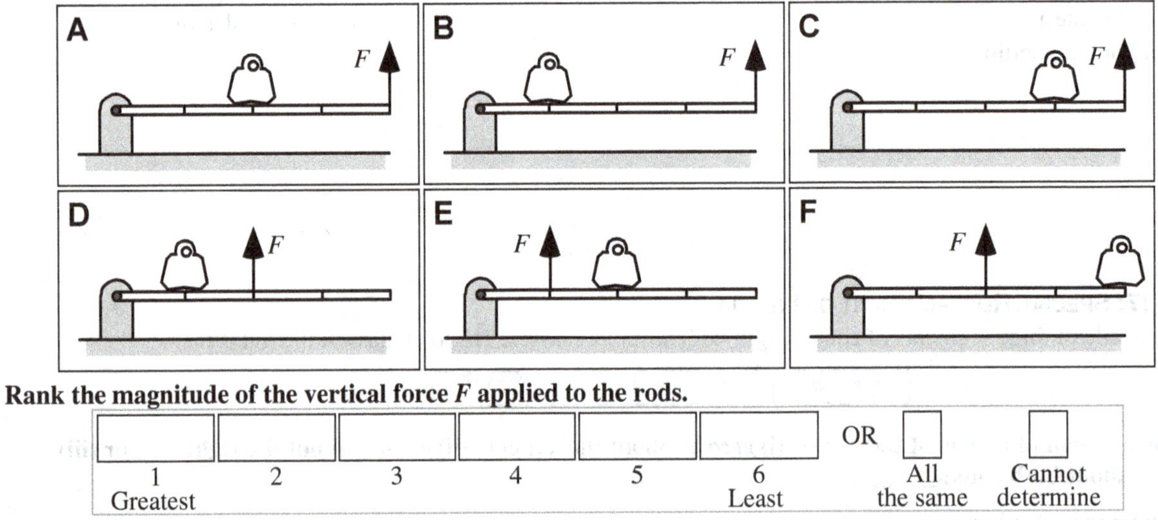

Rank the magnitude of the vertical force *F* applied to the rods.

						OR		
1 Greatest	2	3	4	5	6 Least		All the same	Cannot determine

Explain your reasoning.

B6-LMCT20: HORIZONTAL PIVOTED BOARD WITH LOAD II—FORCE TO HOLD BOARD

A 100-N weight is placed on a massless board a distance L_1 to the left of frictionless pin. A vertical downward force F is applied to the other side of the board a distance of L_2 from the pin as shown. The system is at rest.

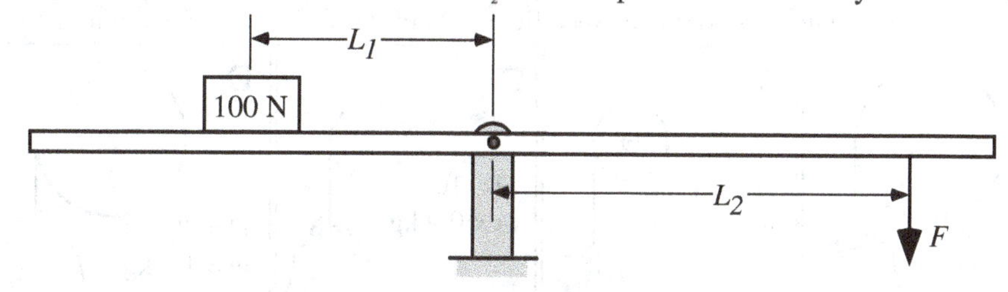

Identify from choices (i)–(v) how each change described below will affect the magnitude of the applied force (*F*) on the right side of the board needed to keep the system in equilibrium.

Compared to the case above, this change will:

(i) ***increase*** the magnitude of the support force (F) on the board.
(ii) ***decrease*** the magnitude of the support force (F) on the board but not to zero.
(iii) ***decrease*** the magnitude of the support force (F) on the board **to zero**.
(iv) ***have no effect*** on the magnitude of the support force (F) on the board.
(v) *have an* ***indeterminate*** effect on the magnitude of the support force (F) on the board.

Each of these modifications is the only change to the initial situation shown in the diagram above.

(a) The 100-N weight is moved to a position closer to the pin. _____
Explain your reasoning.

(b) The support force (*F*) is moved to a position closer to the pin. _____
Explain your reasoning.

(c) The weight is decreased to 50 N. _____
Explain your reasoning.

(d) The support force (F) is moved to the right end of the board. _____
Explain your reasoning.

(e) The board is made longer but the support force (F) remains at the same location. _____
Explain your reasoning.

(f) The 100-N weight and the support force (F) are both moved to positions closer to the pin. _____
Explain your reasoning.

B6-RT21: Hanging Weights on Fixed Disks—Torque

Vertically oriented circular disks have strings wrapped around them. The other ends of the strings are attached to hanging masses. The diameters of the disks, the masses of the disks, and the masses of the hanging masses are given. The disks are fixed and are *not* free to rotate. Specific values of the variables are given in the figures.

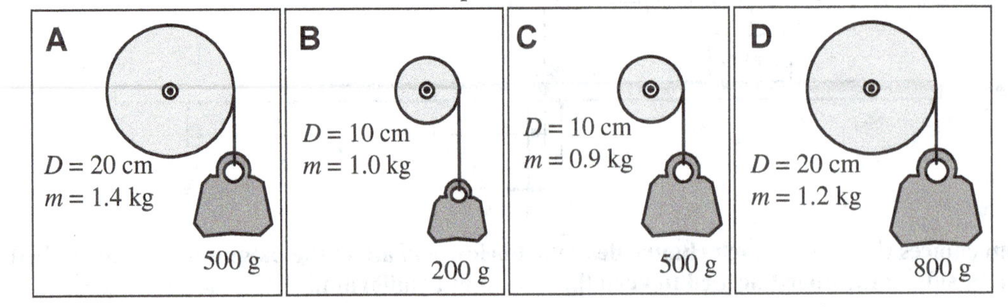

Rank the magnitudes of the torques exerted by the strings about the center of the disks.

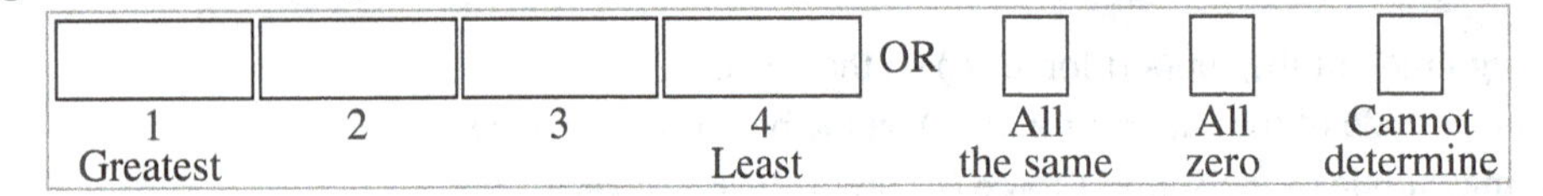

Explain your reasoning.

B6-RT22: Systems of Point Masses—Difficult to Rotate

Each of the ten point masses in each case is identical. The solid line in each figure represents an axis about which the masses are going to be rotated. The point masses are fixed together so that they all maintain the arrangements shown while being rotated.

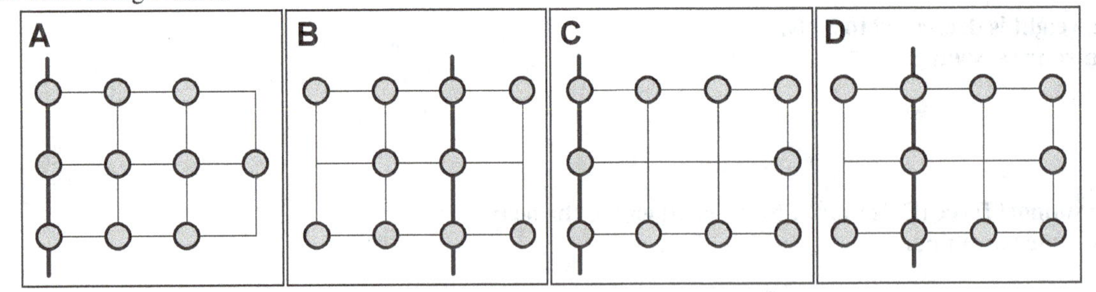

Rank these arrangements on how hard it will be to start the systems rotating.

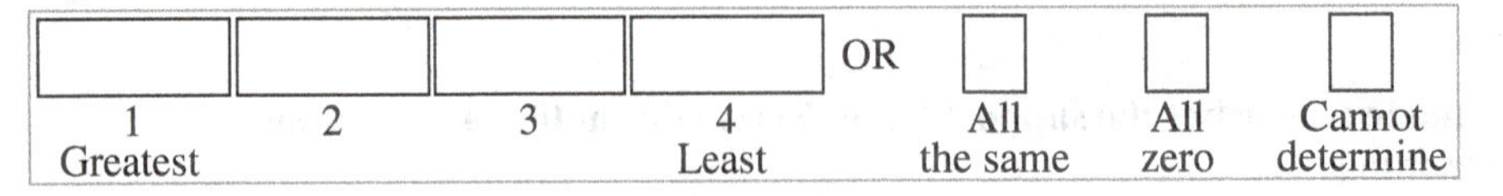

Explain your reasoning.

B6-RT23: Meter Stick with Hanging Mass I—Difficulty Holding

A student is holding a meter stick by one end. A 1,000 g mass is hung on the meter sticks. All of the meter sticks are identical, but the distance along the meter stick at which the 1,000 g mass is hung and the angles at which the student holds the meter stick vary. (Ignore the mass of the meter stick.)

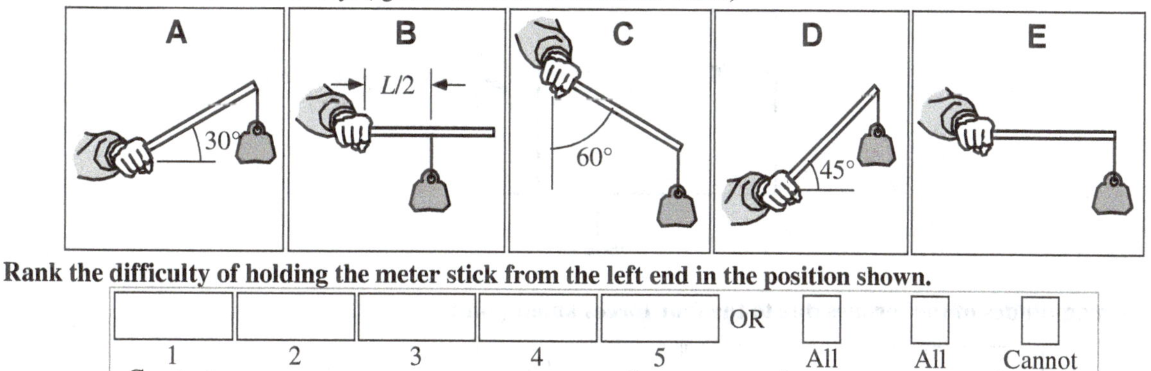

Rank the difficulty of holding the meter stick from the left end in the position shown.

					OR			
1 Greatest	2	3	4	5 Least		All the same	All zero	Cannot determine

Explain your reasoning.

B6-CT24: Horizontal Meter Stick with Two Hanging Masses—Torque

In each case, a student is holding a meter stick horizontally. Each meter stick has a mass attached at the 50 cm mark and another at the 100 cm mark. The meter sticks are identical, and the specific values and locations are given in the figures.

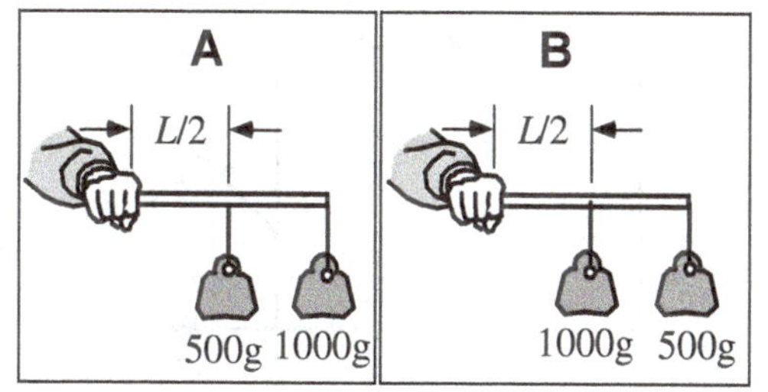

Is the magnitude of the torque by the student on the meter stick (i) *greater* in Case A, (ii) *greater* in Case B, or (iii) *the same* in both cases? _____

Explain your reasoning.

B6-RT25: Four Forces Acting on a Piece of Plywood—Torque

Four 4-Newton forces (*A–D*) act on a 3-m by 4-m piece of plywood as shown.

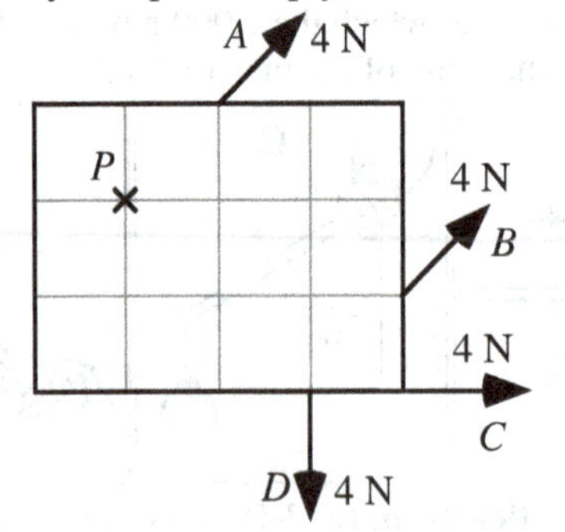

Rank the magnitudes of the torques due to the four forces about point *P*.

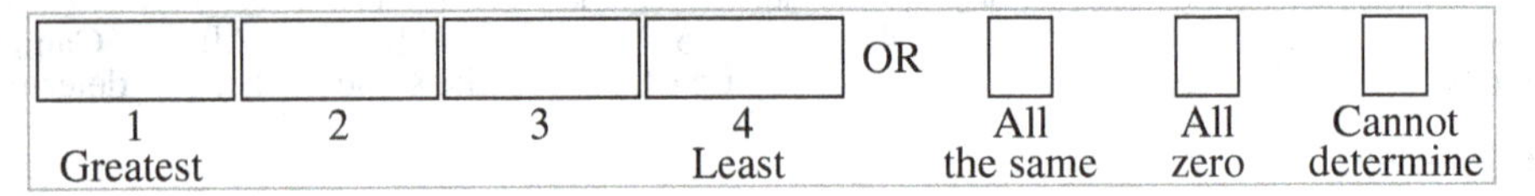

Explain your reasoning.

B6-QRT26: Four Forces Acting on a Piece of Plywood—Rotation Direction

Four 4-Newton forces (*A–D*) act on a 3 m by 4 m piece of plywood that has a pivot point at *P*.

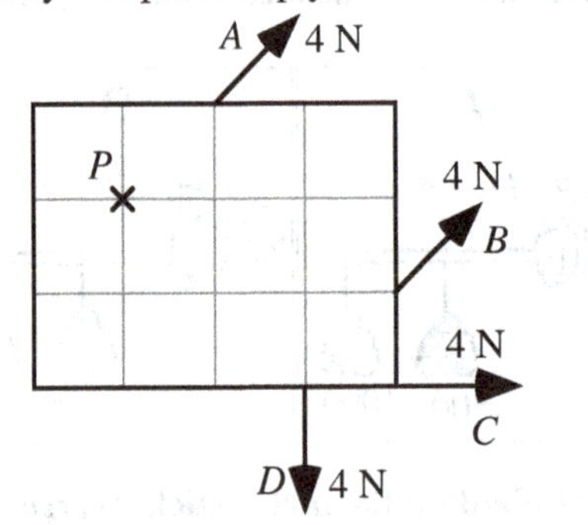

Will the plywood rotate about the pivot point *P* (i) *clockwise*, (ii) *counterclockwise*, or (iii) *not at all*? _____

Explain your reasoning.

B6-BCT27: HOOP ROLLING UP A RAMP—ROTATIONAL ENERGY BAR CHART

A thin hoop or ring with a radius of 2 m is moving so that its center of mass is initially moving at 20 m/s while also rolling without slipping at 10 rad/s along a horizontal surface. It rolls up an incline, coming to rest as shown.

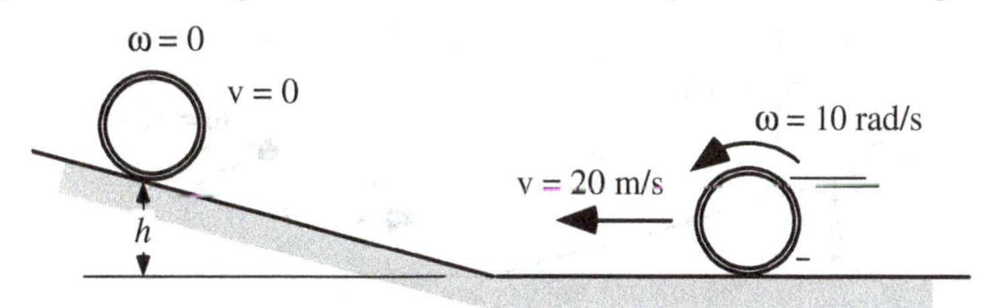

Complete the qualitative energy bar chart below for the earth-hoop system for the time between when the hoop is rolling on the horizontal surface and when it has rolled up the ramp and is momentarily at rest. Put the zero point for the gravitational potential energy at the height of the center of the hoop when it is rolling on the horizontal surface.

Initial system energy				During	Final system energy			
KE_{trans}	KE_{rot}	PE_{grav}	PE_{spring}	W_{ext}	KE_{trans}	KE_{rot}	PE_{grav}	PE_{spring}

0

Bar chart key	
KE_{trans}	Translational kinetic energy
KE_{rot}	Rotational kinetic energy
PE_{grav}	Gravitational potential energy
PE_{spring}	Spring potential energy
W_{ext}	Work done by external forces

Use $g = 10\ m/s^2$ for simplicity

Explain your reasoning.

B6-BCT28: Solid Disk Rolling up a Ramp—Rotational Energy Bar Chart

A solid disk is initially rolling without slipping along a flat, level surface. It then rolls up an incline, coming momentarily to rest as shown.

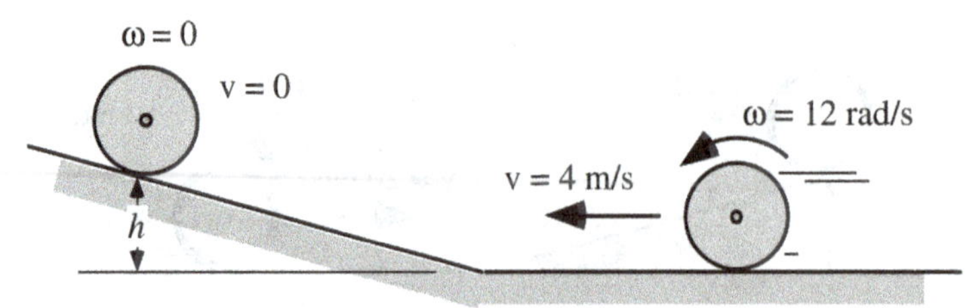

Complete the qualitative energy bar chart below for the earth-disk system for the time between when the disk is rolling on the horizontal and when it has rolled up the ramp and is momentarily at rest. Put the zero point for the gravitational potential energy at the height of the center of the hoop when it is rolling on the horizontal surface.

Initial system energy — During — Final system energy

KE_{trans} KE_{rot} PE_{grav} PE_{spring} W_{ext} KE_{trans} KE_{rot} PE_{grav} PE_{spring}

0

Bar chart key	
KE_{trans}	Translational kinetic energy
KE_{rot}	Rotational kinetic energy
PE_{grav}	Gravitational potential energy
PE_{spring}	Spring potential energy
W_{ext}	Work done by external forces

Use $g = 10\ m/s^2$ for simplicity

Explain your reasoning.

B6-QRT29: SOLID SPHERE ROLLING ALONG A TRACK—LOCATION AT HIGHEST POINT

A solid sphere rolls without slipping along a track shaped as shown at right. It starts from rest at point *A* and is moving vertically when it leaves the track at point *B*.

At its highest point while in the air, will the sphere be (a) *above*, (b) *below*, or (c) *at the same height as* point *A*? _____

Explain your reasoning.

B6-RT30: MOVING DOWN A RAMP—MAXIMUM HEIGHT ON THE OTHER SIDE OF A RAMP

In each case, a 1-kg object is released from rest on a ramp at a height of 2 m from the bottom. All of the spheres roll without slipping, and the blocks slide without friction.

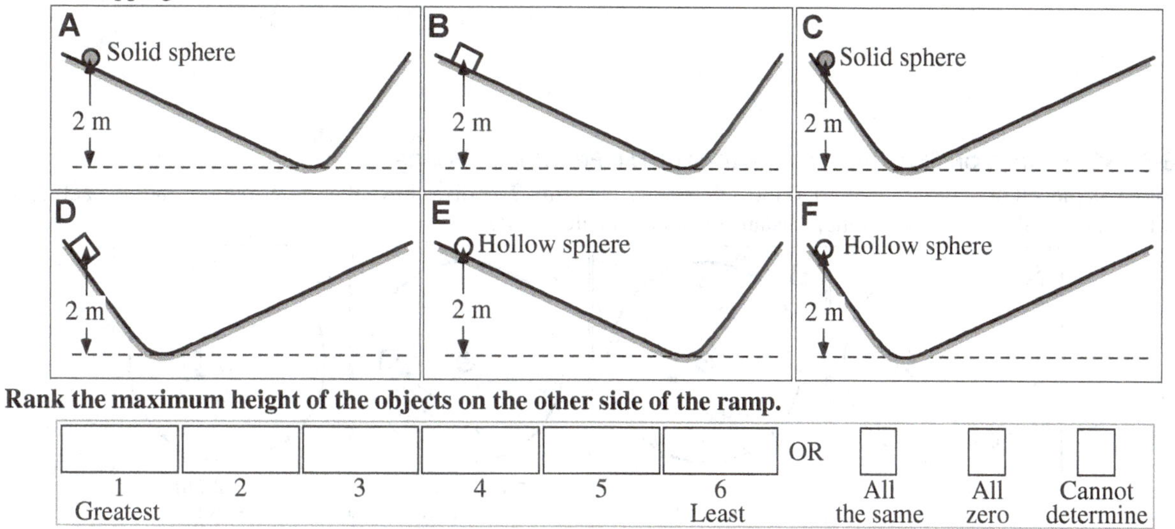

Rank the maximum height of the objects on the other side of the ramp.

____	____	____	____	____	____	OR	☐	☐	☐
1 Greatest	2	3	4	5	6 Least		All the same	All zero	Cannot determine

Explain your reasoning.

B6-RT31: Objects Moving down Ramps—Speed at Bottom

In each case, a 1-kg object is released from rest on a ramp at a height of 2 m from the bottom. All of the spheres roll without slipping, and the blocks slide without friction. The ramps are identical in Cases A and C. The ramps in Cases B and D are identical and are not as steep as the others.

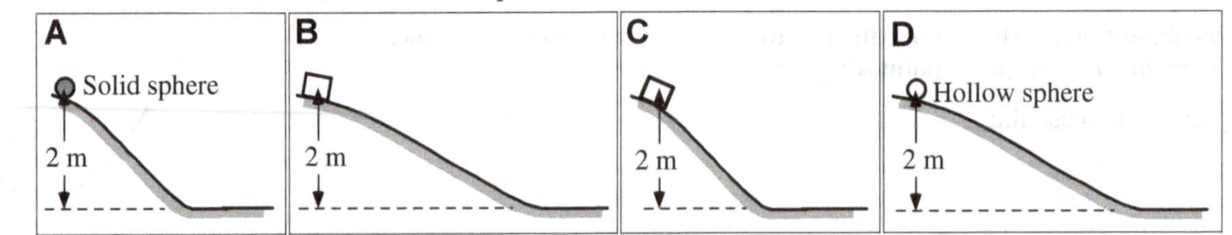

Rank the speed of the objects when they reach the horizontal surface at the bottom of the ramp.

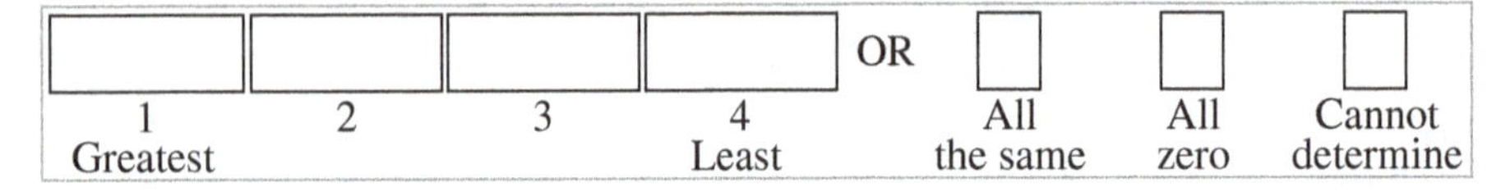

Explain your reasoning.

B6-RT32: Blocks on Rotating Disc—Horizontal Frictional Force

A block is placed on a rotating disc and moves in a circular path. The discs have the same rotation rate in each case, but the masses of the blocks and their distance from the center varies.

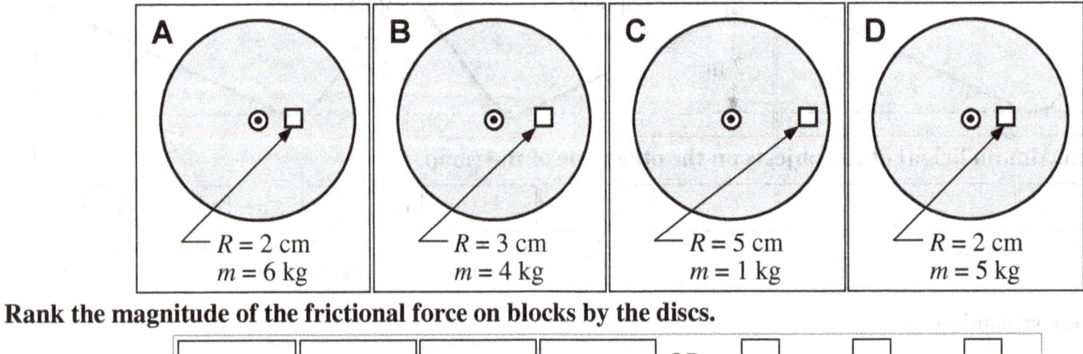

Rank the magnitude of the frictional force on blocks by the discs.

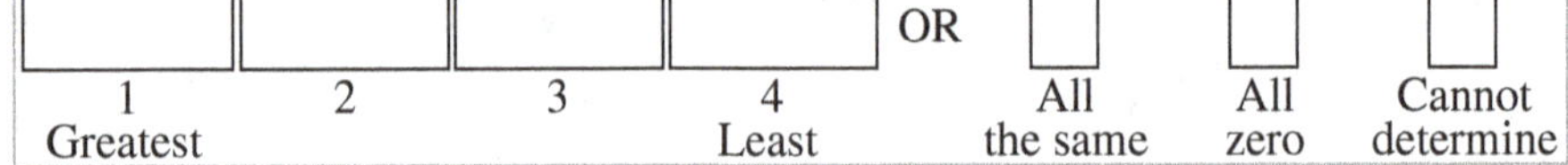

Explain your reasoning.

B7 OSCILLATORY MOTION

B7-RT01: MASS ON HORIZONTAL SPRING SYSTEMS I—OSCILLATION FREQUENCY

A block rests on a frictionless surface and is attached to the end of a spring. The other end of the spring is attached to a wall. Four block–spring systems are considered. The springs are stretched to the right by the distances shown in the figures and then released from rest. The blocks oscillate back and forth. The mass and force constant of the spring are given for each case.

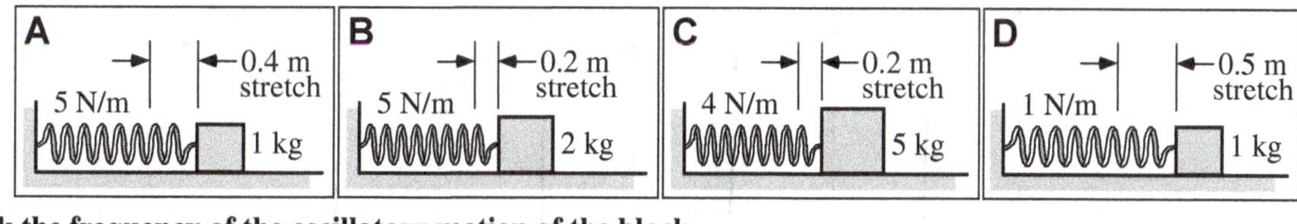

Rank the frequency of the oscillatory motion of the block.

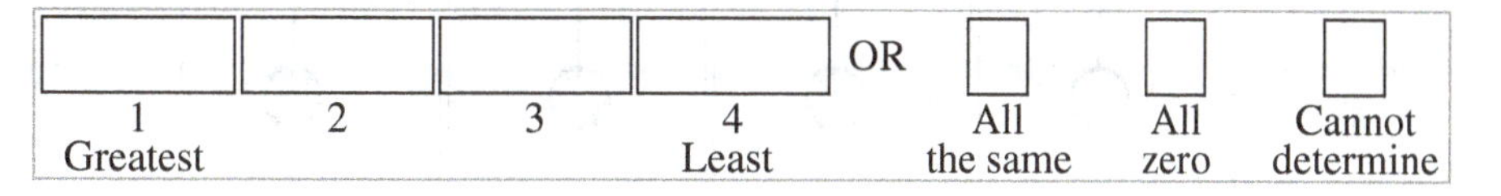

Explain your reasoning.

B7-RT02: SWINGING SIMPLE PENDULA—OSCILLATION FREQUENCY

The simple pendulum shown in Case A consists of a mass M attached to a massless string of length L. If the mass is pulled to one side a small distance and released, it will swing back and forth. Cases B, C, and D are variations of this system.

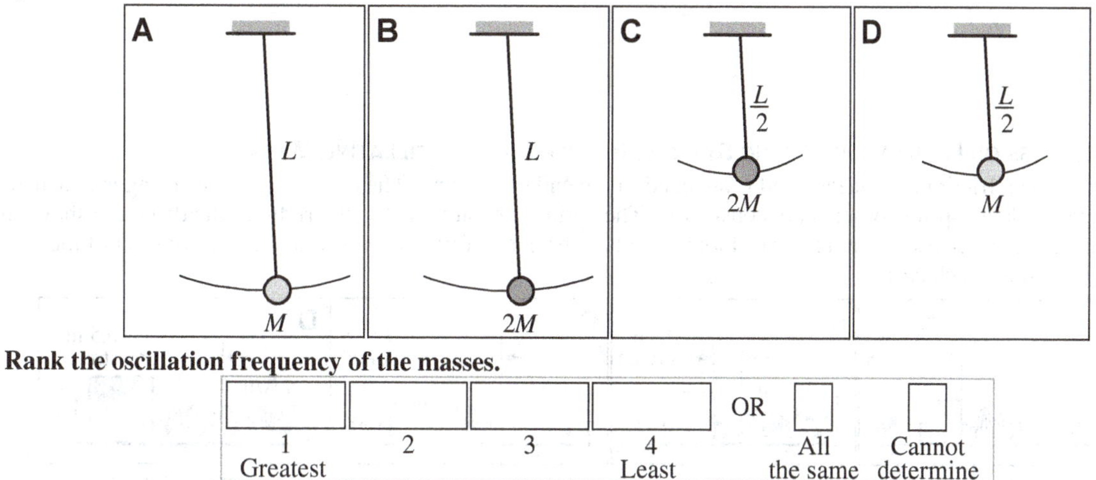

Rank the oscillation frequency of the masses.

				OR		
1 Greatest	2	3	4 Least		All the same	Cannot determine

Explain your reasoning.

B7-RT03: Swinging Sphere on Long Strings—Time for One Swing

Metal spheres are hung on the ends of long strings. The spheres have been pulled to the side and released so that they are swinging back and forth. The mass of the sphere and the frequency of the swing are given in each case.

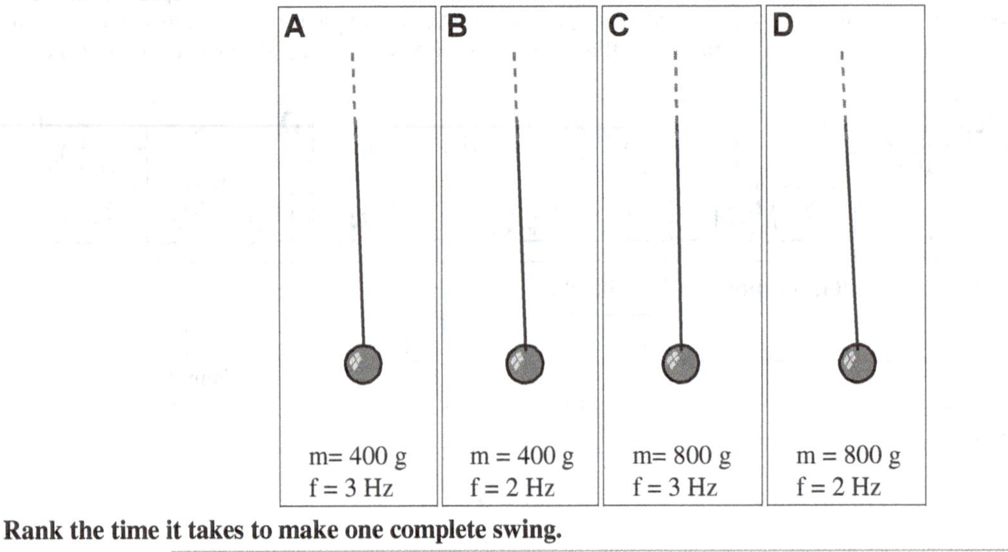

Rank the time it takes to make one complete swing.

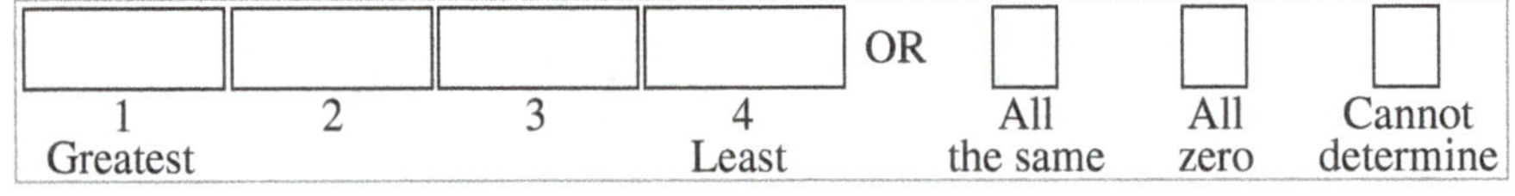

Explain your reasoning.

B7-RT04: Mass on Horizontal Spring Systems II—Period of Oscillating Mass

A block rests on a frictionless surface and is attached to the end of a spring. The other end of the spring is attached to a wall. Four block–spring systems are considered. The springs are stretched to the right by the distances shown in the figures and then released from rest. The blocks oscillate back and forth. The mass and force constant of the spring are given for each case.

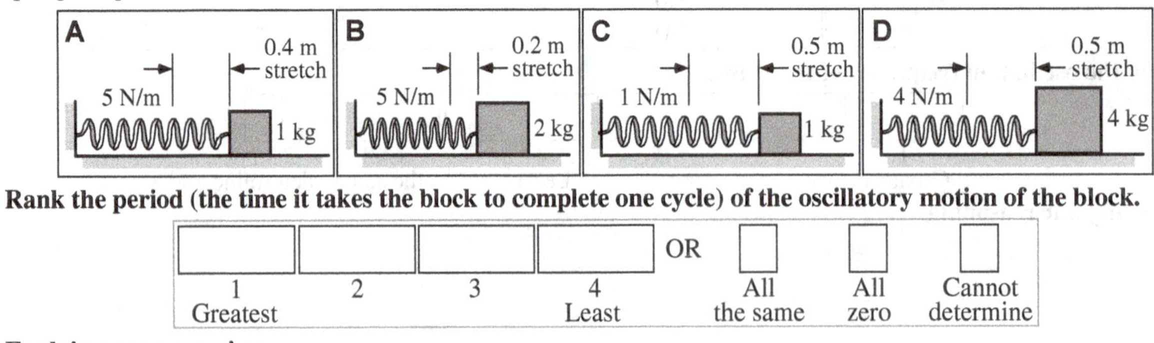

Rank the period (the time it takes the block to complete one cycle) of the oscillatory motion of the block.

				OR			
1 Greatest	2	3	4 Least		All the same	All zero	Cannot determine

Explain your reasoning.

B7-QRT05: POSITION-TIME GRAPH OF A CART ATTACHED TO A SPRING—MASS AND PERIOD

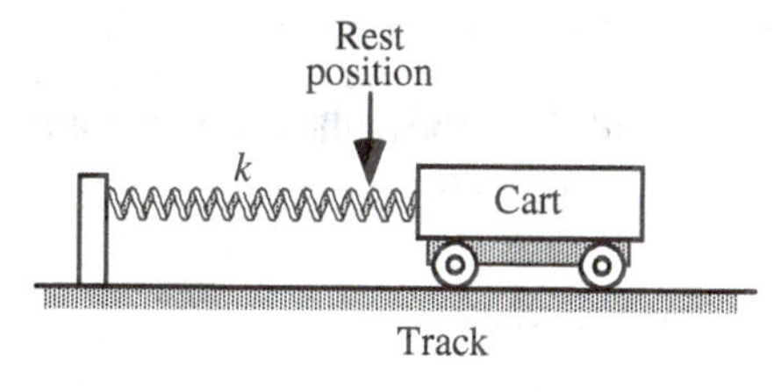

A frictionless cart of mass m is attached to a spring with spring constant k. When the cart is displaced from its rest position and released, it oscillates with a period τ that is given by

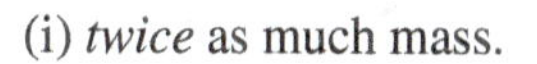

$$\tau = 2\pi\sqrt{m/k}$$

The graph of the position of this cart as a function of time is labeled Experiment A. Graphs for two other experiments that use different masses are shown below this. The same spring is used in all three experiments.

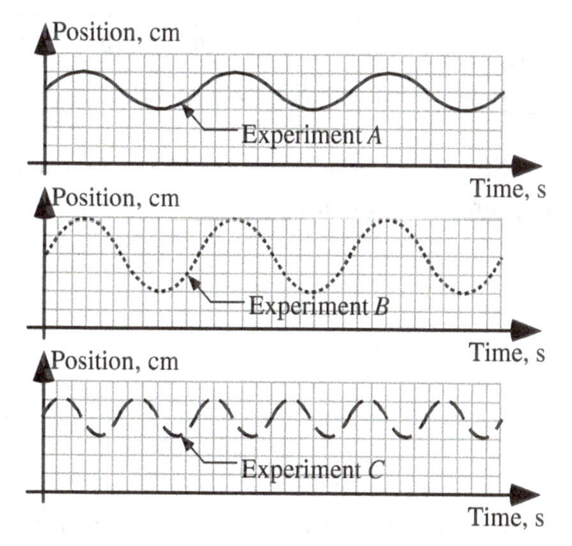

(a) Compared to Experiment *A*, in Experiment *B* the cart has

(i) *twice* as much mass.
(ii) *four times* as much mass.
(iii) *one-half* the mass.
(iv) *one-fourth* the mass.
(v) *the same* mass.

Explain your reasoning.

(b) Compared to Experiment *A*, in Experiment *C* the cart has

(i) *twice* as much mass.
(ii) *four times* as much mass.
(iii) *one-half* the mass.
(iv) *one-fourth* the mass.
(v) *the same* mass.

Explain your reasoning.

(c) Suppose that in a fourth experiment (Experiment *D*), the mass used in Experiment *A* was doubled and the spring was replaced with a spring with spring constant 2*k*. The period in Experiment *D* would be

(i) *the same* as the period in Experiment *A*.
(ii) *double* the period in Experiment *A*.
(iii) *four times* the period in Experiment *A*.
(iv) *one-half* the period in Experiment *A*.
(v) *one-fourth* the period in Experiment *A*.

Explain your reasoning.

B7-CRT06: Velocity-Time Graph—Frequency and Period

A cart attached to a spring is displaced from equilibrium and then released. There is no friction. A graph of velocity as a function of time for the cart is shown. The arrows and signs above the cart indicate the positive and negative directions for the position of the cart.

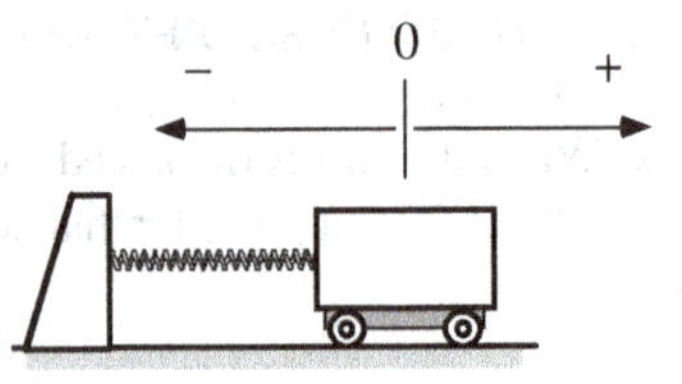

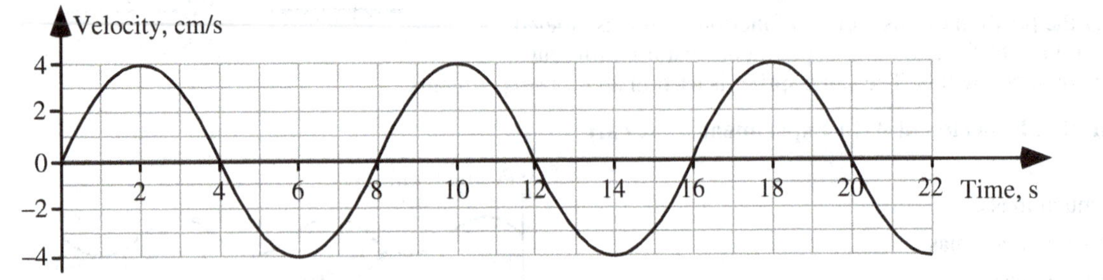

(a) What is the period of the motion for this cart?
Explain your reasoning.

(b) What is the frequency of the motion for this cart?
Explain your reasoning.

(c) In which direction was the cart displaced from equilibrium before it was released?
Explain your reasoning.

B7-SCT07: MASS ON A VERTICAL SPRING—ACCELERATION

A mass is oscillating up and down at the end of a spring. Three students are discussing the acceleration of the mass:

Aileen: *"I think the acceleration of the mass will be largest when it is at the end of its oscillations turning around. That's where the spring is stretched the most."*

Brigitte: *"No, I don't see how that can be. Its velocity is zero at that point, so its acceleration has to be zero also."*

Chandra: *"I disagree. The acceleration is largest when the mass is halfway between the middle and the end because that is where its speed is changing the most."*

With which, if any, of these students do you agree?

Aileen _____ Brigitte _____ Chandra _____ None of them_____

Explain your reasoning.

B7-LMCT08: Mass Connected to a Horizontal Spring—Frequency

A mass-spring system consists of a spring with a spring constant (or stiffness) k and unstretched length L, connected to a cart of mass M resting on a horizontal frictionless surface as shown. If the cart is pulled to one side a small distance and released, it will oscillate back and forth with amplitude A and frequency f.

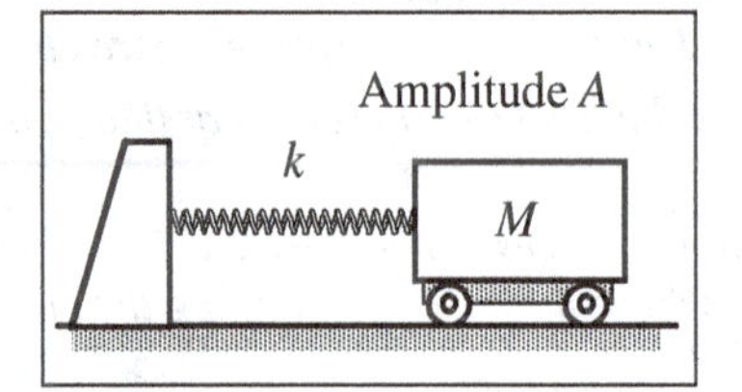

Identify from choices (i)–(iv) how each change described below will affect the frequency of the oscillating mass-spring system.

Compared to the case above, this change will:

(i) ***increase*** the frequency of the system.
(ii) ***decrease*** the frequency of the system.
(iii) ***have no effect*** on the frequency of the system.
(iv) *have an* ***indeterminate*** effect on the frequency of the system.

Each of these modifications is the only change to the initial situation described above.

(a) The mass is increased. ______
Explain your reasoning.

(b) The spring constant or stiffness is increased. ______
Explain your reasoning.

(c) The mass is pulled a little farther and then released. ______
Explain your reasoning.

(d) The spring constant is doubled to $2k$ and the mass is reduced to $M/2$. ______
Explain your reasoning.

(e) The amplitude is increased and the mass is increased. ______
Explain your reasoning.

B7-QRT09: OSCILLATION DISPLACEMENT-TIME GRAPH—KINEMATIC QUANTITIES

A cart attached to a spring is displaced from equilibrium and then released. There is no friction. A graph of displacement as a function of time for the cart is shown. The arrows and signs above the cart indicate the positive and negative directions for the position of the cart.

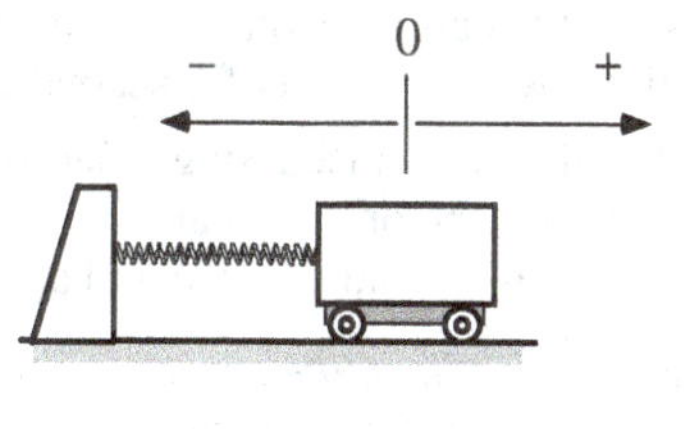

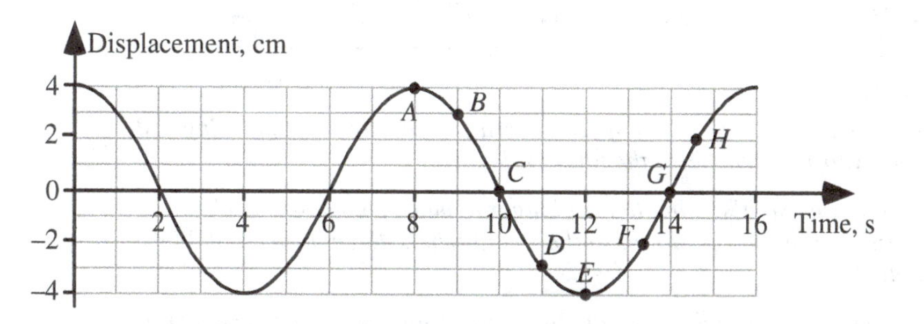

For each question below, choose from the labeled points above, or state "none."

(a) At which point or points is the acceleration positive? _____
Explain your reasoning.

(b) At which point or points does the cart have zero velocity but nonzero net force? _____
Explain your reasoning.

(c) At which point or points is the net force on the cart equal to zero? _____
Explain your reasoning.

(d) At which point or points are the acceleration, velocity, and displacement all positive? ______
Explain your reasoning.

(e) At which point or points is the acceleration nonzero and opposite in sign to the position? _____
Explain your reasoning.

(f) At which point or points is the velocity nonzero and opposite in sign to the acceleration? _____
Explain your reasoning.

B7-SCT10: Horizontal Oscillating Cart—Period

A frictionless cart of mass M is attached to a spring with spring constant k. When the cart is displaced 6 cm from its rest position and released, it oscillates with a period of 2 seconds.

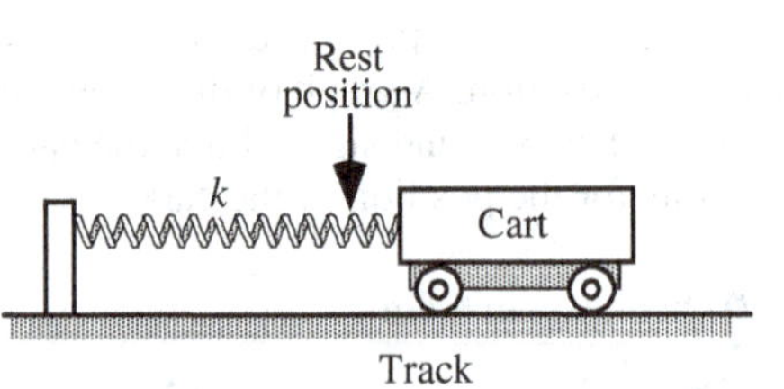

Four students are discussing what would happen to the period of oscillation if the original cart with mass M was displaced 12 cm from its rest position instead of 6 cm and again released:

Adan: *"Since the spring is stretched more, the force will be greater, causing a greater acceleration and greater speeds overall. Since the cart is moving faster, the time will go down, probably to 1 second since the force is doubled."*

Barb: *"The cart has farther to go now, and so it's going to take longer to make a complete cycle. It's going to go farther on both sides of the rest position, so the round-trip is 48 cm instead of 24 cm. The period is going to double."*

Charles: *"The cart has four times as much energy, and the conjugate variable for energy is time according to the Heisenberg uncertainty principle. The energy quadruples when the spring stretch is doubled, and so the time must be only one-quarter as much. The period will be one-half second."*

Dallas: *"Stretching the spring twice as far means that k is twice as big. And the period is 2 pi times the square root of the spring constant divided by the mass. Doubling the spring constant and leaving the mass alone is going to double what's inside the square root, and after we take the square root we get a period of 2 seconds times the square root of 2, or 1.414 seconds."*

With which, if any, of these students do you agree?

Adan _____ Barb _____ Charles _____ Dallas _____ None of them_____

Explain your reasoning.

B7-QRT11: OSCILLATING MASS ON SPRING DISPLACEMENT-TIME GRAPH—DIRECTIONS

A cart attached to a spring is displaced from equilibrium and then released. A graph of displacement as a function of time for the cart is shown. There is no friction. Points are labeled *A–H* in the graph. The arrows and signs above the cart indicate the positive and negative directions for the displacement of the cart.

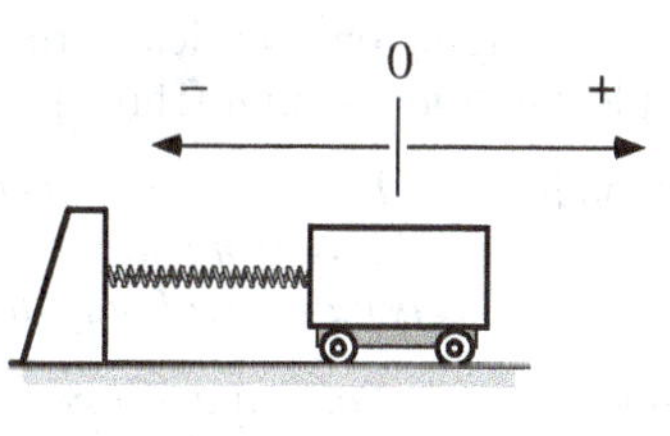

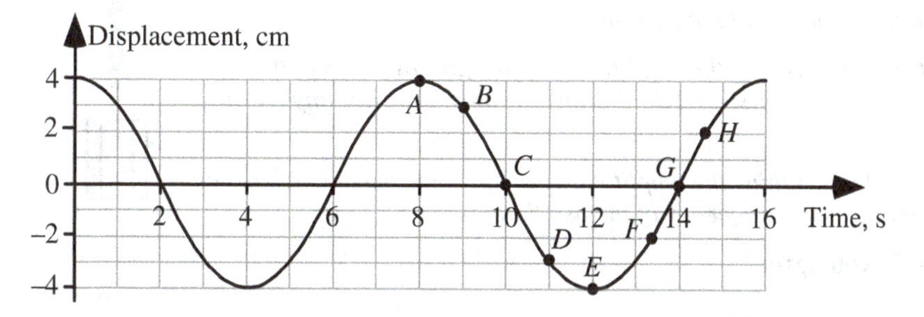

For each labeled point above, identify if the vector quantity listed below is in the positive (+) direction, negative (–) direction, or is zero (0) for no direction.

Point	Acceleration	Velocity	Displacement	Net Force
A				
B				
C				
D				
E				
F				
G				
H				

Explain your reasoning.

B7-SCT12: Mass Oscillating on a Vertical Spring—Energy

A mass hanging on a vertical spring is pulled down a distance *d* and released. The mass undergoes simple harmonic motion. Three physics students make the following contentions about this situation:

Alexandra: *"The maximum kinetic energy of this mass-spring system is fixed by the properties of the system and does not depend on how far down the mass is pulled. How far the mass is pulled will only affect the frequency of the oscillations."*

Bruno: *"No, that can't be right since increasing the amplitude, or how far down it is pulled, increases the potential energy of the system. I don't think the amplitude has any effect on the frequency."*

Chung: *"I agree in part with both of you. I think the amplitude does affect the maximum kinetic energy, but I also think it affects the frequency of the oscillations."*

With which, if any, of these students do you agree?

Alexandra _____ Bruno _____ Chung _____ None of them_____

Explain your reasoning.

B7-QRT13: DISPLACEMENT-TIME GRAPH—ENERGY QUANTITIES

A cart attached to a spring is displaced from equilibrium and then released. A graph of displacement as a function of time for the cart is shown. There is no friction. Points are labeled *A–H* in the graph.

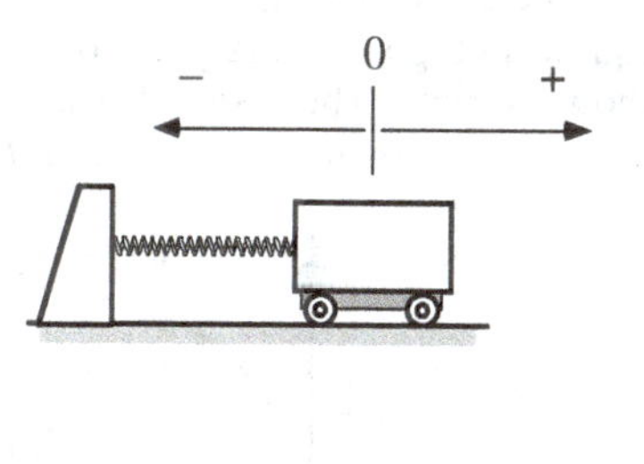

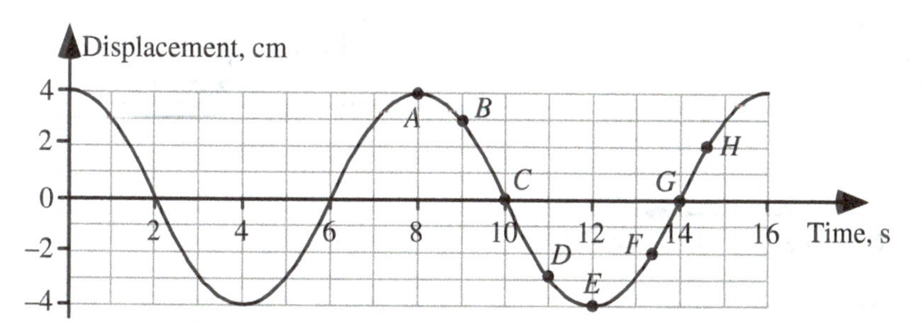

For each question below, choose from the labeled points above or state "none" for the mass-spring-earth system.

(a) At which point or points are the spring potential energy and the cart's kinetic energy both at their maximum values? _____
Explain your reasoning.

(b) At which point or points is the kinetic energy equal to zero? _____
Explain your reasoning.

(c) At which point or points is the total energy at its maximum value? _____
Explain your reasoning.

(d) At which point or points is the spring potential energy negative? _____
Explain your reasoning.

(e) At which point or points is the kinetic energy positive? _____
Explain your reasoning.

(f) At which point or points is the kinetic energy at its maximum value and the spring potential energy at its minimum value? _____
Explain your reasoning.

(g) At which point or points is the kinetic energy at its minimum value and the spring potential energy at its maximum value? _____
Explain your reasoning.

B7-BCT14: Oscillating Mass on Spring Displacement-Time Graph—Energy

A cart attached to a spring is given an initial push, displacing it from its equilibrium position. A graph of displacement as a function of time for the cart is shown at right. The system has a total initial energy of 12 J and there is no friction. Five points are labeled *A*–*E* in the graph.

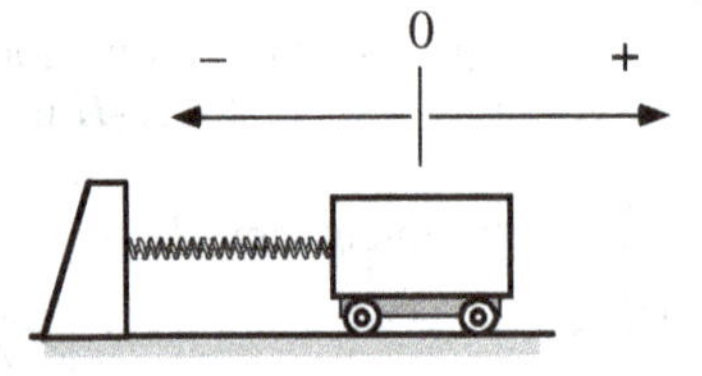

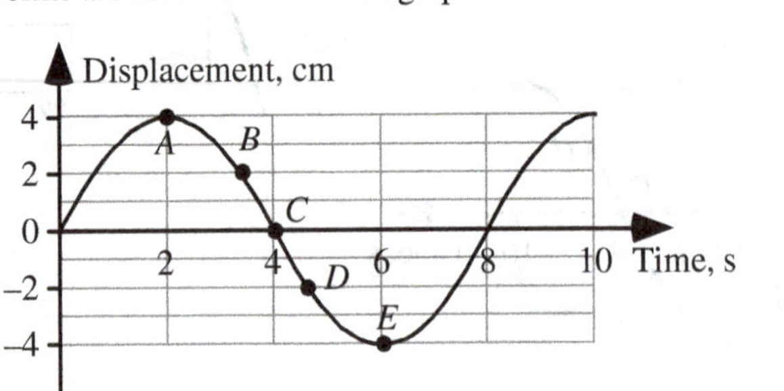

For each labeled point, complete the bar chart below for the kinetic energy and the potential energy for the cart-spring system.

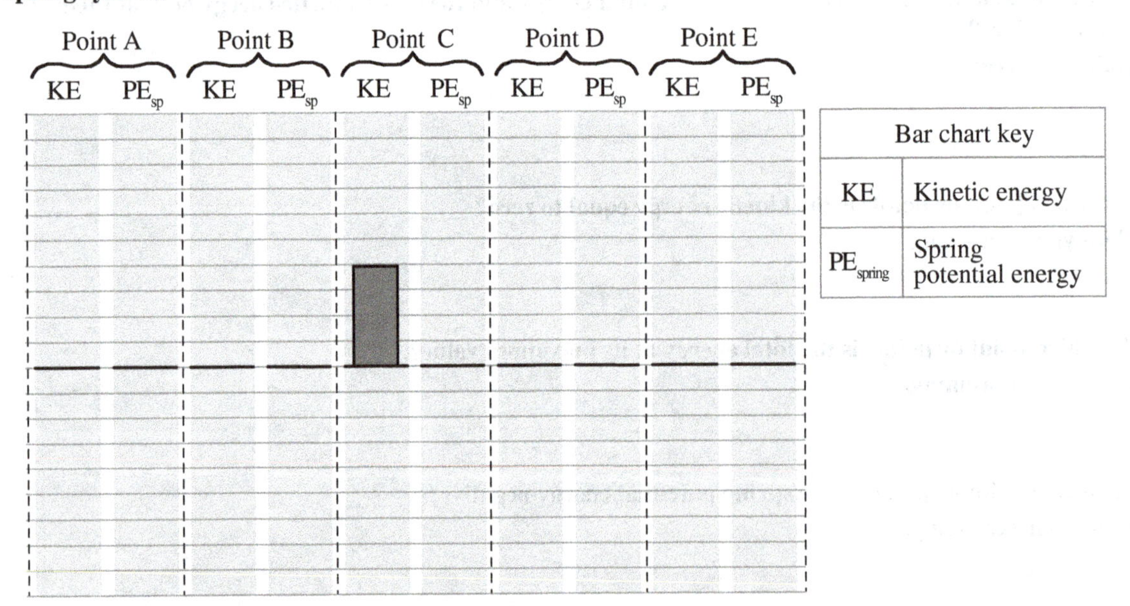

Bar chart key	
KE	Kinetic energy
PE_{spring}	Spring potential energy

Explain your reasoning.

C1 DENSITY

C1-RT01: CUTTING UP A BLOCK—DENSITY

A block of material (labeled A in the diagram) with a width w, height h, and thickness t has a mass of M_o distributed uniformly throughout its volume. The block is then cut into three pieces, B, C, and D, as shown.

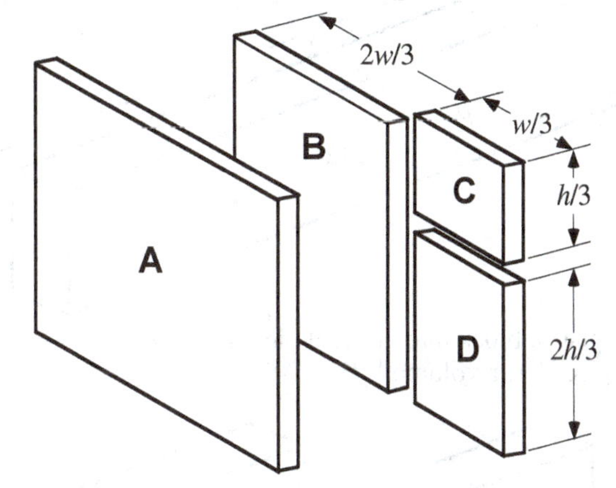

Rank the density of the original block A, piece B, piece C, and piece D.

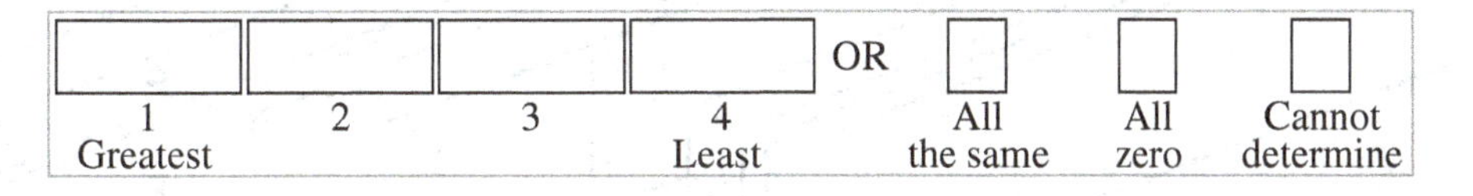

Explain your reasoning.

C1-SCT02: BREAKING UP A BLOCK—DENSITY

A block of material with a width w, height h, and thickness t has a mass of M_o distributed uniformly throughout its volume. The block is then broken into two pieces, A and B, as shown. Three students make the following statements:

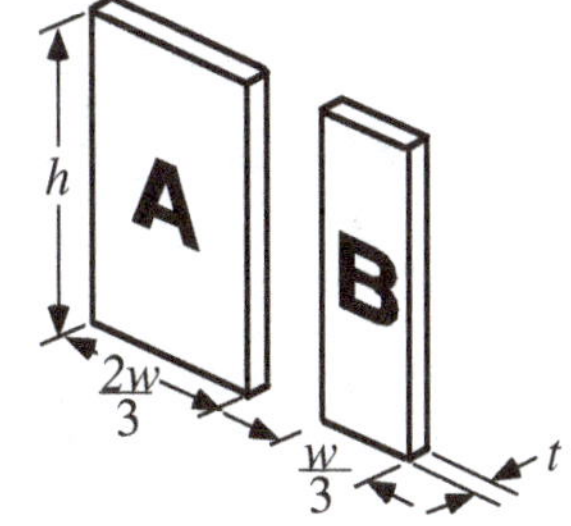

Ajay: *"They both have the same density. It's still the same material."*

Ben: *"The density is the mass divided by the volume, and the volume of B is smaller. Since the mass is uniform and the volume is in the denominator, the density is larger for B."*

Chithra: *"The density of piece A is larger than the density of piece B since A is larger; thus it has more mass."*

With which, if any, of these students do you agree?

Ajay _____ Ben _____ Chithra _____ None of them_____

Explain your reasoning.

C1-QRT03: Slicing up a Block—Mass and Density

The plastic block shown below has a volume V_o and a mass $12M_o$ distributed evenly to give a uniform density ρ_o.

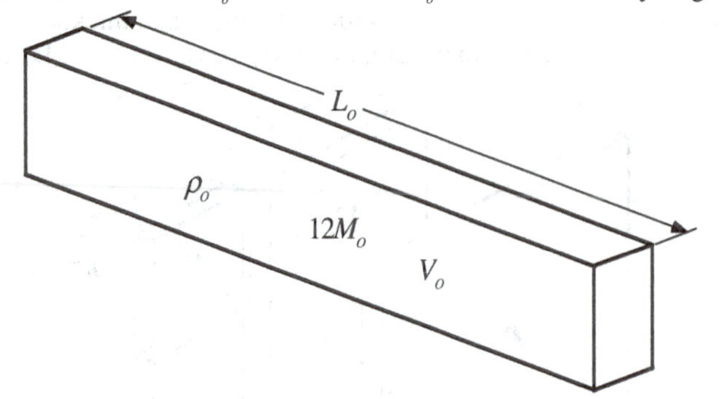

Three possible ways to slice the plastic block into unequal pieces are shown below. In each case, the larger piece has a volume $2V_o/3$ and the smaller piece has a volume $V_o/3$.

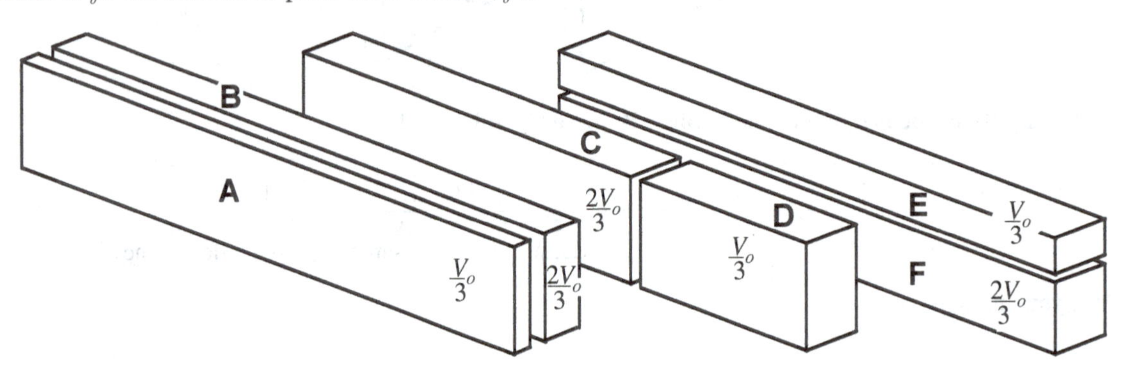

Fill in the table for the mass (in terms of M_o) and density (in terms of ρ_o) of the pieces of the block labeled A– F.

	Mass	Density
Original block		
Piece A		
Piece B		
Piece C		
Piece D		
Piece E		
Piece F		

C1-QRT04: CYLINDERS WITH THE SAME MASS I—VOLUME, AREA, AND DENSITY

Two solid cylinders are shown. Cylinder A has a height H and a radius R, and cylinder B has a height $2H$ and a radius $2R$. Both cylinders have uniform densities and the same mass. Cylinder A has a density ρ_A and volume V_A.

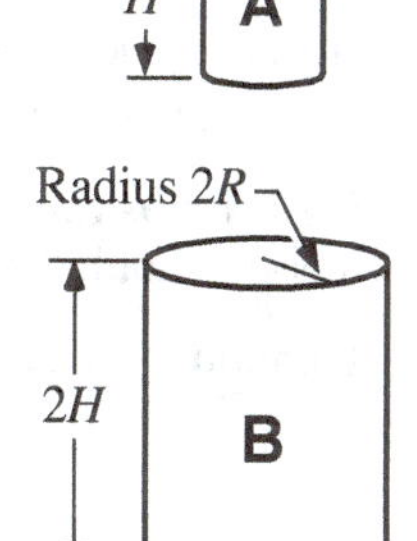

If r is the radius of a cylinder and h is the height, then the volume of the cylinder is $V = \pi r^2 h$, and the surface area is $SA = 2\pi r^2 + 2\pi rh$.

(a) What is the volume of cylinder B in terms of the volume of cylinder A? (Your answer should look like $V_B = n\ V_A$, where n is some number.)

Explain your reasoning.

(b) What is the surface area of cylinder B in terms of the surface area of cylinder A? (Your answer should look like $SA_B = n\ SA_A$, where n is some number.)

Explain your reasoning.

(c) What is the density of cylinder B in terms of the density of cylinder A? (Your answer should look like $\rho_B = n\ \rho_A$, where n is some number.)

Explain your reasoning.

C1-QRT05: Cylinders with the Same Mass II—Volume, Area, and Density

Two solid cylinders are shown. Cylinder A has a height H and a radius R and cylinder B has a height $3H$ and a radius $3R$. Both cylinders have uniform densities and the same mass. Cylinder A has a density ρ_A and volume V_A.

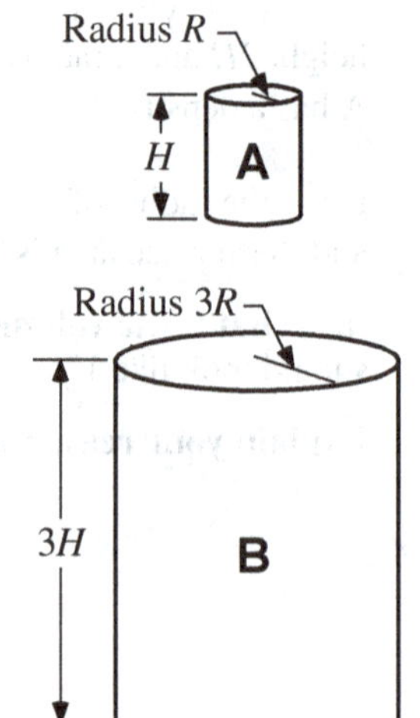

If r is the radius of a cylinder and h is the height, then the volume of the cylinder is $V = \pi r^2 h$, and the surface area is $SA = 2\pi r^2 + 2\pi rh$.

(a) What is the volume of cylinder B in terms of the volume of cylinder A? (Your answer should look like $V_B = n\, V_A$, where n is some number.)

Explain your reasoning.

(b) What is the surface area of cylinder B in terms of the surface area of cylinder A? (Your answer should look like $SA_B = n\, SA_A$, where n is some number.)

Explain your reasoning.

(c) What is the density of cylinder B in terms of the density of cylinder A? (Your answer should look like $\rho_B = n\, \rho_A$, where n is some number.)

Explain your reasoning.

C1-QRT06: CYLINDERS WITH THE SAME MASS III—VOLUME, AREA, AND DENSITY

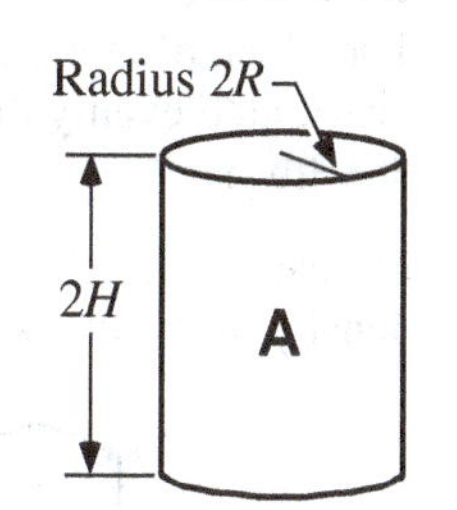

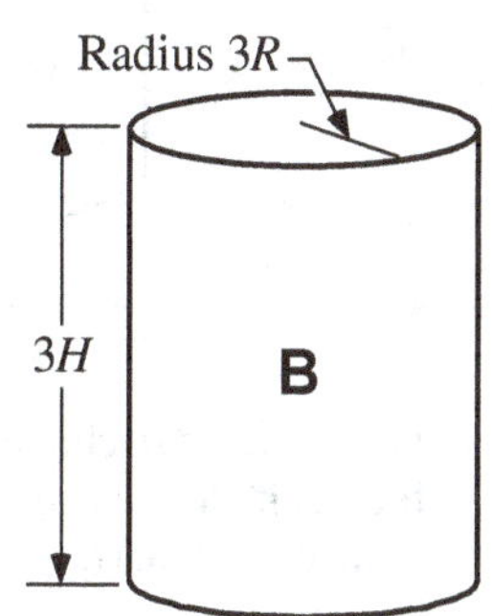

Two solid cylinders are shown. Cylinder A has a height $2H$ and a radius $2R$ and cylinder B has a height $3H$ and a radius $3R$. Both cylinders have uniform densities and the same mass. Cylinder A has a density ρ_A and volume V_A.

If r is the radius of a cylinder and h is the height, then the volume of the cylinder is $V = \pi r^2 h$, and the surface area is $SA = 2\pi r^2 + 2\pi rh$.

(a) What is the volume of cylinder B in terms of the volume of cylinder A? (Your answer should look like $V_B = n\ V_A$, where n is some number.)

Explain your reasoning.

(b) What is the surface area of cylinder B in terms of the surface area of cylinder A? (Your answer should look like $SA_B = n\ SA_A$, where n is some number.)

Explain your reasoning.

(c) What is the density of cylinder B in terms of the density of cylinder A? (Your answer should look like $\rho_B = n\ \rho_A$, where n is some number.)

Explain your reasoning.

C1-BCT07: Four Blocks—Mass and Density

The block of material shown to the right has a volume V_o. An overall mass M_o is distributed evenly throughout the volume of the block so that the block has a uniform density ρ_o.

For each block shown below, the volume is given as well as *either* the mass or the density of the block.

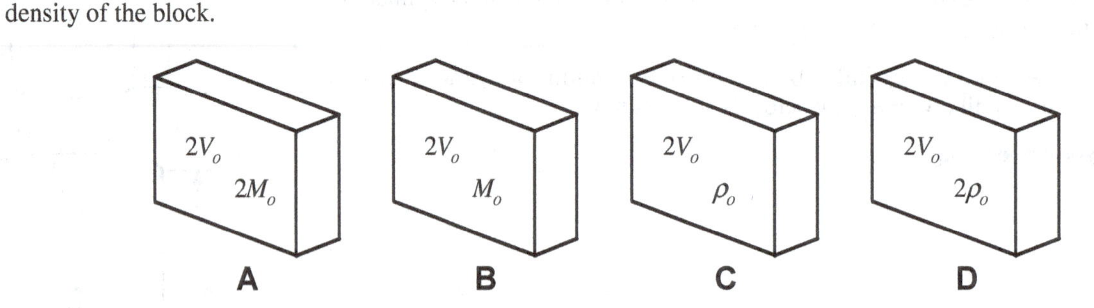

Construct bar charts for the mass and density for the four blocks labeled A–D, and for the pieces of the blocks if they were cut in half labeled A/2–D/2. The mass and density for the original block are shown to set the scale of the chart.

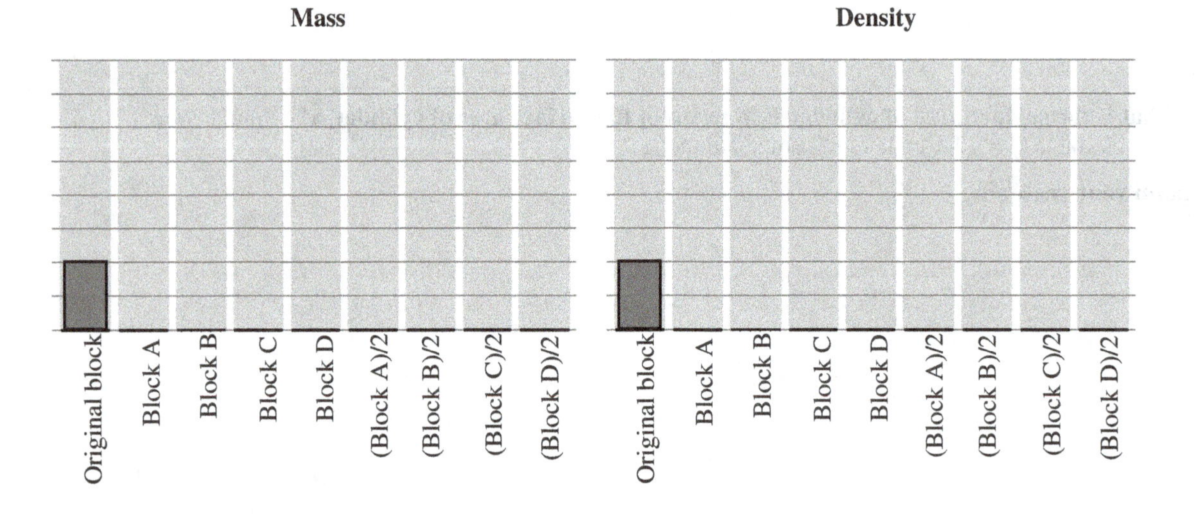

Explain your reasoning.

C1-RT08: CYLINDERS AND CUBES—DENSITY

A cylinder and a cube are carved out of a piece of plastic with uniform density, and a second cylinder and cube are carved out of a piece of metal with uniform density. Dimensions are given for the cylinders and cubes. The mass of the cylinder in Case B is twice the mass of the cylinder in Case A.

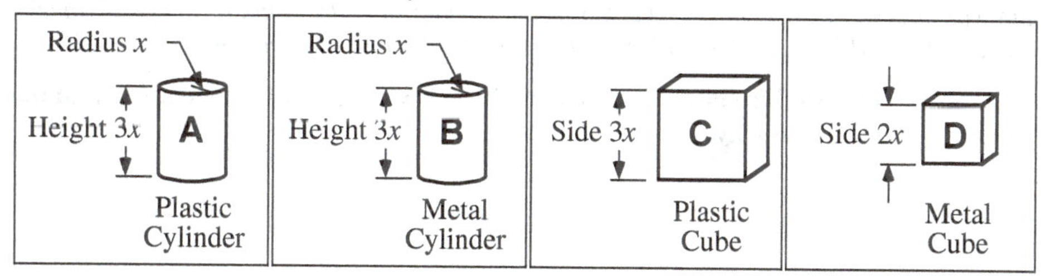

Rank the densities of the objects.

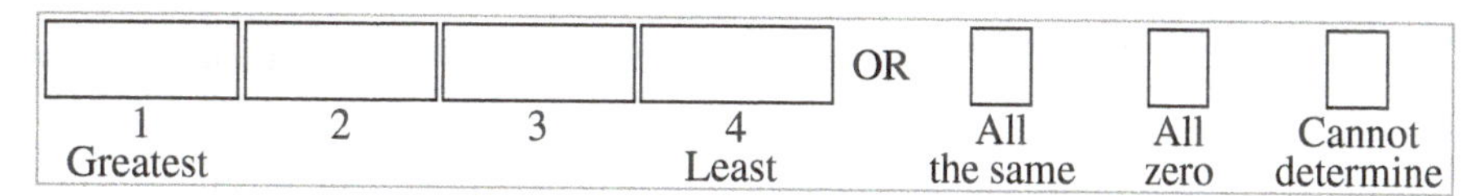

Explain your reasoning.

C1-QRT09: FOUR CUBES—MASS

Of the four cubes shown below, white cubes A and C are made of the same material, and gray cubes B and D are made of the same material. Each cube has a uniform density. The ranking of cube size is C = D > A > B. Cubes A and B have the same mass.

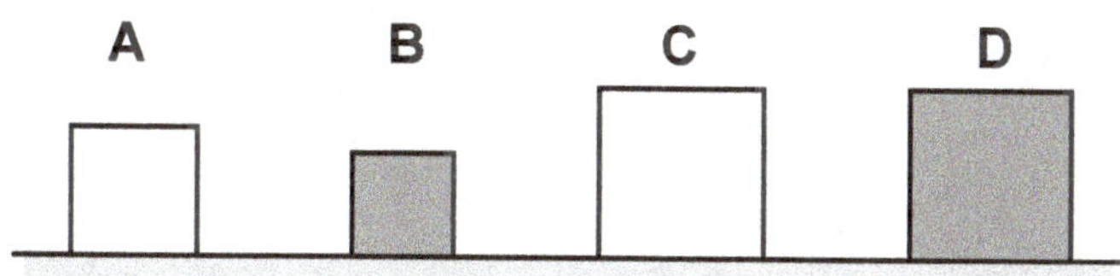

Is the mass of cube C (i) *greater than*, (ii) *less than*, or (iii) *equal to* the mass of cube D?

Explain your reasoning.

C1-WWT10: Pouring Liquid Between Beakers—Density

A liquid in a tall, narrow cylindrical beaker is poured into a wider cylindrical beaker. The liquid only fills the wider beaker to one-fourth its height in the tall beaker. A student makes the following statement:

"When the liquid was poured from the narrow beaker into the wider one, the volume changed. Since no liquid was spilled, all of the liquid is still in the wider beaker, so the density of the liquid must have changed."

What, if anything, is wrong with this statement? If something is wrong, explain the error and how to correct it. If the statement is valid, explain why.

C2 FLUIDS

C2-RT01: BLOCKS SUSPENDED IN LIQUIDS—BUOYANT FORCE

In each case, a block hanging from a string is suspended in a liquid. All of the blocks are the same size, but they have different masses (labeled M_b) because they are made of different materials. All of the containers have the same volume of liquid, but the masses of these liquids vary (labeled M_l) since the liquids are different. The volume of the blocks is one-sixth the volume of the liquids.

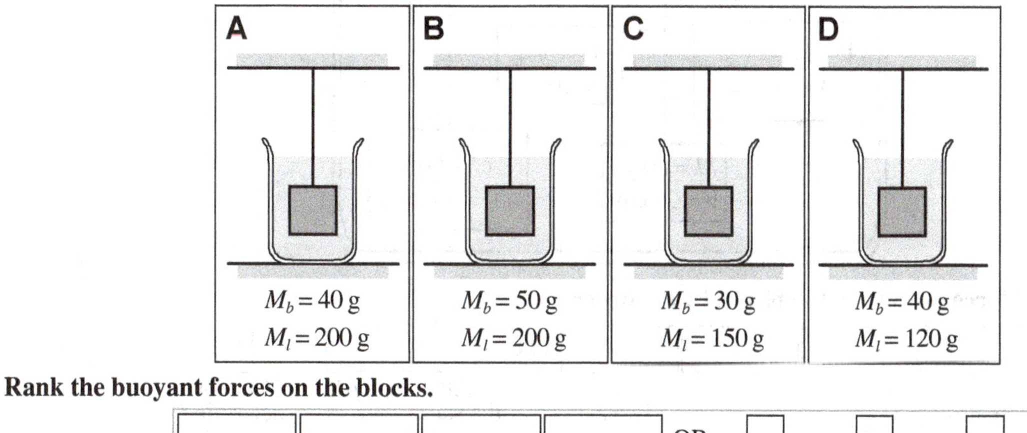

Rank the buoyant forces on the blocks.

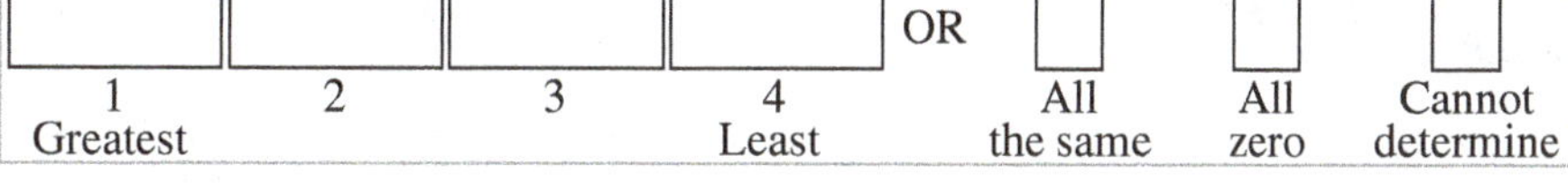

Explain your reasoning.

C2-RT02: BLOCKS SUSPENDED IN LIQUIDS—VOLUME OF LIQUID DISPLACED

In each case, a block hanging from a string is suspended in a liquid. All of the blocks are the same size, but they have different masses (labeled M_b) because they are made of different materials. All of the containers have the same volume of liquid, but the masses of these liquids vary (labeled M_l) since the liquids are different. The volume of the blocks is one-sixth the volume of the liquids.

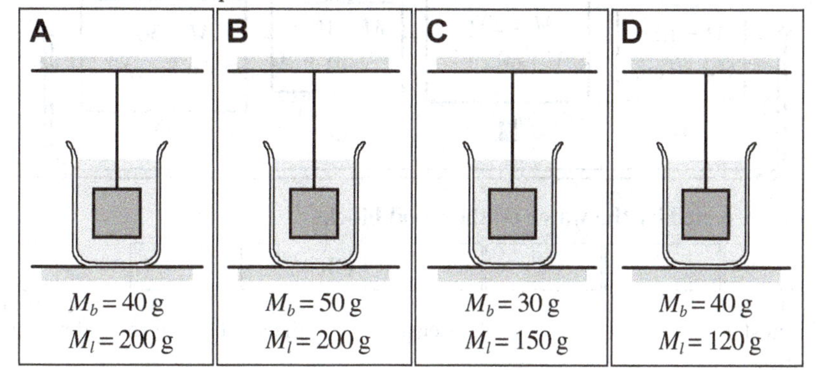

Rank the volume of the liquid displaced by the blocks.

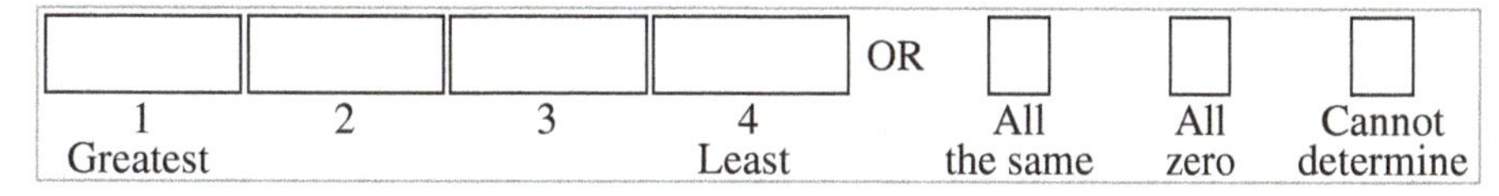

Explain your reasoning.

C2-RT03: Blocks Suspended in Water at Different Depths—Buoyant Force

Blocks that have different masses and volumes are suspended by strings in water. The blocks are at two different depths below the surface as shown.

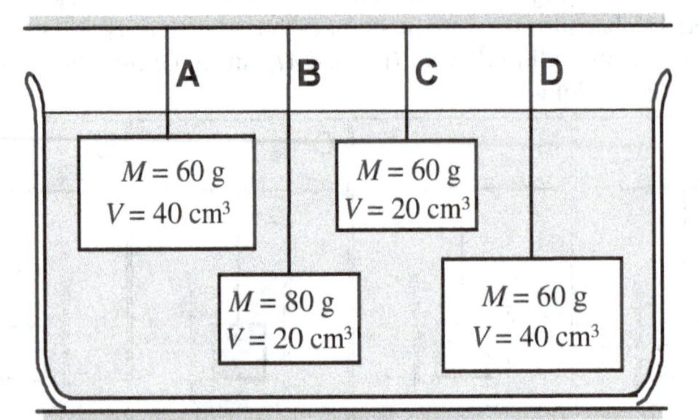

Rank the buoyant force exerted on the blocks by the water.

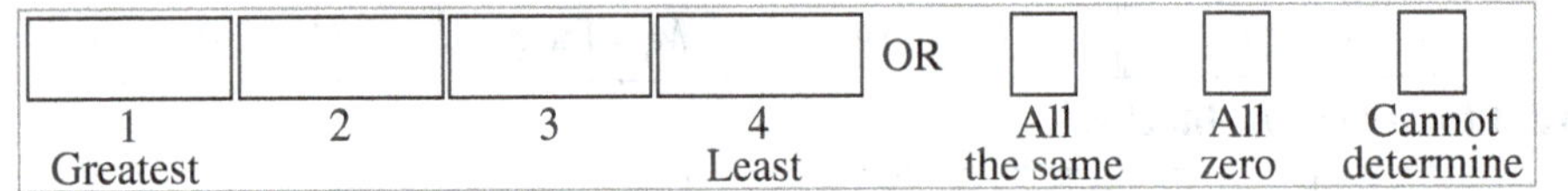

Explain your reasoning.

C2-RT04: Floating Blocks with Different Loads—Buoyant Force

Wood blocks that have different masses and different volumes are floating in water. On top of these blocks are additional masses as shown.

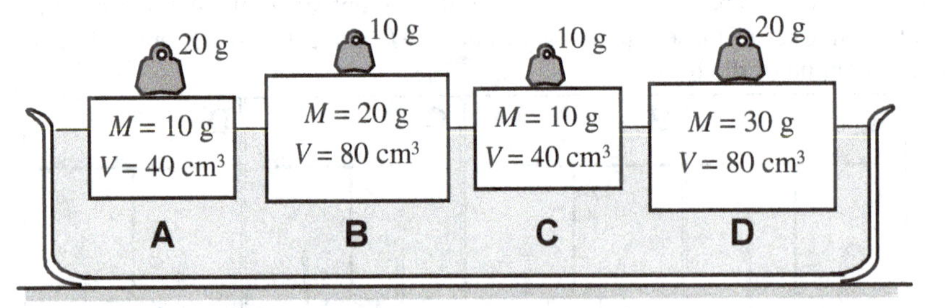

Rank the buoyant force exerted by the water on the wood blocks.

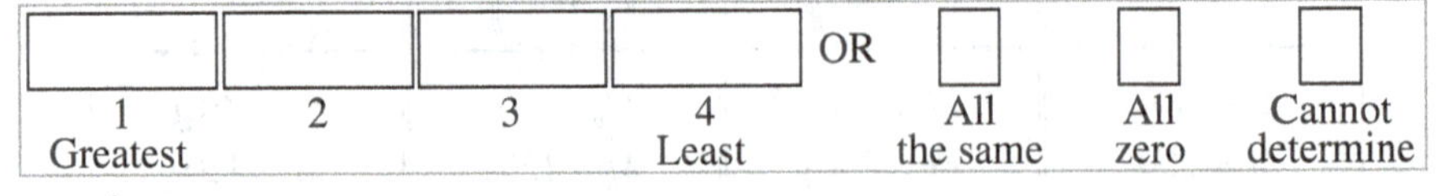

Explain your reasoning.

C2-RT05: BLOCKS SUSPENDED IN LIQUIDS—BUOYANT FORCE

In each case, a block hanging from a string is suspended in a liquid. The blocks are made of different materials and vary in mass and volume as shown. All of the containers have the same volume of an identical liquid.

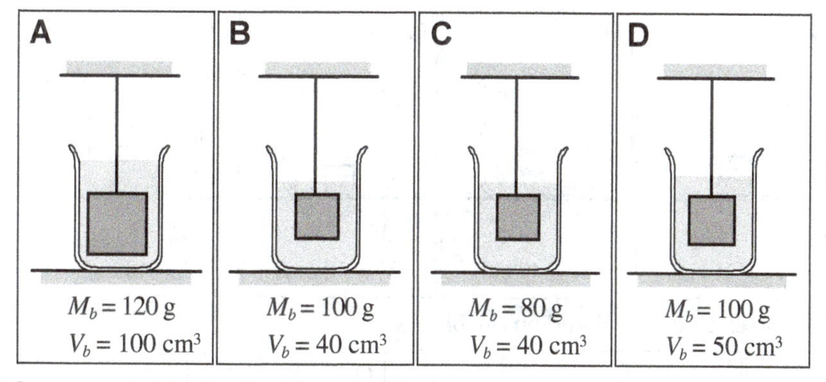

Rank the buoyant force exerted by the liquid on the blocks.

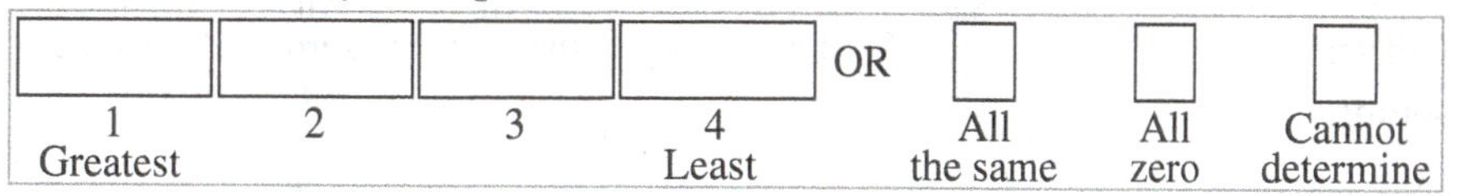

Explain your reasoning.

C2-RT06: BLOCKS FLOATING IN LIQUIDS—BUOYANT FORCE

In each case, a block floats in a liquid. The blocks are made of different materials and vary in mass and volume as shown. All of the containers have the same volume of an identical liquid.

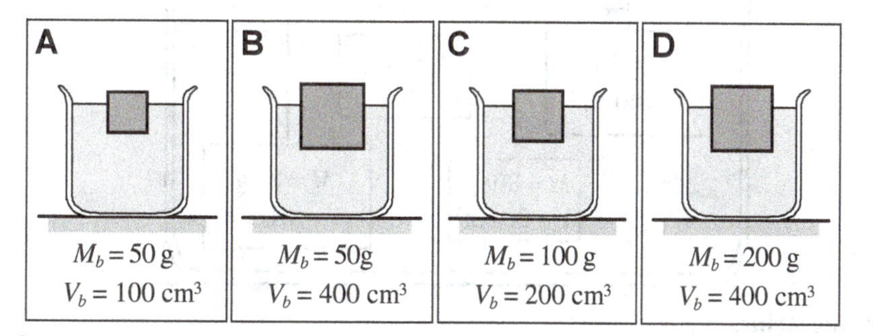

Rank the buoyant force exerted by the liquid on the blocks.

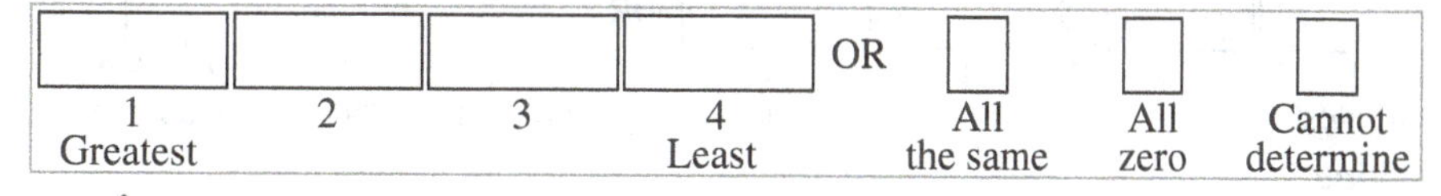

Explain your reasoning.

C2-RT07: Blocks at the Bottom of Liquids—Buoyant Force

In each case, a block is at rest at the bottom of a beaker filled with liquid. The blocks are made of different materials and vary in mass and volume, as shown. The liquid is the same in each beaker, and the liquid levels after the blocks are added are the same for all four beakers.

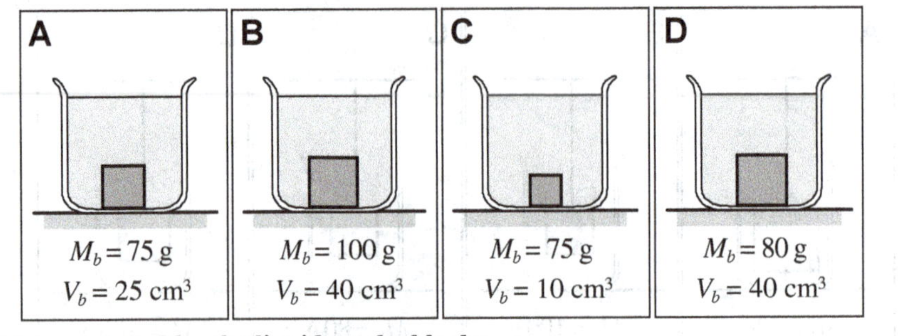

Rank the buoyant force exerted by the liquid on the blocks.

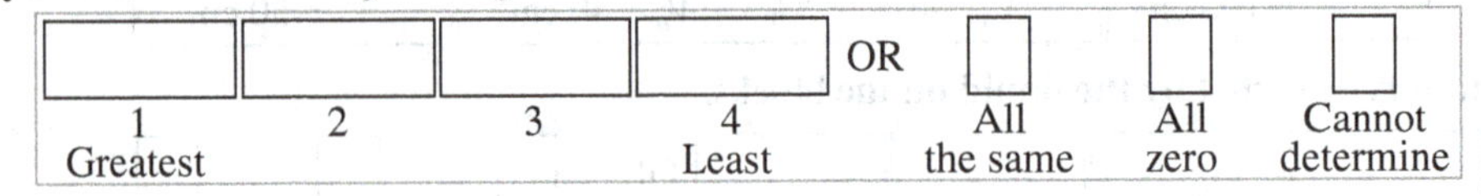

1 Greatest	2	3	4 Least	OR	All the same	All zero	Cannot determine

Explain your reasoning.

C2-RT08: Four Metal Cubes Suspended in Liquids—Tension

Four blocks are suspended from strings in water. Cubes A and C are at the same depth, as are B and D.

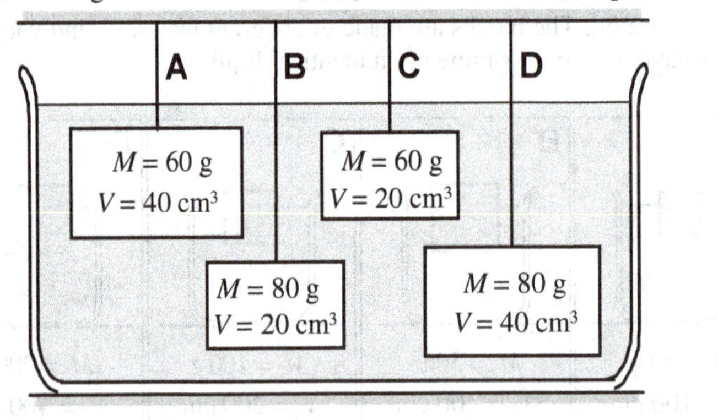

Rank the tensions in the strings.

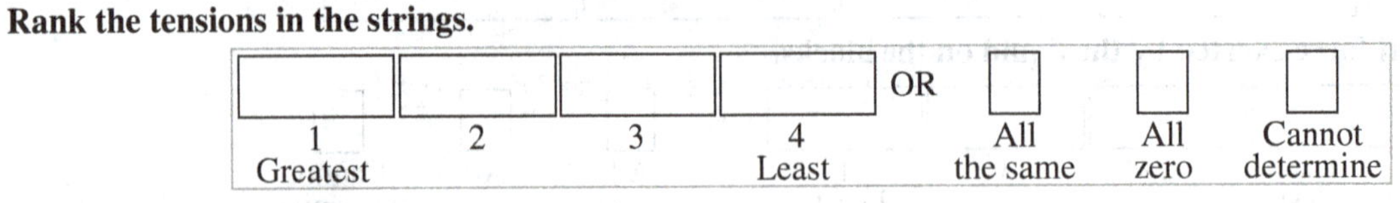

1 Greatest	2	3	4 Least	OR	All the same	All zero	Cannot determine

Explain your reasoning.

C2-RT09: FOUR SUBMERGED CUBES—BUOYANT FORCE

Shown are small cubes that are 10 cm on a side and larger ones that are 12 cm on a side that are submerged in water. Cubes A and B are made of steel ($\rho = 7$ g/cm^3) and cubes C and D are made of aluminum ($\rho = 2.7$ g/cm^3).

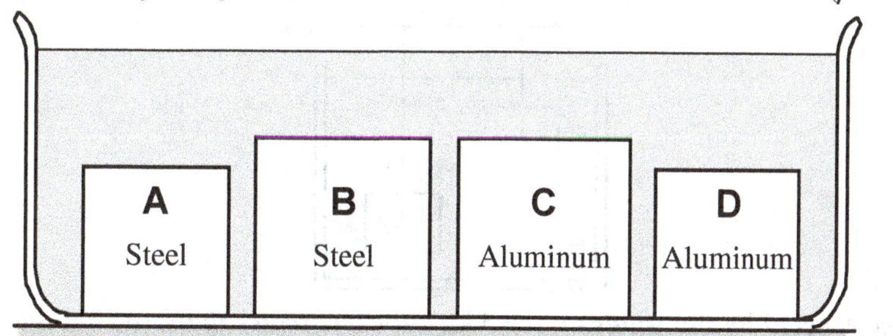

Rank the buoyant force exerted on the cubes by the water.

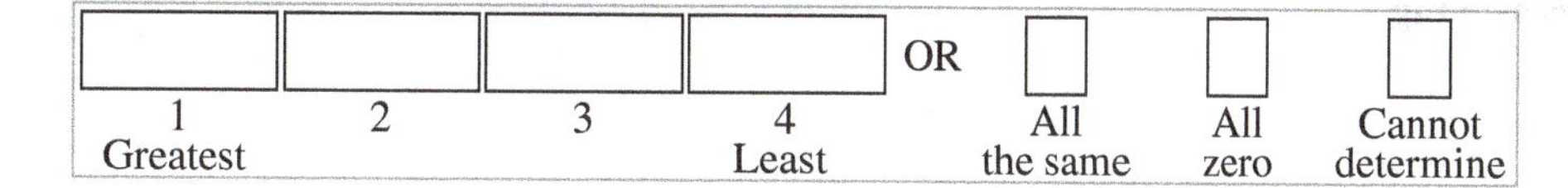

Explain your reasoning.

C2-CT10: TWO FLOATING BLOCKS—BUOYANT FORCE

Two blocks with the same weight but different dimensions are floating in water at different levels. Block A is as tall as block B but is smaller in both other dimensions.

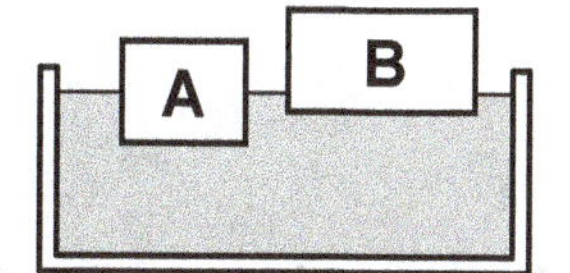

Is the buoyant force exerted by the water on block A (i) *greater than*, (ii) *less than*, or (iii) *equal to* the buoyant force on block B? _____

Explain your reasoning.

C2-CT11: Two Submerged Cubes—Buoyant Force, Tension, and Pressure

Two equal-sized cubes that have different masses are held by strings so that they are submerged in water at different depths.

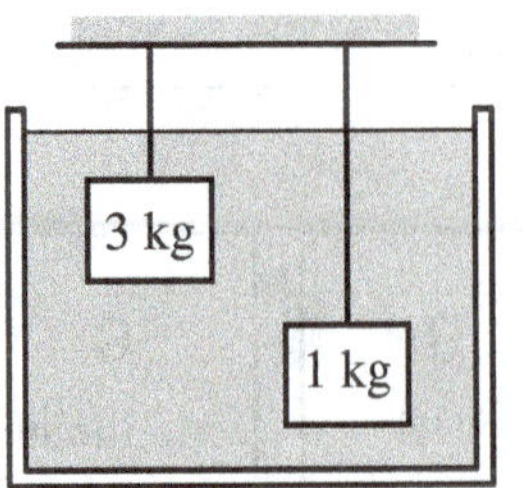

(a) Is the buoyant force exerted by the water on the 3 kg cube (i) *greater than*, (ii) *less than*, or (iii) *equal to* the buoyant force on the 1 kg cube? _____
Explain your reasoning.

(b) Is the tension in the sting holding the 3 kg cube (i) *greater than*, (ii) *less than*, or (iii) *equal to* the tension in the string holding the 1 kg cube? _____
Explain your reasoning.

(c) Is the pressure exerted on the bottom surface of the 3 kg cube by the water (i) *greater than*, (ii) *less than*, or (iii) *equal to* the pressure on the bottom surface of the 1 kg cube? _____
Explain your reasoning.

C2-CT12: FLOATING CUBES—BUOYANT FORCE AND PRESSURE

Two equal-sized cubes are floating in water at different levels.

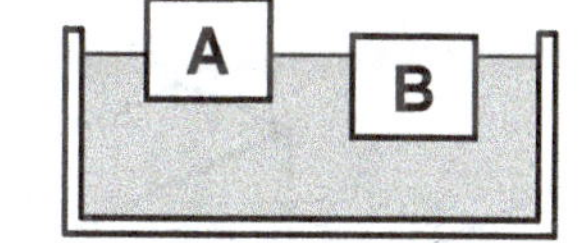

(a) Is the buoyant force exerted by the water on block A (i) *greater than*, (ii) *less than*, or (iii) *equal to* the buoyant force on block B? _____
Explain your reasoning.

(b) Is the weight of block A (i) *greater than*, (ii) *less than*, or (iii) *equal to* the weight of block B? _____
Explain your reasoning.

(c) Is the pressure exerted on the bottom surface of block A (i) *greater than*, (ii) *less than*, or (iii) *equal to* the pressure on the bottom surface of block B? _____
Explain your reasoning.

(d) Is the density of block A (i) *greater than*, (ii) *less than*, or (iii) *equal to* the density of block B? _____
Explain your reasoning.

C2-RT13: Four Rectangular Blocks—Pressure

Four rectangular blocks are made of the same material, with dimensions as shown.

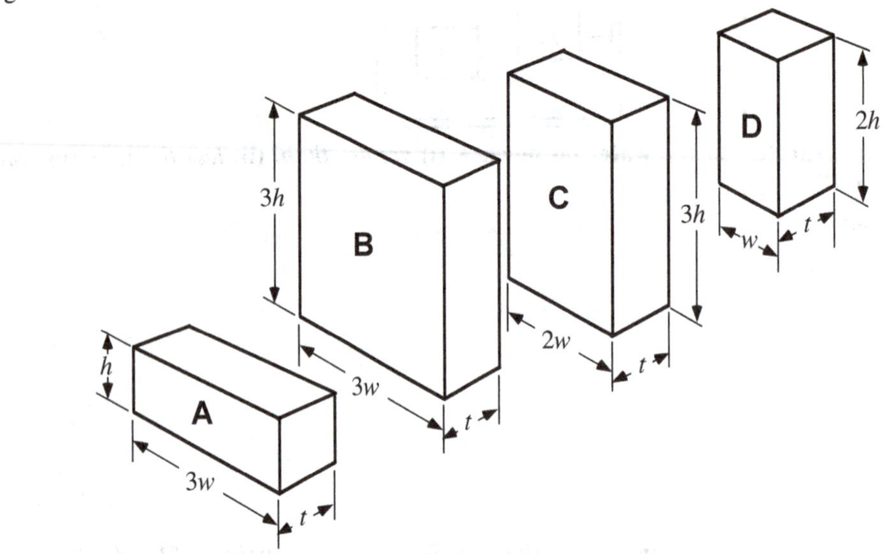

(a) Rank the mass of each block.

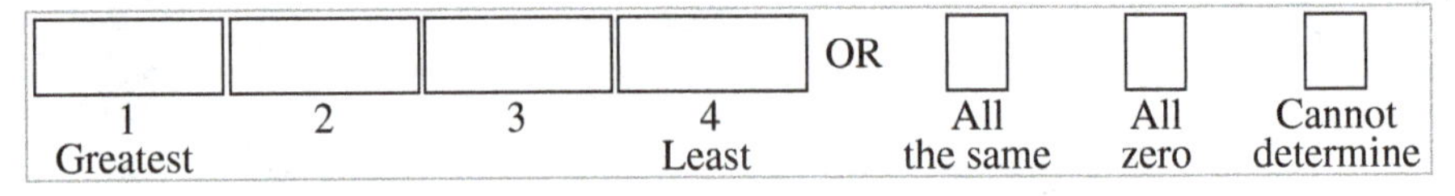

Explain your reasoning.

(b) The blocks are placed as shown onto a table. Rank the pressure exerted by the blocks on the table.

				OR	☐	☐	☐
1 Greatest	2	3	4 Least		All the same	All zero	Cannot determine

Explain your reasoning.

C2-RT14: RECTANGULAR BLOCK—PRESSURE

A rectangular block is at rest on a table. Three faces of the block are labeled A, B, and C. Face A has dimensions 3 cm x 4 cm; face B has dimensions 2 cm x 3 cm; and face C has dimensions 2 cm x 4 cm.

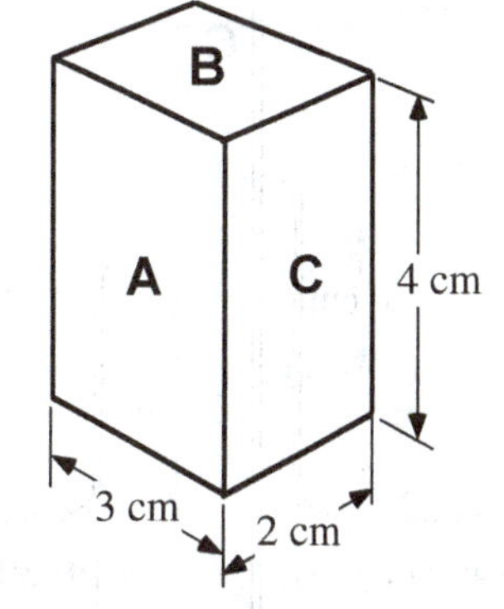

Rank the pressure exerted by the block on the table when it is resting on each labeled face.

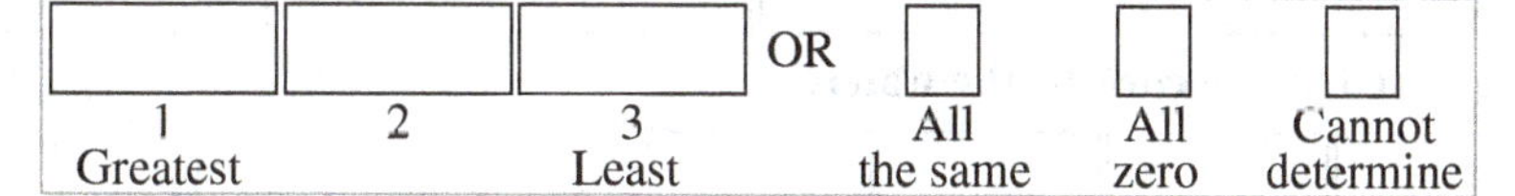

Explain your reasoning.

C2-RT15: BEAKERS OF WATER—PRESSURE ON THE CORK

In each case a beaker is filled with water to the height shown. The diameters of the beakers are also shown. The cylinders have identical holes in their side at the same height above the base. There are corks in all of the holes.

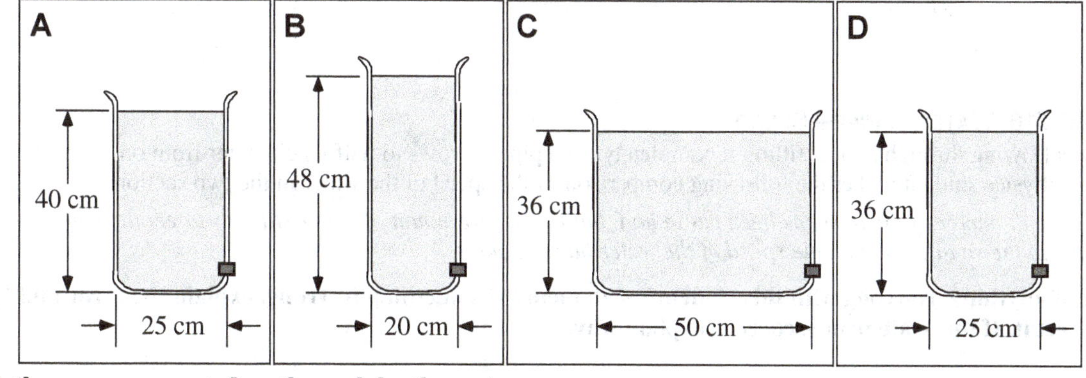

Rank the pressure exerted on the cork by the water.

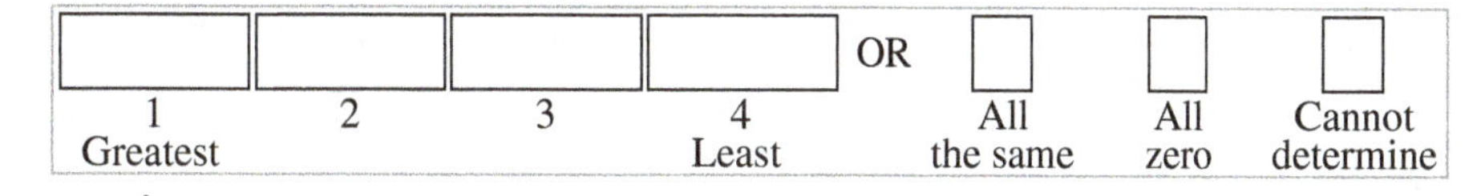

Explain your reasoning.

C2-RT15: Beakers of Water—Cork Pressure

Shown are various beakers filled with water. The diameters of the beakers and the heights of the water to which they are filled are also shown. The beakers have identical holes filled with corks in the side at the specified depths.

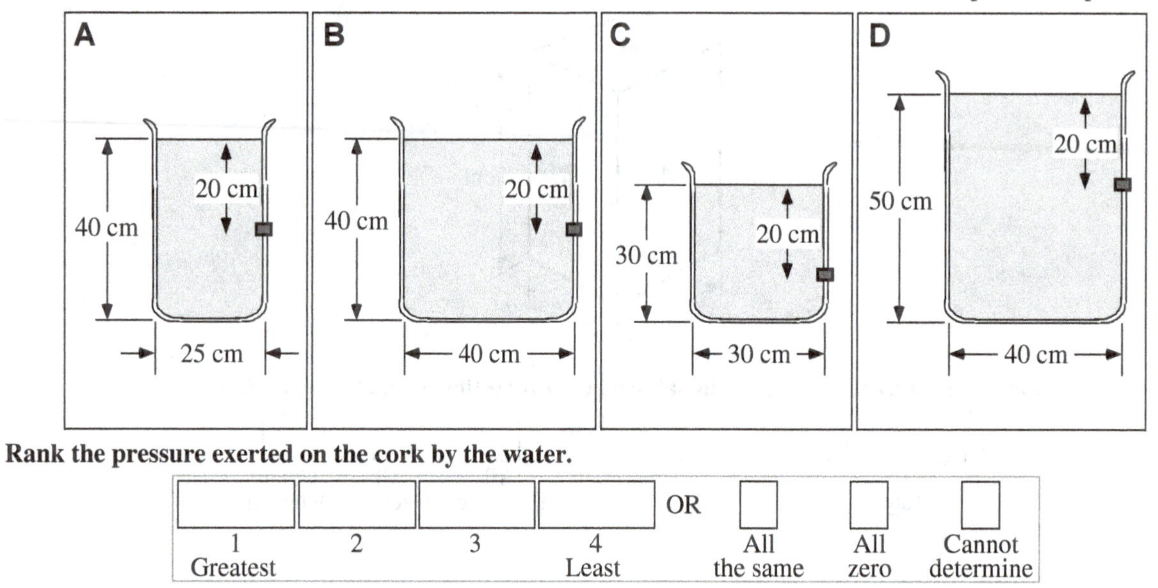

Rank the pressure exerted on the cork by the water.

1 Greatest	2	3	4 Least	OR	All the same	All zero	Cannot determine

Explain your reasoning.

C2-WWT16: Water in Pipe—Speed

Water is flowing through a pipe, filling it completely. The pipe narrows to half its diameter from one section to the next. A physics student makes the following comparison of the speed of the water in the two sections:

"Since the diameter of the pipe has been cut in half, but the same amount of water still has to get through the pipe, in the smaller section the speed of the water has to double."

What, if anything, is wrong with this student's statement? If something is wrong, explain the error and how to correct it. If this statement is correct, explain why.

C3 HEAT AND TEMPERATURE

C3-SCT01: FAHRENHEIT TEMPERATURE CHANGE—CENTIGRADE TEMPERATURE CHANGE

Several students are discussing temperature conversions between Fahrenheit and Centigrade scales. They are considering a temperature change of 45 degrees Fahrenheit (°F).

Ariel: *"Since in going from Fahrenheit to Centigrade we have to use 5/9 (45 °F + 32 °F), a temperature change of 45 °F is a change of 43 °C."*

Brent: *"I think this is a temperature change of 25 °C because all we have to do is take 5/9 (45 °F)."*

Coen: *"No, you are both making this hard. A temperature change of 45 °F is 45 °C since no conversion is needed for changes in temperature, just for specific temperatures."*

With which, if any, of these students do you agree?

Ariel _____ Brent _____ Coen _____ None of them_____

Explain your reasoning.

C3-WWT02: CENTIGRADE TEMPERATURE CHANGE— KELVIN CHANGE

A student discussing a change in temperature states:

"A temperature change of 200 °C is also a 200 degree change in the Kelvin system."

What, if anything, is wrong with this statement? If something is wrong, identify it and explain how to correct it. If the statement is correct, explain why.

C3-WWT03: MIXING LIQUIDS—FINAL TEMPERATURE

A student mixes 100 g of liquid A at a temperature of 80 °C with 100 g of liquid B at a temperature of 20 °C. After the mixture comes to thermal equilibrium, it has a temperature of 40 °C. The student contends:

"If I mix 100 g of liquid A at a temperature of 60 °C with 100 g of liquid B at a temperature of 30 °C, I will also end up with a 200 g mixture at a temperature of 40 °C."

What, if anything, is wrong with the student's contention? If something is wrong, identify it and explain how to correct it. If nothing is wrong, explain the physics behind the student's answer.

C3-WWT04: Boiling Water—Temperature

Two pans of water are being heated on two burners on a stove. In pan A the water is boiling vigorously, but in pan B the water is boiling at a much slower rate. A student contends that:

"The temperature of the water in the pan that is boiling vigorously is a little higher than the other pan because it has definitely reached the boiling point."

What, if anything, is wrong with this student's contention? If something is wrong, identify it and explain how to correct it. If nothing is wrong, explain why the statement is valid.

C3-SCT05: Objects in a Room—Temperature

The objects shown have been sitting untouched on a bedside table overnight. The room they are in has been at a constant temperature of 25 °C.

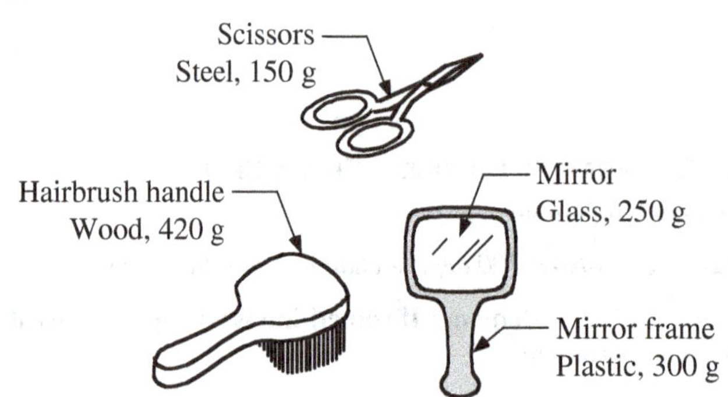

Four students are discussing the temperatures of these objects.

Abigail: *"All of the objects have been sitting there all night. They will be at the same temperature as the room."*

Beto: *"That can't be—if you touch them you can feel the differences. I think the scissors will have the lowest temperature, then the glass of the mirror, then the plastic mirror frame. The wood will be warmest."*

Carlos: *"I agree with Beto that the temperatures will be different, but I think you have to pay attention to the masses as well. The actual temperature depends on the material and the mass, with the more massive objects keeping cooler. It's hard to say whether the scissors will have a lower temperature than the mirror glass, and it's also hard to say whether the plastic will be warmer than the wood, but the mirror and metal will definitely be cooler than the wood and plastic."*

Dave: *"I think the mirror and the mirror frame are going to transfer heat to each other until they are the same temperature, because they are in contact. They'll reach some temperature that is between the cold scissors and the warm brush handle."*

With which, if any, of these students do you agree?

Abigail _____ Beto _____ Carlos _____ Dave _____ None of them_____

Explain your reasoning.

C3-CT06: COMBINING WATER, STEAM, OR ICE—FINAL MASS AND FINAL TEMPERATURE

In three experiments described below, combinations of water (liquid), steam (gas), and ice (solid) are mixed together in an insulated container, and are allowed to reach thermal equilibrium.

First, 40 g of water at 100 °C and 60 g of water at 0 °C are mixed together.

(a) When the mixture reaches thermal equilibrium, will the mass of water be (i) *greater than*, (ii) *less than*, or (iii) *the same as* the sum (100 g) of the two initial masses of water? _____

Explain your reasoning.

(b) When the mixture reaches thermal equilibrium, will the final temperature of the system be (i) *greater than*, (ii) *less than*, or (iii) *equal to* 50 °C? _____

Explain your reasoning.

Next, 50 g of steam at 100 °C and 50 g of water at 80 °C are mixed together in a different insulated container.

(c) When the combination reaches thermal equilibrium, will the mass of water (liquid) be (i) *greater than*, (ii) *less than*, or (iii) *the same as* the sum (100 g) of the two initial masses? _____

Explain your reasoning.

(d) When the mixture reaches thermal equilibrium, will the temperature of the system be (i) *greater than*, (ii) *less than*, or (iii) *equal to* 90 °C? _____

Explain your reasoning.

Finally, 40 g of liquid water at 20 °C and 60 g of ice at 0 °C are mixed together in another insulated container.

(e) When the combination reaches thermal equilibrium, will the mass of water (liquid) be (i) *greater than*, (ii) *less than*, or (iii) *equal to* 100 g? _____

Explain your reasoning.

(f) When the mixture reaches thermal equilibrium, will the temperature of the system be (i) *greater than*, (ii) *less than*, or (iii) *equal to* 10 °C? _____

Explain your reasoning.

C3-CT07: WATER AND ICE—TEMPERATURE

One student at a restaurant asks for water with lots of ice, while another asks for only a little ice. The waiter uses bottled water from the same bottle and ice from the same ice bucket to fill the order. There is still ice left in both glasses when the ice water comes to thermal equilibrium.

Will the temperature of the water in the glass with only a little ice be (a) *greater than*, (b) *less than*, or (c) *the same as* the temperature of the water in the glass with a lot of ice? _____

Explain your reasoning.

C3-CT08: PREPARING COFFEE—TIME TO HEAT

A teacher prepares a cup of instant coffee by heating 200 g of water that was initially at 20 °C with an electric immersion heater placed directly in the cup. It takes 207 seconds to warm the water to 90 °C.

(a) Another teacher with an identical cup uses the same heater to warm up 150 g of water, initially at 20 °C. Is the time taken to heat this second cup of water to 90 °C (i) *greater than*, (ii) *less than*, or (iii) *equal to* 207 seconds? _____

Explain your reasoning.

(b) A third teacher with an identical cup uses the same heater to warm up 200 g of warmer water, initially at 30 °C. Is the time taken to heat this third cup of water to 90°C (i) *greater than*, (ii) *less than*, or (iii) *equal to* 207 seconds? _____

Explain your reasoning.

(c) A fourth teacher with an identical cup uses the same heater to warm up 200 g of colder water, initially at 10 °C. Is the time taken to heat this fourth cup of water to a very warm 80 °C (i) *greater than*, (ii) *less than*, or (iii) *equal to* 207 seconds that it took for the first teacher to heat the water? _____

Explain your reasoning.

C3-TT09: TWO GLASSES OF WATER—AMOUNT OF HEAT

Two glasses each contain 500 mL samples of water. The water in glass A has a temperature of 66 °C, and the water in glass B has a temperature of 94 °C.

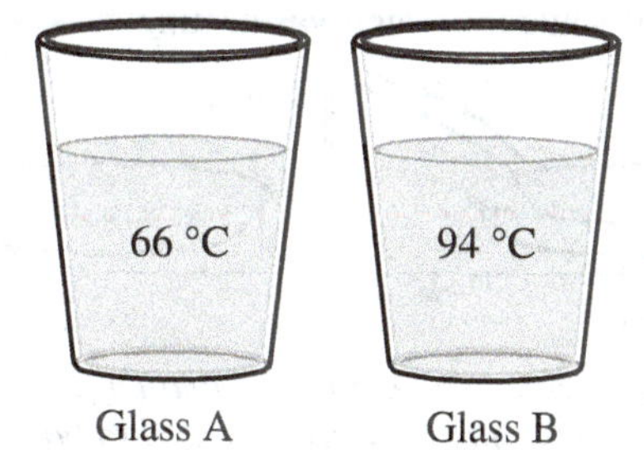

A student makes the following contention:

"Since both glasses have the same amount of water, but Glass B has a higher temperature, Glass B contains more heat."

There is a problem with this student's contention. Identify the problem and explain how to correct it.

C3-CT10: COMBINING TWO GLASSES OF WATER—FINAL TEMPERATURE

Two glasses each contain 500 mL samples of water. The water in glass A has a temperature of 66 °C, and the water in glass B has a temperature of 94 °C. The two glasses are mixed together in a larger glass.

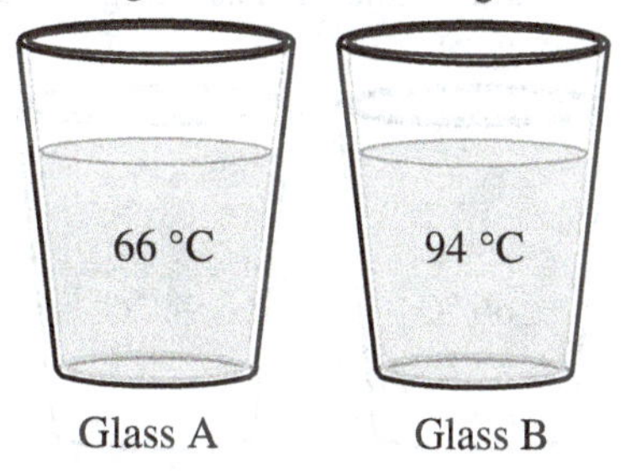

Is the final temperature of the mixture (i) *greater than*, (ii) *less than*, or (iii) *the same as* the average temperature of 80 °C of the two glasses? _____

Explain your reasoning.

C3-TT11: Two Pans of Water—Amount of Heat Added

Two pans each contain the same amount of water. The water in each pan started out at 20 °C, and both pans were heated until the water reached a temperature of 94 °C. The pan in Case A was left open while it was heating, and the pan in Case B had a tight glass lid over it while the water was heating.

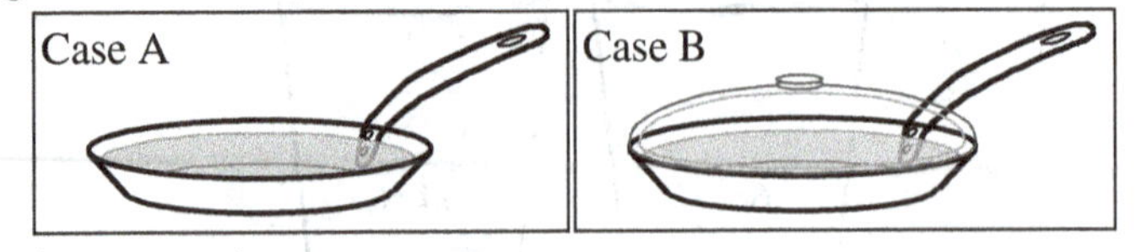

A student makes the following contention:

"Since both pans have the same amount of water and both were heated from the same initial temperature to the same final temperature, the same amount of heat had to be added to both."

There is a problem with the student's contention. Identify the problem and explain how to correct it.

C3-CT12: Combining Two Glasses of Different Liquids—Final Temperature

Two glasses each contain 500 mL samples of liquid. The liquid in Glass A has a temperature of 66 °C, and the liquid in Glass B has a temperature of 94 °C. The liquid in Glass A has a larger specific heat than the liquid in Glass B. The two glasses are mixed together in a larger glass.

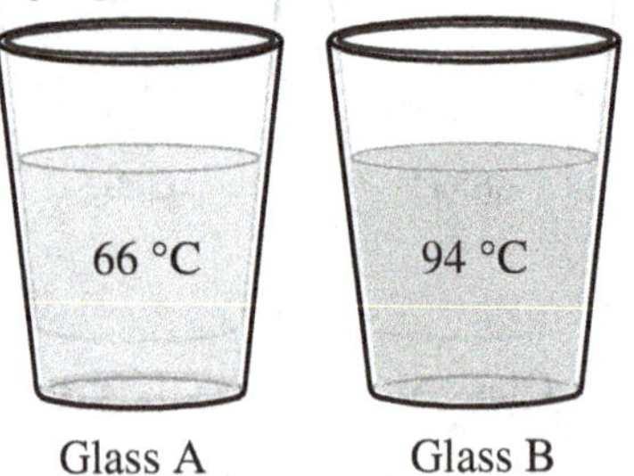

Is the final temperature of the mixture (i) *greater than*, (ii) *less than*, or (iii) *the same as* the average temperature of 80 °C of the two glasses? _____

Explain your reasoning.

C3-CT13: THERMAL ENERGY IN TWO GLASSES OF WATER—TEMPERATURE

(a) Two identical glasses contain equal samples of water. The water in Glass A has twice the thermal (internal) energy as the water in Glass B.

Is the temperature of Glass A (i) *greater than,* (ii) *less than*, or (iii) *the same as* the temperature of Glass B?

______

Explain your reasoning.

(b) Two identical glasses contain different samples of water. Glass A contains 500 mL of water, and Glass B contains 250 mL of water. The water in Glass A has twice the thermal (internal) energy as the water in Glass B.

Is the temperature of Glass A (i) *greater than,* (ii) *less than*, or (iii) *the same as* the temperature of Glass B?

______

Explain your reasoning.

(c) Two different glasses contain different samples of water. Glass A contains 500 grams of water, and Glass B contains 750 grams of water. The water in Glass B has twice the thermal (internal) energy as the water in Glass A.

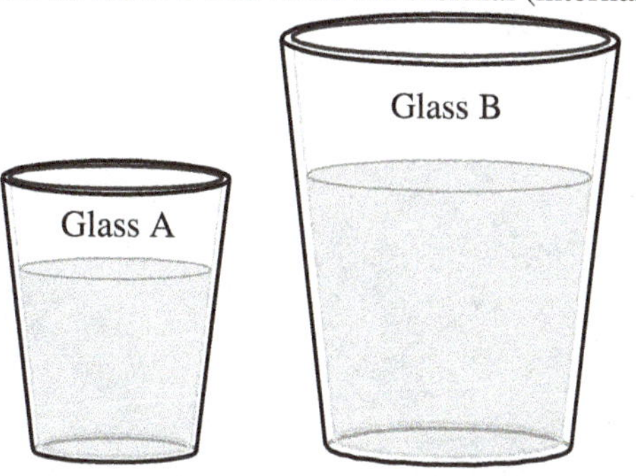

Is the temperature of Glass A (i) *greater than,* (ii) *less than*, or (iii) *the same as* the temperature of Glass B?

______

Explain your reasoning.

C3-CRT14: Melting an Ice Cube—Temperature-Time Graph

An ice cube at a temperature of −20 °C is put in a plastic bowl and placed in a microwave oven, and the oven is turned on. When the oven is turned off, the bowl contains water at a temperature of +60 °C.

Assuming the microwave transferred energy to the ice at a constant rate throughout, draw a graph (below) of the temperature of the H_2O as a function of time. The endpoints of the graph are given.

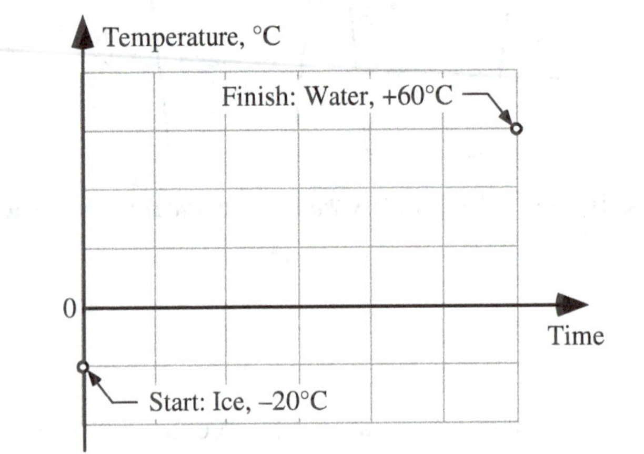

Explain why your graph looks the way it does.

C3-CT15: Heating Ice—Final Temperature

Suppose we heat sample A, which is 150 g of ice at –10 °C, and sample B, which is 150 g of ice at 0 °C, in the same microwave until both samples are water at 30 °C.

Will the time taken to heat sample A be (i) *longer than*, (ii) *shorter than*, or (iii) *equal to* the time to heat sample B? _____

Explain your reasoning.

C3-CT16: MIXING WATER AND/OR ICE—TEMPERATURE

Insulated cylinder A initially contains 150 g of ice at 0 °C, and insulated cylinder B initially contains 150 g of water at 0 °C. To each cylinder, 150 g of water at 80 °C is added, and the contents of the cylinders are stirred and then left to reach thermal equilibrium.

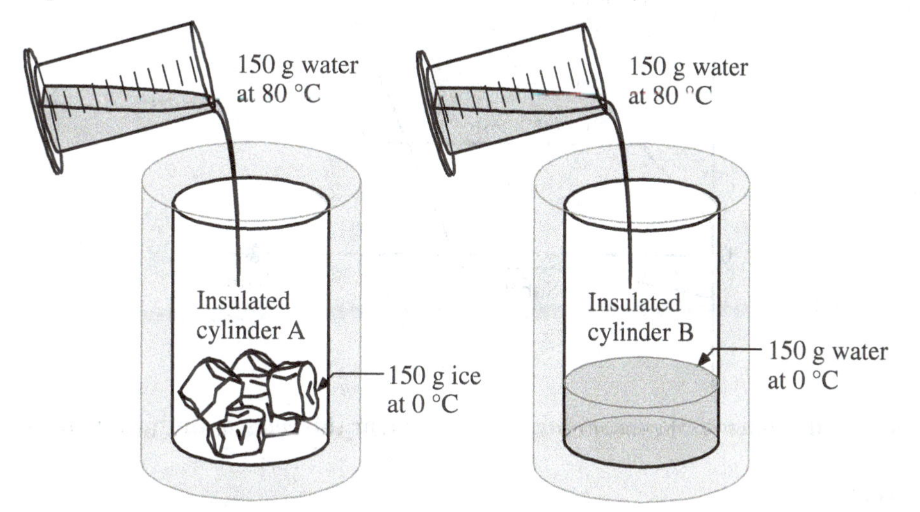

Will the final temperature of the mixture in cylinder A be (i) *greater than*, (ii) *less than*, or (iii) *equal to* the final temperature of the mixture in cylinder B? _____

Explain your reasoning.

C3-CT17: USING A STEEL TAPE AT DIFFERENT TEMPERATURES—ACTUAL LENGTH

A surveyor's 100 m steel tape is calibrated to be precise to 0.1 mm at a temperature of 15 °C. A distance between two points is measured as 63.7300 m with this steel tape on a 40 °C day. The linear temperature coefficient of expansion for steel is α_{Steel} = 11 x 10-6/°C.

Is the actual (or correct) distance (i) *shorter than*, (ii) *longer than*, or (iii) *the same as* the measured distance of 63.7300 m? _____

Explain your reasoning.

C3-CT18: Temperature-Time Graph—Properties of Samples

Samples of two pure substances are heated at a constant rate, and their temperature as a function of time is recorded. Both substances started as solids and melted. The mass of the two samples is the same.

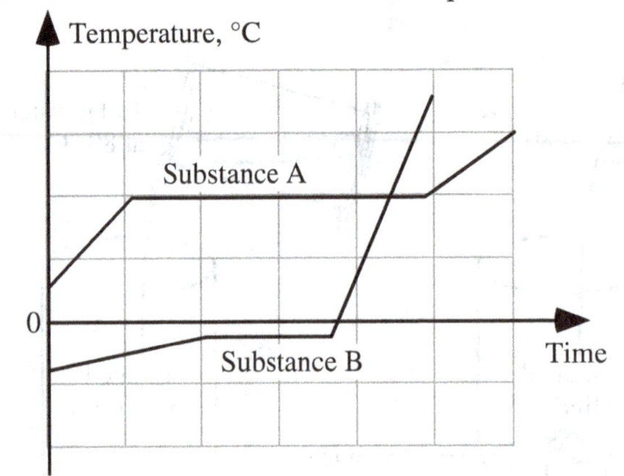

Is the melting point of substance A (i) ***greater than*****, (ii)** ***less than*****, or (iii)** ***equal to*** **the melting point of substance B? _____**

Explain your reasoning.

Is the specific heat of substance A in its solid state (i) ***greater than*****, (ii)** ***less than*****, or (iii)** ***equal to*** **the specific heat of substance B in its solid state? _____**

Explain your reasoning.

Is the latent heat of fusion of substance A (i) ***greater than*****, (ii)** ***less than*****, or (iii)** ***equal to*** **the latent heat of fusion of substance B? _____**

Explain your reasoning.

C3-CT19: Heated Beaker Filled with Glycerin—Overflow

A glass beaker is partially filled with 500 cm^3 of glycerin at 15 °C. The beaker and the glycerin are then heated to 40 °C. The thermal linear coefficient of expansion for the glass is $\alpha_{glass} = 3 \times 10^{-6}$/°C. The thermal volume coefficient of expansion for glycerin is $\beta_{Glycerin} = 5.1 \times 10^{-4}$/°C.

As the beaker and contents are heated, will the glycerin level (i) ***increase*****, (ii)** ***decrease*****, or (iii)** ***remain the same*****? _____**

Explain your reasoning.

C3-RT20: IDEAL GAS SAMPLES—TEMPERATURE I

Four sealed containers hold different amounts of an ideal gas at different temperatures and pressures. The pressure P of the gas is given in each case, as is the number of molecules N and the volume V.

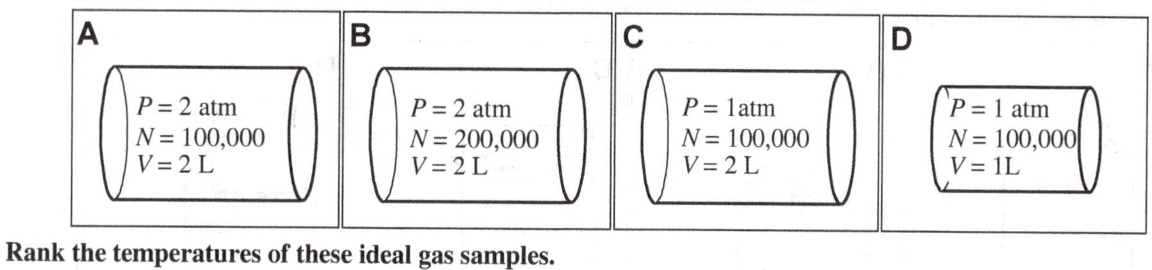

Rank the temperatures of these ideal gas samples.

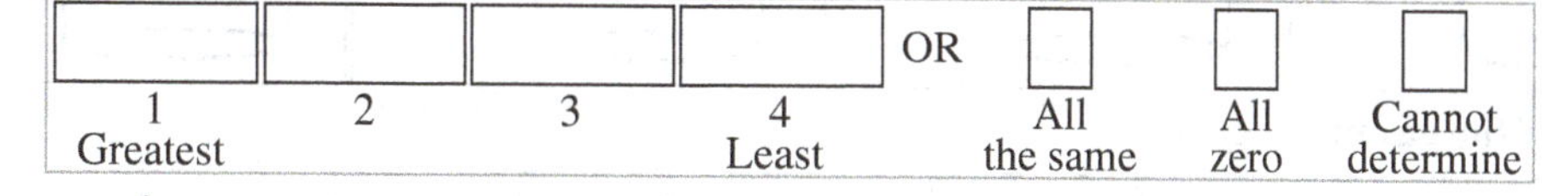

Explain your reasoning.

C3-RT21: IDEAL GAS IN CYLINDERS WITH MOVEABLE PISTONS I—PRESSURE

Cylinders with equal cross-sectional areas contain different volumes of an ideal gas sealed in by pistons. There is a weight sitting on top of each piston. The gas is the same in all four cases and is at the same temperature. The pistons are free to move without friction.

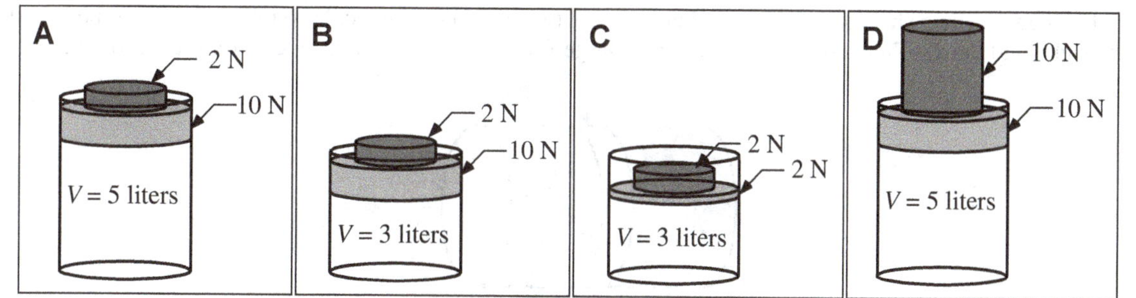

Rank the pressure of the gas in each cylinder.

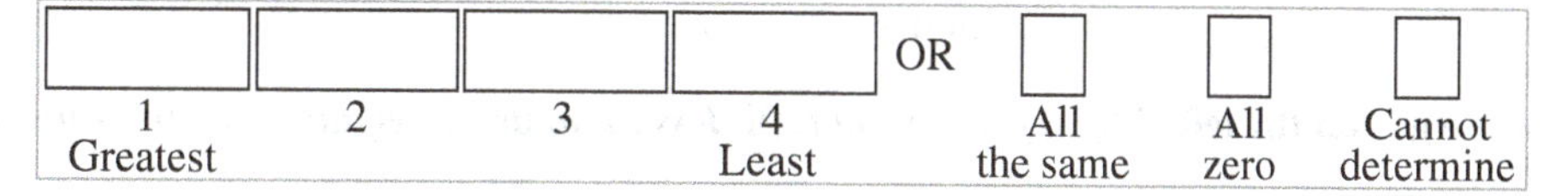

Explain your reasoning.

C3-RT22: Gas in Cylinders with Moveable Pistons—Mass

Cylinders with equal cross-sectional areas contain different volumes of an ideal gas sealed in by pistons. There is a weight sitting on top of each piston. The gas is the same in all four cases and is at the same temperature. The pistons are free to move without friction.

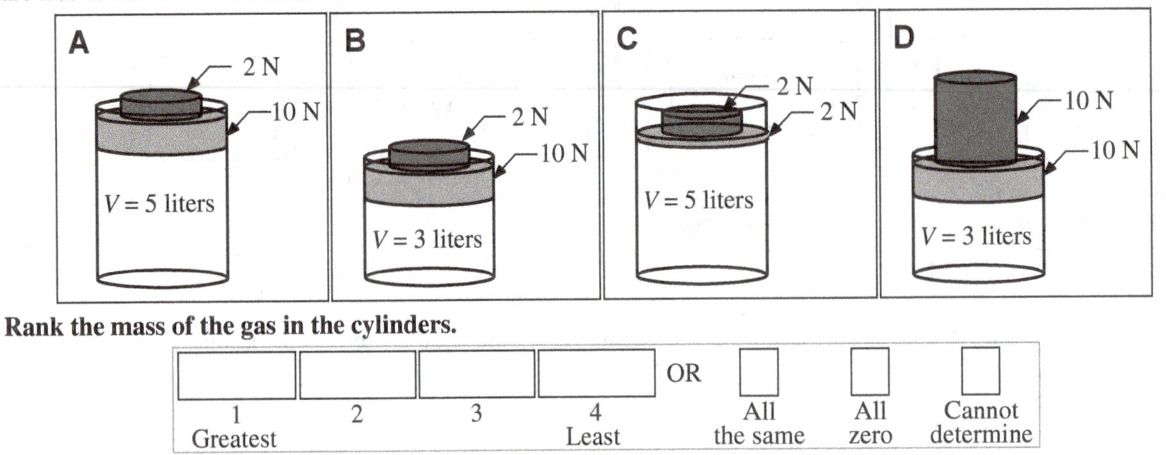

Rank the mass of the gas in the cylinders.

				OR			
1 Greatest	2	3	4 Least		All the same	All zero	Cannot determine

Explain your reasoning.

C3-CT23: Ideal Gases in a Cylinder—Pressure

A cylinder contains two samples of different ideal gases. A piston separating the two gases is free to move without friction. Each gas occupies half of the cylinder, but there are twice as many molecules on the left side of the piston as on the right, and the absolute temperature is twice as large for the gas on the right side as for the gas on the left.

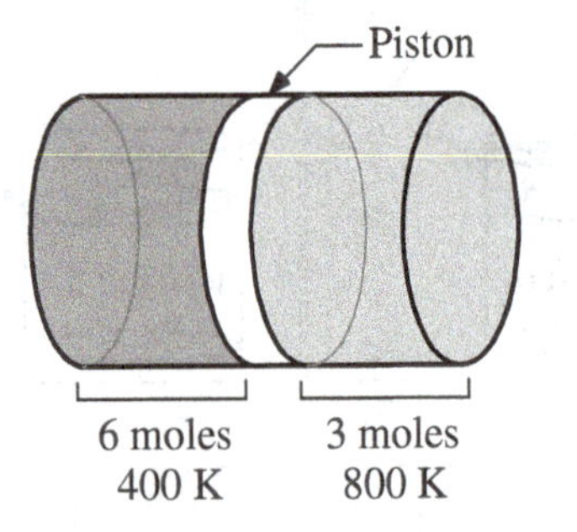

Is the pressure in the gas on the left side (i) *greater than*, (ii) *less than*, or (iii) *equal to* the pressure in the gas on the right side? _____

Explain your reasoning.

C3-CT24: IDEAL GASES IN CYLINDERS—TEMPERATURE

Two cylinders are filled to the same height H with ideal gases. The gases are different, and the cross-sectional areas of the cylinders are different. Both cylinders have pistons that are free to move without friction.

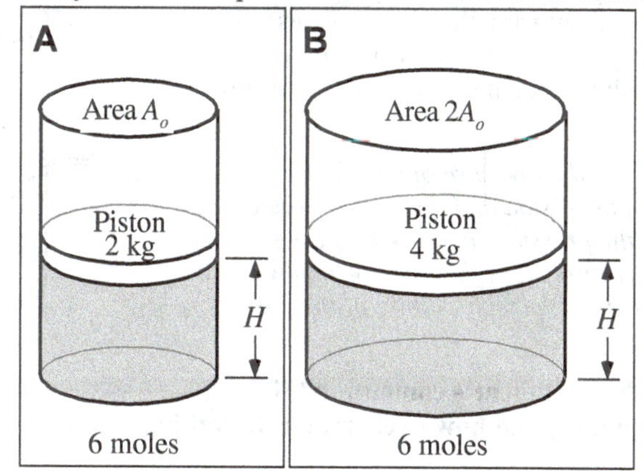

Is the temperature of the gas in cylinder A (i) *greater than*, (ii) *less than*, or (iii) *equal to* the temperature of the gas in cylinder B? _____

Explain your reasoning.

C3-WWT25: IDEAL GAS IN AN INSULATED CYLINDER I—TEMPERATURE CHANGE

An insulated cylinder contains an ideal gas. The cylinder has a piston that is initially locked in place with pins. When the pins are moved outward, the piston can move freely without friction. On top of the piston is an 8 kg metal disc. A student who is asked what will happen to the temperature of the gas after the piston is unlocked makes the following contention:

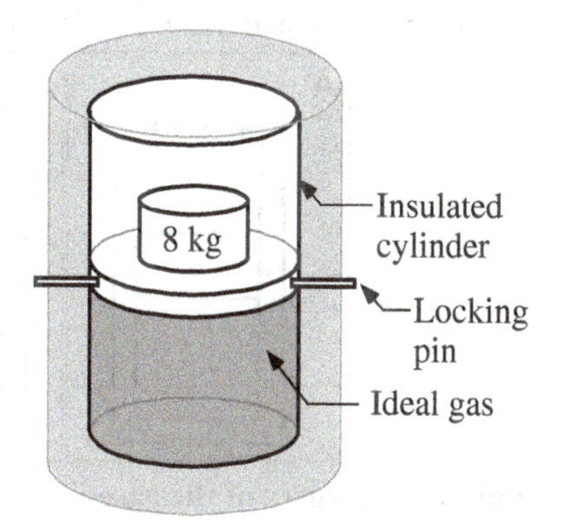

"The piston will move, but we don't know which way. Whether it moves up or down, though, since the cylinder is insulated, the temperature of the gas cannot change during the process."

What, if anything, is wrong with the student's contention? If something is wrong, identify it and explain how to correct it. If nothing is wrong, explain the physics behind the student's answer.

C3-WWT26: Ideal Gas in an Insulated Cylinder II—Temperature Change

An insulated cylinder contains an ideal gas. The cylinder has a piston that is initially locked in place with pins. On top of the piston is an 8 kg metal disc. When the pins are moved outward, the piston moves downward without friction, coming to rest 2 cm below its initial position. A student who observes this piston motion makes the following contention:

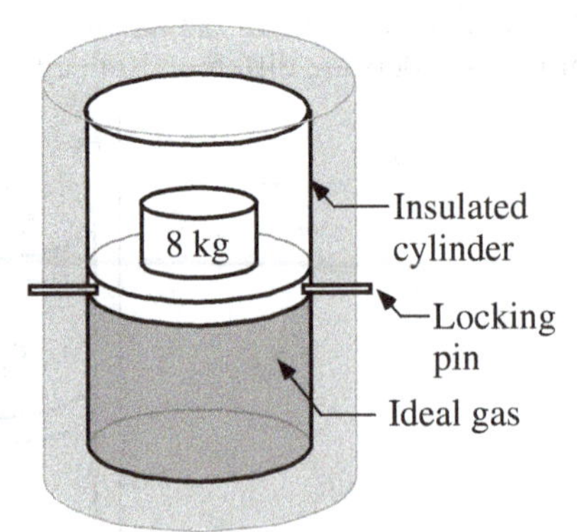

"The piston will move down because the pressure of the gas was too low to support the combined weight of the piston and the metal cylinder. As the volume of the gas decreased, the pressure increased. These two effects compensate, and the temperature stays the same, which we also know since the cylinder is insulated, so the temperature of the gas cannot change during the process."

What, if anything, is wrong with the student's contention? If something is wrong, identify it and explain how to correct it. If nothing is wrong, explain the physics behind the student's answer.

C3-RT27: Ideal Gas in Cylinders with Moveable Pistons II—Pressure

Each cylinder contains an ideal gas trapped by a piston that is free to move without friction. The pistons are at rest. All gases are at the same temperature. The diameter of the cylinder in Case B is twice the diameter of the cylinders in the other cases, and the mass of the piston in Case C is twice the mass in the other cases.

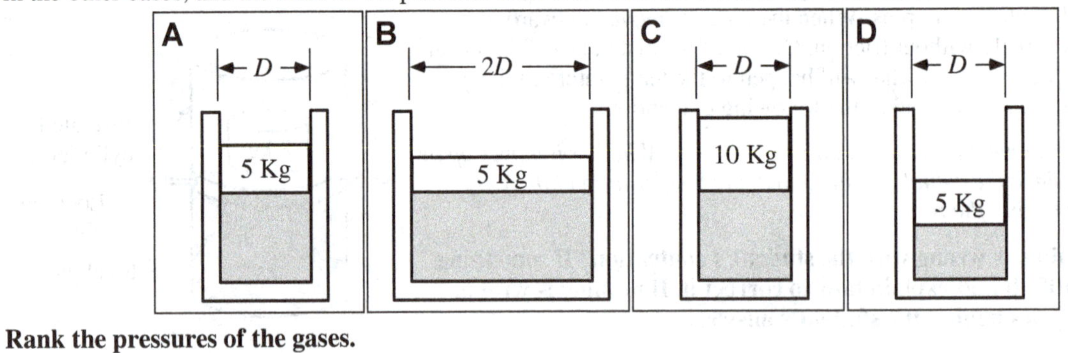

Rank the pressures of the gases.

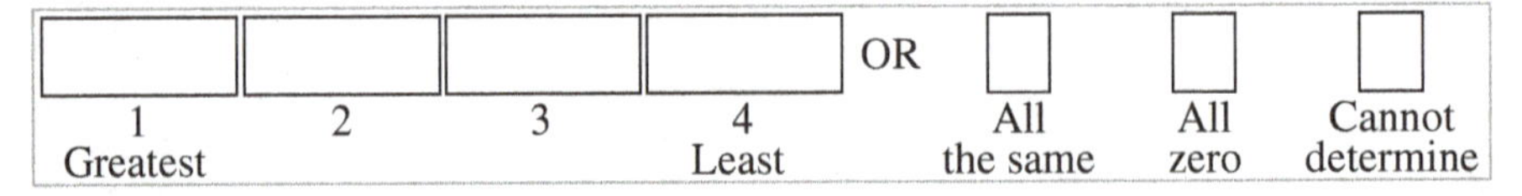

Explain your reasoning.

C3-LMCT28: HEAVY MOVEABLE PISTON IN CONTAINER WITH VOLUME—PRESSURE

An ideal gas is trapped in a container with a moveable, frictionless piston, and a metal disc is placed on top of the piston.

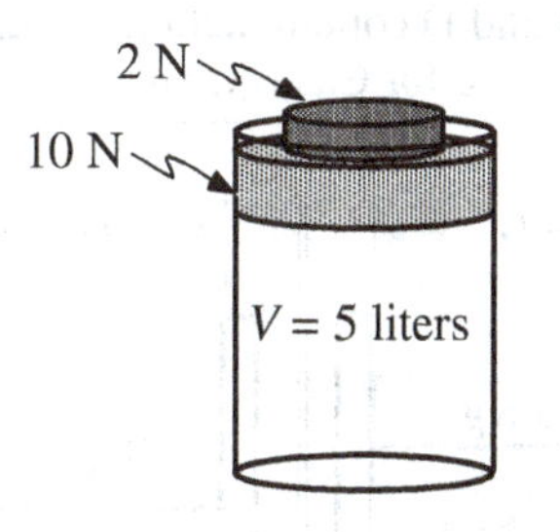

Identify from choices (i) – (iv) how the changes to this system described in (a) to (d) below affect the pressure of the gas in the cylinder.

This change will:

(i) **increase** the pressure of the system.
(ii) **decrease** the pressurc of the system.
(iii) have **no effect** on the pressure of the system.
(iv) have an **indeterminate** effect on the pressure of the system.

All of these changes are made to the initial situation described above. In each case, the volume of gas in the cylinder is the same as above.

(a) The new system has a lighter piston but has the same metal disc placed on it. _____
Explain your reasoning.

(b) The new system has a heavier metal disc but has the same piston. _____
Explain your reasoning.

(c) The new system has a lighter piston and a lighter metal disc. _____
Explain your reasoning.

(d) The new system has a smaller cross-sectional area cylinder and piston, but the weight of the piston is the same and the disc is the same. _____
Explain your reasoning.

C3-RT29: Ideal Gases in Cylinders with a Piston—Pressure

Each cylinder contains an ideal gas trapped by a piston that is free to move without friction. The pistons are at rest. All gases are at the same temperature, and the pistons and cylinders are identical. The cylinders in Cases A and C contain nitrogen, and the cylinders in Cases B and D contain helium, which has fewer grams per mole. The volume of gas is the same for Cases A and B, and the same for Cases C and D.

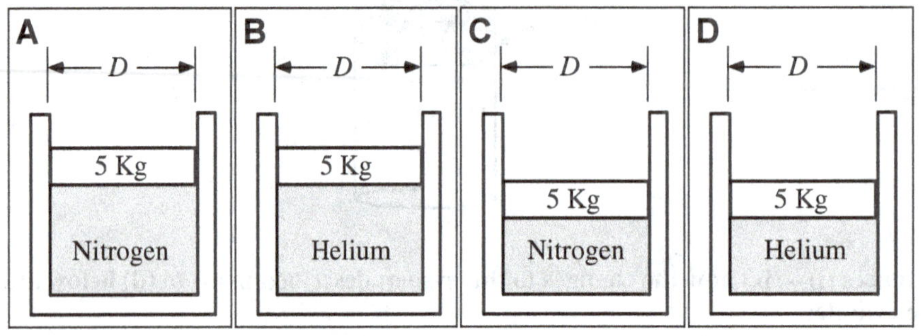

Rank the pressures of the gases in the cylinders.

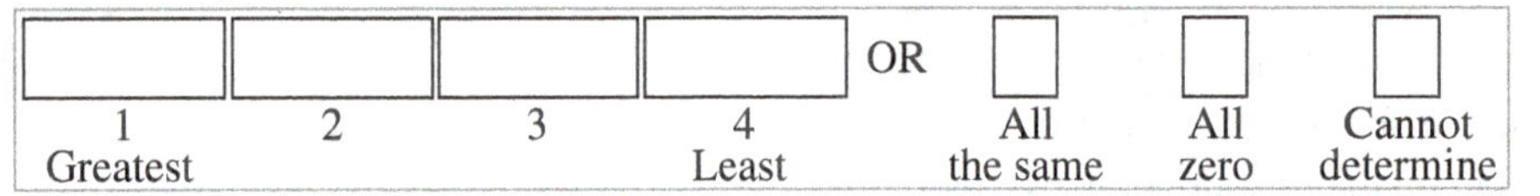

Explain your reasoning.

C3-RT30: Ideal Gases in Cylinders with a Piston—Number of Moles

Each cylinder contains an ideal gas trapped by a piston that is free to move without friction. The pistons are at rest. All gases are at the same temperature, and the pistons and cylinders are identical. The cylinders in Cases A and C contain nitrogen, and the cylinders in Cases B and D contain helium, which has fewer grams per mole. The volume of gas is the same for Cases A and B, and the same for Cases C and D.

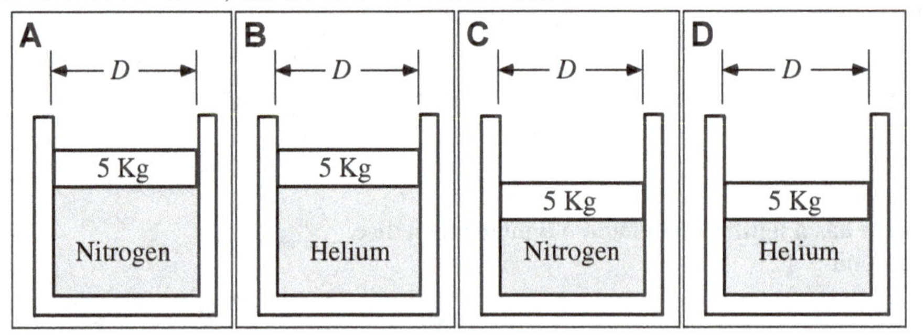

Rank the number of moles of gas in the cylinders.

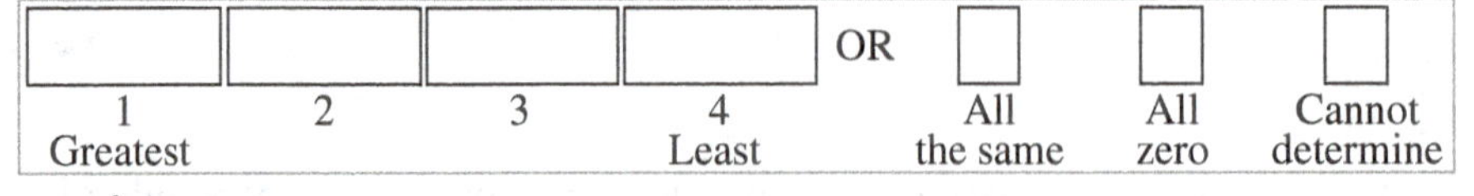

Explain your reasoning.

C3-RT31: IDEAL GASES IN CYLINDERS WITH A PISTON—TEMPERATURE

Each cylinder contains an ideal gas trapped by a piston that is free to move without friction. The pistons are at rest. All gases are at the same temperature, and each cylinder contains the same number of moles of gas. The pistons and cylinders are identical. The cylinders in Cases A and C contain nitrogen, and the cylinders in Cases B and D contain helium, which has fewer grams per mole. The volume of gas is the same for Cases A and B, and the same for Cases C and D.

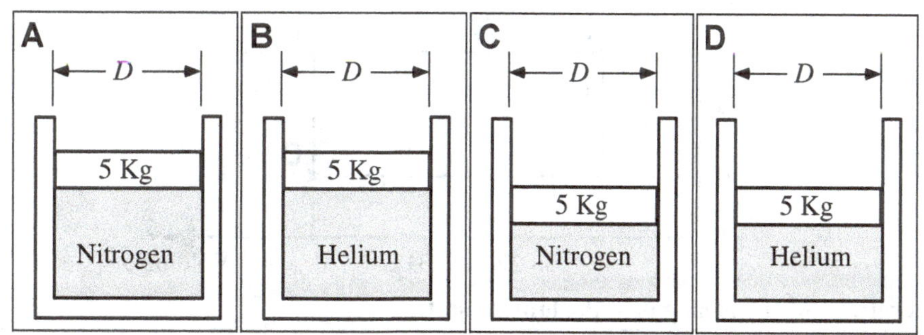

Rank the temperature of the gas in the cylinders.

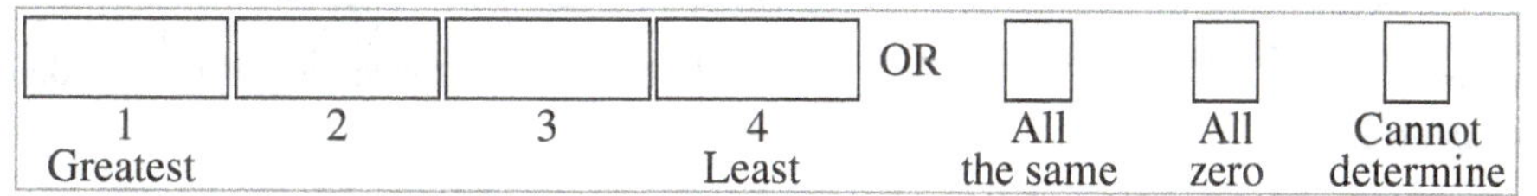

Explain your reasoning.

C3-RT32: PRESSURE-VOLUME GRAPH I—TEMPERATURE IN DIFFERENT STATES

Five points representing five different states of one mole of an ideal gas are labeled on the pressure–volume graph below.

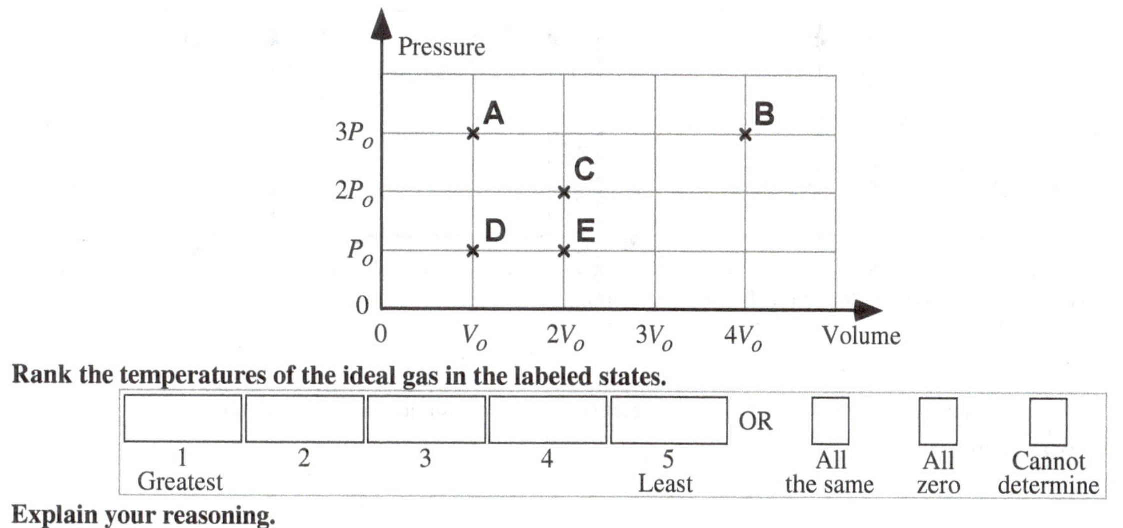

Explain your reasoning.

C3-RT33: Pressure-Volume Graph II—Temperature in Different States

Five points representing five different states of one mole of an ideal gas are labeled on the pressure–volume graph below.

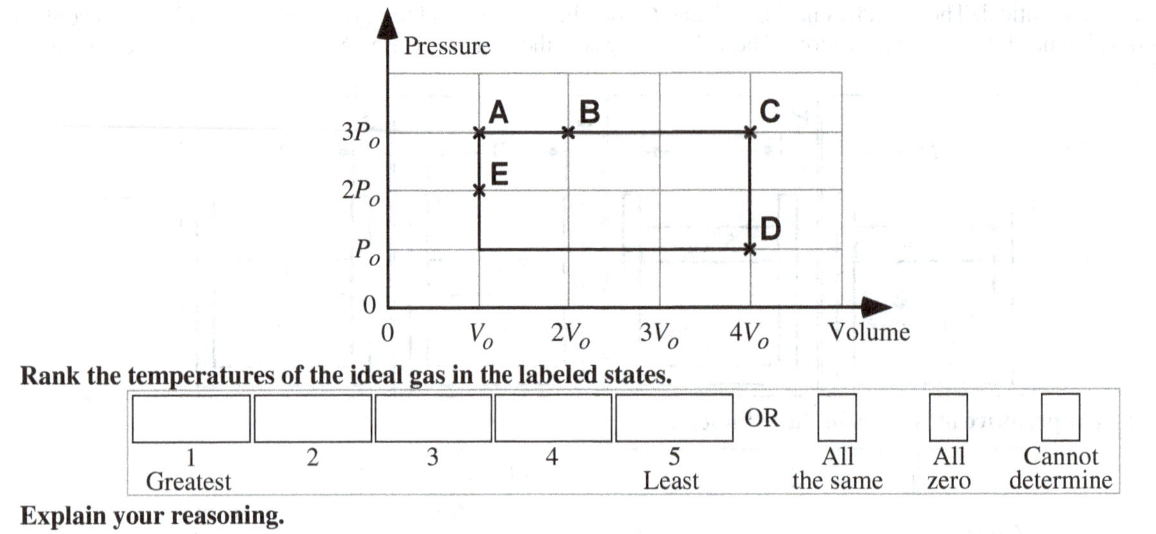

Rank the temperatures of the ideal gas in the labeled states.

1 Greatest | 2 | 3 | 4 | 5 Least | OR | All the same | All zero | Cannot determine

Explain your reasoning.

C3-RT34: Thermodynamic Ideal Gas Processes—Final Temperature

Four thermodynamic processes are illustrated below. These processes are for the same ideal gas starting in the same state (same pressure, volume, temperature, and amount of gas) and ending at the same final volume that is twice the initial volume.

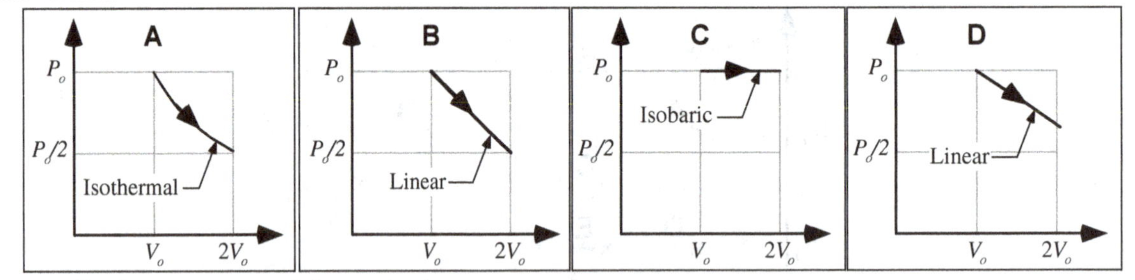

Rank the final temperature of the gas in these processes.

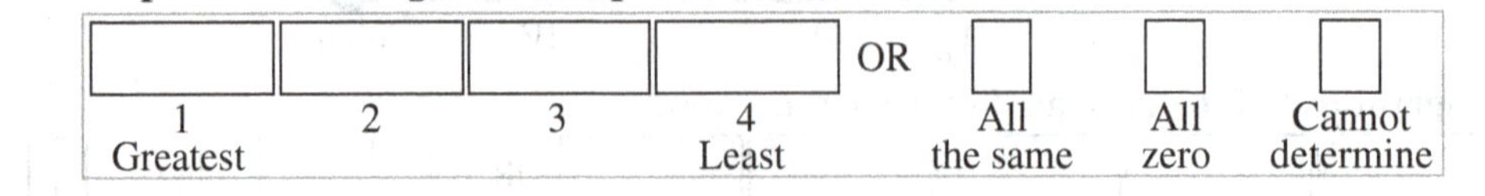

Explain your reasoning.

C3-CRT35: PRESSURE-VOLUME GRAPH—PRESSURE, VOLUME, AND TEMPERATURE BAR CHARTS

An ideal gas trapped in a cylinder expands while its pressure drops. The starting point for this process is labeled *A* and the endpoint is labeled *B* in the graph of pressure versus volume.

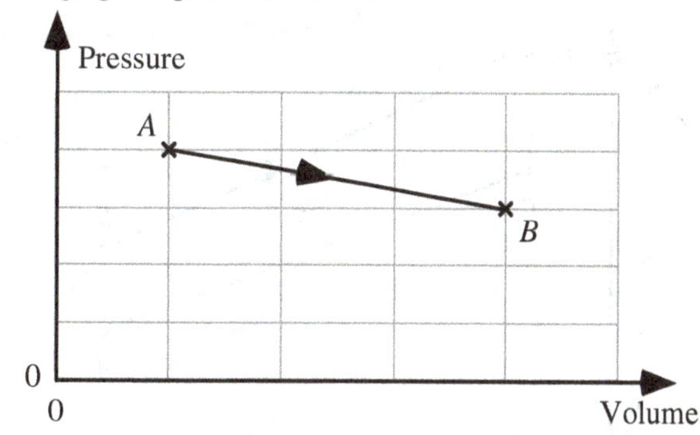

This initial pressure, volume, and temperature are shown in the histograms. Complete the histograms, showing the final pressure, volume, and absolute temperature.

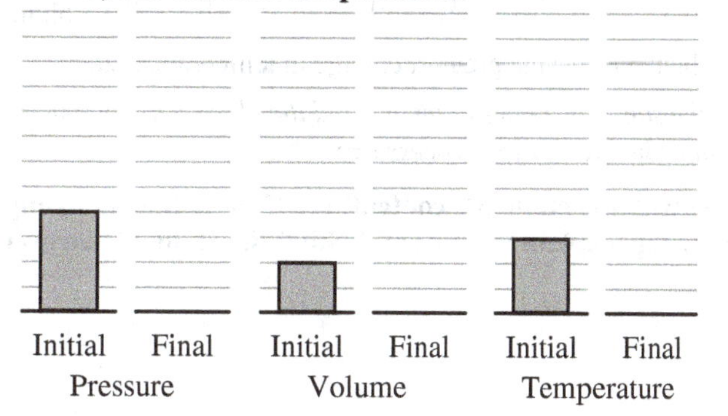

Explain your reasoning.

C3-CT36: PRESSURE–VOLUME GRAPHS FOR EXPANDING GAS—WORK DONE

Two containers of the same number of moles of an ideal gas are taken from the same initial state (same pressure, volume, and temperature) to the same final state by two different paths, as shown.

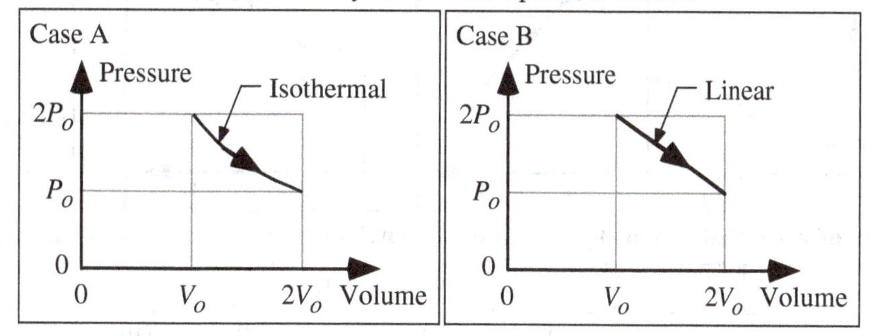

Is the work done by the gas on its surroundings (i) *greater* in Case A, (ii) *greater* in Case B, or (iii) *the same* in both cases? _____

Explain your reasoning.

C3-WWT37: Pressure–Volume Graph for Two Processes—Change in Temperature

Two processes $a \Rightarrow b$ and $c \Rightarrow d$ are shown on the pressure versus volume graph. A sample of an ideal gas is expanded using process $a \Rightarrow b$, and an identical sample is expanded using process $c \Rightarrow d$.

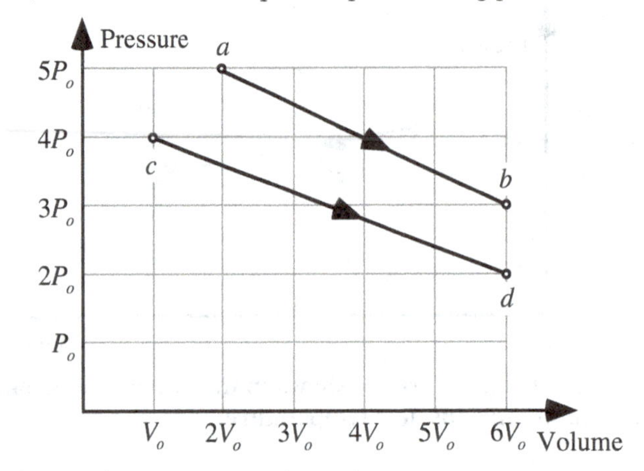

A student trying to decide which gas has the greatest change in temperature states:

"The change in pressure is the same for both processes, and the change in volume is greater for process $c \Rightarrow d$. So the change in temperature must be greater for process $c \Rightarrow d$."

What, if anything, is wrong with the student's contention? If something is wrong, identify it and explain how to correct it. If nothing is wrong, explain the physics behind the student's answer.

C3-RT38: Pressure–Volume Graphs for Various Processes—Work Done by Gas

In each case, the same ideal gas undergoes a thermodynamic process starting in the same state (same pressure, volume, temperature, and amount of gas). The final equilibrium states are different for each case.

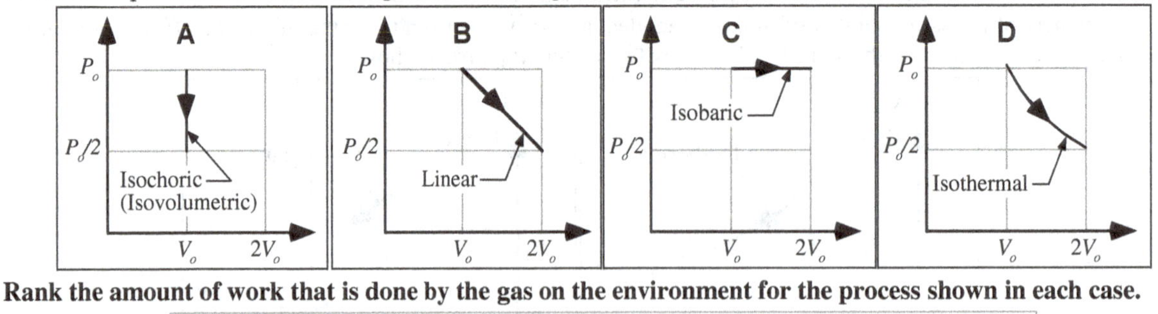

Rank the amount of work that is done by the gas on the environment for the process shown in each case.

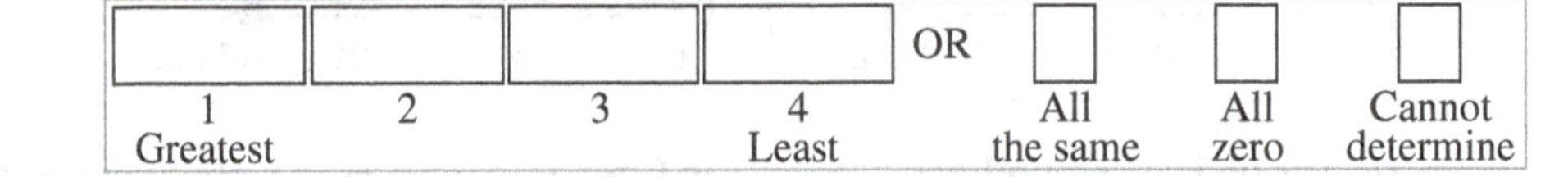

Explain your reasoning.

C3-CT39: ISOTHERMAL AND ADIABATIC IDEAL GAS PROCESSES—WORK, HEAT, AND TEMPERATURE

In both cases below, one mole of an ideal gas is expanded from an initial volume V_o to a final volume $2V_o$. In both cases, the gas is identical and the initial pressure is $2P_o$. The expansion is adiabatic in A and isothermal in B.

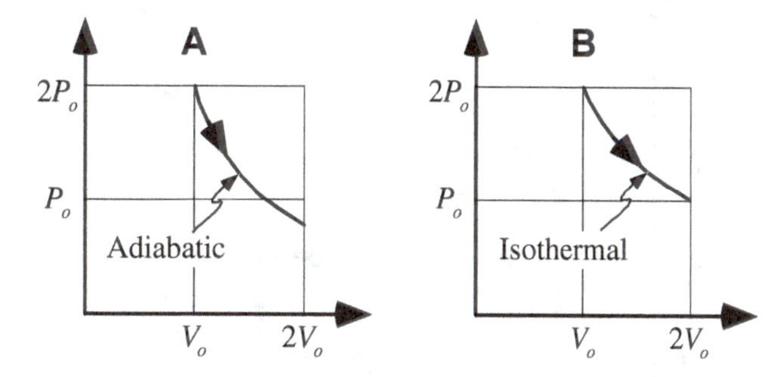

(a) Will the final temperature of the gas be (i) *greater in* Case A, (ii) *greater* in Case B, or (iii) *the same* in both cases? _____

Explain your reasoning.

(b) Will the work done by the gas be (i) *greater* in Case A, (ii) *greater* in Case B, or (iii) *the same* in both cases? _____

Explain your reasoning.

(c) Will the heat exchanged by the gas with its surroundings be (i) *greater* in Case A, (ii) *greater* in Case B, or (iii) *the same* in both cases? _____

Explain your reasoning.

C3-CT40: PRESSURE–VOLUME GRAPH FOR PROCESSES—WORK DONE

Two processes $a \Rightarrow b$ and $c \Rightarrow d$ are shown on the pressure versus volume graph. A sample of an ideal gas is expanded using process $a \Rightarrow b$, and an identical sample is expanded using process $c \Rightarrow d$.

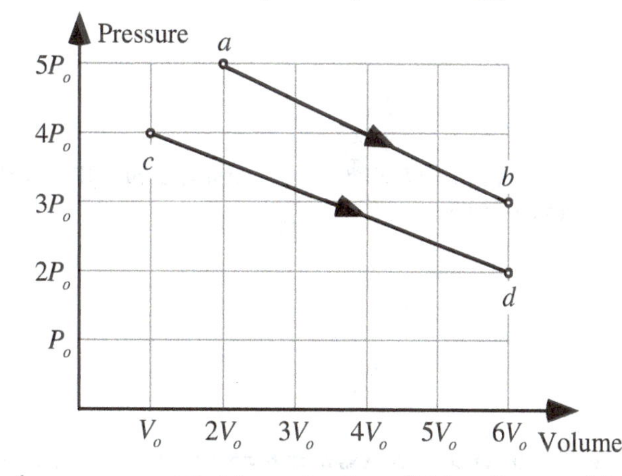

Is the work done by the gas in process a ⇒ b (a) *greater than*, (b) *less than*, or (c) *the same as* the work done by the gas in process c ⇒ d? _____
Explain your reasoning.

C3-CT41: Pressure–Volume Graph for Various Processes—Work

An ideal gas is trapped in a cylinder with a piston. Eight states of the gas are labeled *a–h*, and eight processes between these states are shown as solid lines with arrows.

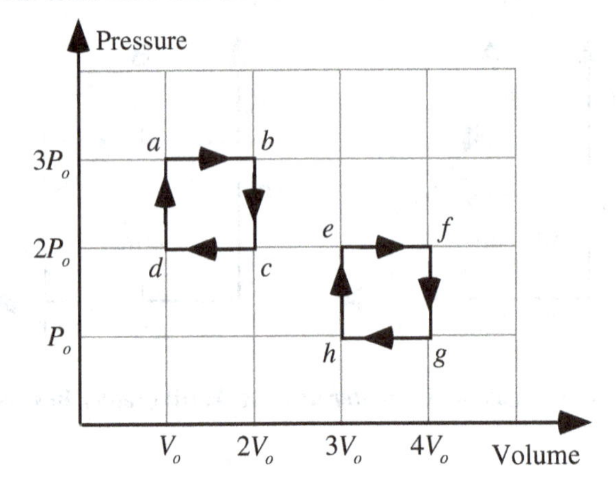

(a) Is work done by the gas for the process *a* ⇒ *b* (i) *greater than*, (ii) *less than*, or (iii) *equal to* the work done by the gas for the process *e* ⇒ *f*? _____

Explain your reasoning.

(b) Is work done by the gas for the process *b* ⇒ *c* (i) *greater than*, (ii) *less than*, or (iii) *equal to* the work done by the gas for the process *f* ⇒ *g*? _____

Explain your reasoning.

(c) Is work done by the gas for the process *a* ⇒ *b* ⇒ *c* ⇒ *d* (i) *greater than*, (ii) *less than*, or (iii) *equal to* the work done by the gas for the process *e* ⇒ *f* ⇒ *g* ⇒ *h*? _____

Explain your reasoning.

(d) Is total work done by the gas for the cyclic process *a* ⇒ *b* ⇒ *c* ⇒ *d* ⇒ *a* (i) *greater than*, (ii) *less than*, or (iii) *equal to* the total work done by the gas for the process *e* ⇒ *f* ⇒ *g* ⇒ *h* ⇒ *e*? _____

Explain your reasoning.

C3-LMCT42: CARNOT HEAT ENGINE I—EFFICIENCY

A Carnot heat engine operating between 727 °C and 127 °C takes in 2,000 J from the hot reservoir and exhausts 800 J to the cold reservoir.

Identify from choices (i)–(iv) how each change (a) to (e) described below will affect the efficiency of the heat engine as compared to the initial efficiency.

This change will cause the *efficiency* of this Carnot heat engine:

(i) to ***increase***.

(ii) to ***decrease***.

(iii) to *remain the **same***.

(iv) to be ***indeterminate***.

All of these modifications are individual changes to the initial situation.

(a) The temperature of the hot reservoir (727 °C) is increased. _____
Explain your reasoning.

(b) The temperature of the cooler reservoir (127 °C) is increased. _____
Explain your reasoning.

(c) Energy taken into the engine (2000 J) is increased. _____
Explain your reasoning.

(d) The temperatures of the hot reservoir and cool reservoir are increased by the same amount. _____
Explain your reasoning.

(e) The temperature of the hot reservoir is increased while the temperature of the cool reservoir is decreased by the same amount. _____
Explain your reasoning.

C3-WWT43: Carnot Engine II—Efficiency

A newly designed engine operating between 727 °C and 127 °C takes in 5,000 J from the hot reservoir and exhausts 1,000 J at the lower temperature. A student states:

"This is a great new engine. The efficiency of a Carnot engine is 82.5% between those temperatures using $(T_h\text{-}T_c)/T_h$, *but this new engine has an efficiency of 80%, which is pretty close to the best possible Carnot engine between those temperatures."*

What, if anything, is wrong with this student's contention? If something is wrong, identify it and explain how to correct it. If nothing is wrong, explain why the statement is valid.

C3-SCT44: Carnot Engine III—Efficiency

A Carnot engine operating between 727 °C and 127 °C takes in 1,000 J from a hot reservoir. Students are discussing the efficiency of this engine:

Albert: *"The efficiency of this engine is 82.5% using* $(T_h\text{-}T_c)/T_h$ *and it does not depend on how much energy is taken from the hot reservoir."*

Ben: *"Using* $(T_h\text{-}T_c)/T_h$ *is the right way to go, but the temperatures have to be in K so the efficiency of this engine is 60%."*

Connie: *"The efficiency of this engine is 80.6% using* $(T_h\text{-}T_c)/T_h$ *after converting the temperatures to Fahrenheit (°F)."*

With which, if any, of these students do you agree?

Albert _____ Ben _____ Connie _____ None of them_____

Explain your reasoning.

D1 ELECTROSTATICS

D1-RT01: ELECTROSCOPE NEAR A CHARGED ROD—ELECTROSCOPE NET CHARGE

A charged rod is brought close to an electroscope that is initially uncharged. In Cases A and B, the rod is positively charged; in Cases C and D, the rod is negatively charged. In Cases A and C, the leaf of the electroscope is deflected the same amount, which is more than it is deflected in Cases B and D.

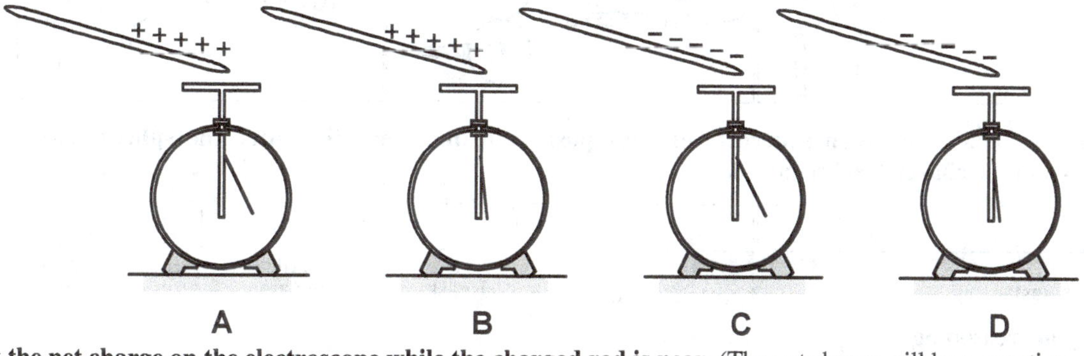

Rank the net charge on the electroscope while the charged rod is near. (The net charge will be a negative value if there is more negative than positive charge on the electroscope.)

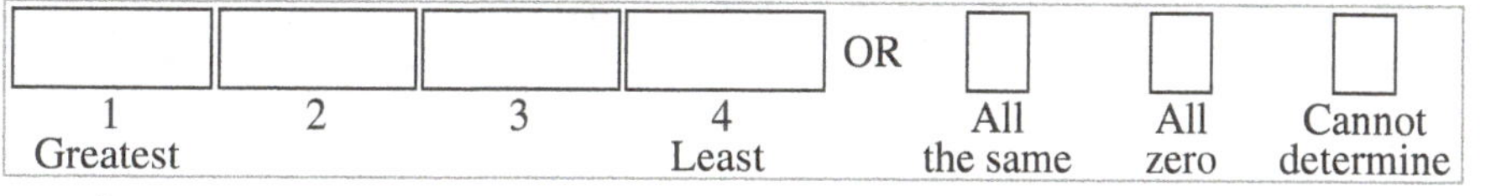

Explain your reasoning.

D1-RT02: TRANSFER OF CHARGE IN CONDUCTORS—CHARGE ON LEFT CONDUCTOR

Two identical conducting spheres are shown with an initial given number of units of charge. The two spheres are brought into contact with each other. After several moments the spheres are separated.

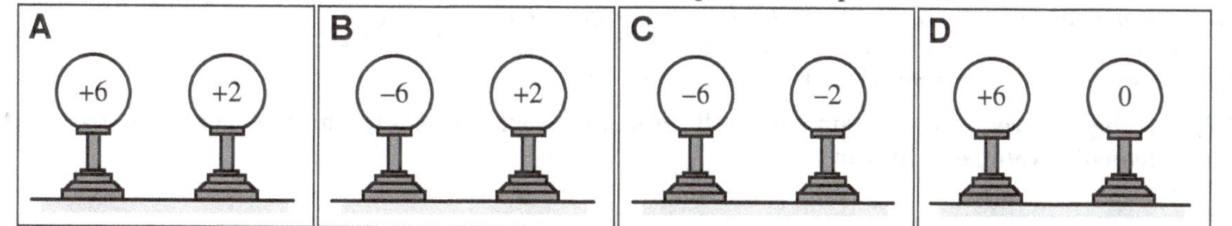

Rank the charge on the left sphere from the highest positive charge to the lowest negative charge after they have been separated. (Note that -6 is lower than -2).

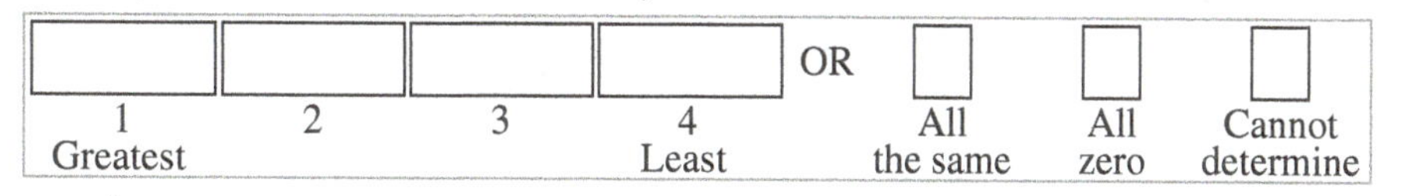

Explain your reasoning.

D1-RT03: Induced Charges near a Charged Rod—Net Charge

A charged rod is moved to the same distance from a pair of uncharged metal spheres as shown. The spheres in each pair are initially in contact, but they are then separated while the rod is still in place. Then the rod is removed.

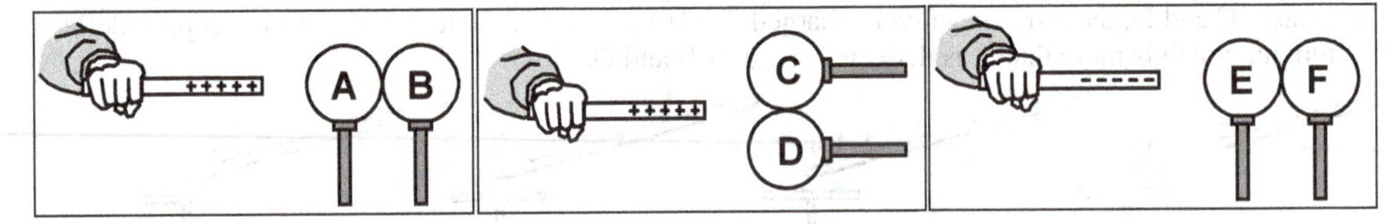

Rank the net charge on each sphere from most positive to most negative after the spheres have been separated and the charged rod removed.

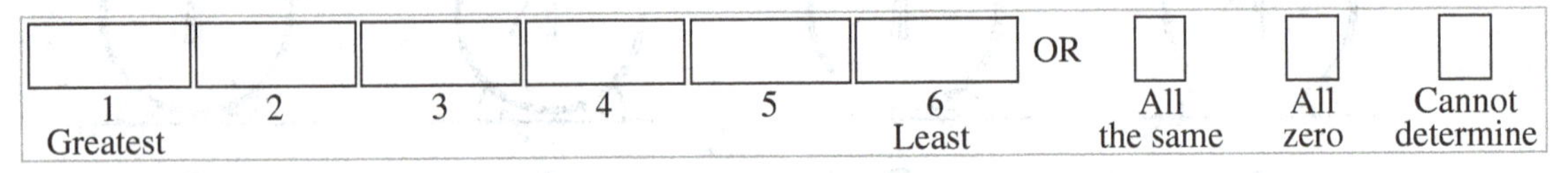

Explain your reasoning.

D1-WWT04: Charged Insulator and a Grounded Conductor—Induced Charge

A charged insulating sphere and a grounded conducting sphere are initially far apart. The charged insulator is then moved near the grounded conductor as shown. A student makes the following statement:

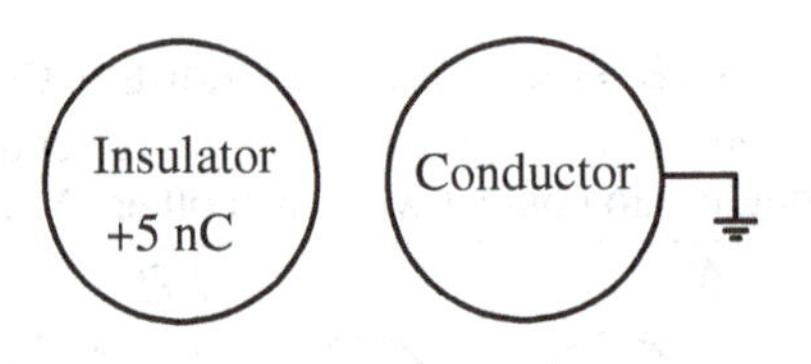

"When the charged insulator is brought close to the grounded conductor, it will cause the negative charges in the conductor to move to the side closest to the insulator. If the charged insulator is taken away, the conductor will be left with a negative charge evenly distributed over its surface."

What, if anything, is wrong with this statement? If something is wrong, explain the error and how to correct it. If the statement is valid, explain why.

D1-QRT05: THREE CONDUCTING SPHERES—CHARGE

Two conducting spheres rest on insulating stands. Sphere B is smaller than Sphere A. Both spheres are initially uncharged and they are touching. A third conducting sphere, C, has a positive charge. It is brought close to (but not touching) Sphere B as shown.

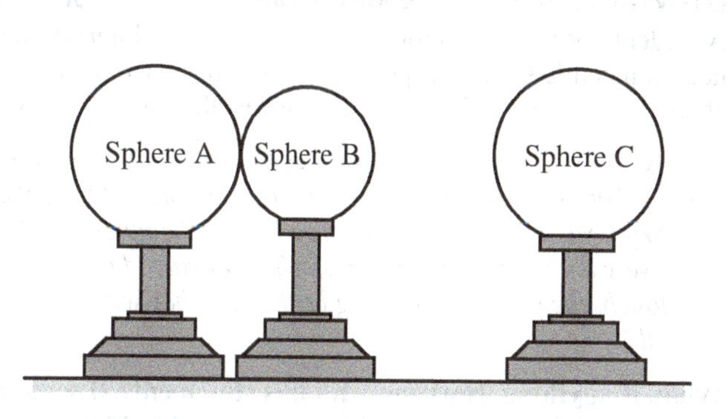

(a) Is the net charge on Sphere A at this time (i) *positive*, (ii) *negative*, or (iii) *zero*? _____
Explain your reasoning.

(b) Is the net charge on Sphere B at this time (i) *positive*, (ii) *negative*, or (iii) *zero*? _____
Explain your reasoning.

(c) Is the magnitude of the net charge on Sphere A (i) *greater than*, (ii) *less than*, or (iii) *equal to* the magnitude of the net charge on Sphere B? _____
Explain your reasoning.

Sphere B is now moved to the right so that it touches Sphere C. As a result of this move:

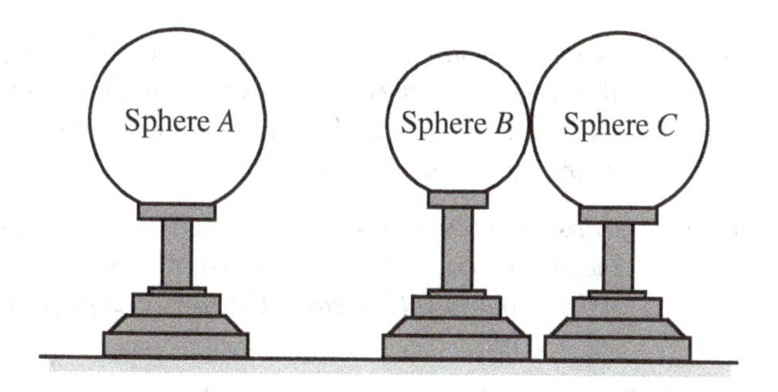

(d) Does the magnitude of the net charge on Sphere A (i) *increase*, (ii) *decrease*, or (iii) *remain the same*? _____
Explain your reasoning.

(e) Does the magnitude of the net charge on Sphere C (i) *increase*, (ii) *decrease*, or (iii) *remain the same*? _____
Explain your reasoning.

D1-WWT06: Uncharged Metal Sphere near a Charged Rod—Charge Distribution

A student observes a demonstration involving an interaction between a neutral metallic sphere suspended from a string and a negatively charged insulating rod. The student makes the following statement:

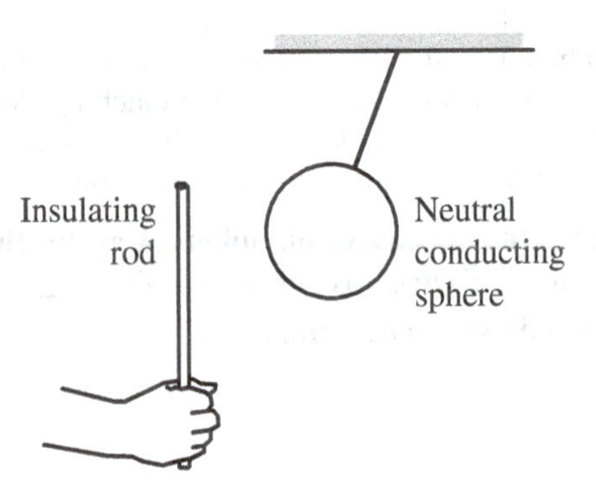

"As the negatively charged rod nears the sphere, it causes the electrons in the sphere to move away from the rod. The side of the sphere nearest to the rod becomes positively charged while the other side becomes negatively charged. So the sphere will be attracted toward the rod. If they touch, the sphere will swing back since they will both become neutral."

What, if anything, is wrong with this statement? If something is wrong, explain the error and how to correct it. If the statement is valid, explain why.

D1-SCT07: Charged Rod and Electroscope—Deflection

A positively charged rod is brought near an electroscope. Even though the rod does not touch the electroscope, the leaf of the electroscope deflects. Below, three students discuss this demonstration.

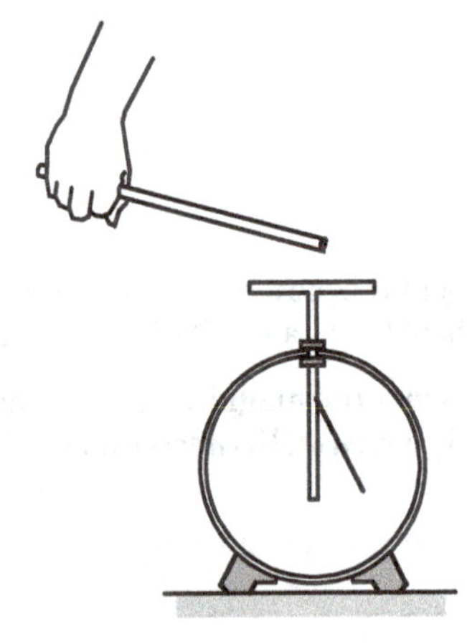

Amadeo: *"There are positive charges that jump from the rod to the plate of the electroscope. Since the electroscope is now charged, the leaf moves out."*

Barun: *"Charges don't have to move from the rod to the plate to deflect. When the rod comes close, electrons in the electroscope move toward the plate. This leaves the bottom of the electroscope positively charged, and the leaf lifts."*

Carmen: *"Positive charges are fixed in place. When the rod is brought close to the electroscope plate, the electrons in the plate are attracted and jump to the rod. This leaves the electroscope positively charged, and the leaf lifts."*

With which of these students do you agree?

Amadeo _____ Barun _____ Carmen _____ None of them _____

Explain your reasoning.

D1-QRT08: CHARGED ROD NEAR ELECTROSCOPE—CHARGE

A student first holds a positively charged rod near the top plate of an electroscope without touching it. The electroscope foil deflects. The electroscope was initially uncharged.

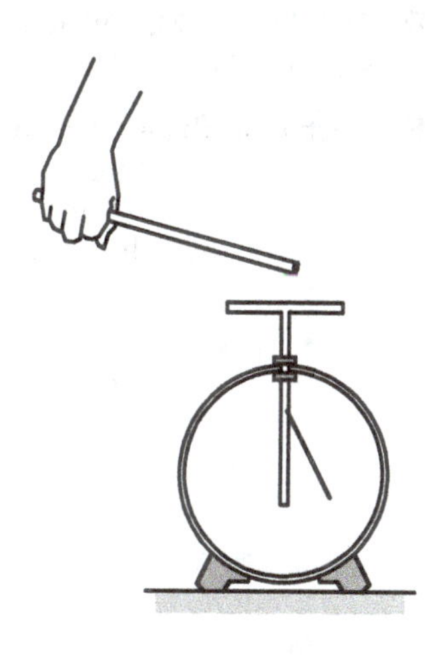

(a) Is the electroscope now (i) ***positively charged*****, (ii)** ***negatively charged*****, or (iii)** ***neutral*****.** _____

Explain your reasoning.

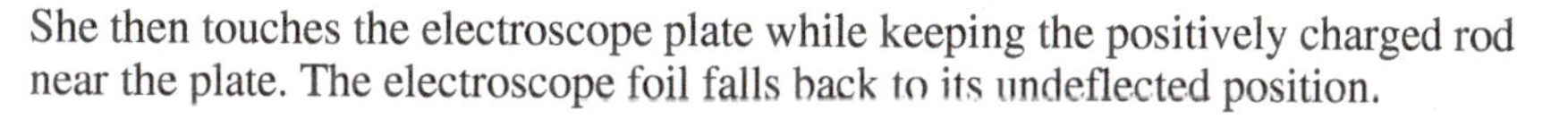

She then touches the electroscope plate while keeping the positively charged rod near the plate. The electroscope foil falls back to its undeflected position.

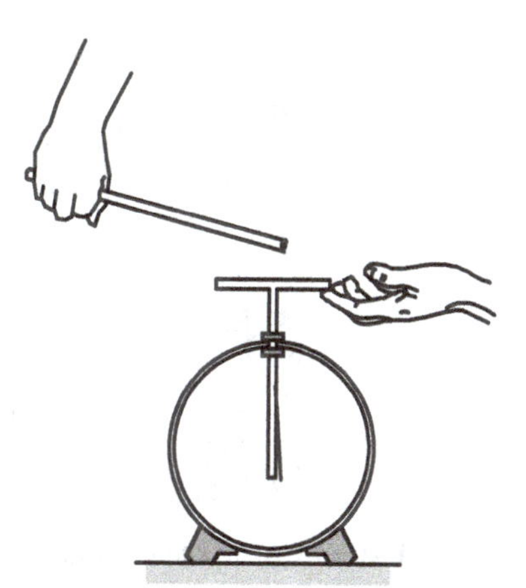

(b) Is the electroscope (i) ***positively charged*****, (ii)** ***negatively charged*****, or (iii)** ***neutral*****.** _____

Explain your reasoning.

While holding the positively charged rod stationary, she removes her hand which is touching the electroscope. Finally, she removes the charged rod.

(c) Is the electroscope (i) ***positively charged*****, (ii)** ***negatively charged*****, or (iii)** ***uncharged*****.** _____

Explain your reasoning.

(d) Will the electroscope foil be (i) ***deflected*** **or (ii)** ***undeflected*****?** _____

Explain your reasoning.

D1-QRT09: Two Charges—Force on Each

In each case shown below, two charges are fixed in place and are exerting forces on each other.

For each case, draw a vector of appropriate length and direction representing the electric force acting on each charge due to the other charge. Draw the vector representing the force *with the length proportional to the magnitude* on the left charge a*bove* that charge; and draw the vector representing the force *with the length proportional to the magnitude* on the right charge below that charge (see the example). For each diagram, use the same scale as the example.

Example: $+Q$ $+Q$

(a) $+2Q$ $+Q$

(b) $-2Q$ $+Q$

(c) $-Q$ $+2Q$

(d) $-2Q$ $+2Q$

(e) $-Q$ $+2Q$

(f) $-2Q$ $+2Q$

Explain your reasoning.

D1-WWT10: Two Negative Charges—Force

Two negatively charged particles are separated by a distance x. The particle on the left has a charge $-Q$ which is three times the charge $-q$ of the particle on the right.

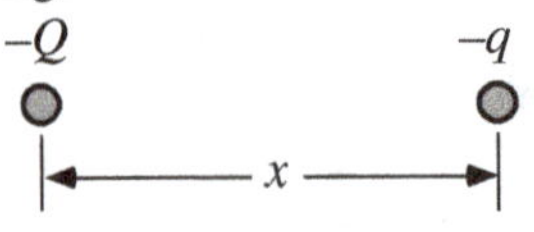

A student makes the following statement:

"Since $F = kQq/x^2$ and Q and q are both negative, the force on Q will be positive. Therefore, the force on Q points to the right."

What, if anything, is wrong with this statement? If something is wrong, explain the error and how to correct it. If the statement is valid, explain why.

D1-WWT11: TWO NEGATIVELY CHARGED PARTICLES—FORCE

A student's diagram for the electric forces acting on two negatively charged ($-Q$ and $-4Q$) particles is shown. Particle A has four times the mass of particle B.

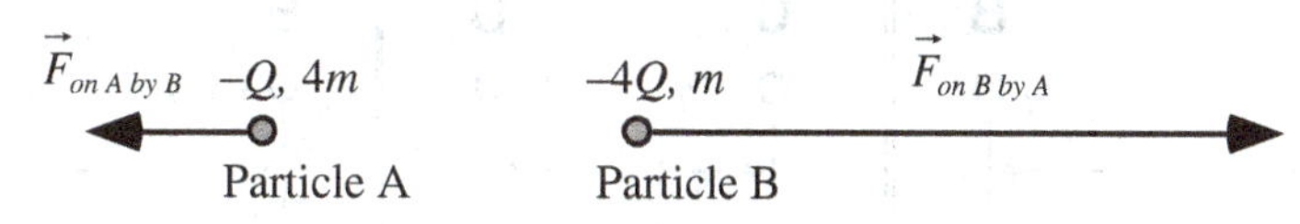

What, if anything, is wrong with this diagram? If something is wrong, explain the error and how to correct it. If the diagram is valid, explain why.

D1-RT12: TWO ELECTRIC CHARGES—ELECTRIC FORCE

In each figure, two charges are fixed in place on a grid, and a point near those particles is labeled P. All of the charges are the same size, Q, but they can be either positive or negative.

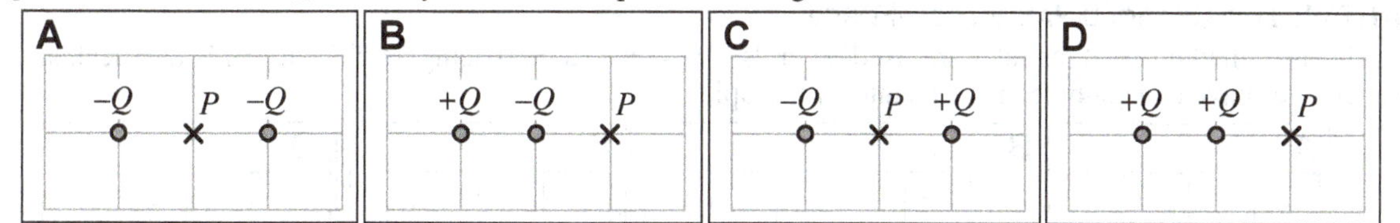

Rank the strength (magnitude) of the electric force on a charge $+q$ that is placed at point P.

☐	☐	☐	☐	OR	☐	☐	☐
1 Greatest	2	3	4 Least		All the same	All zero	Cannot determine

Explain your reasoning.

D1-RT13: Pairs of Point Charges—Attractive and Repulsive Force

The following diagrams show three separate pairs of point charges.

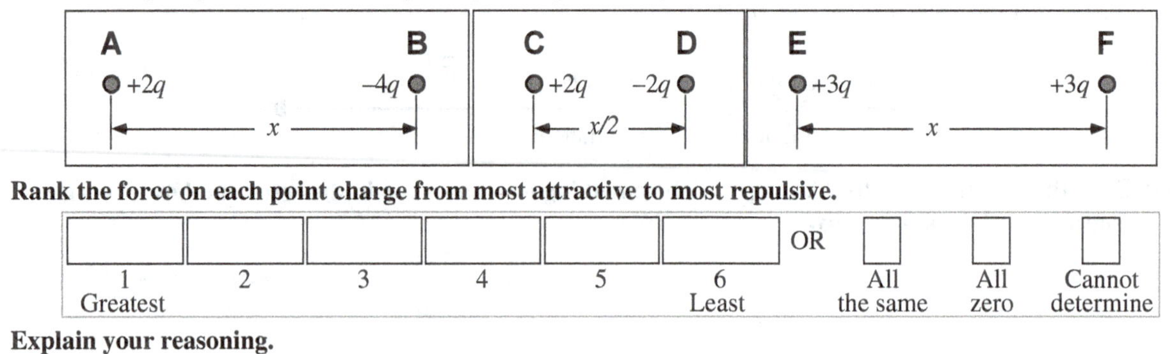

Rank the force on each point charge from most attractive to most repulsive.

						OR			
1 Greatest	2	3	4	5	6 Least		All the same	All zero	Cannot determine

Explain your reasoning.

D1-RT14: Two Charged Particles—Force

In each case, small charged particles are fixed on grids having the same spacing. Each charge q is identical, and all other charges have a magnitude that is an integer multiple of q.

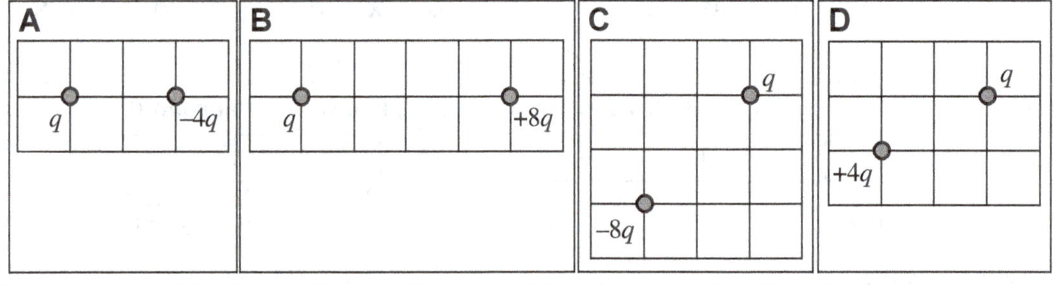

Rank the magnitude of the electric force on the charge labeled q due to the other charge.

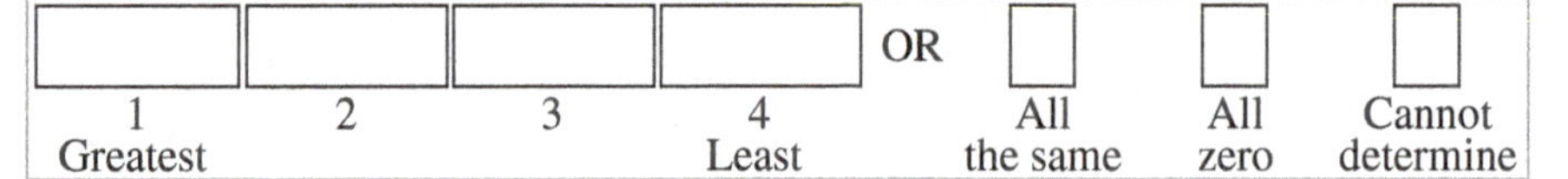

				OR			
1 Greatest	2	3	4 Least		All the same	All zero	Cannot determine

Explain your reasoning.

D1-TT15: Two Charged Particles—Force

Shown below is a student's drawing of the electric forces acting on Particle A (with charge $+Q$ and mass m) and Particle B (with charge $+4Q$ and mass m).

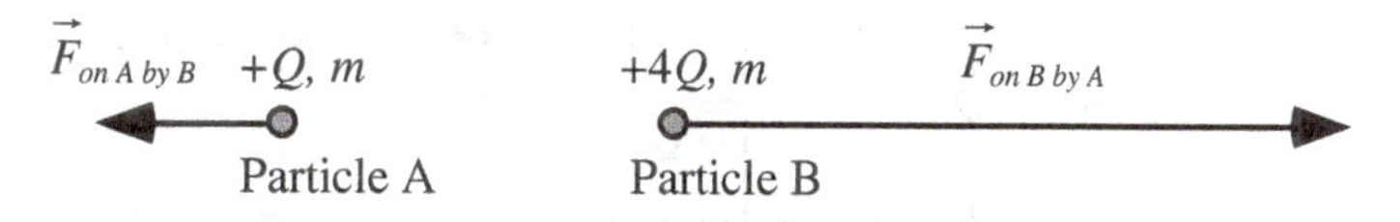

There is something wrong with this diagram. Explain what is wrong and how to correct it.

D1-RT16: Two and Three Charges in a Line—Force

In each case, small charged particles are fixed on grids having the same spacing. Each charge q is identical, and all other charges have a magnitude that is an integer multiple of q.

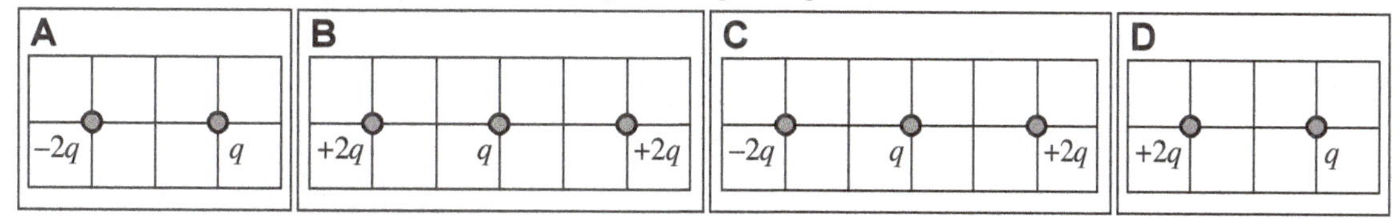

Rank the magnitude of the electric force on the charge labeled q due to the other charges.

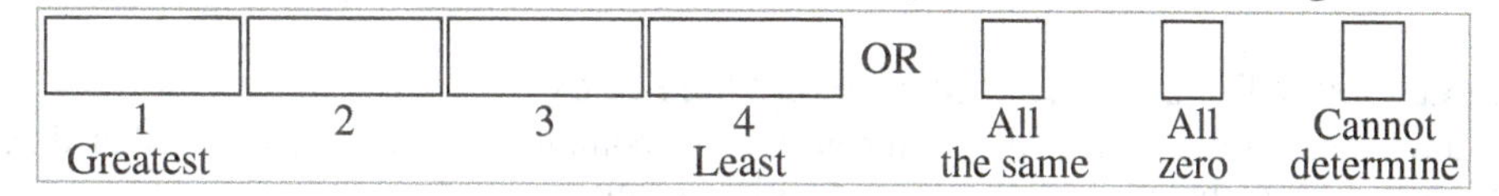

Explain your reasoning.

D1-RT17: CHARGED PARTICLES IN A PLANE—FORCE

In each case, small charged particles are fixed on grids having the same spacing. Each charge q is identical, and all other charges have a magnitude that is an integer multiple of Q.

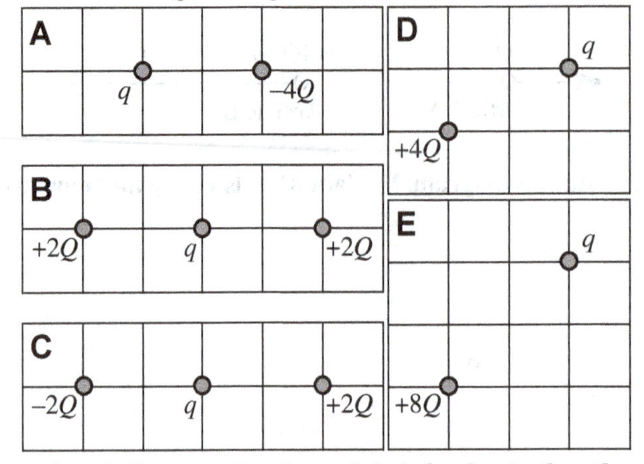

Rank the magnitude of the net electric force on the charge labeled *q* due to the other charges.

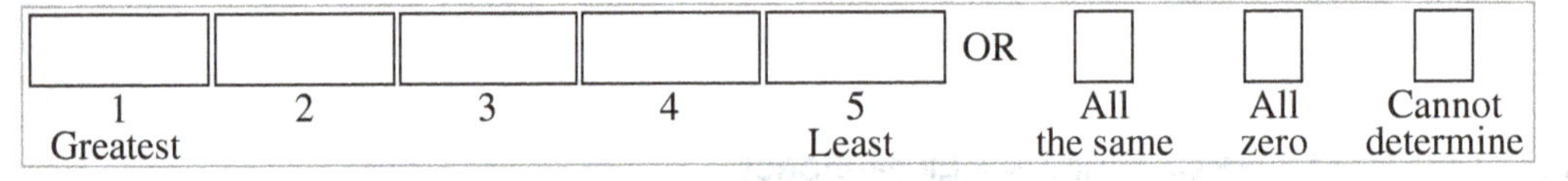

Explain your reasoning.

D1-RT18: THREE LINEAR ELECTRIC CHARGES—ELECTRIC FORCE

In each figure, three charges are fixed in place on a grid, and a point near those particles is labeled P. All of the charges are the same size, Q, but they can be either positive or negative.

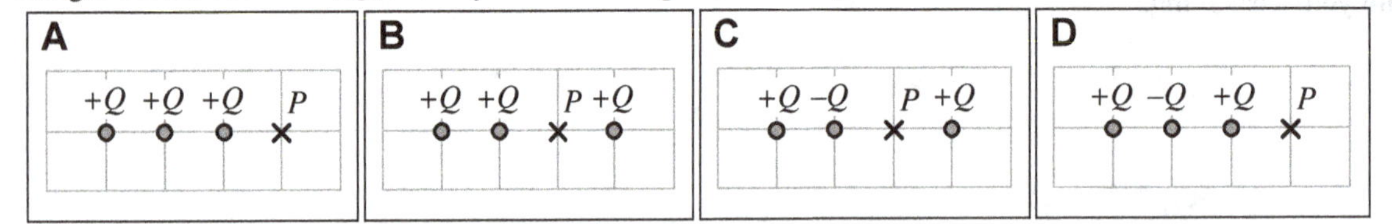

Rank the magnitude of the net electric force on a charge +*q* that is placed at point *P*.

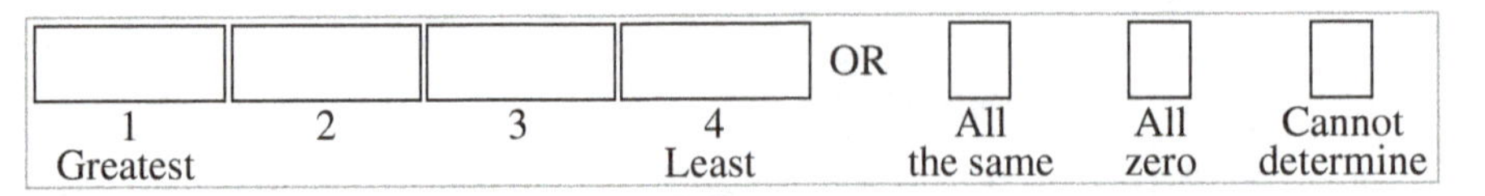

Explain your reasoning.

D1-QRT19: TWO UNEQUAL CHARGES—FORCE

Shown below are two charged particles that are fixed in place. The magnitude of the charge Q is greater than the magnitude of the charge q. A third charge is now placed at one of the points A–E. The net force on this charge due to q and Q is zero.

(**a**) Both q and Q are positive.
At which point A–E is it possible that the third charge was placed? _____
Explain your reasoning.

(**b**) Charge q is positive and charge Q is negative.
At which point A–E is it possible that the third charge was placed? _____
Explain your reasoning.

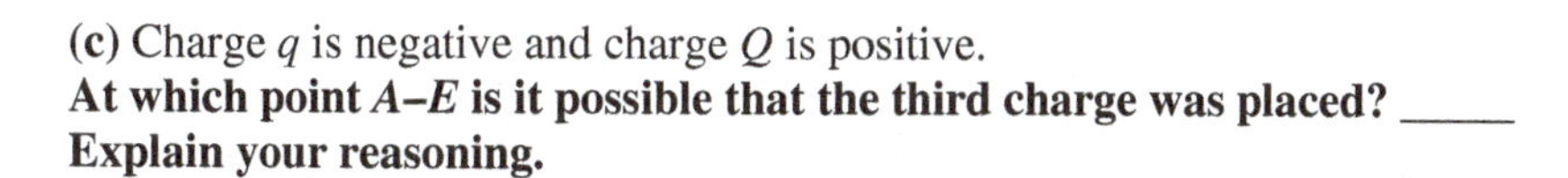

(**c**) Charge q is negative and charge Q is positive.
At which point A–E is it possible that the third charge was placed? _____
Explain your reasoning.

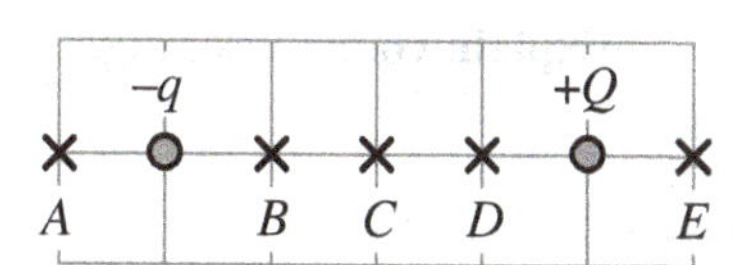

(**d**) Both q and Q are negative.
At which point A–E is it possible that the third charge was placed? _____
Explain your reasoning.

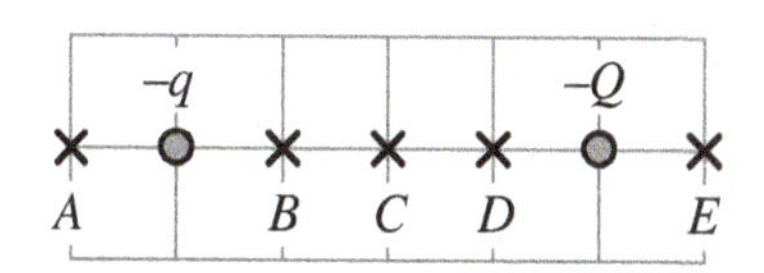

D1-QRT20: Three Charges in a Line I—Force

Three charged particles, *A, B,* and *C,* are fixed in place in a line. Charge *C* is twice as far from charge *B* as charge *A* is. All charges are the same magnitude.

In the chart to the left below, use arrows (← or →) to indicate the direction of the net force on charge *C* due to charges *A* and *B*. If the force is zero, state that explicitly.

In the chart on the right below, use arrows (← or →) to indicate the direction of the net force on charge *B* due to charges *A* and *C*. If the force is zero, state that explicitly.

A	B	C	$\Sigma\vec{F}$ on charge *C*
+	+	+	Direction:
+	+	−	Direction:
+	−	+	Direction:
+	−	−	Direction:
−	+	+	Direction:
−	+	−	Direction:
−	−	+	Direction:
−	−	−	Direction:

A	B	C	$\Sigma\vec{F}$ on charge *B*
+	+	+	Direction:
+	+	−	Direction:
+	−	+	Direction:
+	−	−	Direction:
−	+	+	Direction:
−	+	−	Direction:
−	−	+	Direction:
−	−	−	Direction:

Explain your reasoning.

D1-QRT21: THREE CHARGES IN A LINE II—FORCE

Three charged particles, *A*, *B*, and *C*, are fixed in place in a line. Charge *C* is twice as far from charge *B* as charge *A* is. All charges have different magnitudes.

For each of the following combinations of charge signs, determine whether it is possible for the net electric force on each charge due to the other two charges to be zero.

Charges (A B C)	$\Sigma\vec{F}$ on charge *A*	$\Sigma\vec{F}$ on charge *B*	$\Sigma\vec{F}$ on charge *C*
+ + +	Must be nonzero ☐ Possibly zero ☐	Must be nonzero ☐ Possibly zero ☐	Must be nonzero ☐ Possibly zero ☐
+ + −	Must be nonzero ☐ Possibly zero ☐	Must be nonzero ☐ Possibly zero ☐	Must be nonzero ☐ Possibly zero ☐
+ − +	Must be nonzero ☐ Possibly zero ☐	Must be nonzero ☐ Possibly zero ☐	Must be nonzero ☐ Possibly zero ☐
+ − −	Must be nonzero ☐ Possibly zero ☐	Must be nonzero ☐ Possibly zero ☐	Must be nonzero ☐ Possibly zero ☐
− + +	Must be nonzero ☐ Possibly zero ☐	Must be nonzero ☐ Possibly zero ☐	Must be nonzero ☐ Possibly zero ☐
− + −	Must be nonzero ☐ Possibly zero ☐	Must be nonzero ☐ Possibly zero ☐	Must be nonzero ☐ Possibly zero ☐
− − +	Must be nonzero ☐ Possibly zero ☐	Must be nonzero ☐ Possibly zero ☐	Must be nonzero ☐ Possibly zero ☐
− − −	Must be nonzero ☐ Possibly zero ☐	Must be nonzero ☐ Possibly zero ☐	Must be nonzero ☐ Possibly zero ☐

Explain your reasoning.

D1-QRT22: Three Charges in a Line III—Force

Three charged particles are fixed in place in a line. Charge C is twice as far from charge B as charge A is. It is known that there is no net force on charge C due to charges A and B.

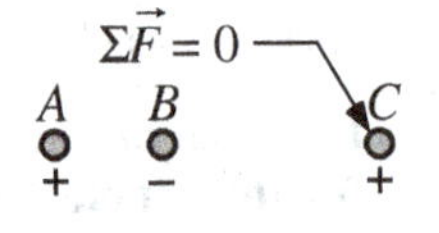

Indicate whether each of the following statements is *true*, *false*, or *cannot be determined*.

	Statement	True	False	Cannot be determined
(a)	Charge A has a greater magnitude than charge C.			
(b)	Charge A has a greater magnitude than charge B.			
(c)	Charge C has a greater magnitude than charge B.			
(d)	Charge A has the same magnitude as charge C.			
(e)	Charge A has the same magnitude as charge B.			
(f)	Charge C has the same magnitude as charge B.			

Explain your reasoning.

Three charged particles, A, B, and C, are fixed in place in a line. Charge C is twice as far from charge B as charge A is. It is known that there is no net force on charge B due to charges A and C.

$\Sigma\vec{F} = 0$

A B C

\+ − +

Indicate whether each of the following statements is *true*, *false*, or *cannot be determined*.

	Statement	True	False	Cannot be determined
(g)	Charge A has a greater magnitude than charge C.			
(h)	Charge A has a greater magnitude than charge B.			
(i)	Charge C has a greater magnitude than charge B.			
(j)	Charge A has the same magnitude as charge C.			
(k)	Charge A has the same magnitude as charge B.			
(l)	Charge C has the same magnitude as charge B.			

Explain your reasoning.

D1-BCT23: THREE CHARGES IN A LINE IV—FORCE

Three charged particles, *A*, *B*, and *C*, are fixed in place in a line. Charge *C* is twice as far from charge *B* as charge *A* is. All charges have the same magnitude.

Construct a bar chart for the net force on charge *B* due to charges *A* and C. Use positive values for net forces directed to the right and negative values for net forces directed to the left. If the force is zero, state that explicitly.

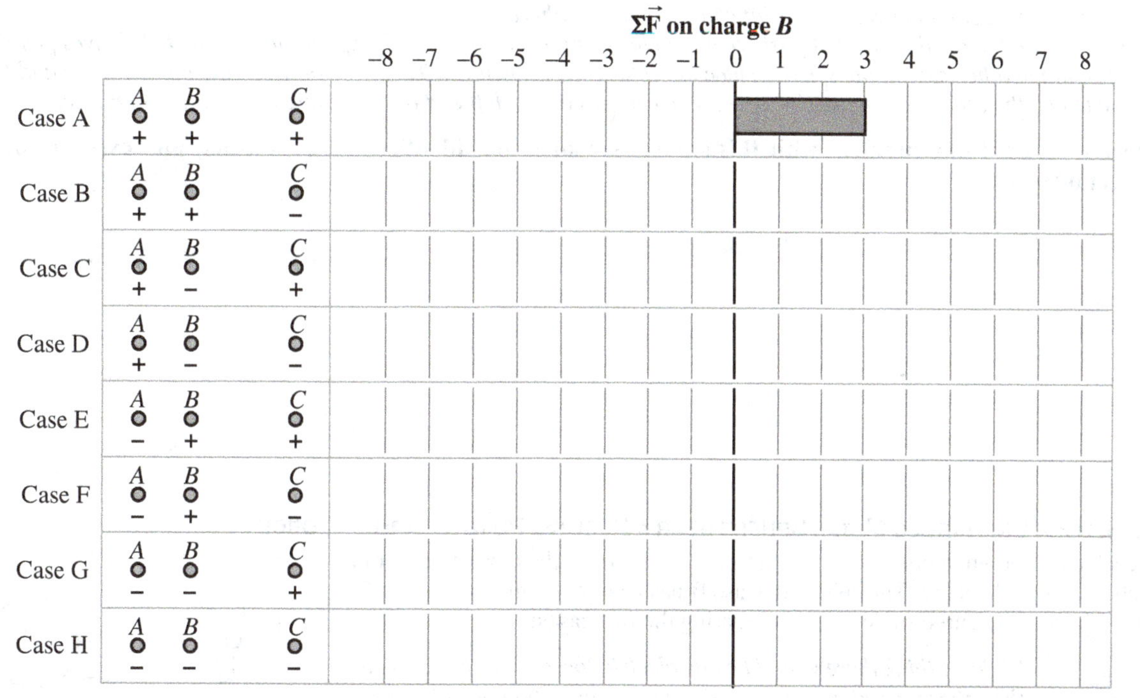

Explain your reasoning.

D1-TT24: Neutral Metal Sphere near a Positive Point Charge—Force

A positive point charge is placed a distance d away from a neutral solid metal sphere.

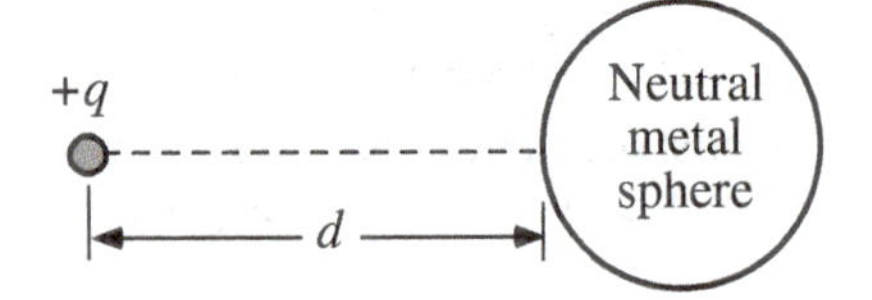

A student makes the following statement about the electric force between the neutral metal sphere and the point charge:

"There is an attraction between the point charge and the sphere. Since the sphere is a conductor, the external positive point charge pulls electrons in the sphere toward it. This leaves positive charges on the other side of the sphere, since the sphere is still neutral. The force between the point charge and the sphere is just the attraction between the negative charges on the left end of the sphere and the point charge."

There is at least one problem with this student's contention. Identify any problem(s) and explain how to correct it/them.

D1-SCT25: Uncharged Metal Sphere near a Positive Point Charge—Force

In each case shown, a point charge $+q$ is a distance d from the closest point of an uncharged metal sphere. The sphere in Case B has a larger diameter than the sphere in case A. Three students are comparing the two cases:

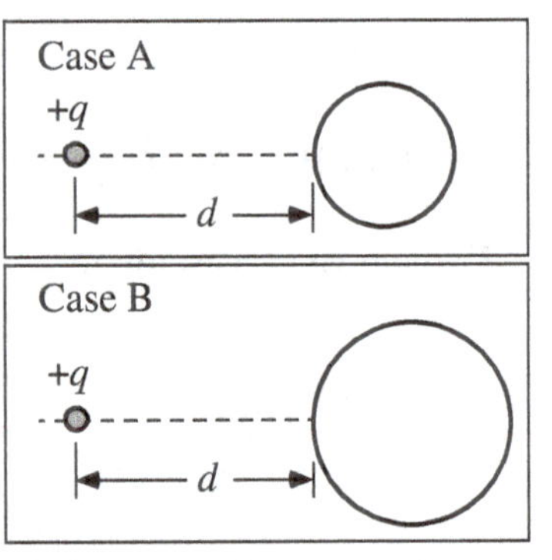

Aaron: *"I don't think there would be any electric forces in either case. Since the sphere has no net charge, there is no attraction or repulsion."*

Bae: *"The forces on the point charges are equal in the two cases. There is an attraction because the point charge will pull the electrons in the sphere toward it. But the distance between the point charge and the electrons is the same in both cases, so the force of attraction is the same."*

Carlota: *"When the electrons are pulled toward the point charge, they leave a pool of positive charges on the other side of the sphere. These positive charges repel the point charge, and this balances the attraction of the electron. The sphere overall is still uncharged, so there is as much positive charge as negative charge, and there is no net force between the objects."*

With which of these students do you agree?

Aaron _____ Bae _____ Carlota _____ None of them _____

Explain your reasoning.

D1-WWT26: NEUTRAL METAL SPHERE NEAR A POSITIVE POINT CHARGE—FORCE

A positive point charge is placed a distance d away from a neutral metal sphere.

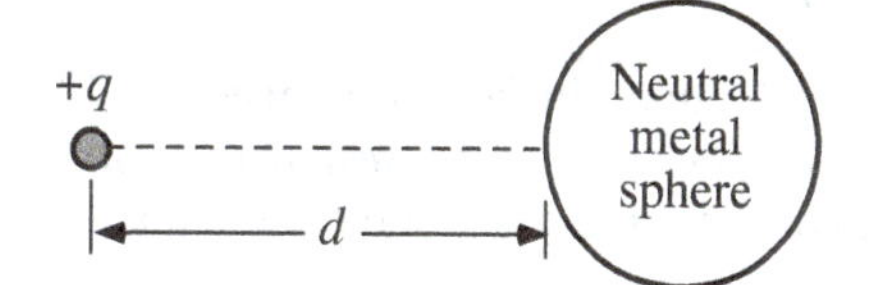

A student makes the following statement:

"The electric force is zero. Coulomb's law states that the electric force between two objects is proportional to the product of the charges. Since the charge of the sphere is zero, and zero times anything gives zero, the force between the point charge and the sphere is zero."

What, if anything, is wrong with this statement? If something is wrong, explain the error and how to correct it. If the statement is valid, explain why.

D1-RT27: NEUTRAL METAL SPHERE NEAR A POINT CHARGE—FORCE

A point charge is placed a distance d away from a neutral metal sphere. The diameters of the spheres in Cases A and C are the same and smaller than the equal diameters in Cases B and D. The point charge is positive for Cases A and B, and negative for Cases C and D.

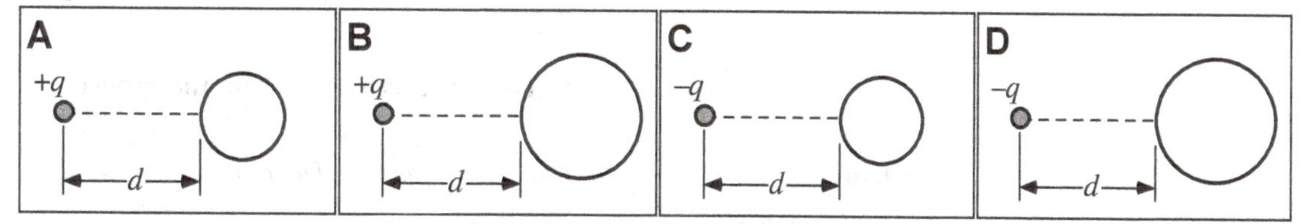

Rank the magnitude of the force exerted on the point charge by the sphere.

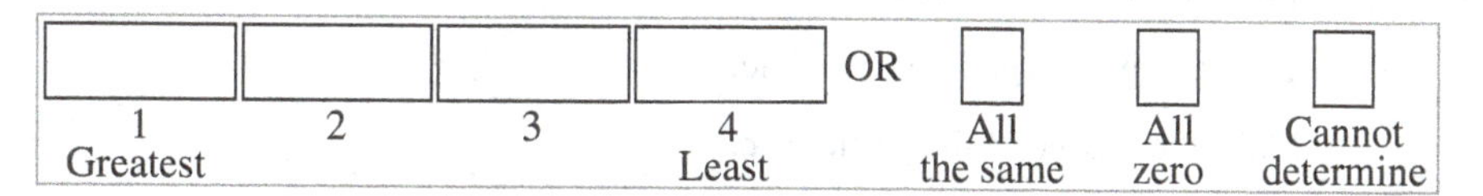

Explain your reasoning.

D1-LMCT28: Neutral Metal Sphere near a Positive Point Charge—Force

A positive point charge is placed a distance x away from the closest surface of a neutral metal sphere that has a diameter D.

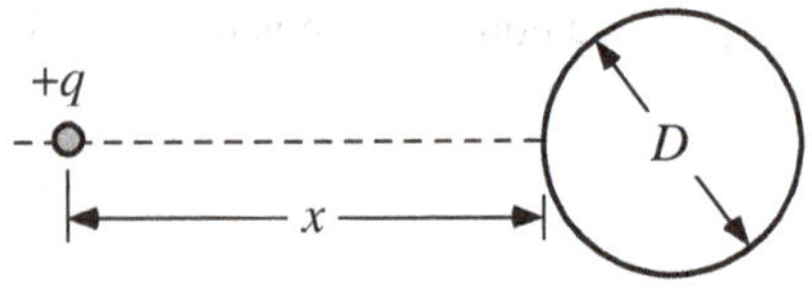

(a) For each change listed, state whether the magnitude of the force exerted on the point charge by the sphere *increases, decreases,* or *remains the same.* (Assume that all of the other given variables remain the same for each change given.)

	Change	Effect on the force exerted on the particle			
		No force	*Increases*	*Decreases*	*Remains the Same*
(a)	Increase the distance x.				
(b)	Increase D, keeping the charge a distance x away.				
(c)	Increase the charge of the particle.				
(d)	Make the charge of the particle $-q$.				
(e)	Add negative charge to the sphere.				

Explain your reasoning.

(b) For each change listed, state whether the magnitude of the force exerted on the sphere by the point charge *increases, decreases,* or *remains the same.* (Assume that all of the other given variables remain the same for each change given.)

	Change	Effect on the force exerted on the sphere			
		No force	*Increases*	*Decreases*	*Remains the Same*
(f)	Increase the distance x.				
(g)	Increase D, keeping the charge a distance x away.				
(h)	Increase the charge of the particle.				
(i)	Make the charge of the particle $-q$.				
(j)	Add negative charge to the sphere.				

Explain your reasoning.

D1-CT29: CONDUCTING CUBE BETWEEN POINT CHARGES—NET FORCE

In both cases, two particles with equal and opposite charges are fixed in place a distance d apart. The cases are identical, except that in Case B an uncharged metal cube is placed between the two particles.

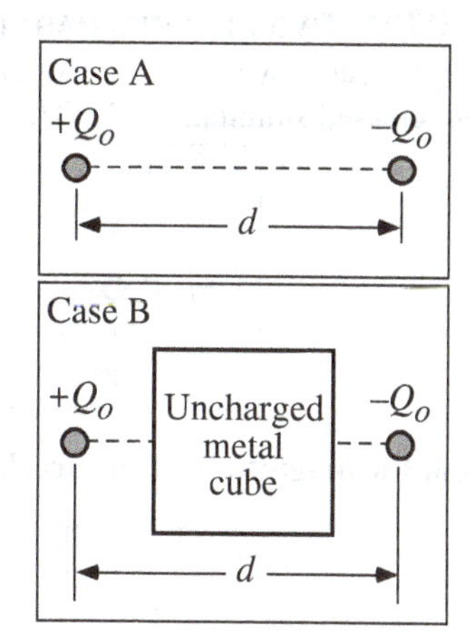

Is the net electric force on the positively charged particle (i) *greater in* Case A, (ii) *greater* in Case B, or (iii) *the same* in both cases? _____

Explain your reasoning.

D1-QRT30: CUBES BETWEEN POINT CHARGES—FORCE EXERTED BY ONE CHARGE ON THE OTHER

In both cases, two equal and opposite charges are fixed in place a distance d apart. The cases are identical, except that in Case B an uncharged metal cube is placed between the two charges.

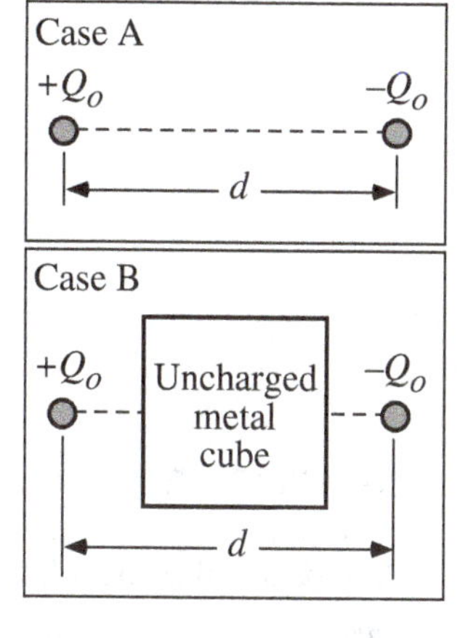

(a) Will the force exerted on the positive charge by the negative charge be (i) *greater* in Case A, (ii) *greater* in Case B, or (iii) *the same* in both cases? _____

Explain your reasoning.

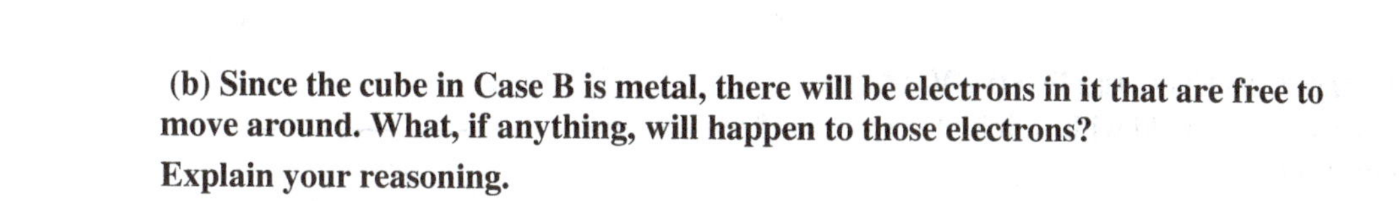

(b) Since the cube in Case B is metal, there will be electrons in it that are free to move around. What, if anything, will happen to those electrons?

Explain your reasoning.

Now the uncharged metal cube in Case B is replaced with an uncharged plastic cube, keeping everything else exactly the same.

(c) Will the force exerted on the positively charged particle by the negatively charged particle be (i) *greater* in Case A, (ii) *greater* in case B, or (iii) *the same* in both cases? _____

Explain your reasoning.

(d) Since the cube is plastic, there will be no electrons in it that are free to move around, but the molecules can become polarized (i.e., the electrons move closer on average to one end of the molecule and the protons move closer to the other). Will the plastic cube exert a force on the positive charge?

Explain your reasoning.

D1-RT31: Two Charged Particles—Acceleration

In each case shown, a particle of charge $+q$ is placed a distance d from a particle of charge $+4q$. The particles are then released simultaneously. The masses of the particles are indicated in the diagram.

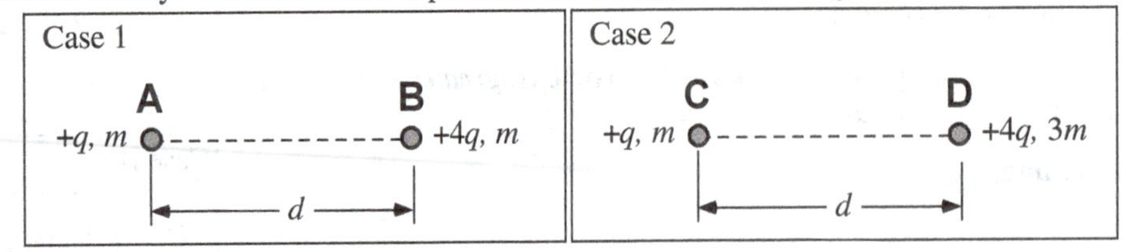

Rank the magnitude of the acceleration of each particle just after it is released.

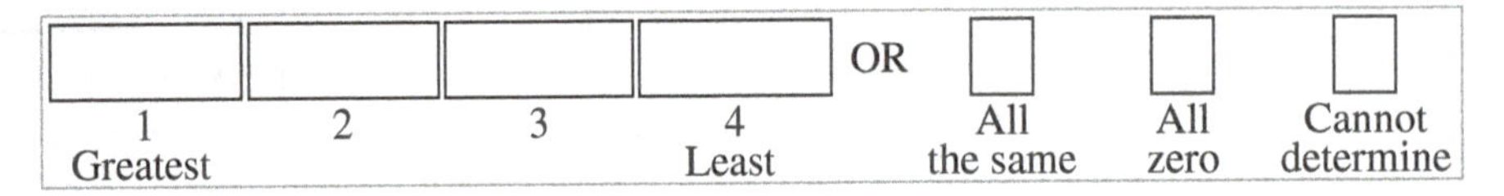

Explain your reasoning.

D1-WWT32: Electron in a Uniform Electric Field—Velocity

An electron is placed in a uniform electric field with an initial velocity of 5 m/s as shown. A student makes the following statement:

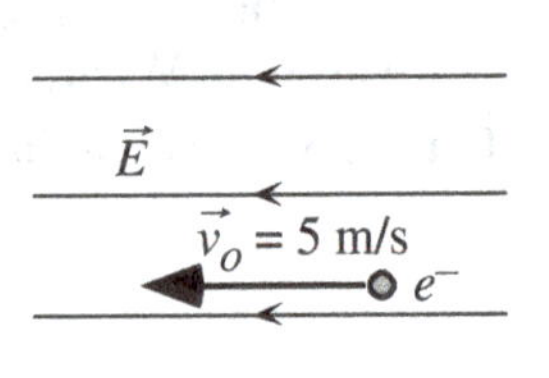

"The electron will continue to move in the same direction at a constant velocity because it is moving in the same direction as the electric force on it; since the electric field is constant, the force on the electron is constant."

What, if anything, is wrong with this statement? If something is wrong, explain the error and how to correct it. If the statement is valid, explain why.

D1-LMCT33: POSITIVE CHARGE IN A UNIFORM ELECTRIC FIELD—ELECTRIC FORCE

A particle with a charge $+q$ is placed in a uniform electric field.

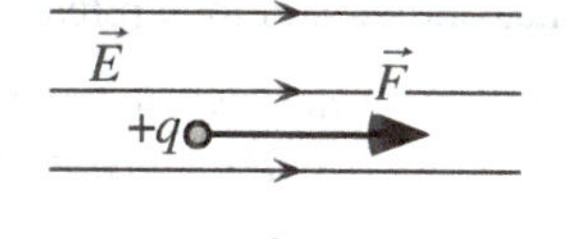

Identify from choices (i)–(vi) how each change described in (a) to (e) will affect the electric force on the particle.

This change will:

(i) change only the **direction** of the electric force.
(ii) **increase** the magnitude of the electric force.
(iii) **decrease** the magnitude of the electric force.
(iv) **increase** the magnitude and change the **direction** of the electric force.
(v) **decrease** the magnitude and change the **direction** of the electric force.
(vi) **not affect** the electric force.

All of these modifications are changes to the initial situation shown in the diagram.

(a) The charge q on the particle is doubled. _____
Explain your reasoning.

(b) The sign of the charge q on the particle is changed to the opposite sign. _____
Explain your reasoning.

(c) The particle is given a push, causing a leftward initial velocity. _____
Explain your reasoning.

(d) The magnitude of the uniform electric field is halved. _____
Explain your reasoning.

(e) The direction of the uniform electric field is rotated 90° clockwise. _____
Explain your reasoning.

D1-SCT34: Electron in a Uniform Electric Field—Electric Force

Consider the following statements about the motion of an electron placed at rest in a uniform electric field as shown and then released:

Anna: *"Since the electron is negative, it will move downward. Since the field is uniform, it will move at a constant velocity proportional to the strength of the electric field."*

Brooke: *"The electron will accelerate upward because particles move in the direction of the electric field, which points upward."*

Chico: *"The electron will move downward because it is a negative particle. The force acting on it will be opposite the direction of the electric field. It will move with a constant acceleration."*

With which of these students do you agree?

Anna _____ Brooke _____ Chico _____ None of them _____

Explain your reasoning.

D1-SCT35: Two Negatively Charged Particles—Acceleration

Two negatively charged particles labeled *A* and *B* are separated by a distance *x*. The particles have different charges and masses as shown.

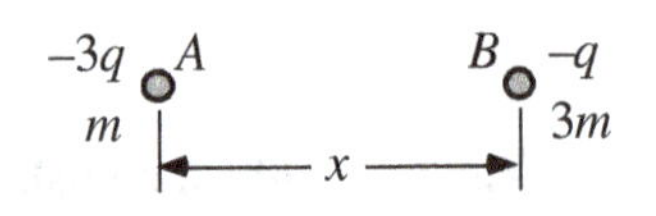

Three students are discussing what will happen just after the particles are released.

Antonio: *"The magnitude of the force that A exerts on B will be the same as the magnitude of the force that B exerts on A. Since A has less mass, it will have a larger acceleration."*

Brenda: *"The magnitude of the force on A by B is greater than the magnitude of the force on B by A since B has more mass. So A will have the largest acceleration."*

Cho: *"A has more charge but it has less mass. The larger mass of B is exactly compensated for by the larger charge of A. The acceleration of both will be the same."*

With which of these students do you agree?

Antonio _____ Brenda _____ Cho _____ None of them _____

Explain your reasoning.

D1-RT36: THREE CHARGED PARTICLES ARRANGED IN A TRIANGLE—FORCE

In each case, three charged particles are fixed in place at the vertices of an equilateral triangle. The triangles are all the same size.

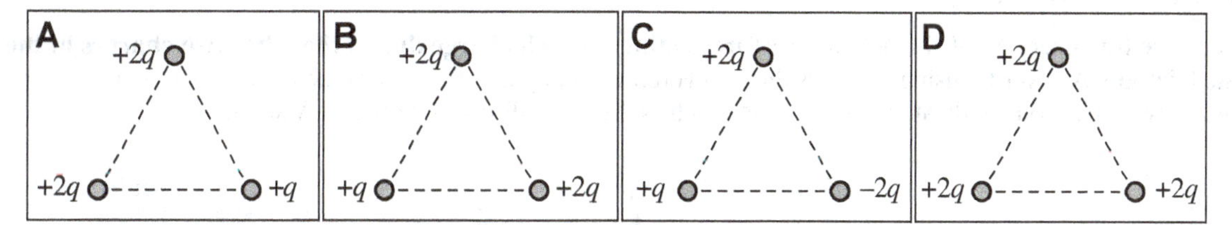

Rank the magnitude of the net electric force on the lower-left particle.

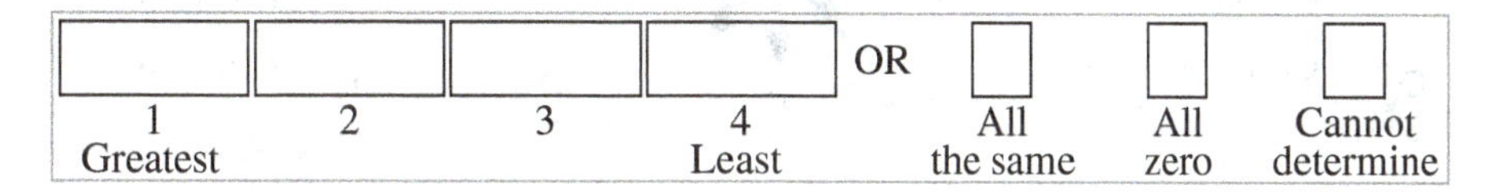

Explain your reasoning.

D1-QRT37: FORCE DIRECTION ON THREE CHARGES IN AN EQUILATERAL TRIANGLE—FORCE

Three charges are fixed at the vertices of each of the equilateral triangles shown below. All charges have the same magnitude. Only charge 1 is positive.

Determine the direction of the net electric force acting on each charge due to the other two charges in the same triangle. Answer by using letters A through L representing directions from the choices below.

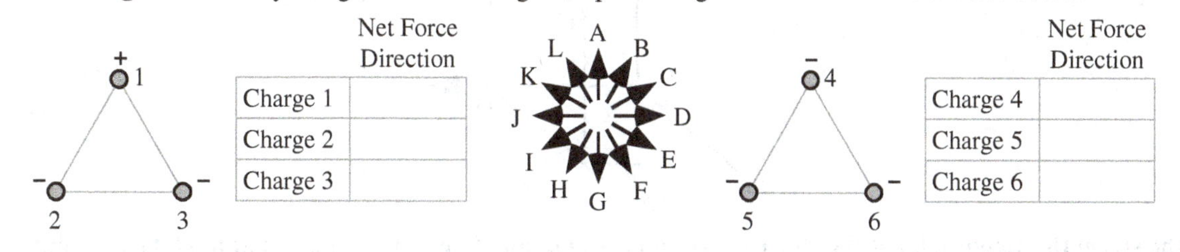

Explain your reasoning.

D1-QRT38: Force Direction on Three Charges in a Right Triangle—Force

Three charges are fixed at the vertices of each of the right isosceles triangles shown below. All charges have the same magnitude. Only charge 1 is positive.

Determine the direction of the net electric force acting on each charge due to the other two charges in the same triangle. Answer by using letters A through H representing directions from the choices below. If the angle is between two directions, indicate both directions such as AB for a direction between A and B.

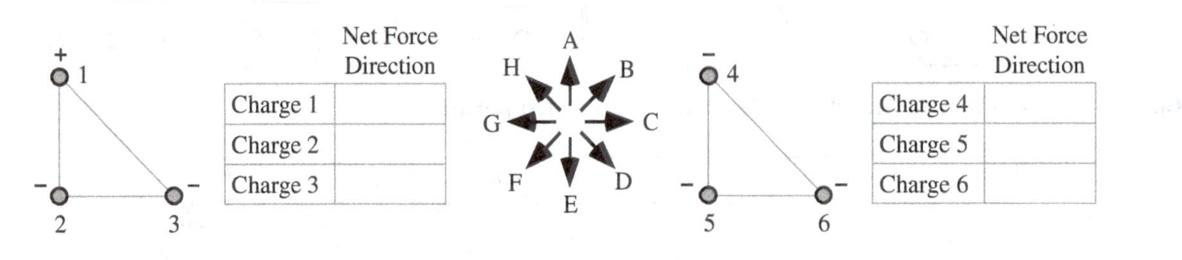

	Net Force Direction
Charge 1	
Charge 2	
Charge 3	

	Net Force Direction
Charge 4	
Charge 5	
Charge 6	

Explain your reasoning.

D1-RT39: Near a Point Charge—Electric Force at Three-Dimensional Locations

There is a positive point charge $+q$ located at (0, 0, 0) in the three-dimensional region below. Within that region are points located on the corners of a cube as shown.

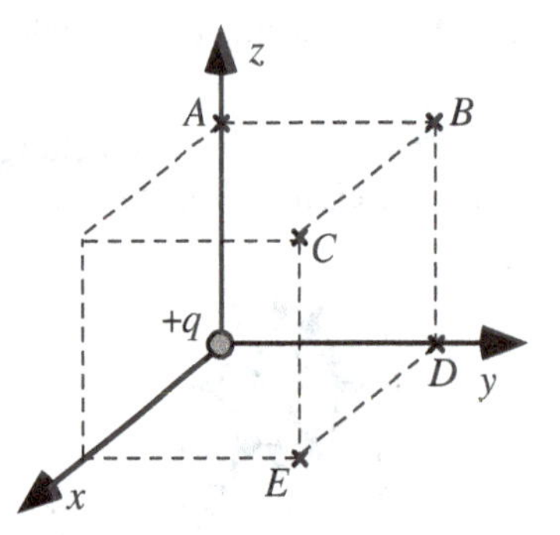

Rank the strength (magnitude) of the electric force on a $+3q$ point charge if it is placed at the labeled points.

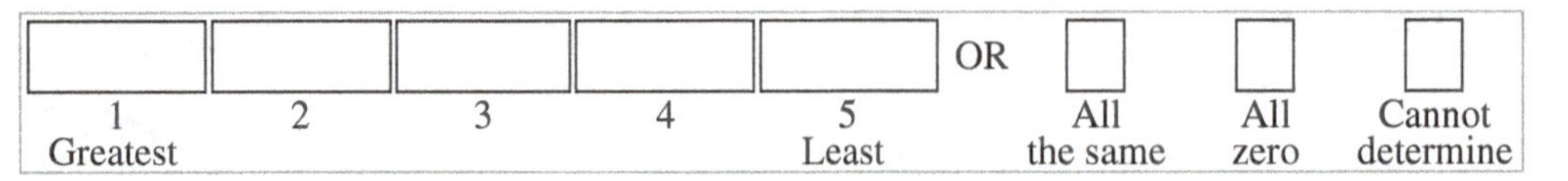

Explain your reasoning.

D1-WBT40: Forces on Three Charges Along a Line—Charge Location

Three charges are fixed in place along a line. All three charges have the same magnitude, but they may have different signs. Shown below are diagrams showing the forces exerted on each charge by the other two charges.

In each case, the sign of one of the charges is shown, as well as its position along a dashed line. **Indicate the signs of the other two charges and their approximate positions on the dashed line.**

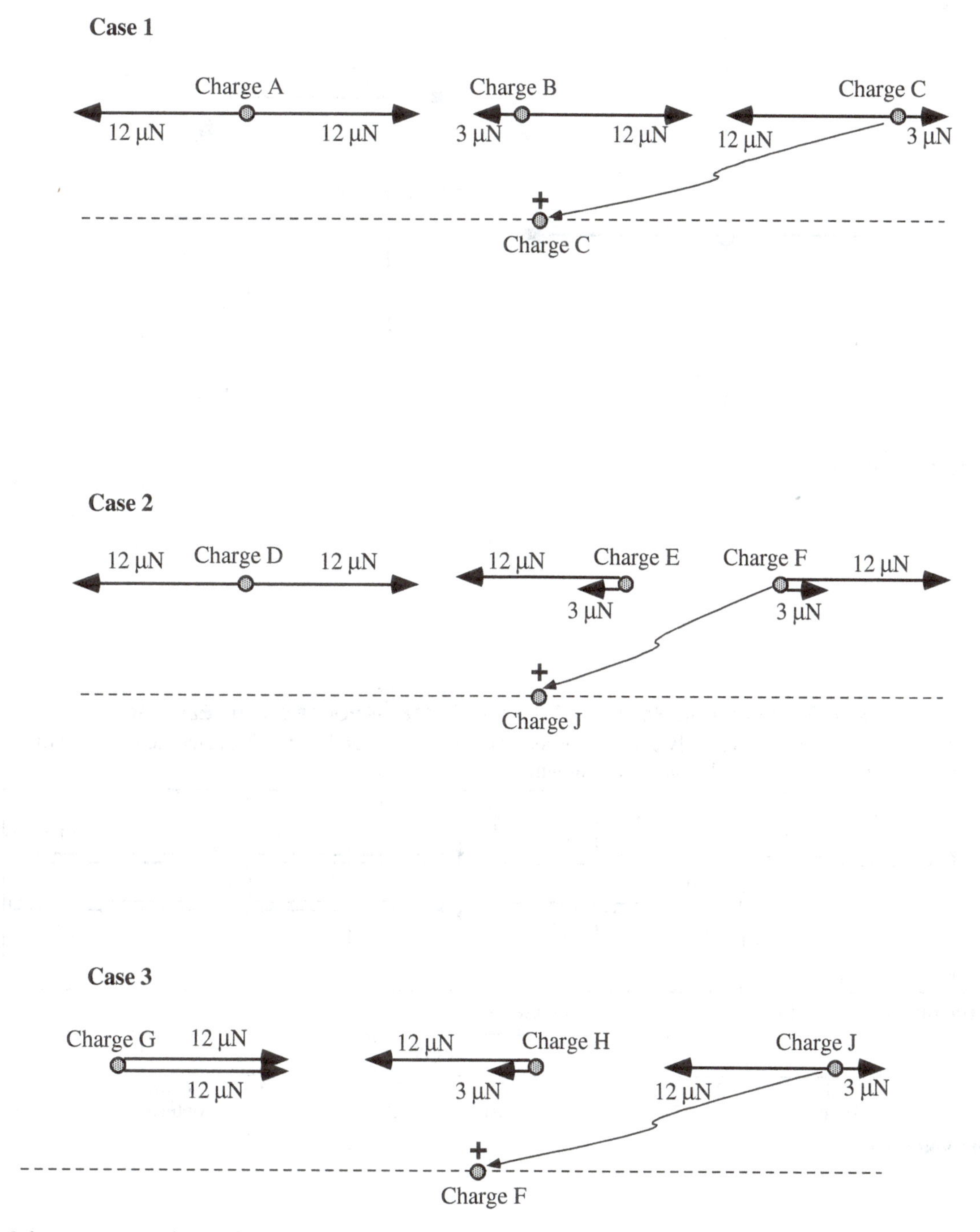

Explain your reasoning.

D1-WBT41: FORCES ON THREE CHARGES IN TWO DIMENSIONS—CHARGE LOCATIONS

Three charged particles are fixed to a grid and are exerting electric forces on one another. Particles *A* and *B* have a charge $+2q$, and particle *C* has a charge $-q$. The diagrams at the right, below, show the electric forces exerted on each particle due to the other two particles.

Particle *B* is shown fixed at the origin of a grid. **On the grid, indicate the positions of particles *A* and *C* relative to particle *B*.**

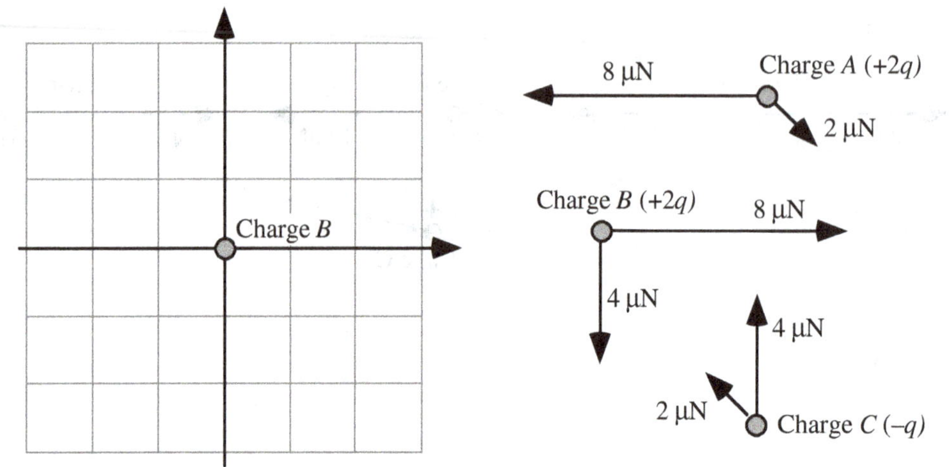

Explain your reasoning.

D1-RT42: ELECTRON BETWEEN TWO PARALLEL CHARGED PLATES—FORCE ON THE ELECTRON

In each case, an electron is momentarily at rest between two parallel charged plates. The electric potential of each plate and the separations between the plates are shown.

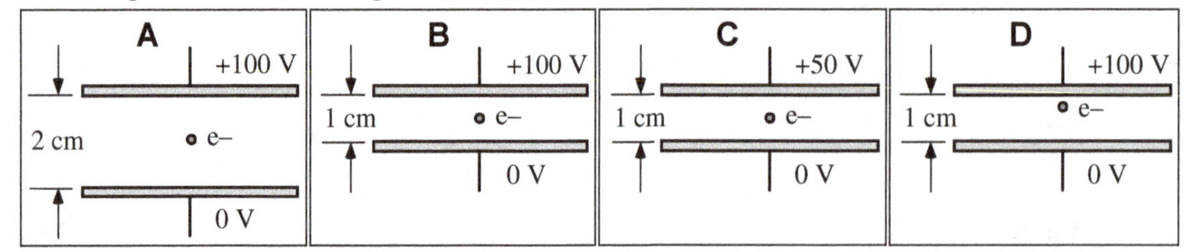

Rank the magnitude of the force exerted on the electron.

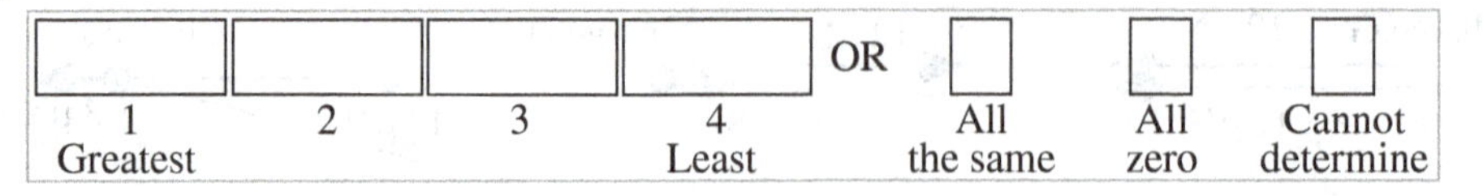

Explain your reasoning.

D1-RT43: SUSPENDED CHARGES IN AN ELECTRIC FIELD—ANGLE

A charged sphere is suspended from a string in a uniform electric field directed horizontally. There is an electric force on the sphere to the right and a gravitational force pointing downward. As a result, the sphere hangs at an angle θ from the vertical. Combinations of sphere mass and electric charge are listed in the chart for four cases, all in the same uniform electric field.

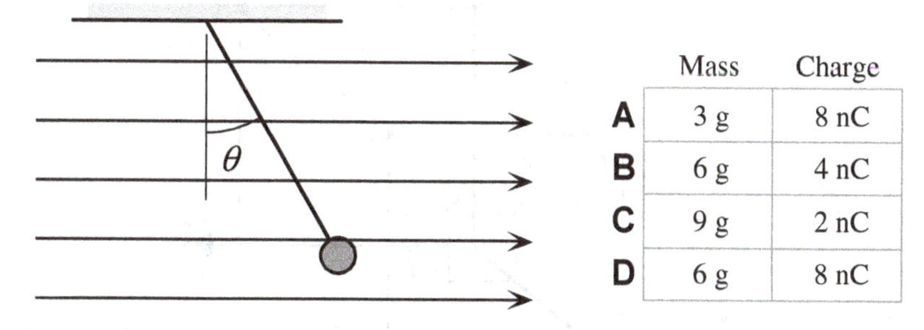

	Mass	Charge
A	3 g	8 nC
B	6 g	4 nC
C	9 g	2 nC
D	6 g	8 nC

Rank the angle θ that the string forms with the vertical for these different spheres.

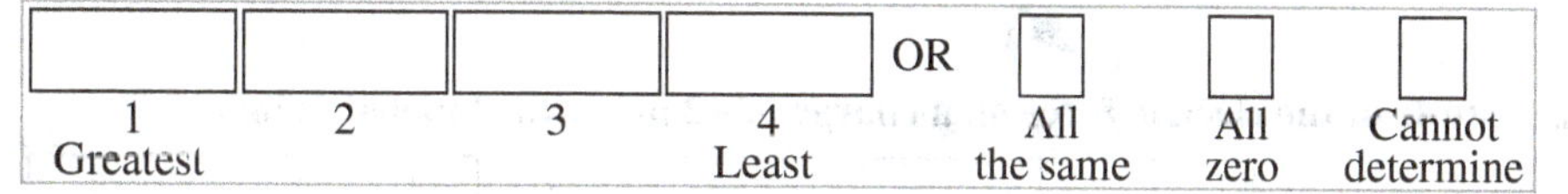

Explain your reasoning.

D1-RT44: UNIFORM ELECTRIC FIELD—ELECTRIC FORCE ON CHARGE

A large region of space has a uniform electric field in the +x direction (⇒). At the point (0,0) m, the electric field magnitude is 30 N/C.

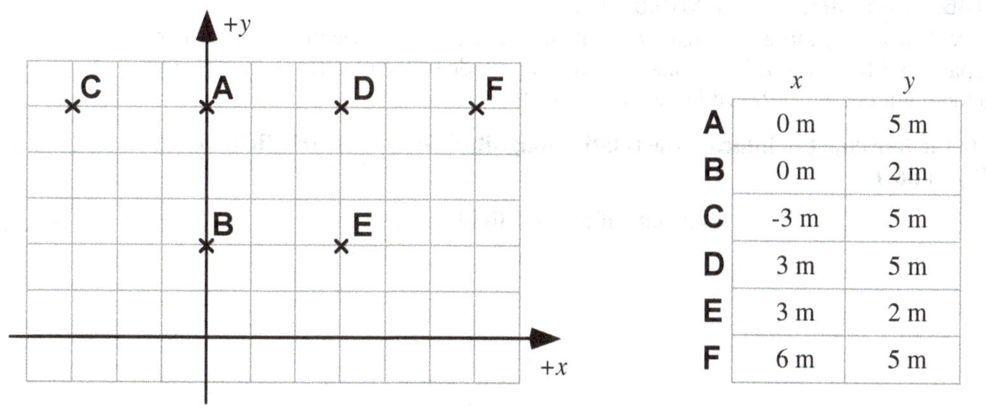

	x	y
A	0 m	5 m
B	0 m	2 m
C	-3 m	5 m
D	3 m	5 m
E	3 m	2 m
F	6 m	5 m

Rank the strength (magnitude) of the electric force on a +5 μC charge when it is placed at rest at each of the labeled points.

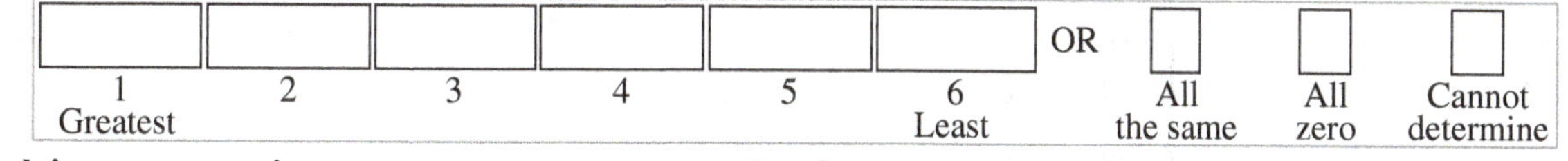

Explain your reasoning.

D1-RT45: Uniform Electric Field—Electric Force at Three-Dimensional Locations

All the labeled points are within a region of space with a uniform electric field. The electric field points toward the top of the page (that is, in the positive z-direction).

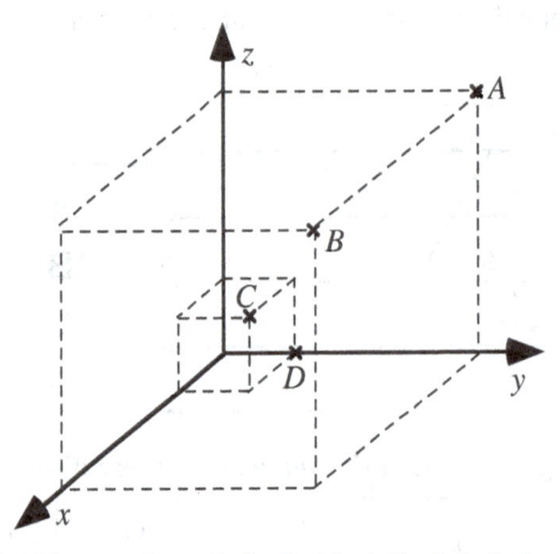

Rank the magnitude of the electric force on a charge of +2 μC at the labeled points.

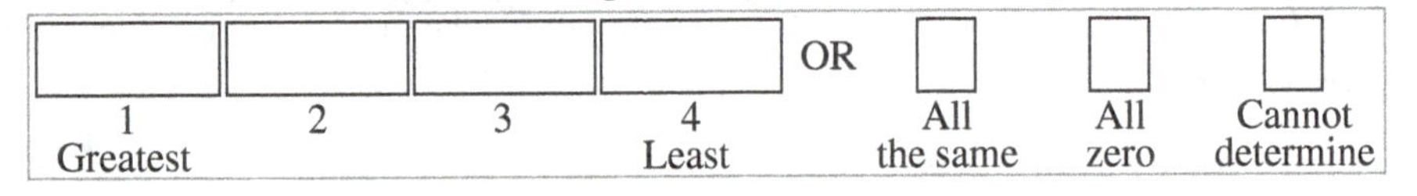

Explain your reasoning.

D1-BCT46: Point Charge—Electric Field

Points P, R, S, and T lie close to a positive point charge. The concentric circles shown are equally spaced with radii of r, $2r$, $3r$, and $4r$. The magnitude of the electric field at point P due to the point charge is shown in the bar chart below.

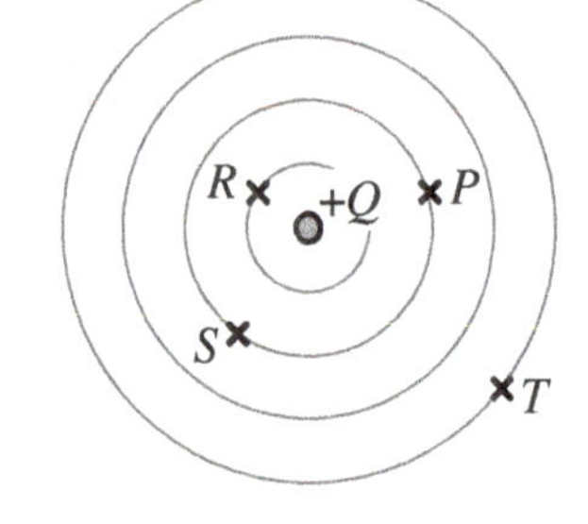

Complete the bar chart to indicate the relative magnitude of the electric field at points R, S, and T.

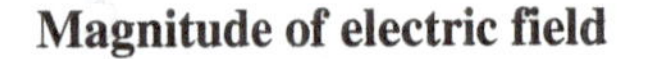

P R S T

D1-RT47: TWO ELECTRIC CHARGES—ELECTRIC FIELD ALONG A LINE

In each figure, two charges are fixed in place on a grid, and a point near those particles is labeled *P*. All of the charges are the same size, *Q*, but they can be either positive or negative.

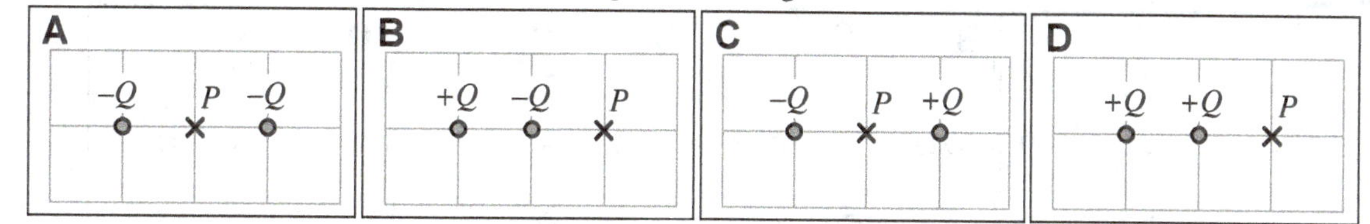

Rank the magnitude of the electric field at point *P*.

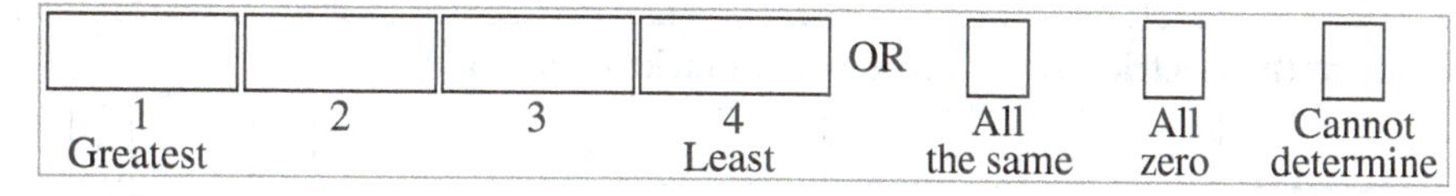

Explain your reasoning.

D1-SCT48: THREE CHARGES IN A LINE—ELECTRIC FIELD

Shown are two cases where three charges are placed in a row. Three students are comparing the electric field that exerts a force on the middle charge in the diagrams.

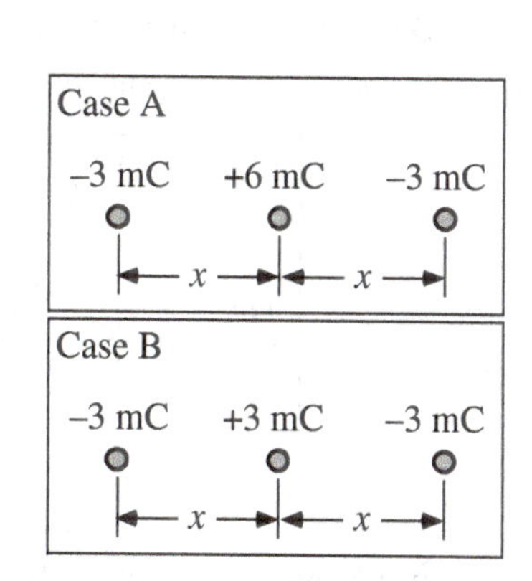

Adrianna: *"All three charges contribute by the principle of superposition. So the field is going to be greatest in case A since the contributions due to the three charges will be greatest."*

Brandon: *"I think it's a bogus question. The field at that point is undefined because there is a charge there."*

Catalina: *"I don't think that's right. The field that exerts a force on the middle charge is the field due to the other two charges because a charge cannot feel it own field. Since those other two charges don't change, the field acting on the middle charge is the same in both cases."*

With which of these students do you agree?

Adrianna ______ Brandon _____ Catalina _____ None of them _____

Explain your reasoning.

D1-RT49: Four Point Charges in Two Dimensions—Electric Field

In each case, four charged particles, each with a charge magnitude Q, are fixed on grids. The cases are identical except for the signs of the charges.

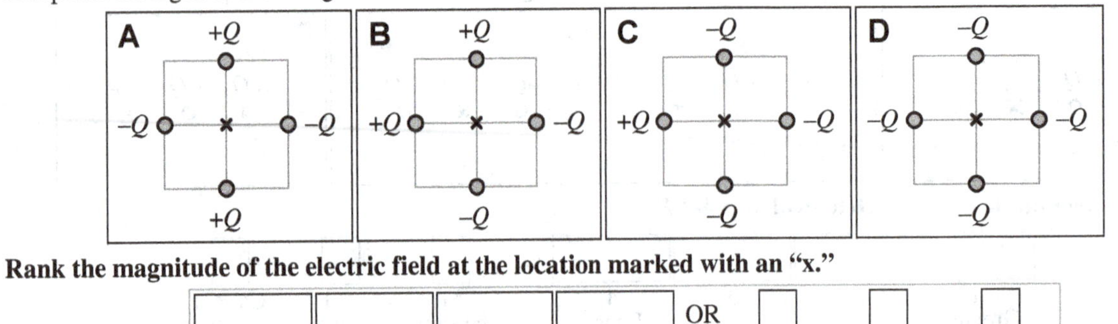

Rank the magnitude of the electric field at the location marked with an "x."

1 Greatest	2	3	4 Least	OR	All the same	All zero	Cannot determine

Explain your reasoning.

D1-RT50: Six Charges in Three Dimensions—Electric Field

In each case, six point charges are all the same distance from the origin as shown. All charges are either $+Q$ or $-Q$.

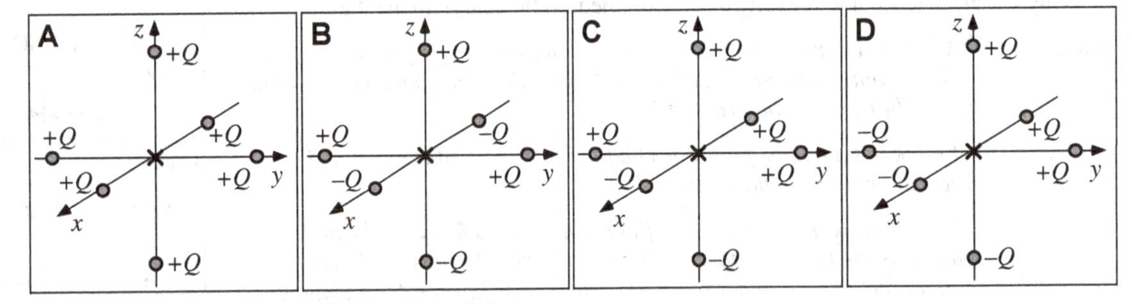

Rank the magnitude of the electric field at the origin.

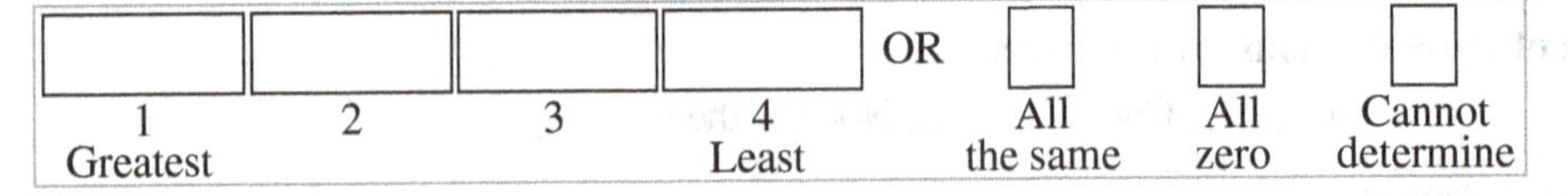

Explain your reasoning.

D1-TT51: POTENTIAL NEAR TWO CHARGES—ELECTRIC FIELD

Two equal magnitude electric charges are separated by a distance d. The electric potential at the midpoint between these two charges is zero. A student considering this situation says:

"The electric field at the midpoint between the two charges will be zero also, since the two charges are opposite in sign, so the fields will be equal but opposite, and add to zero."

There is something wrong with the student's statement. **Identify any problem(s) and explain how to correct it/them.**

D1-CT52: POTENTIAL NEAR CHARGES—ELECTRIC FIELD

In each case, a point midway between equal magnitude electric charges is identified. The signs of these charges are not given. The electric potential at this midpoint is $2V_0$ in both cases, where V_0 is the potential due to a single positive charge.

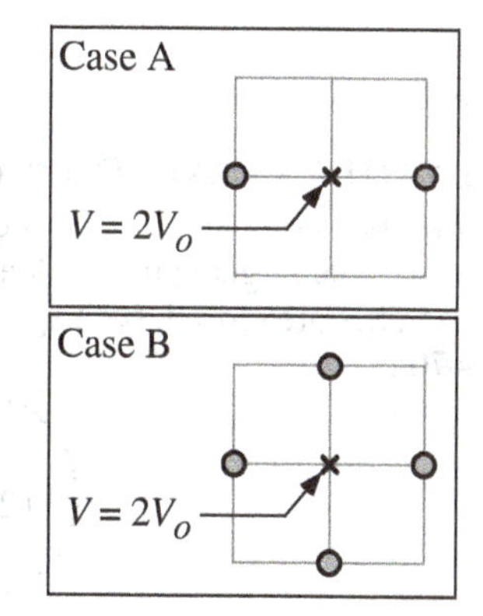

Is the magnitude of the electric field at the midpoint (i) *greater* in Case A, (ii) *greater* in Case B, or (iii) *the same* in both cases? _____

Explain your reasoning.

D1-SCT53: Charged Insulators connected with a Switch—Charge

Two solid, insulating spheres are connected by a wire and a switch. The spheres are the same size, but they have different initial charges.

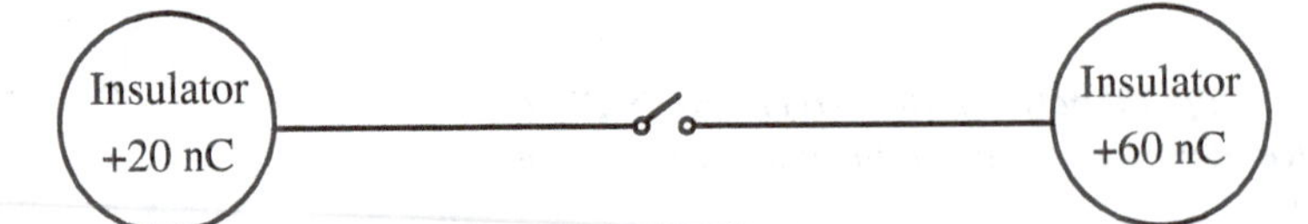

Three students are discussing what would happen if the switch was closed.

Arturo: *"Since the spheres are the same size, charge will move until there is an equal charge of 40 nC on each."*

Beth: *"I agree, but since they are insulators, the charge will move very slowly. Eventually there will be the same charge of 40 nC on each, but it will take a long time, perhaps 5 to 10 minutes.*

Caitlin: *"No, since they are insulators the charge cannot move. It doesn't matter whether the switch is open or closed."*

With which of these students do you agree?

Arturo _____ Beth _____ Caitlin _____ None of them _____

Explain your reasoning.

D1-RT54: Pairs of Connected Charged Conductors—Charge

Two pairs of charged, isolated, conducting spheres are connected with wires and switches. The spheres are very far apart. The larger spheres (A and B) are identical, and the smaller spheres (C and D) are identical. Before the switches are closed, both spheres on the left have a charge of +20 nC, and both spheres on the right have a charge of +70 nC.

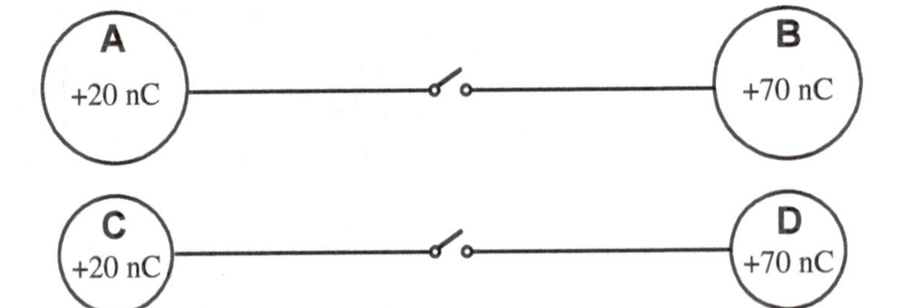

Rank the electric charge on the spheres after the switches are closed.

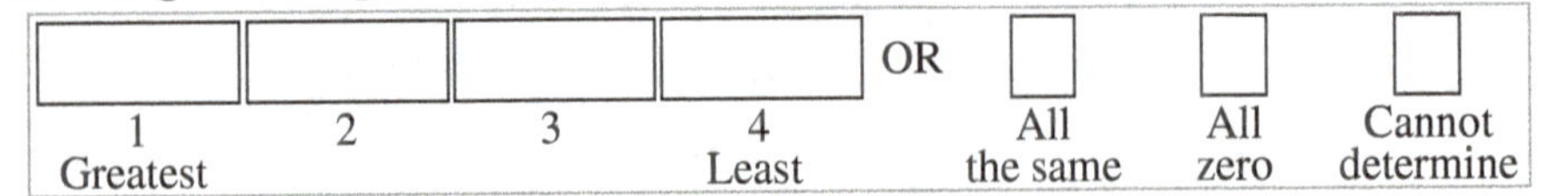

Explain your reasoning.

D1-RT55: FOUR CHARGES IN TWO DIMENSIONS—ELECTRIC POTENTIAL

In each situation shown below, small charged particles are fixed on grids having the same spacing. Each charge Q on this page has the same magnitude with the signs indicated in the diagrams.

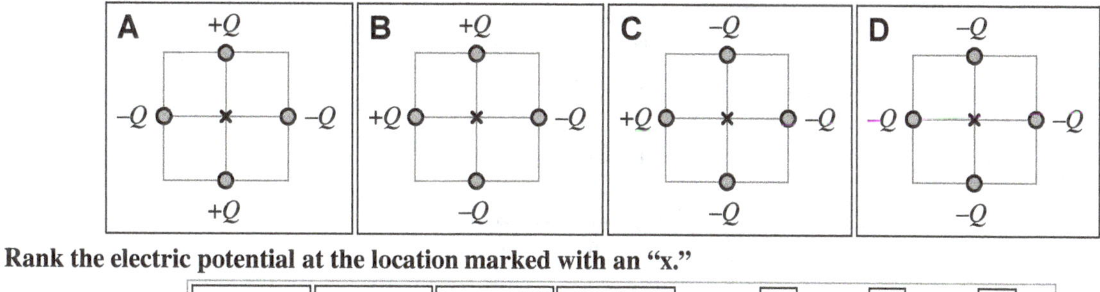

Rank the electric potential at the location marked with an "x."

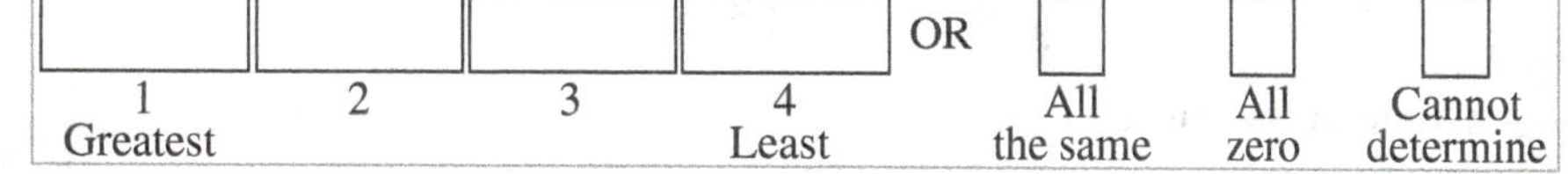

Explain your reasoning.

D1-RT56: POINTS NEAR A PAIR OF EQUAL OPPOSITE CHARGES—ELECTRIC POTENTIAL

Two equal and opposite charges are fixed to a grid at the locations shown. Four points in the vicinity of these charges are labeled *A–D*.

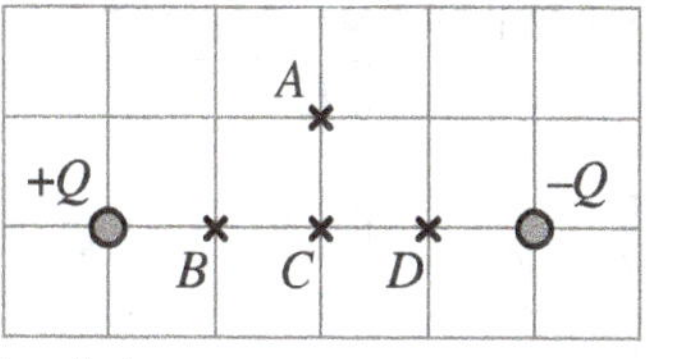

Rank the electric potential at the labeled points.

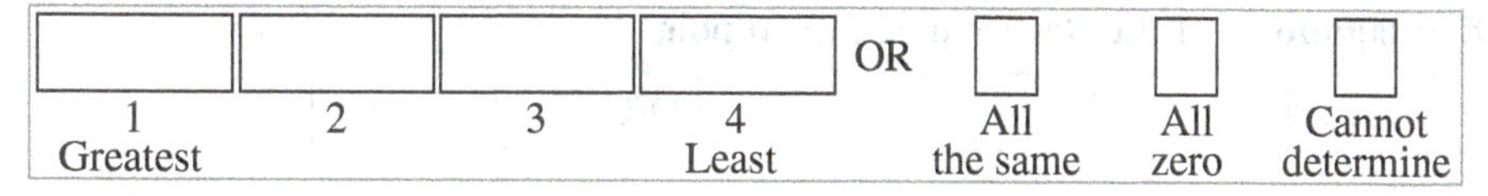

Explain your reasoning.

D1-RT57: Near a Point Charge—Electric Potential at Three-Dimensional Locations

There is a positive point charge $+q$ located at (0, 0, 0) as shown in the three-dimensional region below. Within that region are points located on the corners of a cube as shown.

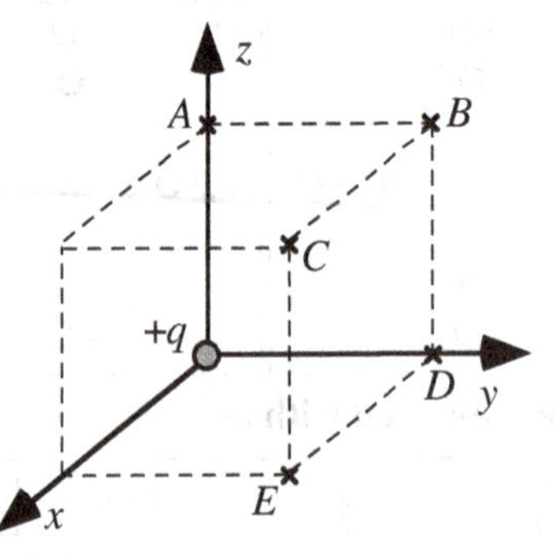

Rank the electric potential at the labeled points.

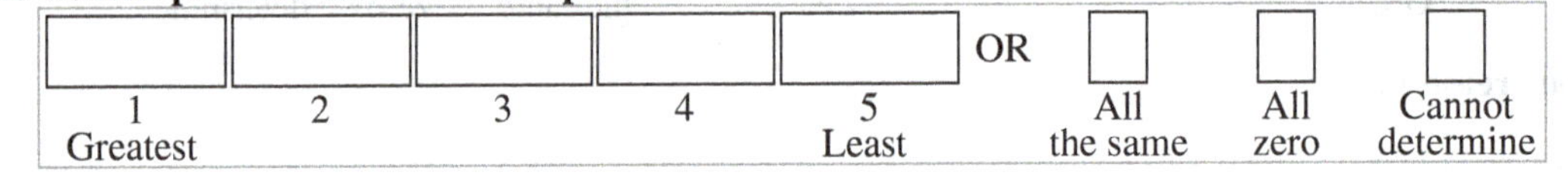

Explain your reasoning.

D1-RT58: Two Electric Charges—Electric Potential

In each figure, two charges are fixed in place on a grid, and a point near those particles is labeled P. All of the charges are the same size, Q, but they can be either positive or negative.

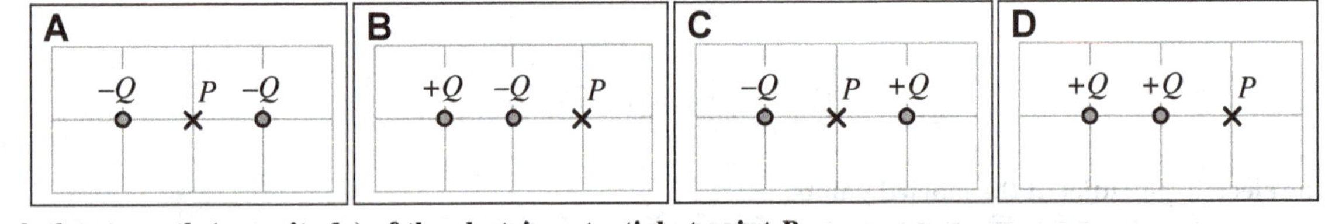

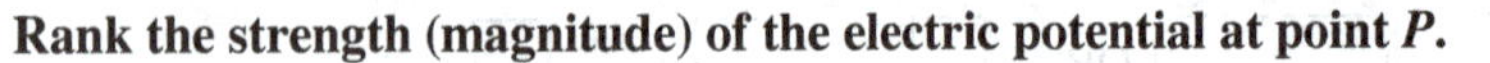

Rank the strength (magnitude) of the electric potential at point *P*.

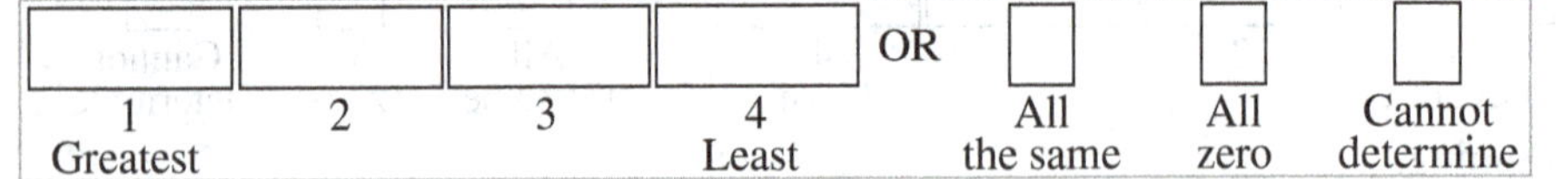

Explain your reasoning.

D1-LMCT59: FOUR CHARGES IN TWO DIMENSIONS—FIELD AND POTENTIAL

Four identical point charges are fixed at the same distance from point *P*. The charges are either $+Q$ or $-Q$.

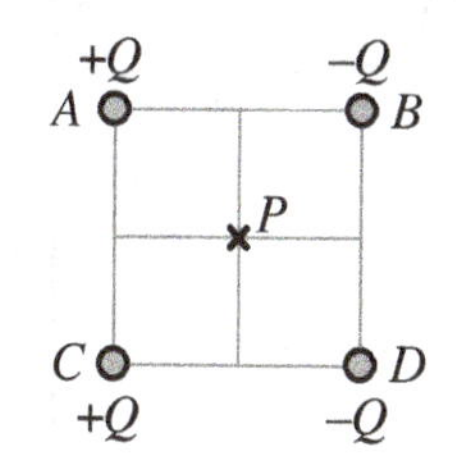

Each action described is made to the situation shown in the diagram (*i.e.*, "Change sign of charge *D*" means that charges *A, C,* and *D* will be positive and charge *B* will be negative).

For each modification:

- **Indicate whether the magnitude of the electric field at the origin (i) *increases*, (ii) *decreases*, or (iii) *remains the same*.**
- **Indicate whether the electric potential at the origin (i) *increases*, (ii) *decreases*, or (iii) *remains the same*.** (Use the convention that the electric potential is zero far from the charges.)
- **Indicate the direction of the electric field at the origin after the modification.**

	Modification	Electric field	Electric potential	Electric field direction
(a)	Change the sign of charge *A*.			
(b)	Change the sign of charge *B*.			
(c)	Change the sign of charge *C*.			
(d)	Change the sign of charge *D*.			
(e)	Change the signs of charges *B* and *D*.			
(f)	Exchange charges *A* and *B*.			
(g)	Exchange charges *A* and *D*.			

Explain your reasoning.

D1-RT60: Uniform Electric Field—Potential Difference

Two parallel plates that have been charged create a uniform electric field of 30 N/C between the plates.

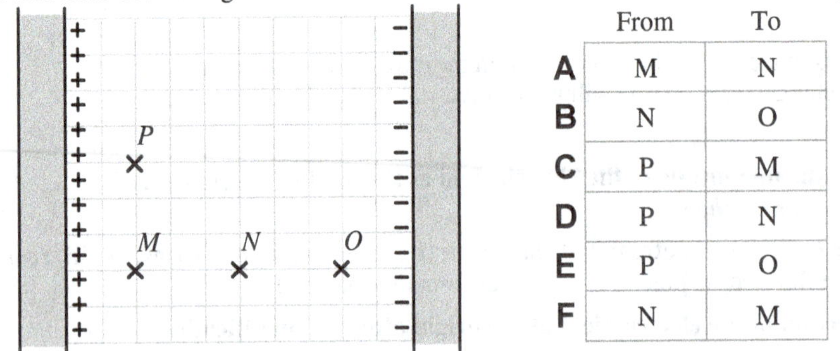

	From	To
A	M	N
B	N	O
C	P	M
D	P	N
E	P	O
F	N	M

Rank the electrical potential differences of all the different combinations listed between the four points ***M*** **at (2, 0) m;** ***N*** **at (5, 0) m;** ***O*** **at (8, 0) m; and** ***P*** **at (2, 3) m within this region. (Positive values are larger than negative values.)**

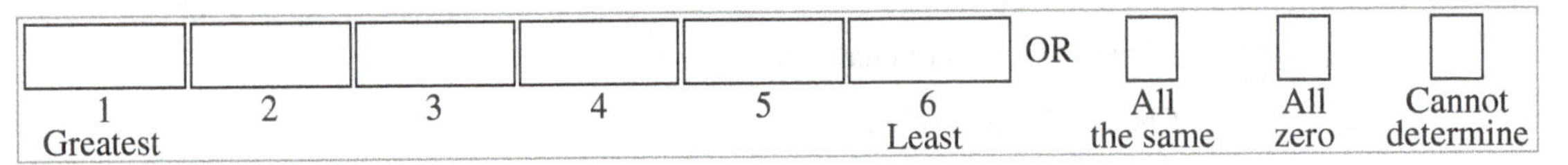

Explain your reasoning.

D2 CIRCUITS

D2-RT01: CARBON RESISTORS—RESISTANCE

Four different resistors are created from the same piece of carbon. The length and the diameter of each resistor are shown.

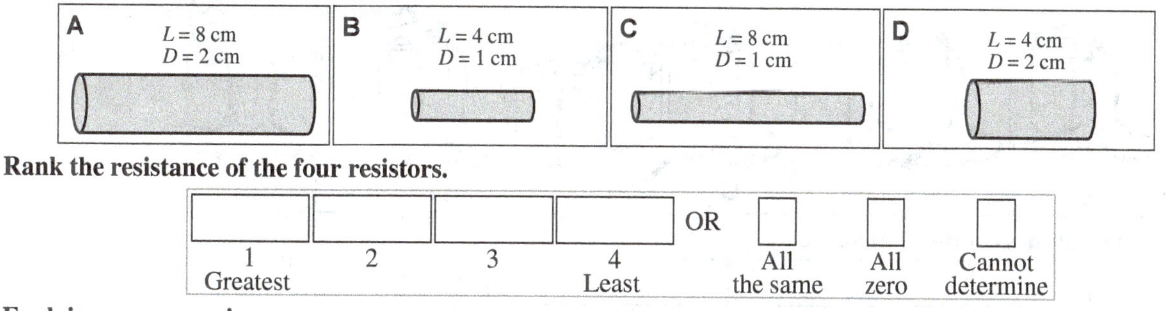

Rank the resistance of the four resistors.

				OR			
1 Greatest	2	3	4 Least		All the same	All zero	Cannot determine

Explain your reasoning.

D2-WWT02: BATTERIES AND LIGHT BULBS—BULB BRIGHTNESS

All of the batteries in the circuits shown are identical, as are the light bulbs. A student comparing the brightness of the bulbs in these circuits states:

"Bulbs E and C are the brightest since they have three batteries, then bulbs B and D since they have two batteries, and the least bright one is A, since there is only one battery. The more batteries, the brighter the bulb, and it does not matter how they are connected."

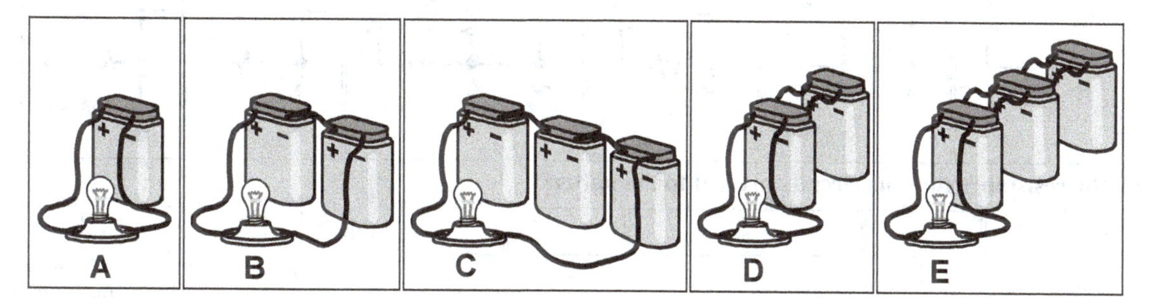

What, if anything, is wrong with this statement? If something is wrong, explain the error and how to correct it. If the statement is correct, explain why.

D2-RT03: Batteries and Light Bulbs—Bulb Brightness

Identical ideal batteries are connected in different arrangements to identical light bulbs as shown.

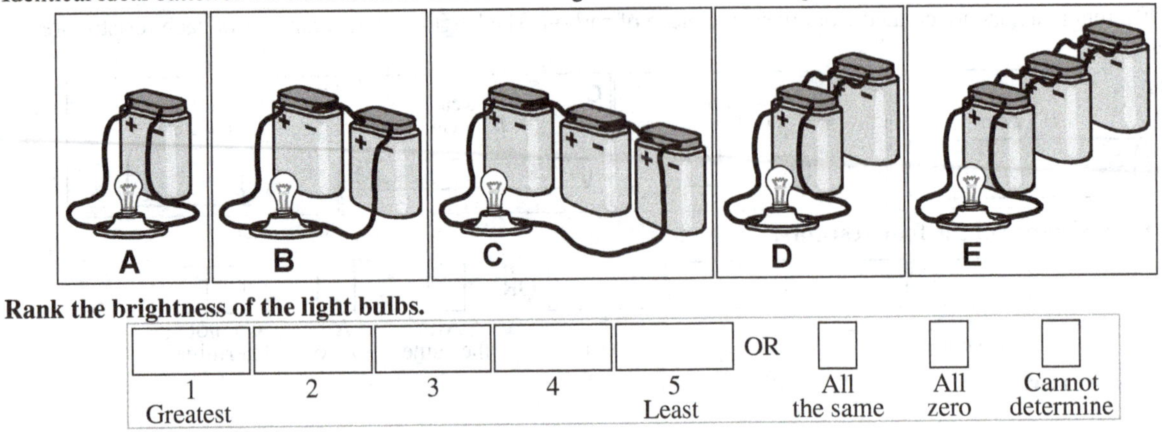

Rank the brightness of the light bulbs.

1 Greatest	2	3	4	5 Least	OR	All the same	All zero	Cannot determine

Explain your reasoning.

D2-RT04: Simple Resistor Circuits I—Resistance

All of the resistors and batteries are identical in the circuits shown.

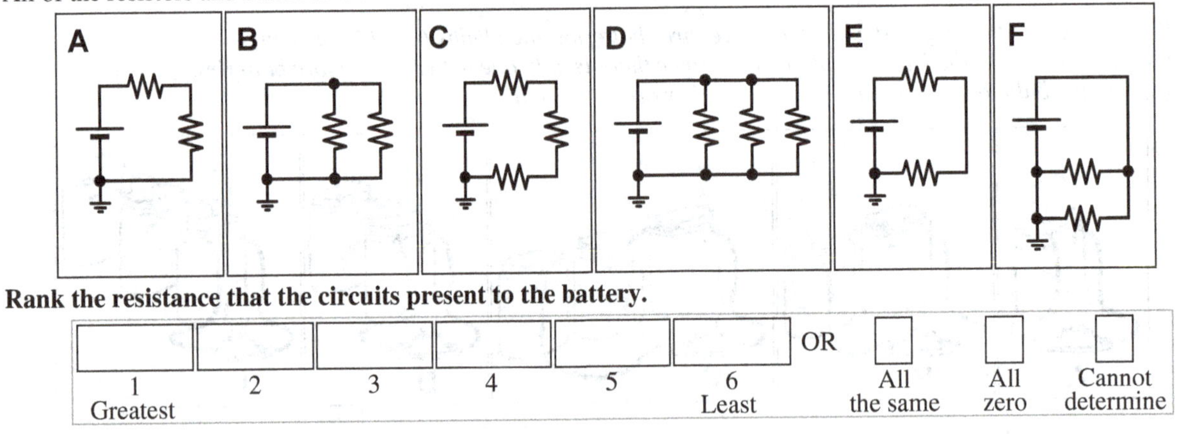

Rank the resistance that the circuits present to the battery.

1 Greatest	2	3	4	5	6 Least	OR	All the same	All zero	Cannot determine

Explain your reasoning.

D2-RT05: SIMPLE LIGHT BULB CIRCUITS I—BULB BRIGHTNESS

All of the bulbs in the circuits below are identical, as are all of the batteries.

For the three items below, rank the brightness of the bulb labeled *X*.

(a)

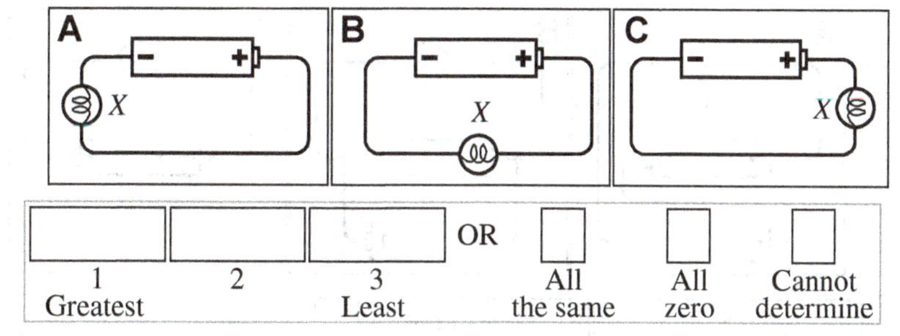

Explain your reasoning.

(b)

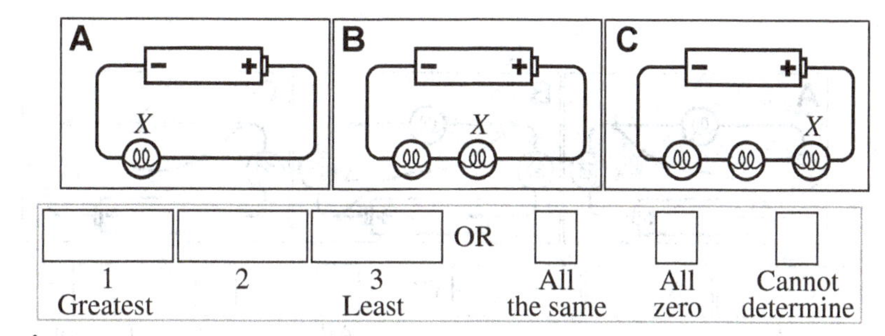

Explain your reasoning.

(c)

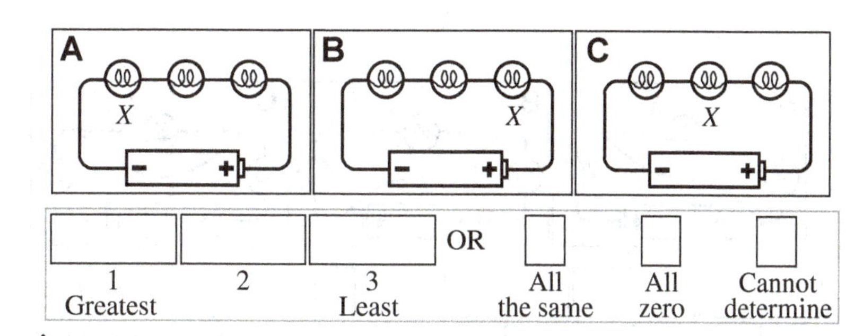

Explain your reasoning.

D2-RT06: Simple Light Bulb Circuits II—Bulb Brightness

All of the bulbs in the circuits below are identical, as are all of the batteries.

In each of the items below rank the brightness of the bulb labeled *X*.

(a)

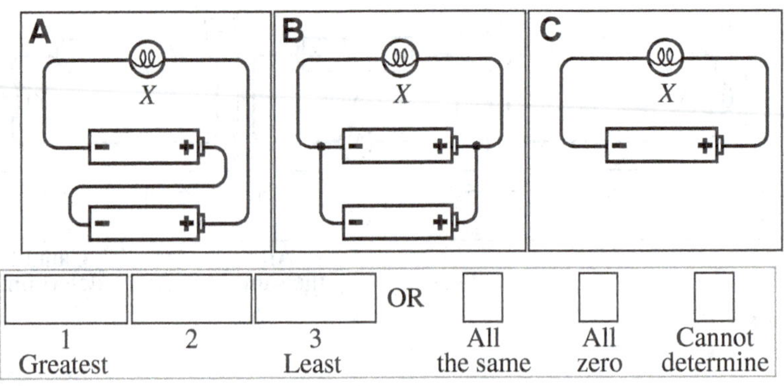

Explain your reasoning.

(b)

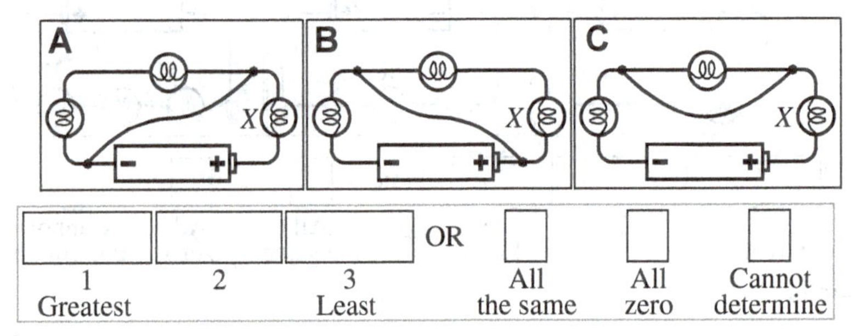

Explain your reasoning.

(c)

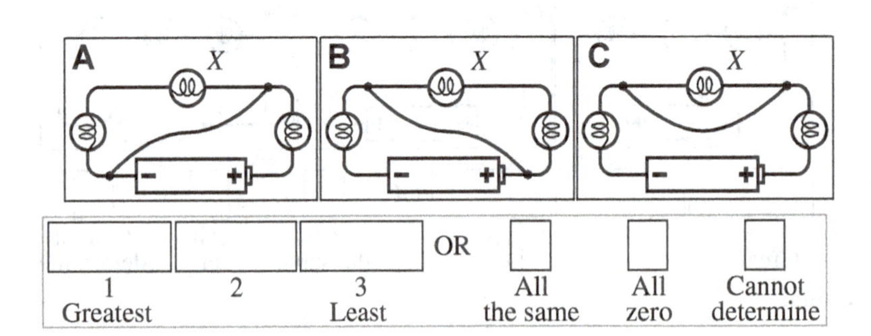

Explain your reasoning.

D2-RT07: SIMPLE LIGHT BULB CIRCUITS I—AMMETER READING

All of the bulbs in the circuits below are identical, as are all of the batteries.

For the two items below rank the current measured by the ammeter.

(a)

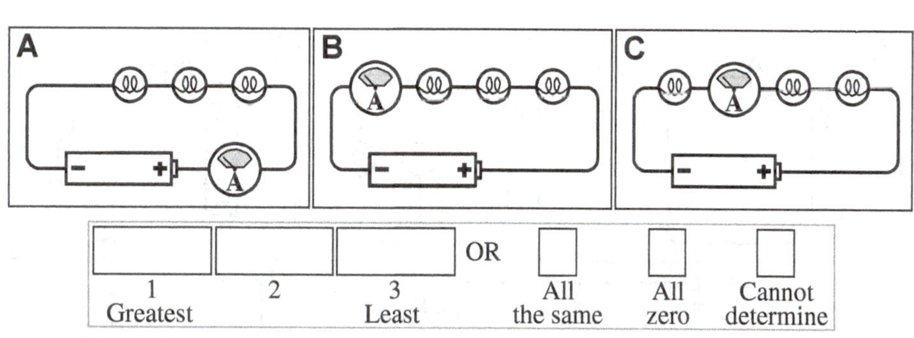

Explain your reasoning.

(b)

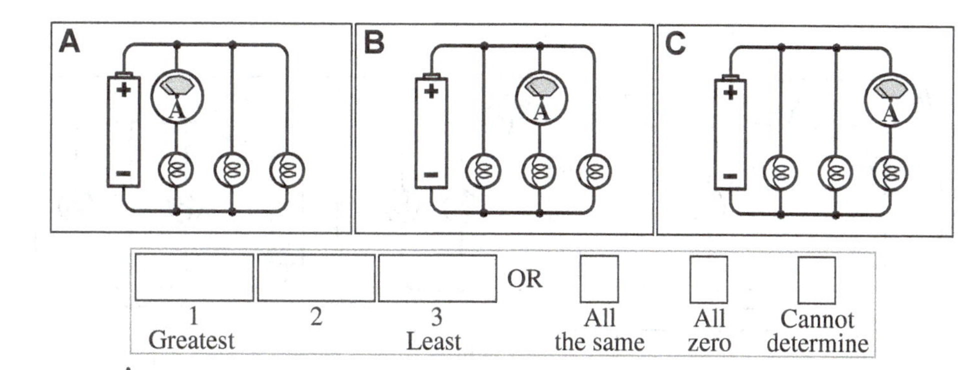

Explain your reasoning.

D2-CT08: CIRCUIT WITH TWO LIGHT BULBS—CURRENT IN BULB

A battery is connected to a circuit with two bulbs and a switch as shown.

When the switch is closed, does the current in bulb *A* (a) *increase,* (b) *decrease,* or (c) *remain the same*? _____

Explain your reasoning.

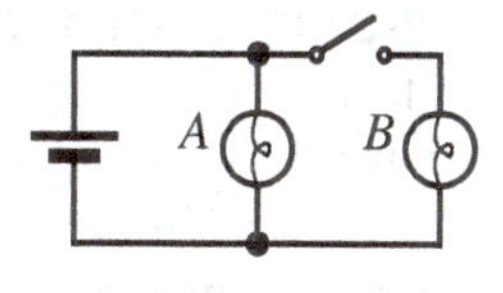

D2-RT09: Simple Light Bulb Circuits II—Ammeter Reading

All of the bulbs in the circuits below are identical, as are all of the batteries.

For the two items below rank the current measured by the ammeter.

(a)

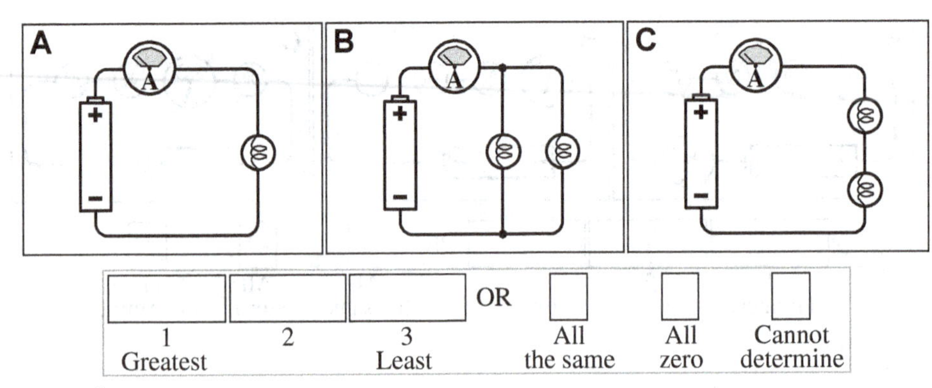

Explain your reasoning.

(b)

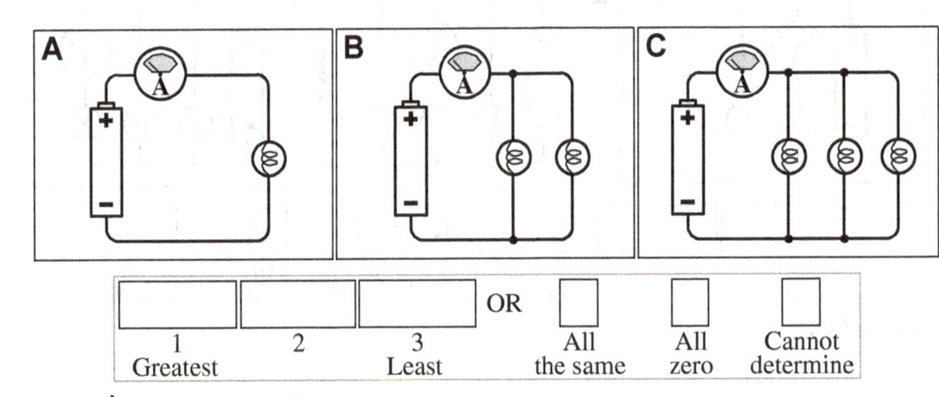

Explain your reasoning.

D2-CT10: Circuit with Two Light Bulbs—Current in Battery

A battery is connected to a circuit with two bulbs and a switch as shown.

When the switch is closed, does the current in the battery (i) *increase*, (ii) *decrease*, or (iii) *remain the same*? _____

Explain your reasoning.

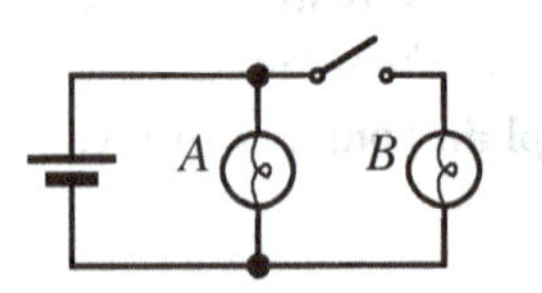

D2-RT11: SIMPLE LIGHT BULB CIRCUITS I—POTENTIAL DIFFERENCE BETWEEN TWO POINTS

All of the bulbs in the circuits below are identical, as are all of the batteries.

In each item below, rank the magnitude of the potential difference between points *M* and *N*.

(a)

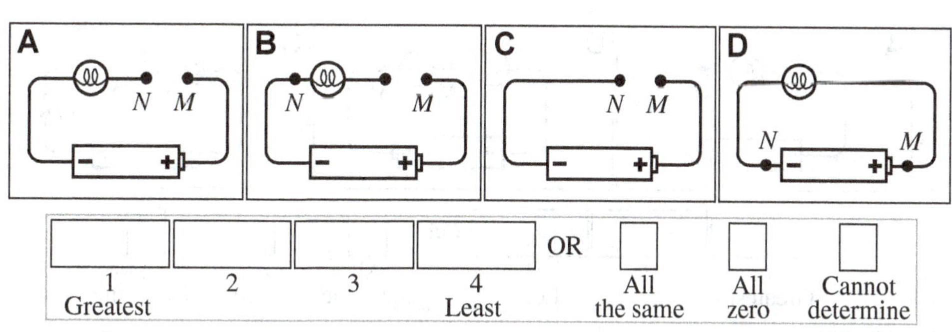

Explain your reasoning.

(b)

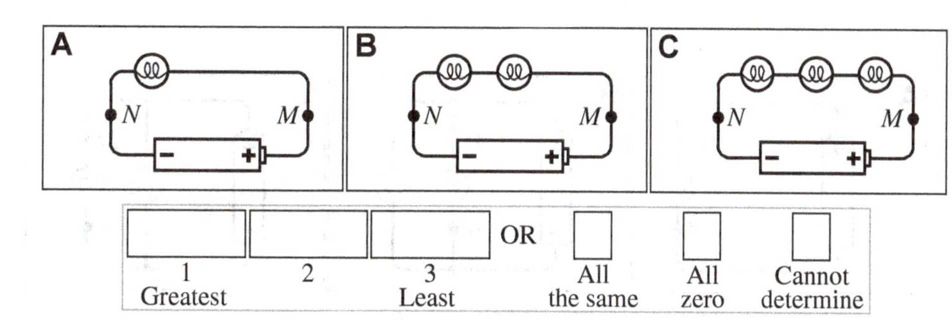

Explain your reasoning.

(c)

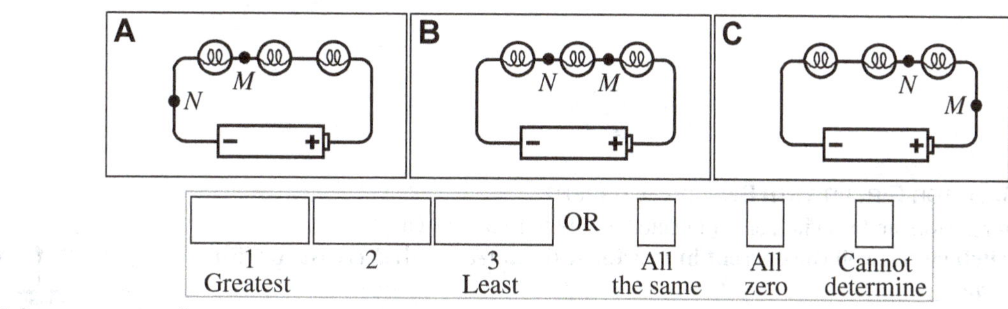

Explain your reasoning.

D2-RT12: Simple Light Bulb Circuits II—Potential Difference between Two Points

All of the bulbs in the circuits below are identical, as are all of the batteries.

For the two items below, rank the magnitude of the potential difference between points *M* and *N*.

(a)

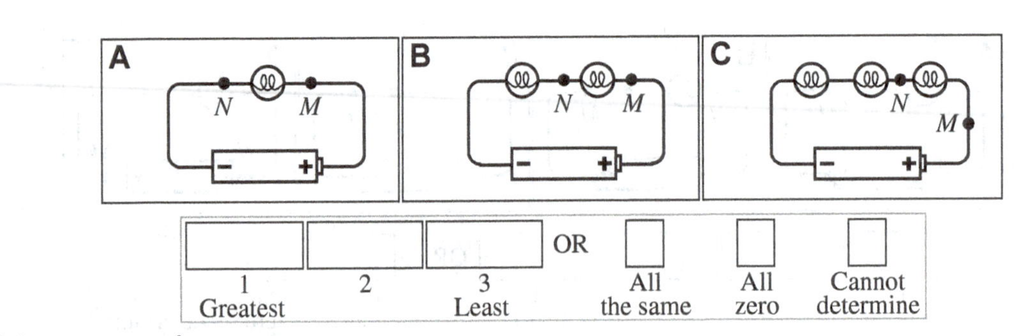

Explain your reasoning.

(b)

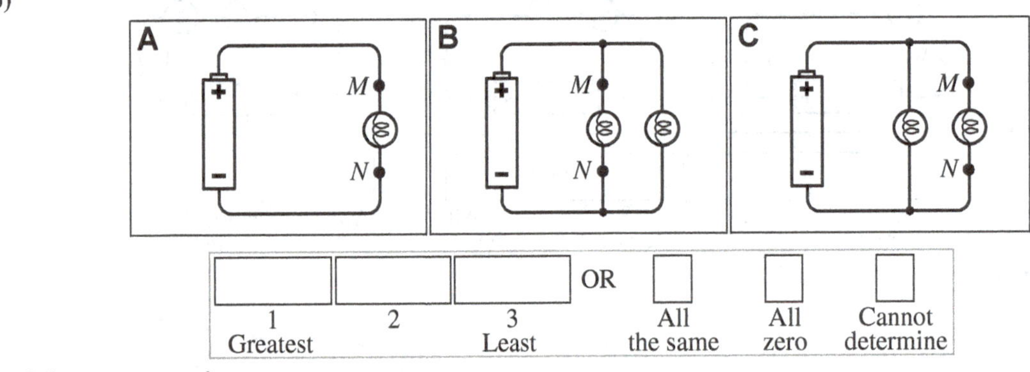

Explain your reasoning.

D2-CT13: Resistor Circuit with Switch—Current

Five identical resistors and a switch are connected to a battery as shown.

When the switch closes, will the current in resistor *A* (i) *increase*, (ii) *decrease*, or (iii) *remain the same*? _____

Explain your reasoning.

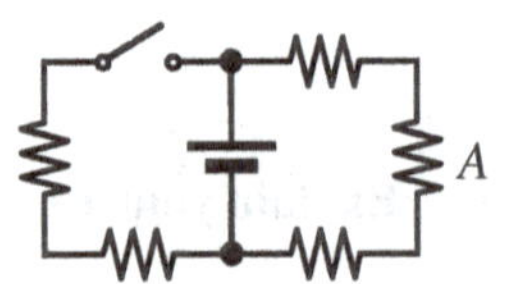

D2-RT14: SIMPLE LIGHT BULB CIRCUITS III—POTENTIAL DIFFERENCE BETWEEN TWO POINTS

All of the bulbs in the circuits below are identical, as are all of the batteries.

For the two items below, rank the magnitude of the potential difference between points *M* and *N*.

(a)

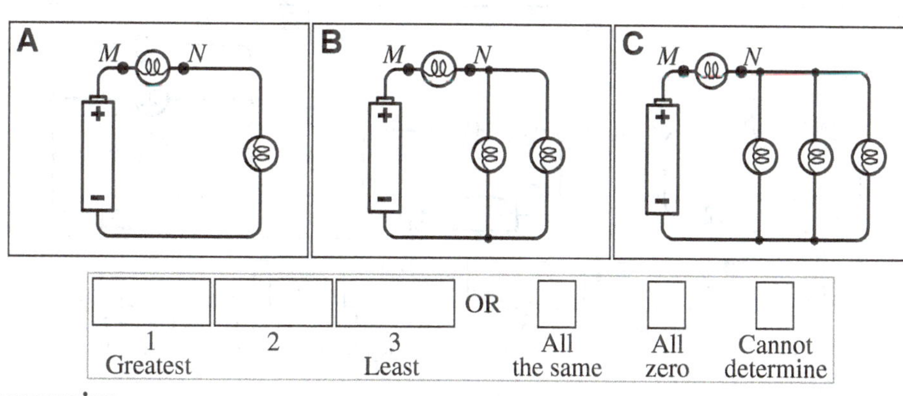

Explain your reasoning.

(b)

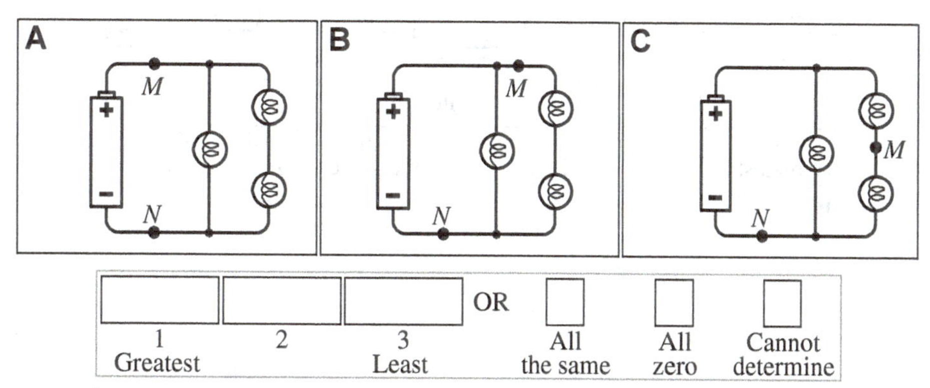

Explain your reasoning.

D2-RT15: Simple Light Bulb Circuits I—Current in Battery

All of the bulbs in the circuits below are identical, as are all of the batteries.

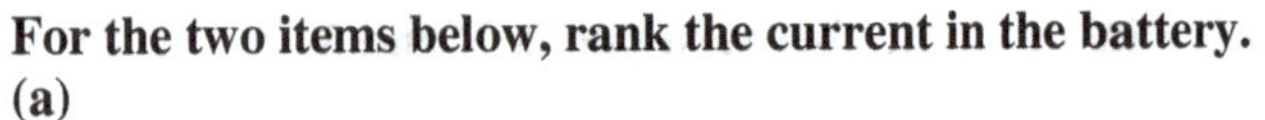

For the two items below, rank the current in the battery.

(a)

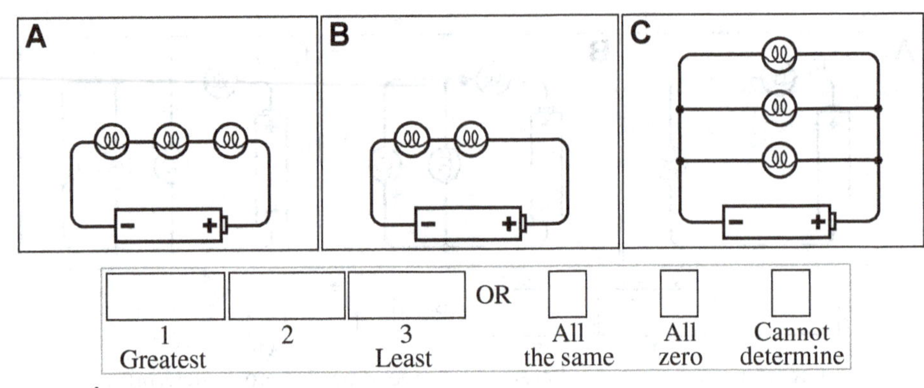

Explain your reasoning.

(b)

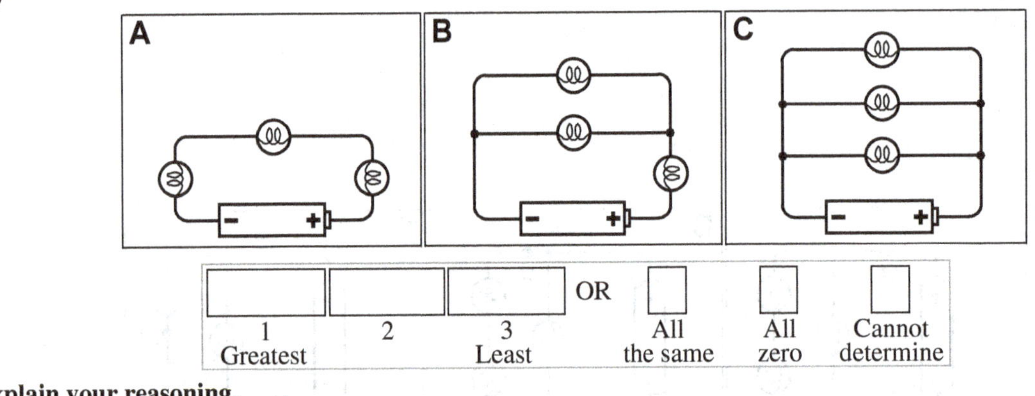

Explain your reasoning.

D2-CT16: Circuit with Two Batteries—Bulb Brightness

Two identical ideal batteries, a switch, and a bulb are connected as shown.

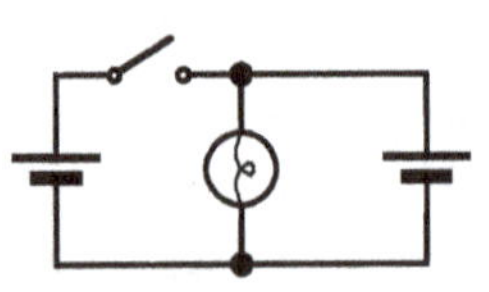

When the switch closes, will the brightness of the bulb (i) *increase*, (ii) *decrease*, or (iii) *remain the same*? _____

Explain your reasoning.

D2-RT17: SIMPLE LIGHT BULB CIRCUITS II—CURRENT IN BATTERY

All of the bulbs in the circuits below are identical, as are all of the batteries.

For the two items below, rank the current in the battery.

(a)

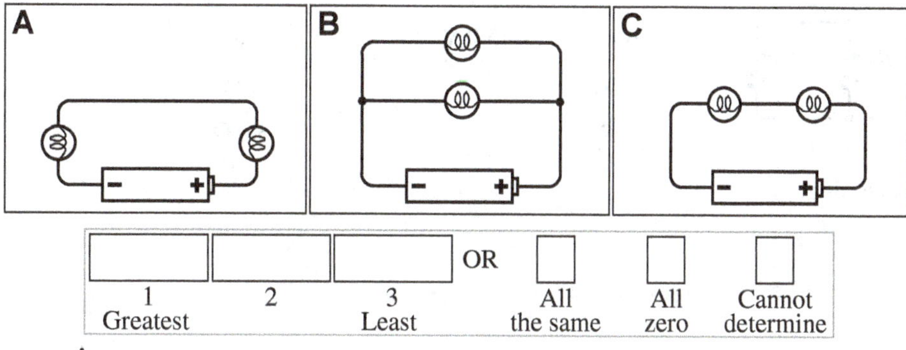

Explain your reasoning.

(b)

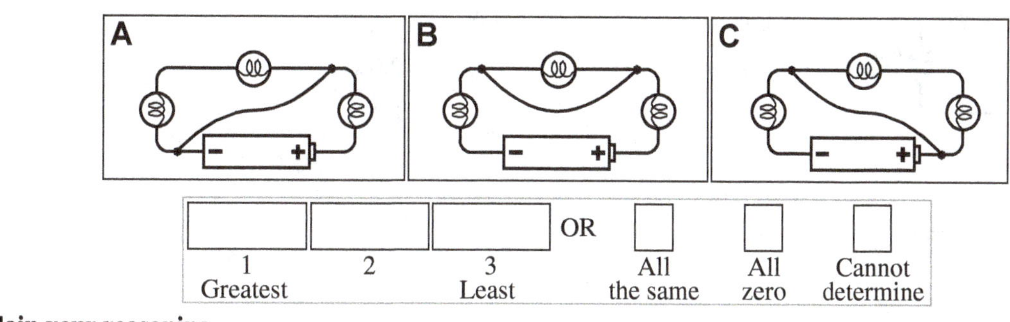

Explain your reasoning.

D2-CT18: LIGHT BULB CIRCUIT WITH SWITCH—CURRENT IN BULB

Three light bulbs and a switch are connected to a battery as shown.

When the switch is closed, will the current in bulb *A* (i) *increase*, (ii) *decrease*, or (iii) *remain the same*? _____

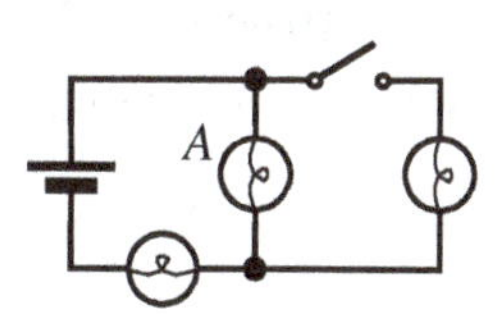

Explain your reasoning.

D2-QRT19: Two Resistor Circuits—Current, Resistance, and Voltage Drop Chart

For items (a) and (b) below complete the table, showing the value of the currents in and voltages across all elements.

(a) The resistance values for this circuit are given in the table, as is the battery voltage.

	ΔV	I	R
Battery	15.0 V		
R_1			5.0 Ω
R_2			3.0 Ω

Explain your reasoning.

(b) The resistance values for this circuit are given in the table, as is the current in the battery.

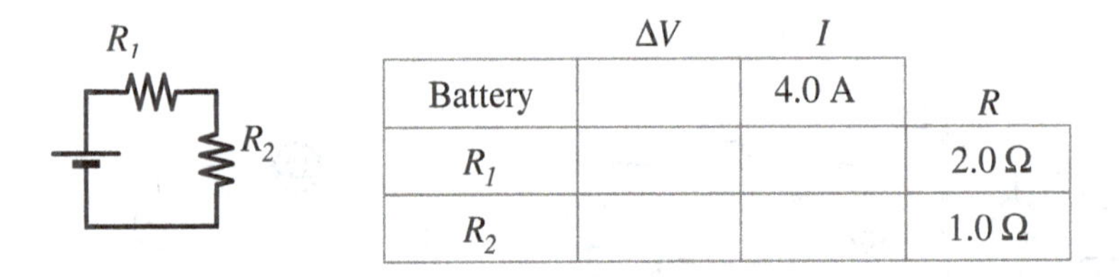

	ΔV	I	R
Battery		4.0 A	
R_1			2.0 Ω
R_2			1.0 Ω

Explain your reasoning.

D2-CT20: Two Light Bulbs in a Circuit—Bulb Brightness

Two identical light bulbs are connected to a battery as shown.

Is bulb *A* (i) *brighter than,* (ii) *dimmer than,* or (iii) *the same brightness as* bulb *B?* _____

Explain your reasoning.

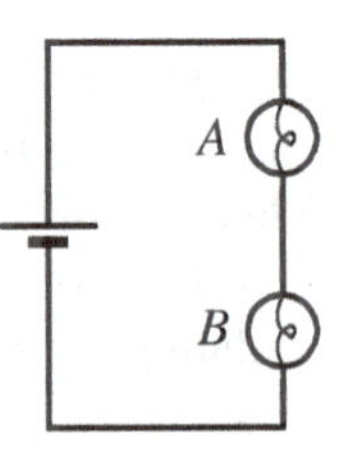

D2-LMCT21: TWO RESISTORS IN PARALLEL—BATTERY CURRENT

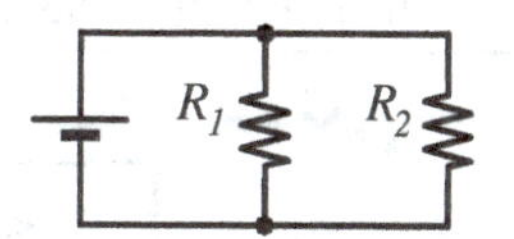

For the circuit shown, identify, from choices (i)–(iv), how each change described below will affect the current in the battery.

This change will:

(i) ***increase*** the current in the battery.
(ii) ***decrease*** the current in the battery.
(iii) ***have no effect*** on the current in the battery.
(iv) have an effect on the current in the battery that ***cannot be determined***.

All of these modifications are changes to the initial situation that is shown.

(a) The resistance in R_1 is reduced. _____
Explain your reasoning.

(b) The resistance in R_2 is reduced. ______
Explain your reasoning.

(c) The resistance in R_1 and R_2 are increased by the same amount. ______
Explain your reasoning.

(d) The resistance in R_1 is reduced, and in R_2 it is increased. ______
Explain.

D2-CT22: CIRCUIT WITH THREE RESISTORS—CURRENT

Three resistors are connected to a battery as shown. Two points in the circuit are labeled *A* and *B*.

Is the current at point *A* (i) *greater than*, (ii) *less than*, or (iii) *equal to* the current at point *B*? _____

Explain your reasoning.

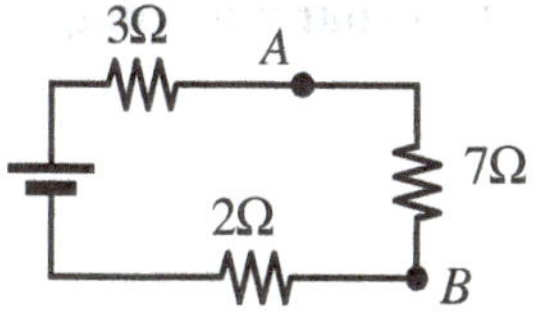

D2-RT23: Simple Resistor Circuits I—Current

All of the resistors in the circuits shown are identical, as are all of the batteries.

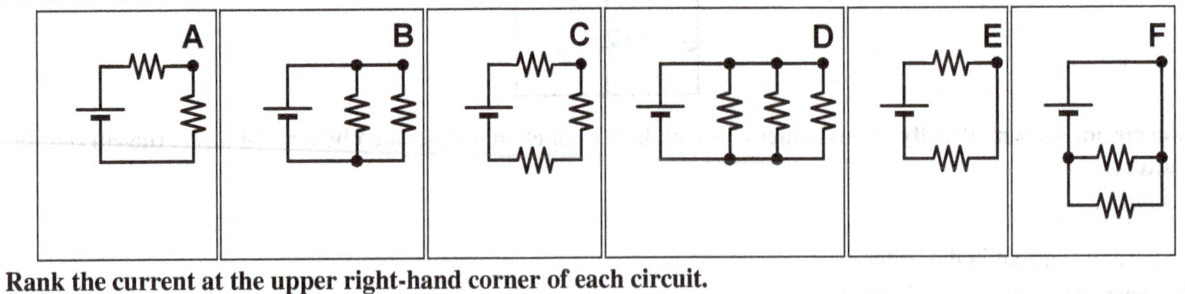

Rank the current at the upper right-hand corner of each circuit.

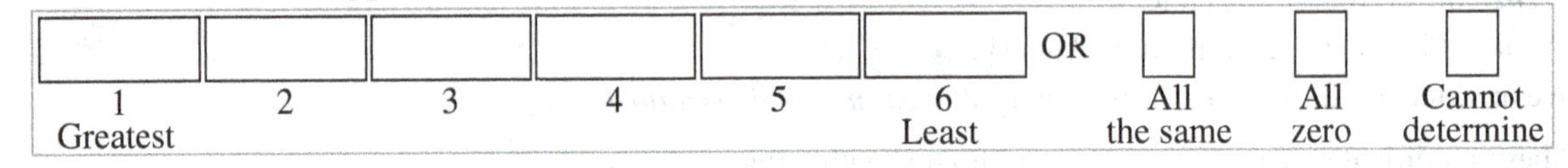

Explain your reasoning.

D2-RT24: Simple Resistor Circuits with a Ground—Voltage

All of the resistors in the circuits below are identical, as are all of the batteries.

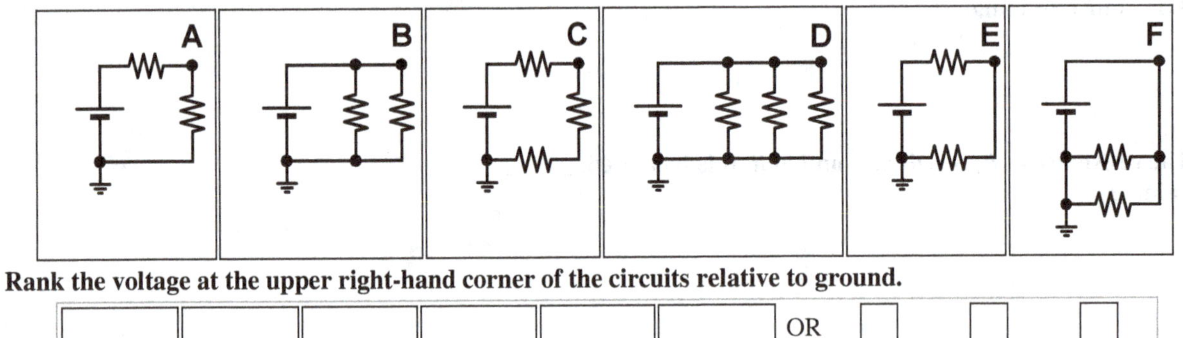

Rank the voltage at the upper right-hand corner of the circuits relative to ground.

Explain your reasoning.

D2-RT25: SIMPLE RESISTOR CIRCUITS II—CURRENT

All of the resistors in the circuits below are identical. Three of the circuits contain 6-volt batteries and three contain 12-volt batteries.

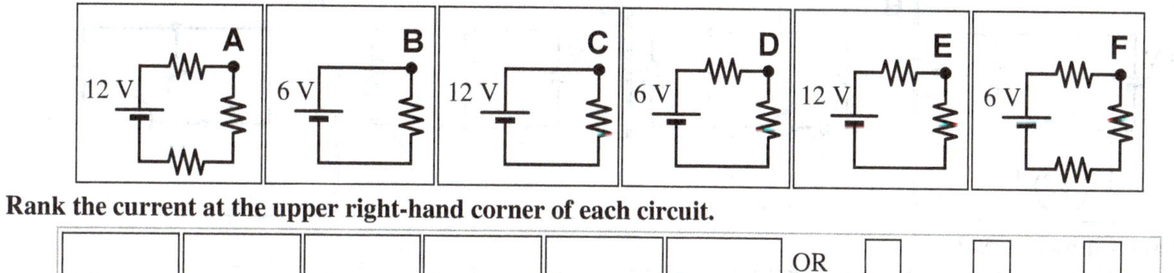

Rank the current at the upper right-hand corner of each circuit.

						OR			
1 Greatest	2	3	4	5	6 Least		All the same	All zero	Cannot determine

Explain your reasoning.

D2-RT26: SIMPLE RESISTOR CIRCUITS WITH A GROUND—VOLTAGE DROP

The following circuits contain either a 6-volt or a 12-volt battery and one or more identical resistors.

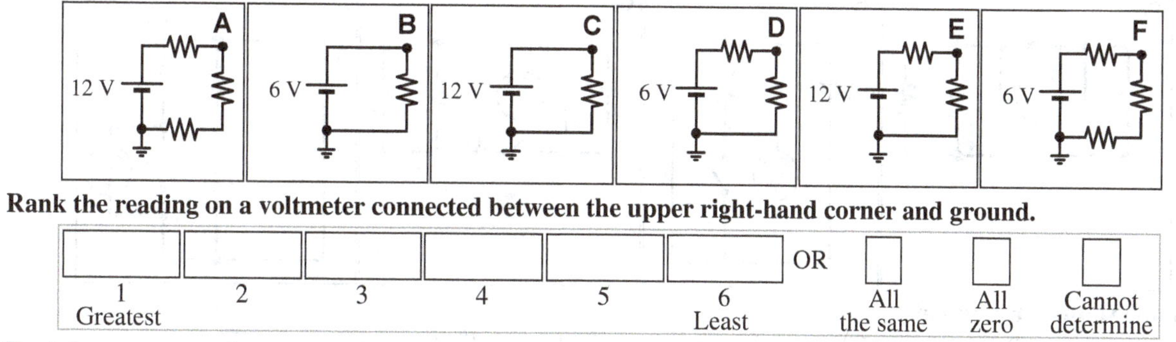

Rank the reading on a voltmeter connected between the upper right-hand corner and ground.

						OR			
1 Greatest	2	3	4	5	6 Least		All the same	All zero	Cannot determine

Explain your reasoning.

D2-RT27: Parallel Circuits I—Voltmeter Readings across Open Switches

All of the resistors in the circuits below are identical. The switch in each case is open.

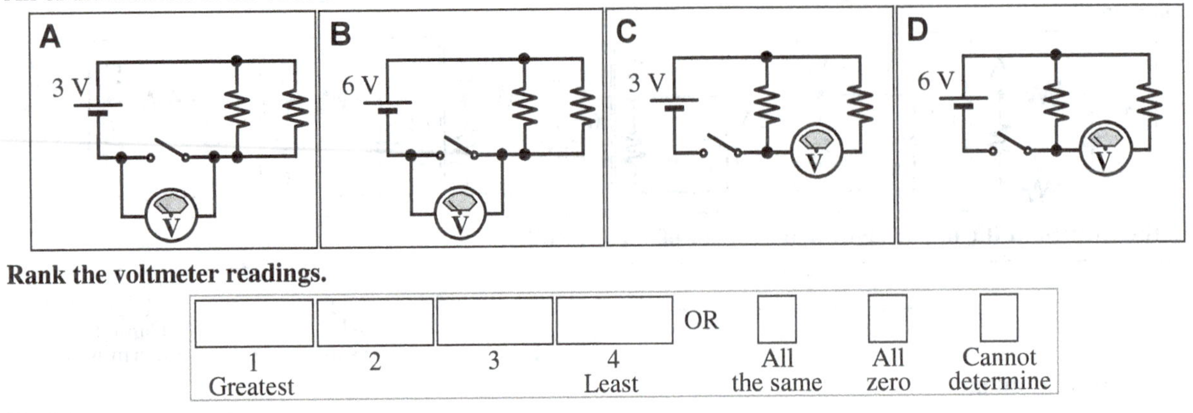

Rank the voltmeter readings.

1 Greatest	2	3	4 Least	OR	All the same	All zero	Cannot determine

Explain your reasoning.

D2-RT28: Parallel Circuits II—Voltmeter Readings across Open Switches

All of the resistors in the circuits below are identical, as are all of the batteries. The switch in each case is open.

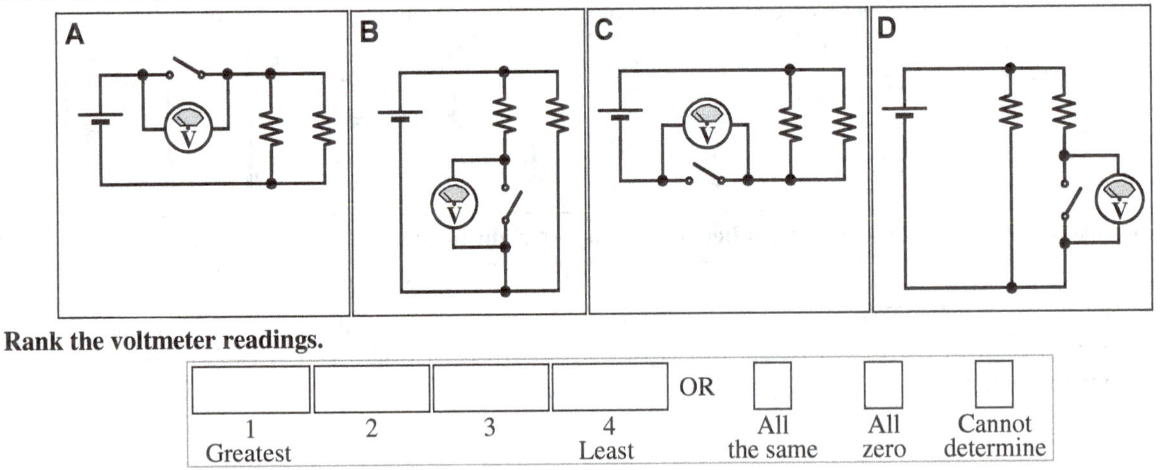

Rank the voltmeter readings.

1 Greatest	2	3	4 Least	OR	All the same	All zero	Cannot determine

Explain your reasoning.

D2-RT29: CIRCUIT WITH TWO SWITCHES—AMMETER READINGS

The circuit contains a battery, two switches, five identical resistors, and an ammeter. Four possible switch configurations (open or closed) for the circuit are shown in the table.

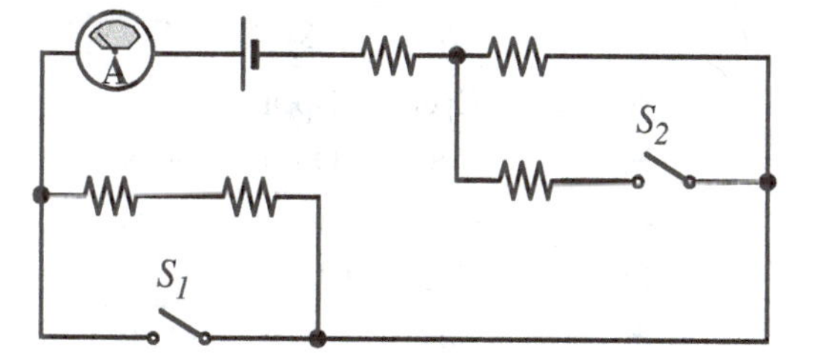

Configuration	Switch S_1	Switch S_2
A	Open	Open
B	Open	Closed
C	Closed	Open
D	Closed	Closed

Rank the ammeter reading for the four configurations.

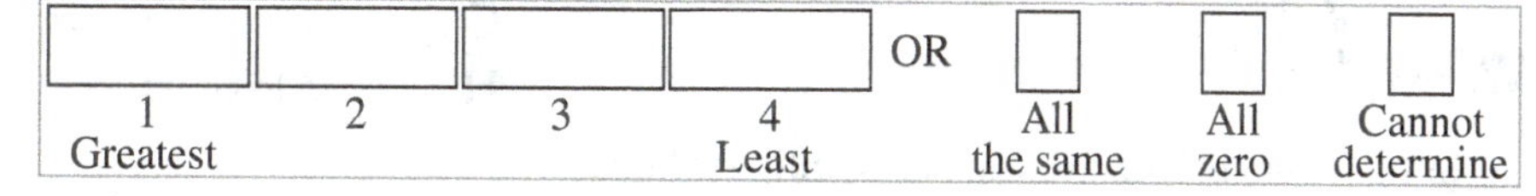

Explain your reasoning.

D2-RT30: CIRCUIT WITH TWO SWITCHES—VOLTMETER READINGS

The circuit contains a battery, two switches, five identical resistors, and a voltmeter. Four possible switch configurations (open or closed) for the circuit are shown in the table.

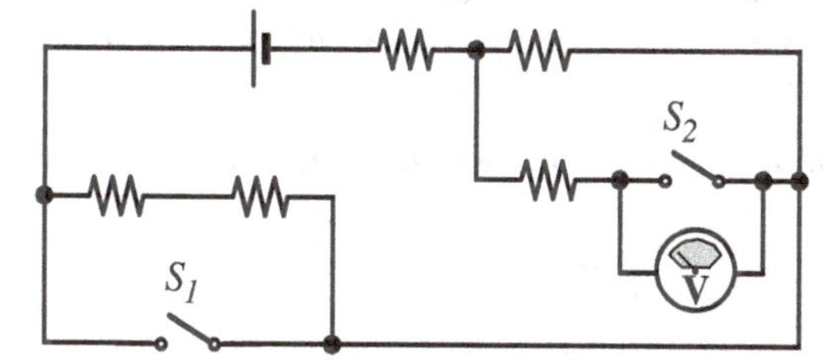

Configuration	Switch S_1	Switch S_2
A	Open	Open
B	Open	Closed
C	Closed	Open
D	Closed	Closed

Rank the voltmeter reading for the four configurations.

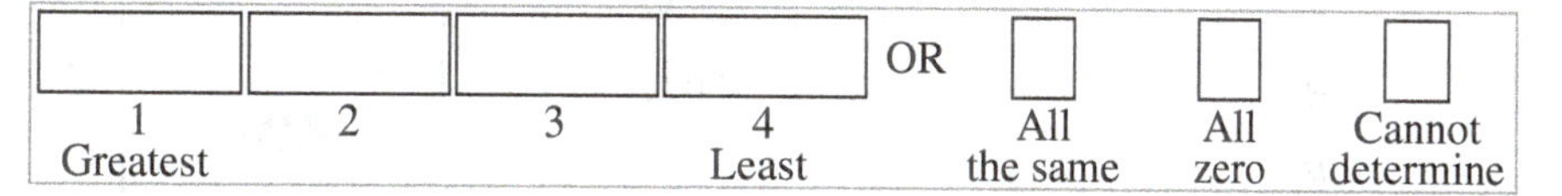

Explain your reasoning.

D2-RT31: CIRCUIT WITH THREE SWITCHES—AMMETER READINGS

The circuit contains a battery, three switches, six identical resistors, and an ammeter. Eight possible switch configurations (open or closed) for the circuit are shown in the table.

Configuration	S_1	S_2	S_3
A	Open	Open	Open
B	Open	Open	Closed
C	Open	Closed	Open
D	Open	Closed	Closed
E	Closed	Open	Open
F	Closed	Open	Closed
G	Closed	Closed	Open
H	Closed	Closed	Closed

Rank the current in the ammeter for these switch configurations.

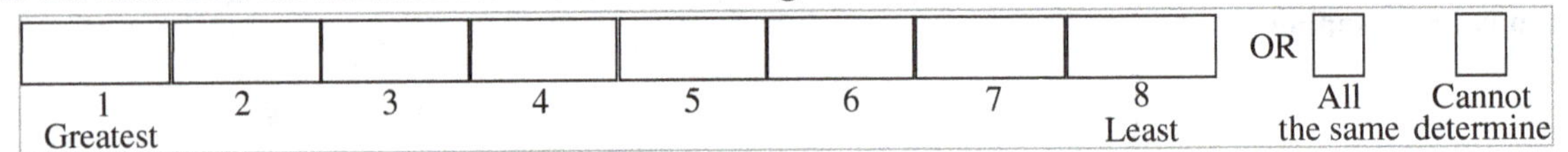

Explain your reasoning.

D2-QRT32: FIVE RESISTOR CIRCUITS—CURRENT, RESISTANCE, AND VOLTAGE DROP

Four of the five resistance values for this circuit are given in the table, as is the battery voltage and the current in resistor R_3.

Complete the table, showing the value of R_1 and the currents in and voltages across all elements.

	ΔV	I	R
Battery	72.0 V		
R_1			
R_2			2.0 Ω
R_3		4.0 A	5.0 Ω
R_4			1.0 Ω
R_5			3.0 Ω

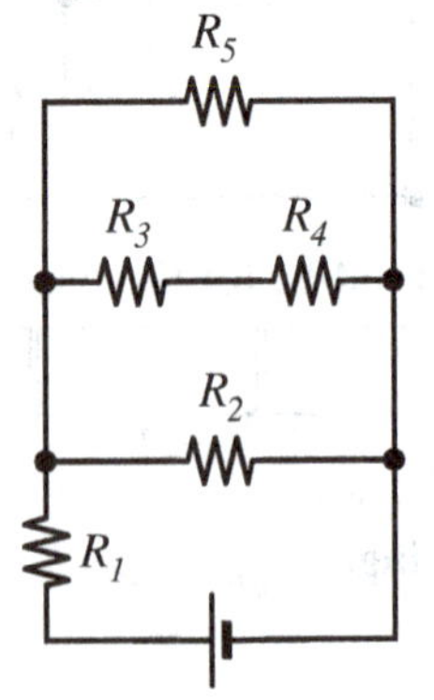

Explain your reasoning.

D2-CT33: LIGHT BULBS CIRCUIT WITH THREE BATTERIES—BULB BRIGHTNESS

Three identical ideal batteries, a switch, and two bulbs are connected as shown.

When the switch closes, will the brightness of bulb A (i) *increase*, (ii) *decrease*, or (iii) *remain the same*? _____

Explain your reasoning.

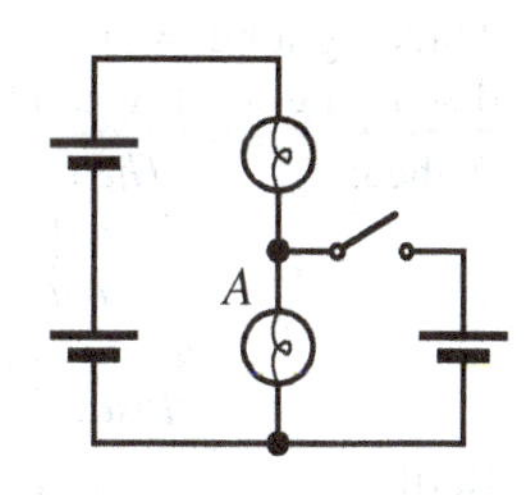

D2-CT34: LIGHT BULBS CIRCUIT WITH SWITCH—BRIGHTNESS OF BULBS

Two light bulbs and a switch are connected to a battery as shown.

(a) When the switch is closed, will the brightness of bulb B (i) *increase*, (ii) *decrease*, or (iii) *remain the same*? _____

Explain your reasoning.

(b) When the switch is closed, will the brightness of bulb A (i) *increase*, (ii) *decrease*, or (iii) *remain the same*?

_____

Explain your reasoning.

D2-WWT35: CIRCUIT WITH TWO RESISTORS—CURRENT

A battery is connected to a circuit containing two resistors and a switch as shown. A student states:

"When the switch closes, the current in resistor A goes down, because resistor A now has to share the current from the battery with resistor B."

What, if anything, is wrong with this statement? If something is wrong, identify it and explain how to correct it. If this statement is correct, explain why.

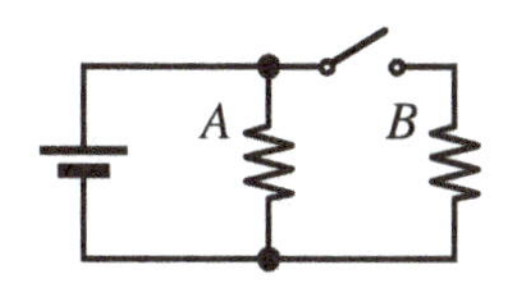

D2-SCT36: Light Bulb Circuit with Switch—Bulb Brightness

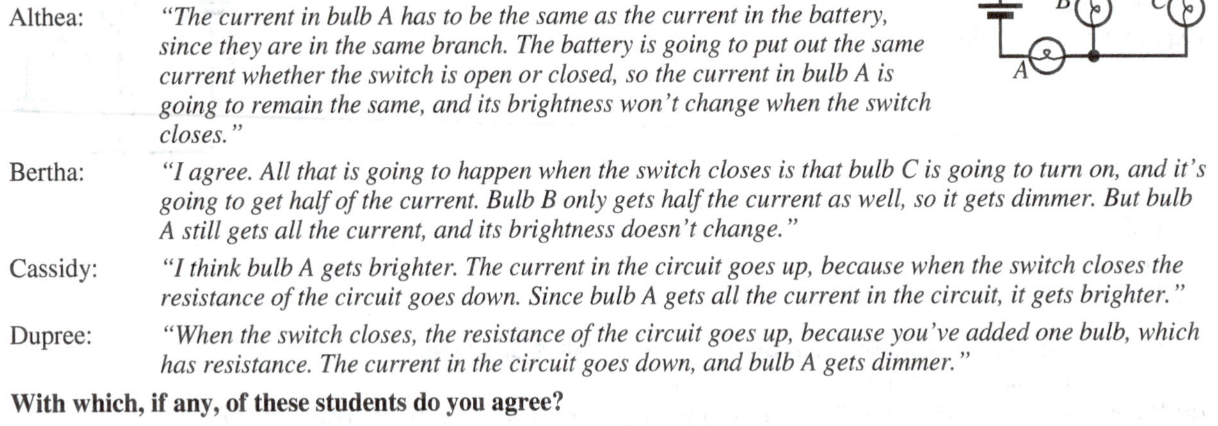

Three light bulbs and a switch are connected to a battery as shown. Four students are discussing what would happen to the brightness of bulb *A* when the switch closes:

Althea: *"The current in bulb A has to be the same as the current in the battery, since they are in the same branch. The battery is going to put out the same current whether the switch is open or closed, so the current in bulb A is going to remain the same, and its brightness won't change when the switch closes."*

Bertha: *"I agree. All that is going to happen when the switch closes is that bulb C is going to turn on, and it's going to get half of the current. Bulb B only gets half the current as well, so it gets dimmer. But bulb A still gets all the current, and its brightness doesn't change."*

Cassidy: *"I think bulb A gets brighter. The current in the circuit goes up, because when the switch closes the resistance of the circuit goes down. Since bulb A gets all the current in the circuit, it gets brighter."*

Dupree: *"When the switch closes, the resistance of the circuit goes up, because you've added one bulb, which has resistance. The current in the circuit goes down, and bulb A gets dimmer."*

With which, if any, of these students do you agree?

Althea _____ Bertha _____ Cassidy _____ Dupree _____ None of them_____

Explain your reasoning.

D2-WWT37: Circuit with Two Resistors—Current

A battery is connected to a circuit containing two resistors as shown. A student states:

"Using Ohm's law, the current is the voltage divided by the resistance, so when you have a bigger resister, you have a smaller current. In this case, resistor B is a larger resistance than A, so it will have a smaller current."

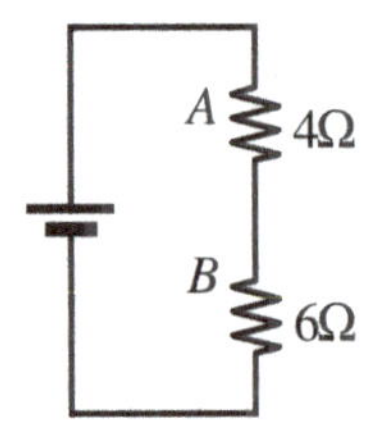

What, if anything, is wrong with this statement? If something is wrong, identify it and explain how to correct it. If this statement is correct, explain why.

D2-WWT38: CIRCUIT WITH FOUR RESISTORS—CURRENT RANKING

A battery is connected to a circuit containing four identical resistors as shown. A student states:

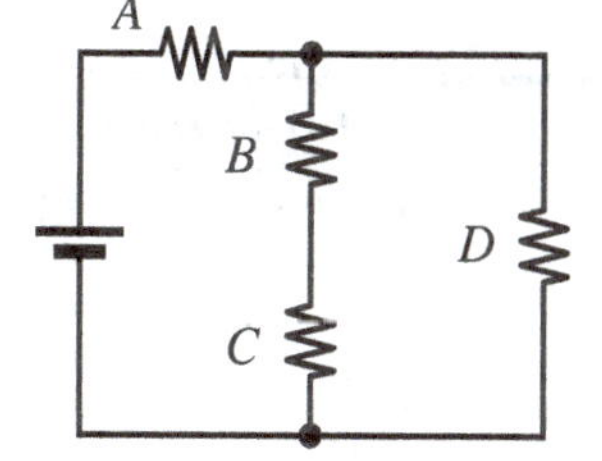

"All of the current from the battery goes through resistor A. At the junction after A the current splits up. Half of the current goes through resistor D, and the other half is shared by resistors B and C. So resistor A has the most current, followed by resistor D, followed by resistors B and C, which have the same current."

What, if anything, is wrong with this statement? If something is wrong, identify it and explain how to correct it. If this statement is correct, explain why.

D2-WWT39: CIRCUIT WITH THREE BULBS—VOLTAGES

A 24-volt battery is in a circuit containing three bulbs as shown. A voltmeter across bulb *A* measures 18 Volts. A student states:

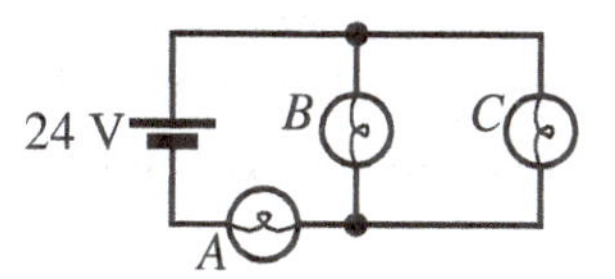

"Since bulbs B and C are identical, then they will each have the same current and the same voltage across them. The sum of the voltages across the three bulbs must add to the battery voltage. So bulb B has a voltage of 3 volts across it, and bulb C also has a voltage of 3 volts."

What, if anything, is wrong with this statement? If something is wrong, identify it and explain how to correct it. If this statement is correct, explain why.

D2-CT40: Four Light Bulbs Circuit with Switch—Effect of Closing Switch

A battery is connected to four identical bulbs and a switch as shown.

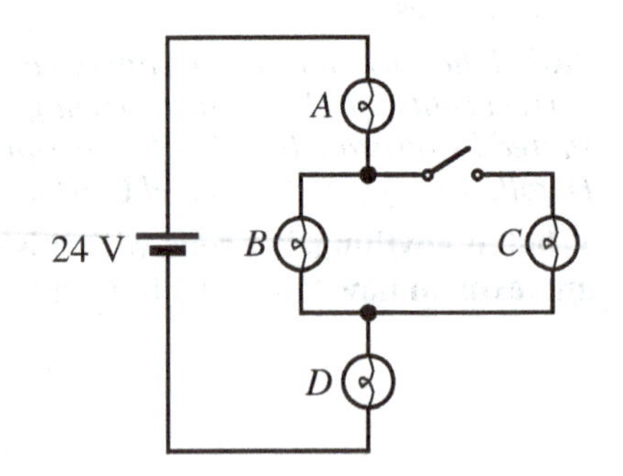

(a) When the switch is closed, does the brightness of bulb *C* (i) *increase*, (ii) *decrease*, or (iii) *remain the same*? _____

Explain your reasoning.

(b) When the switch is closed, does the current in the battery (i) *increase*, (ii) *decrease*, or (iii) *remain the same*? _____

Explain your reasoning.

(c) When the switch is closed, does the brightness of bulb *A* (i) *increase*, (ii) *decrease*, or (iii) *remain the same*? _____

Explain your reasoning.

(d) When the switch is closed, is bulb *D* (i) *brighter* than bulb *A*, (ii) *dimmer* than bulb *A*, or (iii) *the same* brightness as bulb *A*? _____

Explain your reasoning.

(e) When the switch is closed, does the brightness of bulb *D* (i) *increase*, (ii) *decrease*, or (iii) *remain the same*? _____

Explain your reasoning.

(f) When the switch is closed, does the brightness of bulb *B* (i) *increase*, (ii) *decrease*, or (iii) *remain the same*? _____

Explain your reasoning.

D2-SCT41: FOUR RESISTOR CIRCUIT I—CURRENT

In the circuit shown, the sizes of the resistors vary as $R_3 > R_1 > R_2 > R_4$. Four students discussing the currents in this circuit make the following statements:

Ajay: *"I think the current in R_1 will be the largest because all of the current from the battery goes through it."*

Belen: *"Right, and after R_1 the current splits into two parts at the junction. The current through R_2, R_3, and R_4 will all be the same because there are two branches in the circuit and each branch will get half of the current."*

Ciara: *"From Ohm's law, current is biggest where resistance is smallest. I think the current through R_2 will be largest because that branch has the lowest resistance in the circuit."*

Damaris: *"Also using Ohm's law, I think the current in R_3 will be the smallest because R_3 has the largest resistance. The current in R_4 will be largest, because that resistor has the smallest resistance."*

Efren: *"The current in R_3 will be the same as the current in R_4 because they are in the same branch."*

With which, if any, of these students do you agree?

Ajay _____ Belen _____ Ciara _____ Damaris _____ Efren _____ None of them _____

Explain your reasoning.

D2-SCT42: FOUR RESISTOR CIRCUIT II—CURRENT

In the circuit shown, the sizes of the resistors vary as $R_3 > R_2 > R_4 > R_1$. Four students discussing the currents in this circuit make the following statements:

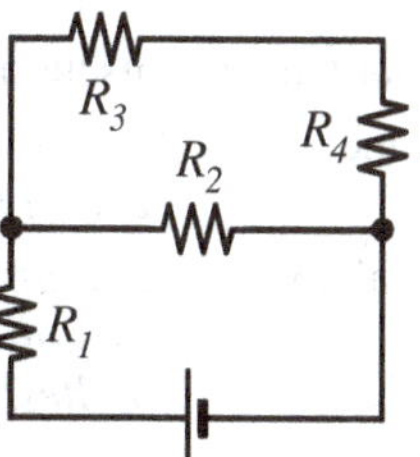

Ali: *"I think the current in R_1 will be the largest because all of the current from the battery goes through it."*

Ben: *"I think the current through R_2, R_3, and R_4 will all be the same because there are two branches in the circuit and each branch will get half of the current."*

Clyde: *"Well I disagree with Ben. I think the current in R_2 will be larger than the current in R_3 and R_4. The currents in the branches depend on the resistances of the branches."*

Dar: *"The only thing I am sure about is that the current in R_3 will be the same as that in R_4 because they are in the same branch."*

With which, if any, of these students do you agree?

Ali _____ Ben _____ Clyde _____ Dar _____ None of them _____

Explain your reasoning.

D2-SCT43: Four Resistor Circuit III—Potential Difference

In the circuit shown, the sizes of the resistors vary as $R_3 > R_1 > R_2 > R_4$. Four students discussing the potential differences in this circuit make the following statements:

Anselma: *"I think the potential difference across R_1 will be the largest because all of the current from the battery goes through it, and it is not the smallest resistance in the circuit."*

Brooke: *"I think the potential difference through R_2 will be largest because that branch will have the larger current of the two branches in the circuit."*

Chandra: *"I am not sure about the potential difference across R_1, but I think the potential differences across the two horizontal branches will be the same."*

Deangelo: *"I'm pretty sure the potential difference across R_3 will be larger than the potential difference across R_4 because R_3 has a larger resistance than R_4."*

Eloy: *"I think the two horizontal branches have the same potential difference as the battery since they are in parallel with the battery."*

With which, if any, of these students do you agree?

Anselma _____ Brooke _____ Chandra _____ Deangelo _____ Eloy _____ None of them _____

Explain your reasoning.

D2-SCT44: Six Resistor Circuit—Current

In the circuit pictured below the sizes of the resistors vary as

$$R_3 > R_5 > R_1 > R_2 > R_4 > R_6$$

Four students discussing the currents in this circuit make the following statements:

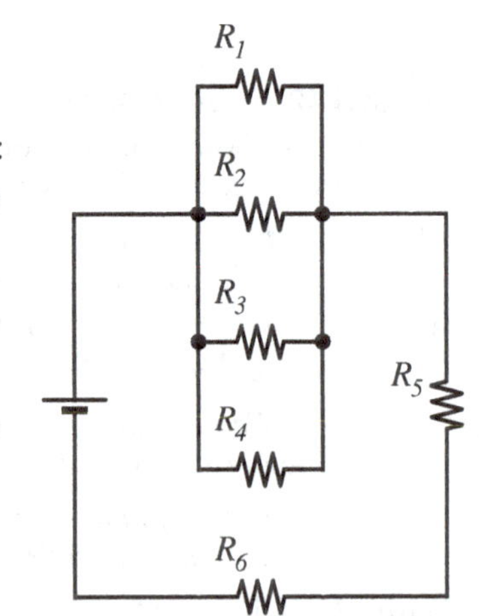

Anne: *"I think the current in R_6 and R_5 will be the largest because all of the current from the battery goes through both of those resistors."*

Benicio: *"I think the current through R_6 will be the smallest because that resistor is the last one in the circuit to get the current, and it is the smallest resistor."*

Celestine: *"I am not sure about the largest current, but I think the current in R_3 will be the lowest because R_3 has the largest resistance."*

Dulce: *"The only thing I am sure about is that the current across R_6 will be the largest because it is the smallest resistor in the circuit."*

With which, if any, of these students do you agree?

Anne _____ Benicio _____ Celestine _____ Dulce _____ None of them _____

Explain your reasoning.

D2-CT45: FOUR RESISTOR CIRCUITS III—CURRENT

For these two circuits, consider the current in the resistor R_1 closest to the battery.

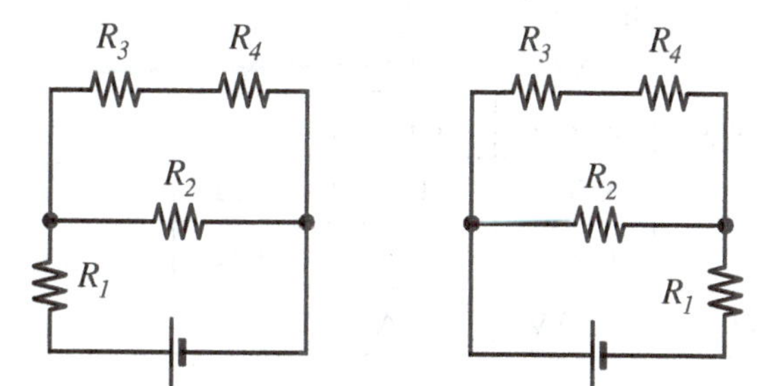

Will the current in R_1 be (i) *larger* in the circuit on the left, (ii) *smaller* in the circuit on the left, or (iii) *equal* in both circuits? _____
Explain your reasoning.

D2-CT46: FOUR RESISTOR CIRCUITS III—POTENTIAL DIFFERENCE

For these two circuits, consider the potential difference across the resistor R_1 closest to the battery.

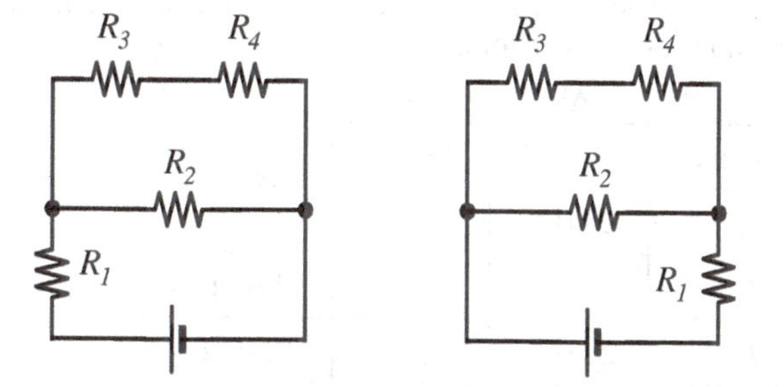

Will the potential difference across R_1 be (i) *larger* in the circuit on the left, (ii) *smaller* in the circuit on the left, or (iii) *equal* in both? _____
Explain your reasoning.

D2-WBT47: Three Resistors Circuit Chart I—Circuit

A circuit contains three resistors and a battery. The chart gives the currents in each element, the potential difference across each element, and the resistance values of the resistors.

	ΔV	I	R
Battery	36.0 V	3.0 A	
R_1	9.0 V	3.0 A	3.0 Ω
R_2	15.0 V	3.0 A	5.0 Ω
R_3	12.0 V	3.0 A	4.0 Ω

Draw an electric circuit that is consistent with the values of this chart. Label the resistors.

D2-WBT48: Three Resistors Circuit Chart II—Circuit

A circuit contains three resistors and a battery. The chart gives the currents in each element, the potential difference across each element, and the resistance values of the resistors.

	ΔV	I	R
Battery	24.0 V	16.0 A	
R_1	24.0 V	8.0 A	3.0 Ω
R_2	24.0 V	6.0 A	4.0 Ω
R_3	24.0 V	2.0 A	12.0 Ω

Draw an electric circuit that is consistent with the values of this chart. Label the resistors.

D2-WBT49: THREE RESISTORS CIRCUIT CHART III—CIRCUIT

A circuit contains three resistors and a battery. The chart gives the currents in each element, the potential difference across each element, and the resistance values of the resistors.

	ΔV	I	R
Battery	18.0 V	6.0 A	
R_1	6.0 V	6.0 A	1.0 Ω
R_2	12.0 V	2.0 A	6.0 Ω
R_3	12.0 V	4.0 A	3.0 Ω

Draw an electric circuit that is consistent with the values of this chart. Label the resistors.

D2-WBT50: THREE RESISTORS CIRCUIT CHART IV—CIRCUIT

A circuit contains three resistors and a battery. The chart gives the currents in each element, the potential difference across each element, and the resistance values of the resistors.

	ΔV	I	R
Battery	12.0 V	4.0 A	
R_1	3.0 V	3.0 A	1.0 Ω
R_2	9.0 V	3.0 A	3.0 Ω
R_3	12.0 V	1.0 A	12.0 Ω

Draw an electric circuit that is consistent with the values of this chart. Label the resistors.

D2-WBT51: Four Resistors Circuit Chart I—Circuit

A circuit contains four resistors and a battery. The chart gives the currents in each element, the potential difference across each element, and the resistance values of the resistors.

	ΔV	I	R
Battery	60.0 V	5.0 A	
R_1	20.0 V	2.0 A	10.0 Ω
R_2	40.0 V	2.0 A	20.0 Ω
R_3	45.0 V	3.0 A	15.0 Ω
R_4	15.0 V	3.0 A	5.0 Ω

Draw an electric circuit that is consistent with the values of this chart. Label the resistors.

D2-WBT52: Four Resistors Circuit Chart II—Circuit

A circuit contains four resistors and a battery. The chart gives the currents in each element, the potential difference across each element, and the resistance values of the resistors.

	ΔV	I	R
Battery	36.0 V	9.0 A	
R_1	12.0 V	6.0 A	2.0 Ω
R_2	12.0 V	3.0 A	4.0 Ω
R_3	24.0 V	6.0 A	4.0 Ω
R_4	24.0 V	3.0 A	8.0 Ω

Draw an electric circuit that is consistent with the values of this chart. Label the resistors.

D2-WBT53: FIVE RESISTORS CIRCUIT CHART—CIRCUIT

A circuit contains five resistors and a battery. The chart gives the currents in each element, the potential difference across each element, and the resistance values of the resistors.

	ΔV	I	R
Battery	30.0 V	9.0 A	
R_1	12.0 V	2.0 A	6.0 Ω
R_2	12.0 V	3.0 A	4.0 Ω
R_3	12.0 V	4.0 A	3.0 Ω
R_4	18.0 V	3.0 A	6.0 Ω
R_5	18.0 V	6.0 A	3.0 Ω

Draw an electric circuit that is consistent with the values of this chart. Label the resistors.

D2-WBT54: SIX RESISTORS CIRCUIT CHART I—CIRCUIT

A circuit contains six resistors and a battery. The chart gives the currents in each element, the potential difference across each element, and the resistance values of the resistors.

	ΔV	I	R
Battery	39.0 V	5.0 A	
R_1	6.0 V	3.0 A	2.0 Ω
R_2	6.0 V	2.0 A	3.0 Ω
R_3	10.0 V	5.0 A	2.0 Ω
R_4	8.0 V	4.0 A	2.0 Ω
R_5	8.0 V	1.0 A	8.0 Ω
R_6	15.0 V	5.0 A	3.0 Ω

Draw an electric circuit that is consistent with the values of this chart. Label the resistors.

D2-WBT55: Six Resistors Circuit Chart II—Circuit

A circuit contains six resistors and a battery. The chart gives the currents in each element, the potential difference across each element, and the resistance values of the resistors.

	ΔV	I	R
Battery	62.0 V	10.0 A	
R_1	12.0 V	3.0 A	4.0 Ω
R_2	12.0 V	1.0 A	12.0 Ω
R_3	12.0 V	4.0 A	3.0 Ω
R_4	12.0 V	2.0 A	6.0 Ω
R_5	20.0 V	10.0 A	2.0 Ω
R_6	30.0 V	10.0 A	3.0 Ω

Draw an electric circuit that is consistent with the values of this chart. Label the resistors.

D2-WBT56: Seven Resistors Circuit Chart—Circuit

A circuit contains seven resistors and a battery. The chart gives the currents in each element, the potential difference across each element, and the resistance values of the resistors.

	ΔV	I	R
Battery	64.0 V	12.0 A	
R_1	20.0 V	5.0 A	4.0 Ω
R_2	20.0 V	2.0 A	10.0 Ω
R_3	20.0 V	5.0 A	4.0 Ω
R_4	24.0 V	12.0 A	2.0 Ω
R_5	8.0 V	4.0 A	2.0 Ω
R_6	8.0 V	8.0 A	1.0 Ω
R_7	12.0 V	12.0 A	1.0 Ω

Draw an electric circuit that is consistent with the values of this chart. Label the resistors.

D3 MAGNETISM

D3-QRT01: ELECTRIC CHARGE NEAR A BAR MAGNET—FORCE DIRECTION

A charged particle is placed so that it is at rest near one pole of a bar magnet. All of the charges are the same distance from the magnet. The strength of the magnetic field due to the bar magnet at the location of the particle is given.

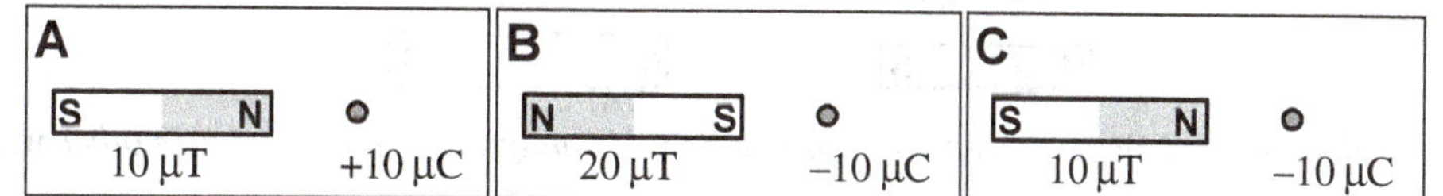

For each case, draw the direction of the force on the magnet and on the charged particle.

	Direction of force on the magnet	Direction of force on the particle
Case A		
Case B		
Case C		

Explain your reasoning.

D3-SCT02: CHARGED ROD NEAR A SUSPENDED BAR MAGNET—ROTATION

A bar magnet is suspended by a string. With the magnet held in place, a charged rod is brought close to the point of suspension of the magnet as shown. The suspended bar magnet is then released so that it is free to rotate. Students are discussing what will happen when the magnet is free to rotate.

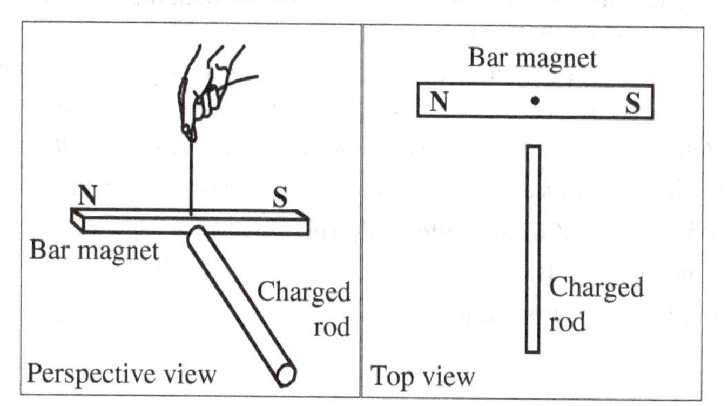

Aaron: *"I think the magnet will rotate clockwise when viewed from above. The north pole will be repelled by the positive charge, and the south pole will be attracted."*

Ben: *"I think the magnet will just sit there. A north pole is not like a positive charge."*

Carl: *"It's more complicated than that. The magnet will induce a north pole on the charged rod because the magnetic field will cause the electron spins to align. Since there are then two north poles close to each other, the magnet will rotate clockwise."*

With which of these students do you agree?

Aaron _____ Ben _____ Carl_____ None of them _____

Explain your reasoning.

D3-SCT03: Two Magnets—Force

Three students are discussing the strengths (or magnitudes) of the forces between two permanent magnets. The smaller magnet is moving to the right. The larger magnet is stronger than the smaller magnet.

Alejandro: *"The velocities and magnet strengths don't matter. The magnets will attract each other with equal strength."*

Bernardo: *"No, the stronger magnet will push more than the weaker one because it has a stronger field."*

Cecilia: *"I don't think we can compare the strength of the forces unless we know the velocity of the smaller magnet."*

With which of these students do you agree?

Alejandro_____ Bernardo_____ Cecilia_____ None of them _____

Explain your reasoning.

D3-SCT04: Electric Charge near a Bar Magnet—Force Direction

Consider the following students' statements about the magnetic force on a positively charged particle placed at rest near a permanent magnet.

Aurelia: *"A positively charged particle placed near the north pole of a permanent magnet will experience a repulsive force because the north pole acts like a positive charge."*

Ben: *"I think it will experience an attractive force, but not because it is a magnet."*

Chila: *"Since it's not moving, I think it won't experience any electromagnetic force."*

With which of these students do you agree?

Aurelia _____ Ben ______ Chila _____ None of them _____

Explain your reasoning.

D3-RT05: CHARGE WITHIN A UNIFORM MAGNETIC FIELD—MAGNETIC FORCE

In each case, charged particles are in a magnetic uniform magnetic field. All magnetic fields have the same strength.

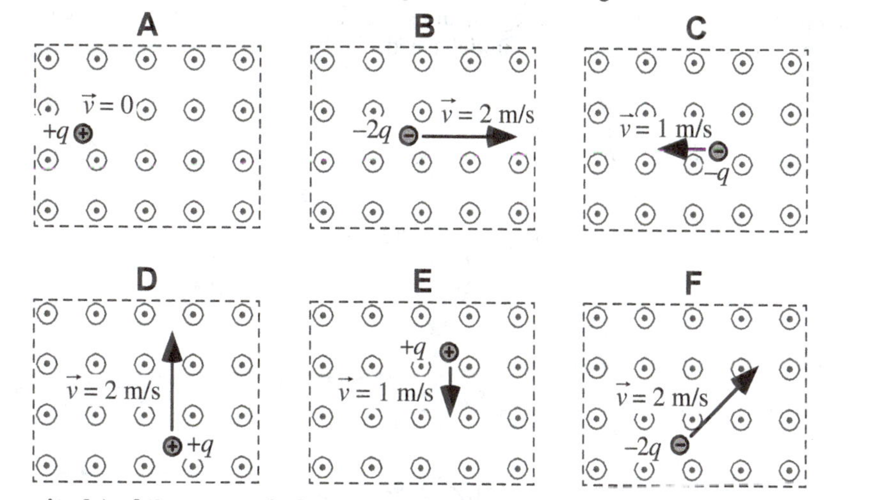

Rank the strength (magnitude) of the magnetic force on each charge.

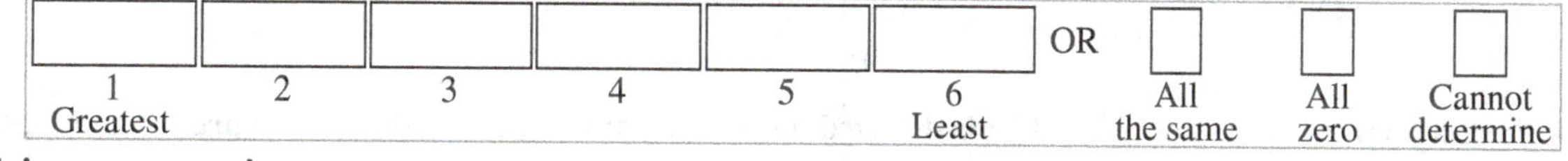

Explain your reasoning.

D3-WWT06: CHARGED PARTICLES AND A UNIFORM MAGNETIC FIELD—DIRECTION OF MOTION

Two particles that have the same mass and electric charge enter the same uniform magnetic field traveling at the same speed. The distance between the two particles is so great that they do not affect each other.

A student makes the following statement:

"*These particles will travel in circular paths of equal radius.*"

What, if anything, is wrong with this statement? If something is wrong, explain the error and how to correct it. If the statement is valid, explain why.

D3-RT07: Moving Charge Path—Direction and Strength of the Magnetic Field

In each case, the shaded region contains a uniform magnetic field that may point either into the page or out of the page. A charged particle moves through the region along the path indicated. All of the charged particles have the same mass and enter the region with the same initial speed.

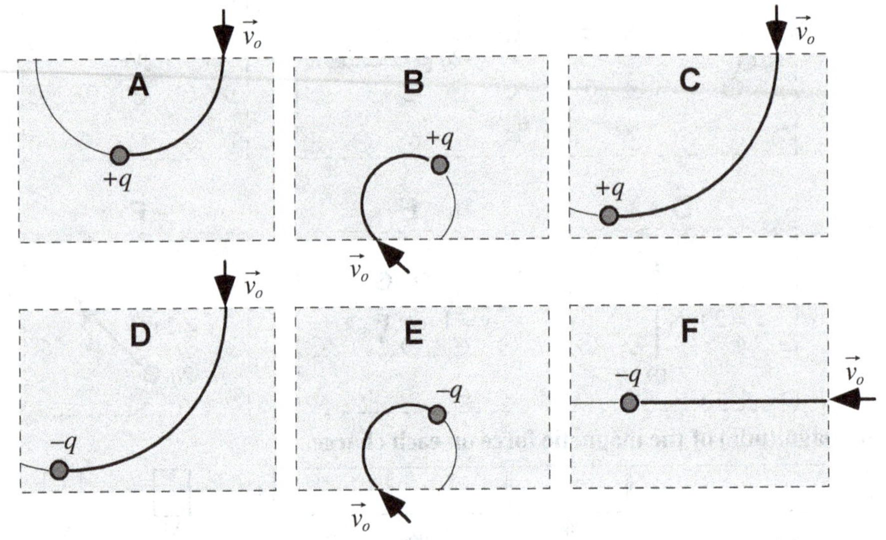

Rank the magnetic field in the region. Fields directed out of the page (considered positive) are ranked higher than fields directed into the page (considered negative).

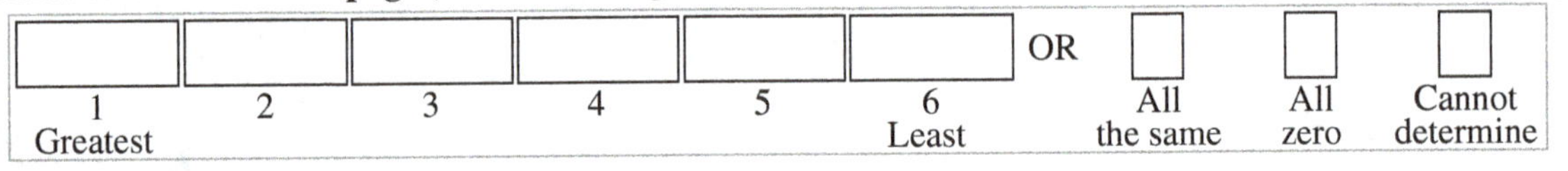

Explain your reasoning.

D3-RT08: Charged Particle and a Uniform Magnetic Field—Force

In each case, a charged particle is moving in a uniform magnetic field. The particle charge and the strength of the field vary among the four cases. The particles all have the same mass, and they were all given the same initial speed.

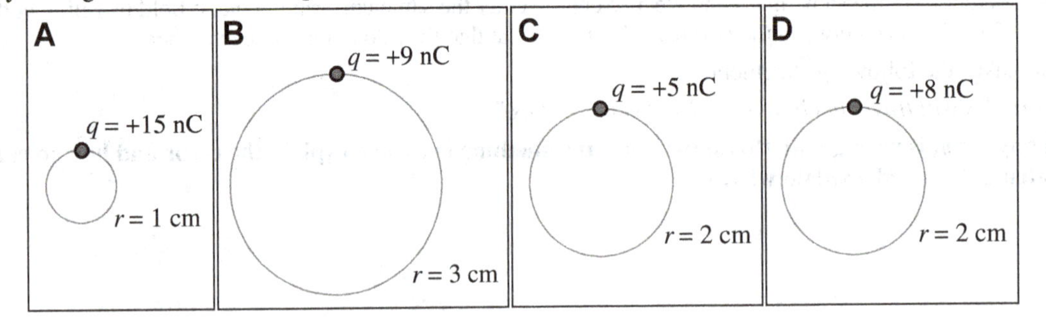

Rank the magnitude of the force on each charge.

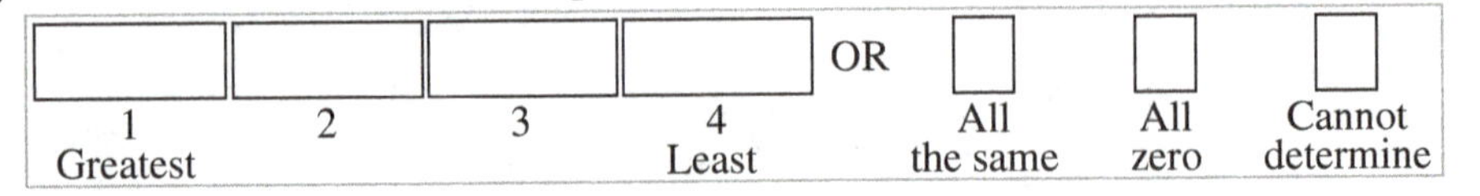

Explain your reasoning.

D3-QRT09: CHARGED PARTICLE AND A UNIFORM MAGNETIC FIELD—PATH

The dark quarter-circle indicates the path of a negatively charged particle as it passes through a region containing a uniform magnetic field.

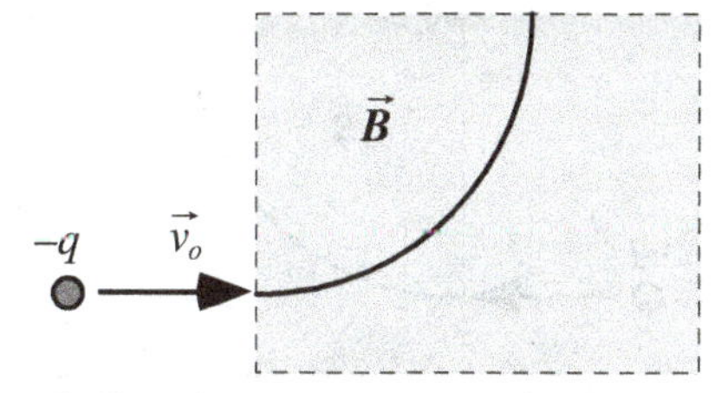

(a) What is the direction of the magnetic field in the shaded region?
Explain your reasoning.

(b) If we double the speed of the particle, how will the path change?
Explain your reasoning.

(c) If we double the magnitude of the uniform magnetic field, how will the path change?
Explain your reasoning.

(d) If we replace the original particle with a negative particle of twice the charge and the same mass, how will the path change?
Explain your reasoning.

(e) If we replace the original particle with a positive particle of the same mass and same magnitude charge as the original negative charge, how will the path change?
Explain your reasoning.

(f) If we replace the original particle with a negative particle of twice the mass and the same charge, how will the path change?
Explain your reasoning.

D3-LMCT10: Moving Charge within a Uniform Magnetic Field—Force

A positively charged particle moving at a constant speed is entering a region in which there is a uniform magnetic field. The particle follows the curved path shown.

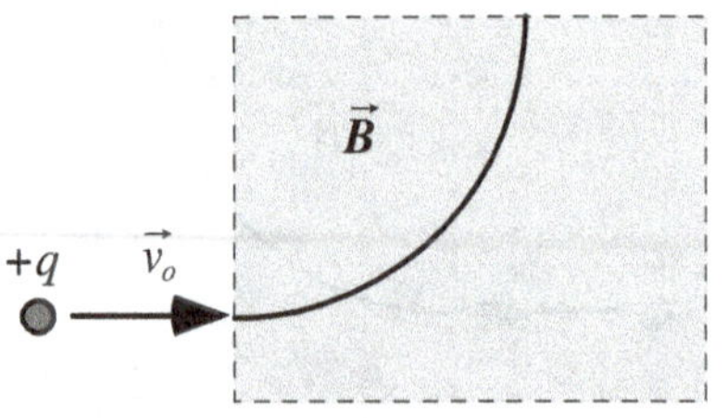

A number of changes to this initial situation are described in (a)–(f) below. Select from choices (i)–(vi) how each change will affect the magnetic force on the particle shortly after it enters the magnetic field.

This change will:

(i) alter only the **direction** of the force on the particle.
(ii) only **increase** the magnitude of the magnetic force on the particle.
(iii) only **decrease** the magnitude of the magnetic force on the particle.
(iv) alter **both the magnitude and direction** of the magnetic force on the particle.
(v) **not affect** the magnetic force on the particle.
(vi) cause the magnetic force on the particle to be **zero**.

Each change below refers to the initial situation described above:

(a) The +*q* particle is replaced by a +2*q* particle. _____
Explain your reasoning.

(b) The +*q* particle is replaced by a • *q* particle. _____
Explain your reasoning.

(c) The +*q* particle is replaced by a neutral particle. _____
Explain your reasoning.

(d) The particle enters the region moving at a slower initial velocity. _____
Explain your reasoning.

(e) The magnetic field is one-third its original strength. _____
Explain your reasoning.

(f) The direction of the magnetic field is parallel to the particle's initial velocity. _____
Explain your reasoning.

D3-RT11: BARS MOVING IN MAGNETIC FIELDS—FORCE ON ELECTRONS

Three straight bars are moving at different speeds through a region in which there is a uniform magnetic field. The bars are made of the same conducting material and are moving in a plane perpendicular to the magnetic field. The bars have different lengths and are moving at constant but different speeds, as shown. Within the bars, four electrons are located and labeled.

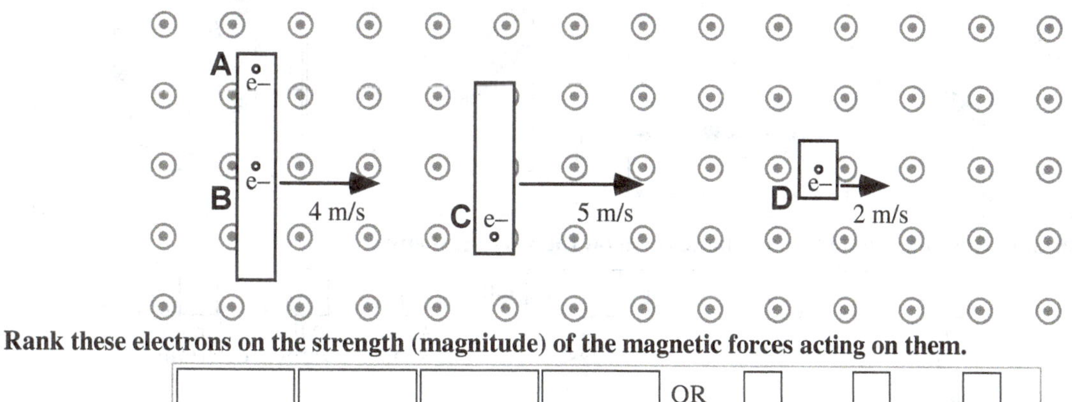

Rank these electrons on the strength (magnitude) of the magnetic forces acting on them.

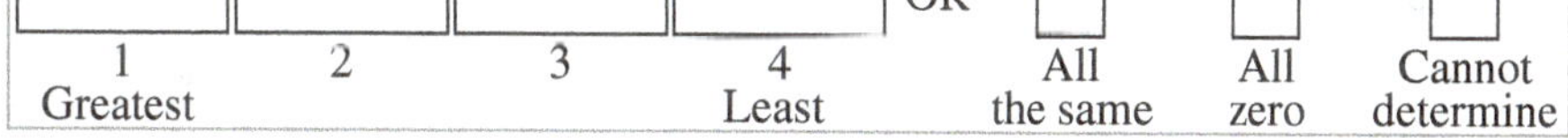

Explain your reasoning.

D3-RT12: MOVING CHARGES IN UNIFORM MAGNETIC FIELD—ACCELERATION

Moving charged particles are released at point P (2 m, 2 m) in a region of space with a uniform magnetic field in the $+x$ direction. All these particles have the same mass, and they are released one at a time into the field with the given velocities.

$+y$

× P (2 m, 2 m)

$+x$

Case	Charge	Speed	Direction
A	+5 mC	3 m/s	$+x$
B	+5 mC	3 m/s	$-x$
C	−10 mC	5 m/s	$+y$
D	−10 mC	5 m/s	$-y$

Rank the magnitude of the initial acceleration of the charged particles as they are released from *P*.

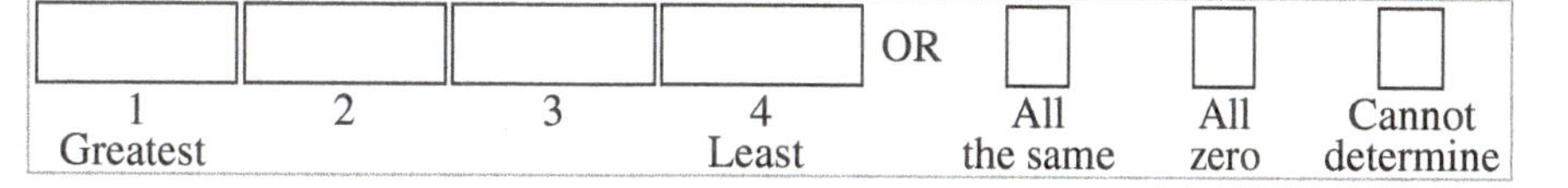

Explain your reasoning.

D3-RT13: Current–Carrying Wire in a Uniform Magnetic Field—Magnetic Force

The figures below show identical current-carrying wire segments in identical uniform magnetic field regions. All the magnetic field regions are the same width and height.

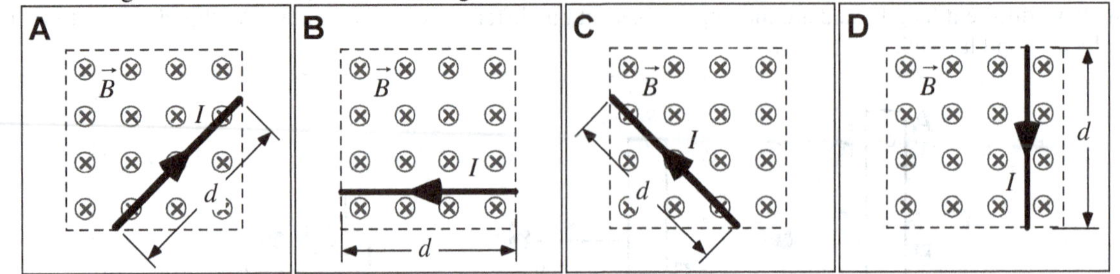

Rank the strength (magnitude) of the magnetic force on the wire segments.

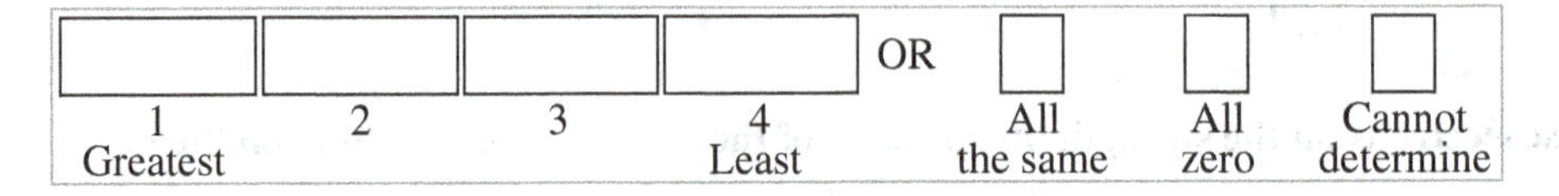

Explain your reasoning.

D3-TT14: Path of a Moving Electron in a Uniform Magnetic Field

An electron is moving to the right at a velocity v when it enters a region containing a uniform magnetic field pointing into the paper. The path of the electron in the magnetic field is shown.

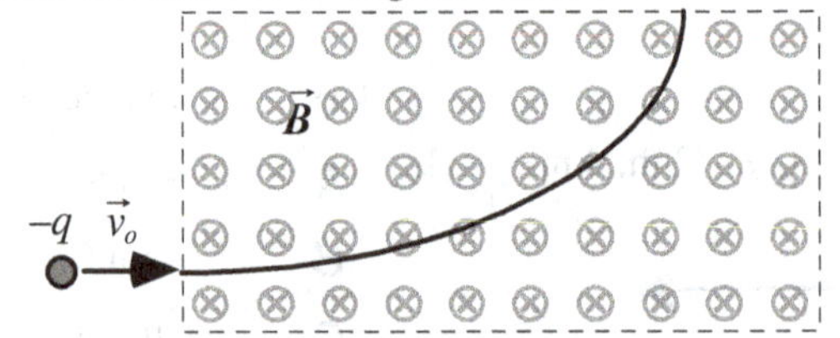

There is at least one error in the diagram. **Identify all errors and explain how to correct them.**

D3-QRT15: STRAIGHT CURRENT–CARRYING WIRE IN A UNIFORM MAGNETIC FIELD—MAGNETIC FORCE

A current-carrying straight wire segment is in a uniform magnetic field directed into the paper. There are connecting wires running parallel to the magnetic field that are not shown.

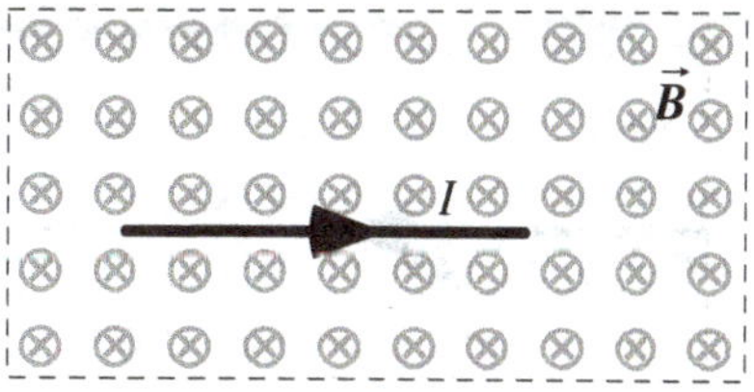

(a) What is the direction of the magnetic force acting on the wire segment due to the uniform magnetic field?

Explain your reasoning.

(b) What would the direction of the magnetic force acting on the wire segment be if the direction of the uniform magnetic field were out of the paper?

Explain your reasoning.

(c) What would happen to the magnitude of the magnetic force acting on the wire segment if the wire segment were longer but still completely within the uniform magnetic field?

Explain your reasoning.

(d) What would happen to the direction of the magnetic force acting on the wire segment if the direction of current in the wire segment were reversed?

Explain your reasoning.

(e) What would happen to the magnitude of the magnetic force acting on the wire segment if the wire segment were moved (without changing its orientation) so that its length was half-in and half-out of the uniform magnetic field region?

Explain your reasoning.

D3-LMCT16: Current in a Uniform Magnetic Field—Magnetic Force

A section of straight wire within a magnetic field is conducting a current to the right. The external magnetic field is uniform and directed into the paper.

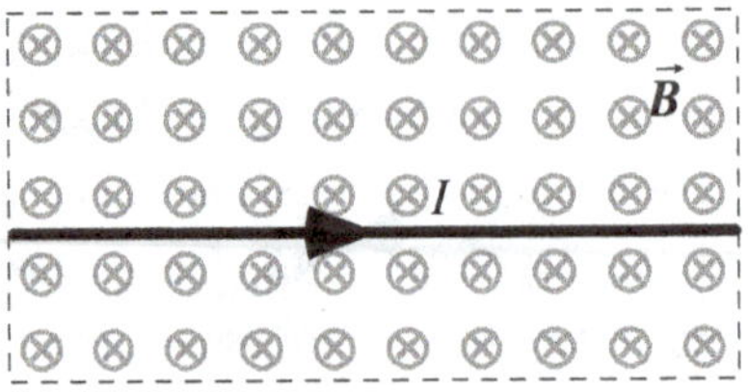

A number of changes to the initial force are described in (a)-(e) below. Select from choices (i)-(vii) the possible causes of the change in the force.

This change could be caused by:

(i) **increasing** the current.
(ii) **decreasing** (but not to zero) the current.
(iii) **reversing the direction** of current.
(iv) **increasing** the strength of the magnetic field.
(v) **decreasing** (but not to zero) the strength of the magnetic field.
(vi) **reversing the direction** of the magnetic field.
(vii) none of these.

If more than one choice is correct, please indicate *all* correct choices for the answer.

Each change below refers to the initial situation described above:

(a) The magnetic force on the wire is larger and in the same direction. _____
Explain your reasoning.

(b) The magnetic force on the wire is larger and in the opposite direction. _____
Explain your reasoning.

(c) The magnetic force on the wire is smaller and in the same direction. _____
Explain your reasoning.

(d) The magnetic force magnitude remains the same, but the direction changes. _____
Explain your reasoning.

(e) The magnetic force on the wire is zero. _____
Explain your reasoning.

D3-QRT17: STRAIGHT CURRENT–CARRYING WIRE—MAGNETIC FIELD NEARBY

The figure below shows a point P near a long current-carrying wire.

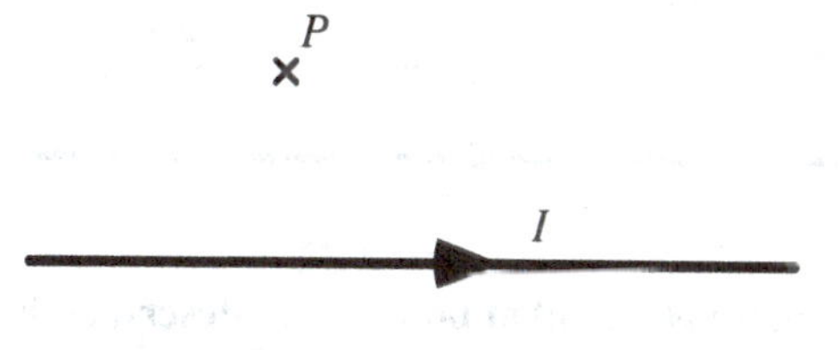

(a) What is the direction of the magnetic field at point P due to the current in the wire? Explain your reasoning.

(b) What would the direction of the magnetic field at point P be if the current in the wire were reversed? Explain your reasoning.

(c) What would happen to the magnetic field at point P if the current in the wire were increased? Explain your reasoning.

(d) What would happen to the magnetic field at P if point P were farther away from the wire? Explain your reasoning.

D3-WWT18: CURRENT–CARRYING WIRE—MAGNETIC FIELD DIRECTION

A long, straight wire is conducting a current whose direction is pointed out of the paper toward you. A student makes the following statement:

"The magnetic field generated by this wire points straight out from the wire."

What, if anything, is wrong with this statement? If something is wrong, explain the error and how to correct it. If the statement is valid, explain why.

D3-LMCT19: Long Wire with a Current—Magnetic Field Nearby

Point P is located above a long straight wire that has a current to the right.

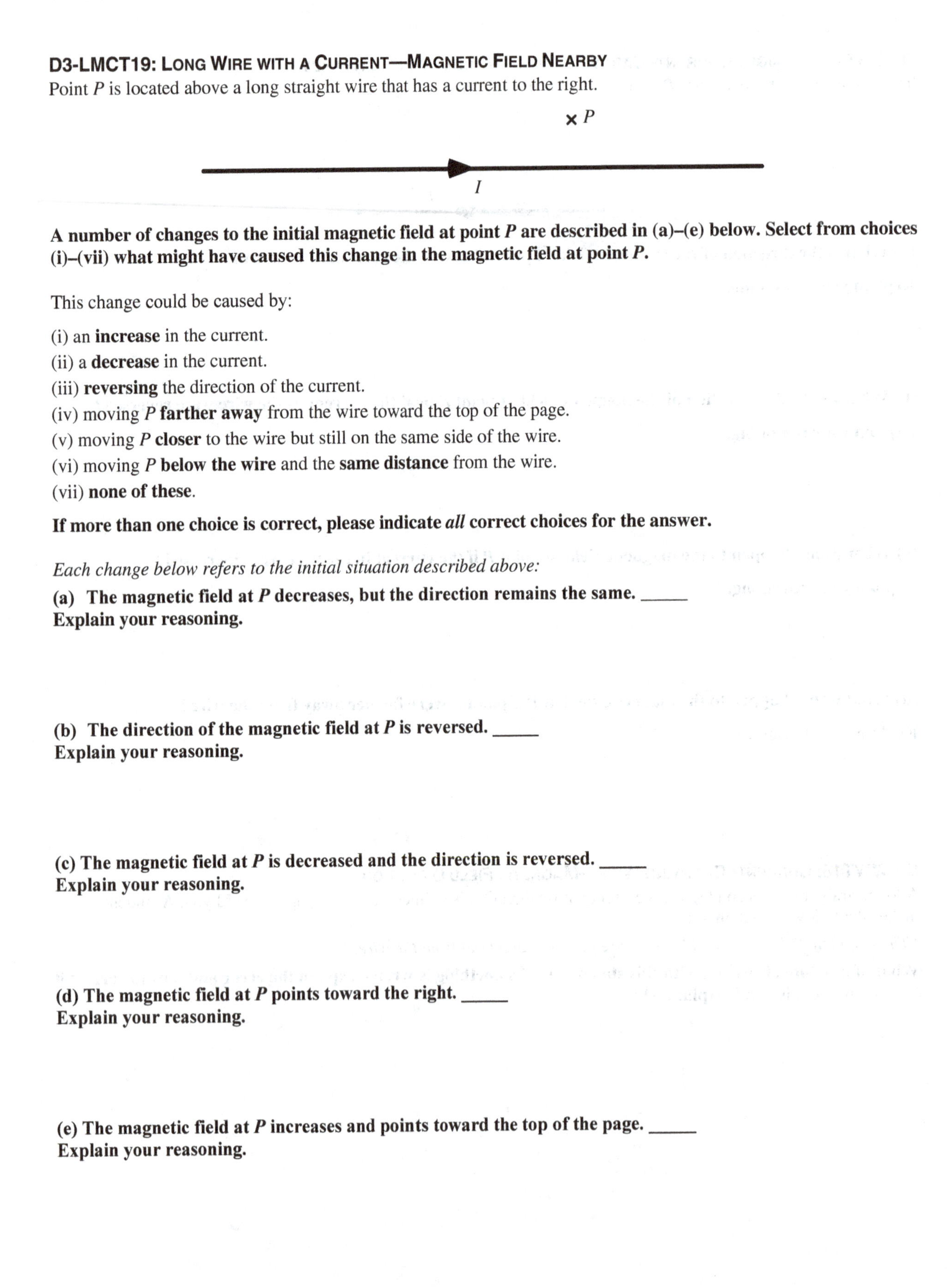

A number of changes to the initial magnetic field at point P are described in (a)–(e) below. Select from choices (i)–(vii) what might have caused this change in the magnetic field at point P.

This change could be caused by:

(i) an **increase** in the current.
(ii) a **decrease** in the current.
(iii) **reversing** the direction of the current.
(iv) moving P **farther away** from the wire toward the top of the page.
(v) moving P **closer** to the wire but still on the same side of the wire.
(vi) moving P **below the wire** and the **same distance** from the wire.
(vii) **none of these**.

If more than one choice is correct, please indicate *all* correct choices for the answer.

Each change below refers to the initial situation described above:

(a) The magnetic field at P decreases, but the direction remains the same. _____
Explain your reasoning.

(b) The direction of the magnetic field at P is reversed. _____
Explain your reasoning.

(c) The magnetic field at P is decreased and the direction is reversed. _____
Explain your reasoning.

(d) The magnetic field at P points toward the right. _____
Explain your reasoning.

(e) The magnetic field at P increases and points toward the top of the page. _____
Explain your reasoning.

D3-BCT20: STRAIGHT CURRENT–CARRYING WIRE—MAGNETIC FIELD NEARBY

A long, straight conducting wire has a current in the $+x$-direction.

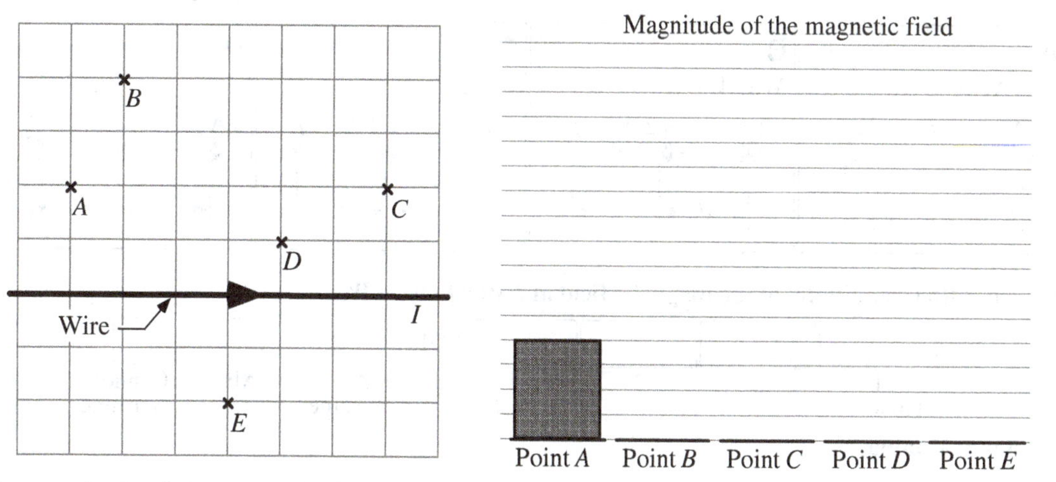

Show the magnitude of the magnetic field at the various points shown in the diagram on the bar chart to the right, above. The magnitude at point A is given in the chart.
Explain your reasoning.

D3-TT21: STRAIGHT CURRENT–CARRYING WIRE—MAGNETIC FIELD NEARBY

A student draws the following diagram representing the magnetic field generated by a straight current-carrying wire.

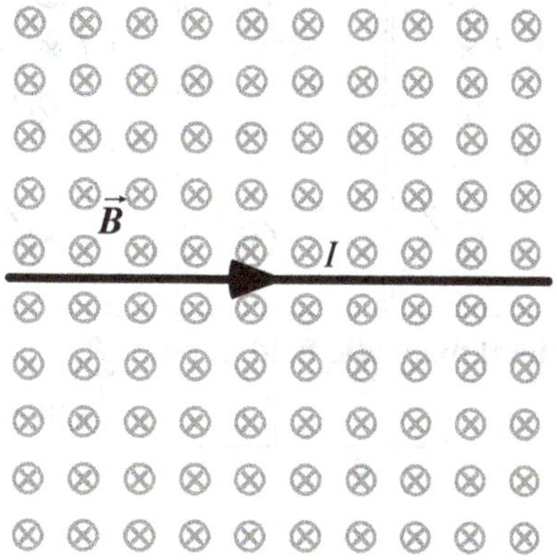

There is at least one problem with the diagram. **Identify any problems and explain how to correct them.**

D3-RT22: Current–Carrying Straight Wires—Magnetic Field Nearby

In these cases, long, straight wires that are perpendicular to the page are carrying electric currents into the page.

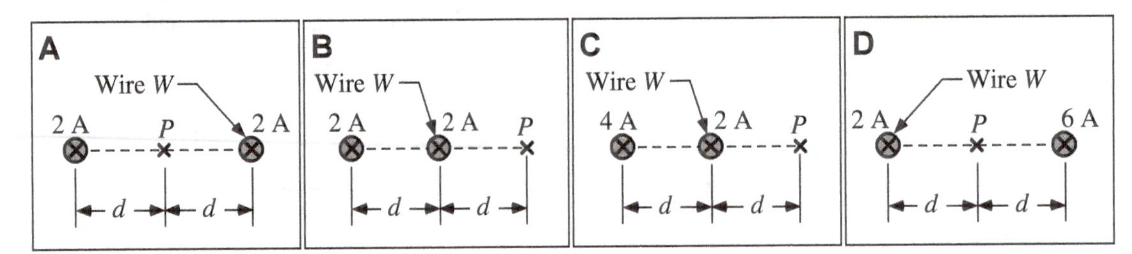

Rank the strength (magnitude) of the magnetic field at *P* due to wire *W*.

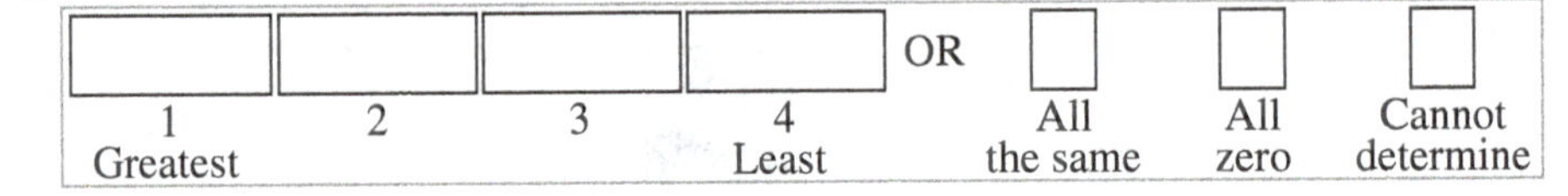

Explain your reasoning.

D3-RT23: Three Parallel Current–Carrying Wires I—Magnetic Field Nearby

In these cases, three long, straight parallel wires carry currents either into or out of the page. Point P is midway between the top two wires. The current has the same magnitude in all wires.

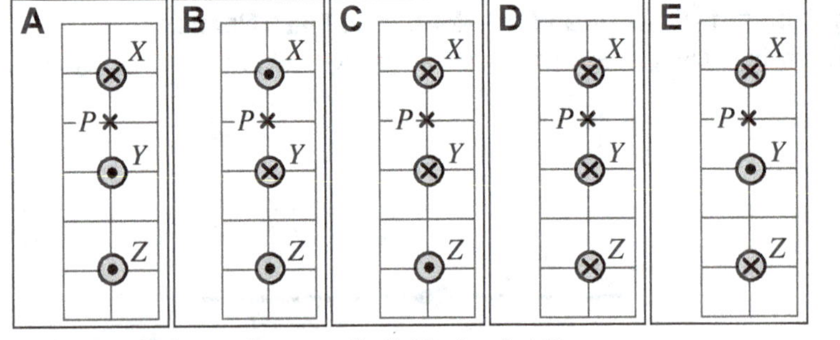

Rank the strength (magnitude) of the total magnetic field at point *P*.

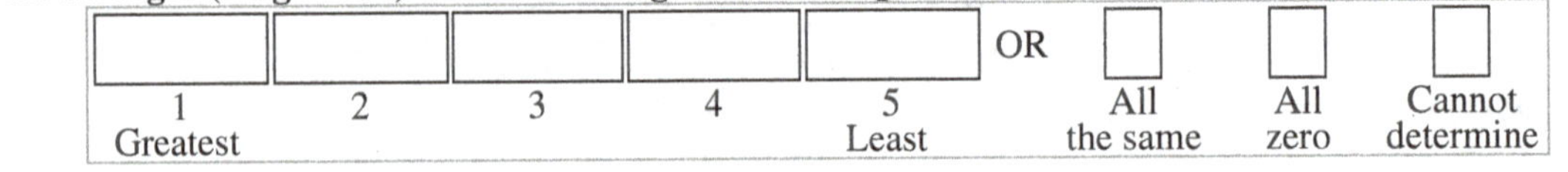

Explain your reasoning.

D3-RT24: CURRENTS AT CORNERS OF SQUARES—MAGNETIC FIELD AT CENTER

Current-carrying wires are positioned at the corners of a square. All of the currents have the same magnitude, but some are into the page and some are out of the page.

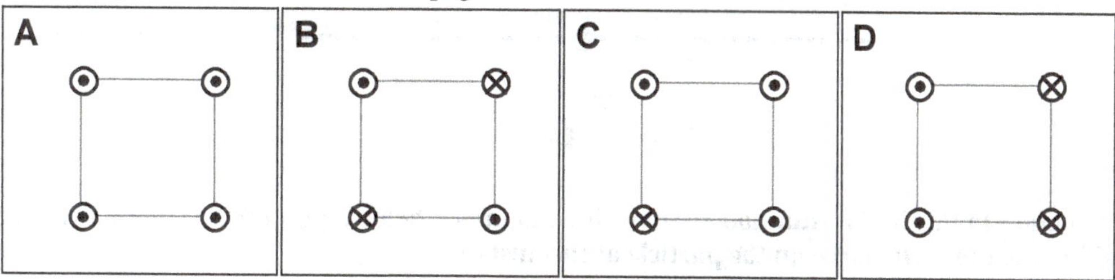

Rank the magnitude of the net magnetic field at the center of the square.

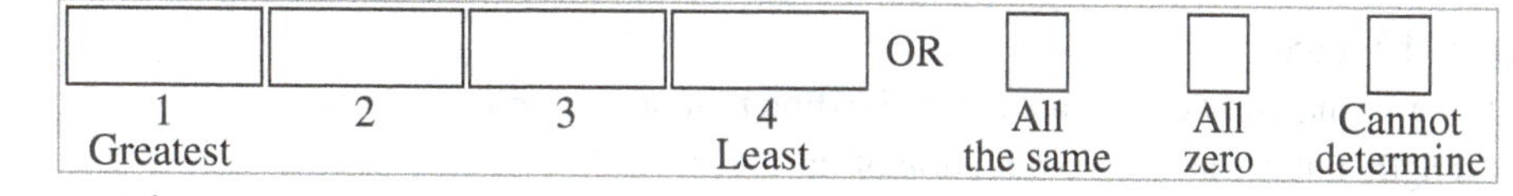

Explain your reasoning.

D3-RT25: PAIRS OF LONG CURRENT-CARRYING WIRES—MAGNETIC FIELD NEARBY

In these cases, the two parallel wires have the same magnitude current perpendicular to the plane of the page. The direction of the current in each wire is shown.

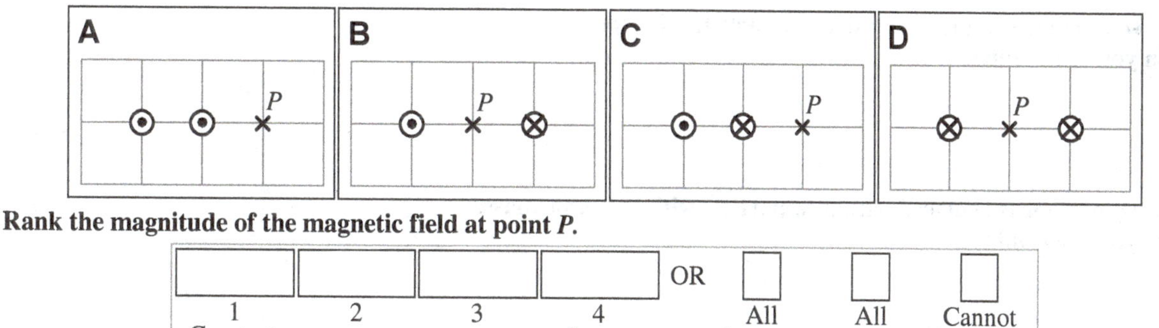

Rank the magnitude of the magnetic field at point *P*.

OR

1 Greatest | 2 | 3 | 4 Least | All the same | All zero | Cannot determine

Explain your reasoning.

D3-LMCT26: CHARGE MOVING ALONG STRAIGHT WIRE—MAGNETIC FORCE

At the instant shown, a particle with a charge of $+q$ is a distance d from a long, straight wire and is moving parallel to the wire.

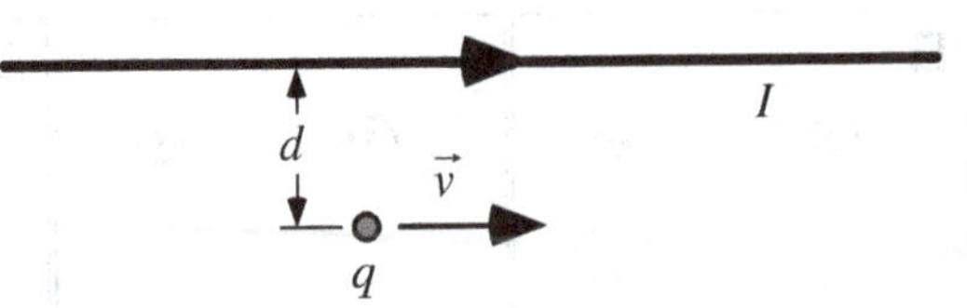

A number of changes to this initial situation are described in (a)–(f) below. Select from choices (i)–(v) how each change will affect the magnetic force on the particle at this instant.

This change will:

(i) have **no effect** on the force.
(ii) **increase** the strength (magnitude), but not affect the direction of the force.
(iii) **decrease** the strength, but not affect the direction of the force.
(iv) **alter the direction**, but not affect the strength of the force.
(v) **alter both** the strength and direction of the force on the particle.

Each change below refers to the initial situation described above:

(a) The current in the wire is doubled. _____
Explain your reasoning.

(b) The direction of the current in the wire is reversed. _____
Explain your reasoning.

(c) The +q particle is replaced with a +2*q* particle with the same mass. _____
Explain your reasoning.

(d) The +q particle is replaced with a –*q* particle with the same mass. _____
Explain your reasoning.

(e) The +*q* particle is replaced with a particle having triple the mass and the same charge. _____
Explain your reasoning.

(f) The charged particle's initial velocity is toward the bottom of the page. _____
Explain your reasoning.

D3-TT27: MOVING AWAY POSITIVE CHARGE NEAR STRAIGHT CURRENT–CARRYING WIRE—FORCE

At the instant shown, a positively charged particle has a velocity that is perpendicular to a current-carrying wire.

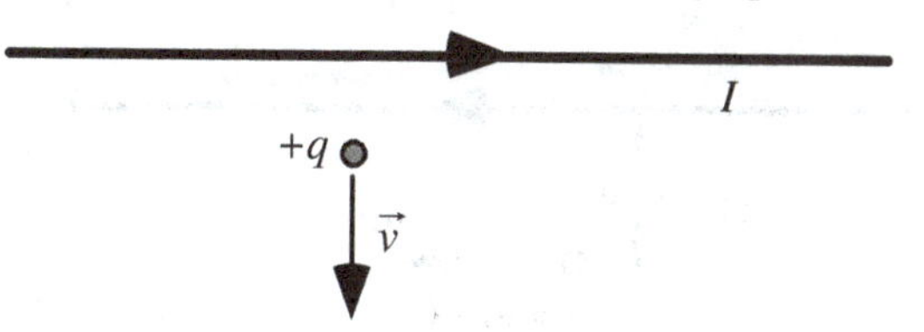

A student makes the following statement:

"The force exerted on the charged particle by the magnetic field is zero because the velocity is parallel to the magnetic field produced by the wire."

There is at least one problem with this student's contention. Identify any problems and explain how to correct them.

D3-RT28: MOVING CHARGE ALONG A STRAIGHT CURRENT–CARRYING WIRE—ACCELERATION

Four charged particles have been projected parallel to identical current-carrying wires. The particles have the same mass and are projected with the same initial speed.

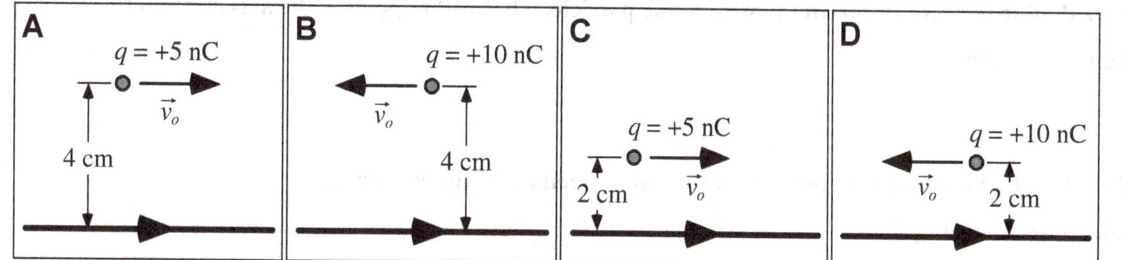

Rank the magnitude of the acceleration of each charge at the instant shown.

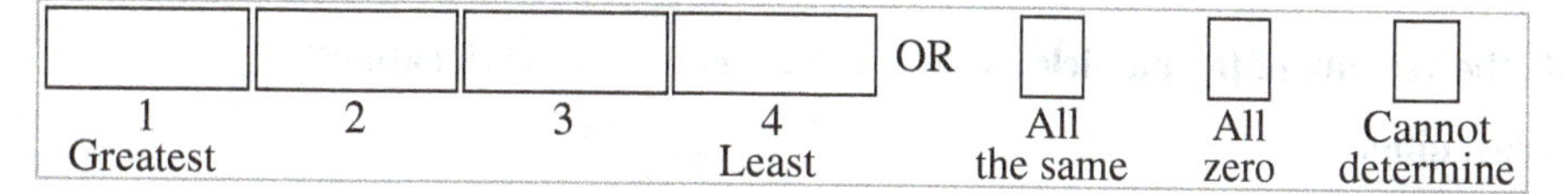

Explain your reasoning.

D3-QRT29: Moving Charge near a Straight Current–Carrying Wire—Acceleration

At the instant shown, a particle with charge of +7 nC is moving at 3 m/s parallel to a long, straight wire that has a current of 8 A.

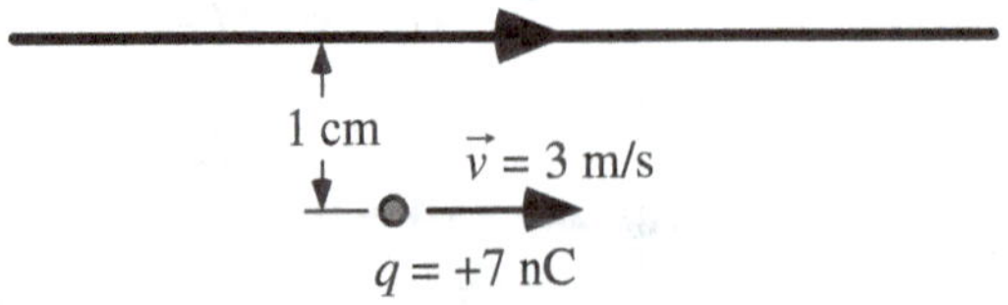

1 cm
$\vec{v}$ = 3 m/s
q = +7 nC

(a) What is the direction of acceleration of the charged particle?

Explain your reasoning.

(b) If we double the charge on the particle, what will happen to the acceleration?

Explain your reasoning.

(c) If we replace the charge with a negative charge of the same mass and same magnitude charge as the original charge, what will happen to the acceleration?

Explain your reasoning.

(d) If we double the distance from the wire to the particle, what will happen to the acceleration?

Explain your reasoning.

(e) If we double the mass of the particle, what will happen to the acceleration?

Explain your reasoning.

(f) If we double the velocity of the particle, what will happen to the acceleration?

Explain your reasoning.

(g) If we reduce the magnitude of the current, what will happen to the acceleration?

Explain your reasoning.

(h) If we reverse the direction of the current, what will happen to the acceleration?

Explain your reasoning.

D3-SCT30: THREE PARALLEL CURRENT–CARRYING WIRES—FORCE

There is a 2 A current in wires X, Y, and Z. There is no magnetic force on wire Y. The distance between adjacent wires is the same. Three students are discussing this situation.

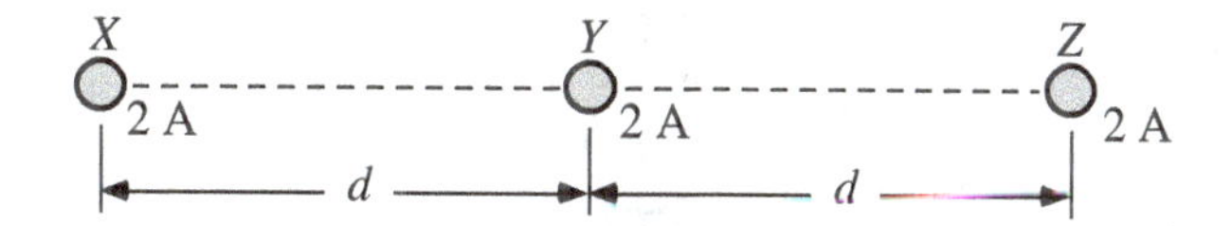

Arcadio: *"For the force on wire Y to cancel, the currents in X and Z must be in opposite directions."*

Bruce: *"No, for the forces on wire Y to cancel, the current in X and Z must be in the same direction, but opposite the current direction in Y."*

Carolina: *"All we care about is the current in X and Z being in the same direction. They don't have to be opposite Y."*

With which of these students do you agree?

Arcadio _____ Bruce _____ Carolina _____ None of them _____

Explain your reasoning.

D3-SCT31: MOVING MAGNET AND CIRCULAR LOOP—FORCE

Three students are comparing the forces on a permanent magnet and a current-carrying circular loop of wire. The mass of the magnet is much larger than the mass of the loop.

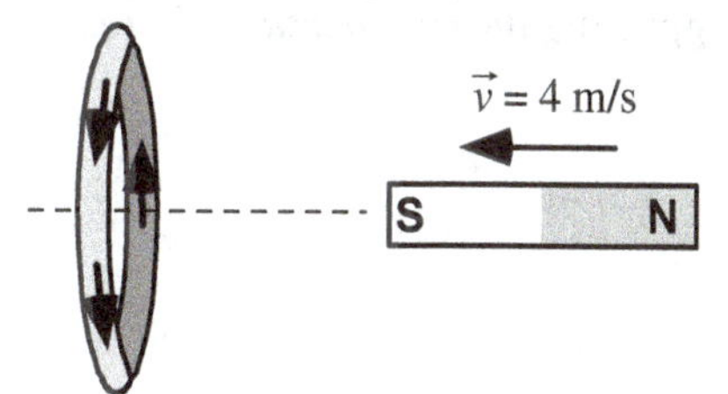

Amador: *"The coil will push or pull on the magnet just as hard as the magnet pushes or pulls on the coil."*

Barbara: *"I think the magnet has to push harder on the coil than the coil pushes on the magnet because the magnet is more massive than the wire."*

Charlene: *"I think the magnet will push or pull on the coil but the coil will not push or pull on the magnet at all because the coil is not a magnet."*

With which of these students do you agree?

Amador _____ Barbara _____ Charlene _____ None of them _____

Explain your reasoning.

D3-QRT32: Suspended Permanent Magnet and Circular Coil—Scale Reading

A small, permanent magnet is suspended by a spring balance above the center of a circular coil of wire that is sitting on a balance. A large current is now introduced into the coil, causing the magnet to be attracted to the coil.

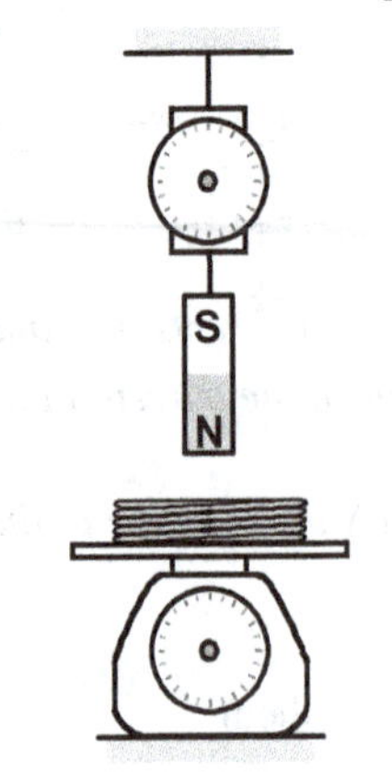

(a) Will the reading on the upper spring balance *increase*, *decrease*, or *stay the same*? Explain your reasoning.

(b) Will the reading on the balance supporting the coil *increase*, *decrease*, or *stay the same*? Explain your reasoning.

(c) Compare the sizes of the changes that will be observed in parts (a) and (b). Explain your reasoning.

D3-RT33: PAIRS OF EQUAL CURRENT ELECTROMAGNETS—FORCE

In each case, a pair of electromagnets is shown placed close together. The electromagnets are made with identical wires wrapped around identical iron cores, but the number of turns of wire varies. Each wire has the same current, and the pairs of electromagnets are separated by the same distance. Carefully observe the orientation of the coil and direction of current flow.

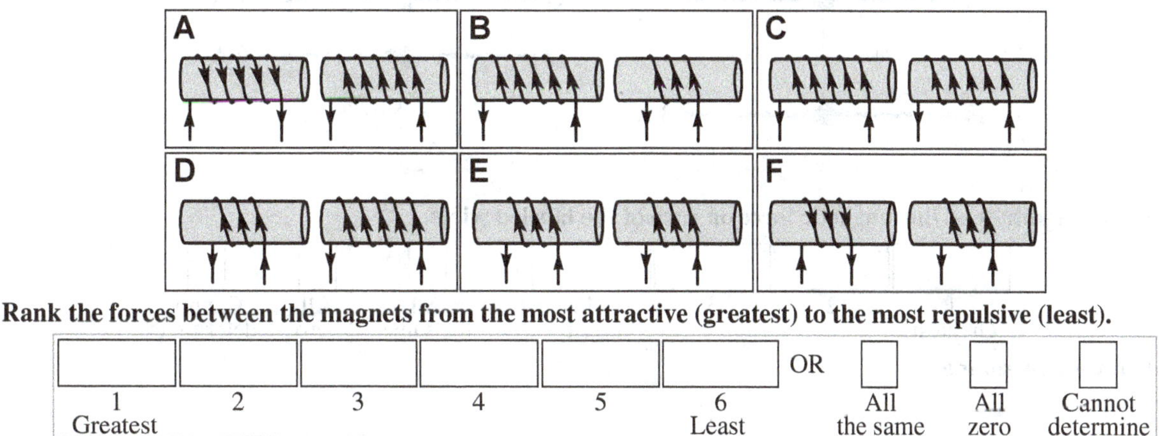

Rank the forces between the magnets from the most attractive (greatest) to the most repulsive (least).

1 Greatest	2	3	4	5	6 Least	OR	All the same	All zero	Cannot determine

Explain your reasoning.

D3-QRT34: CHARGE NEAR A CIRCULAR CURRENT LOOP—MAGNETIC FORCE DIRECTION

An electrically charged particle is placed at rest near a circular current-carrying loop of wire, along the centerline of the loop. All of the charges have the same magnitude. In Cases A and D, the particle is positively charged, while in Cases B and C they are negative. The currents in all the loops are in the same direction.

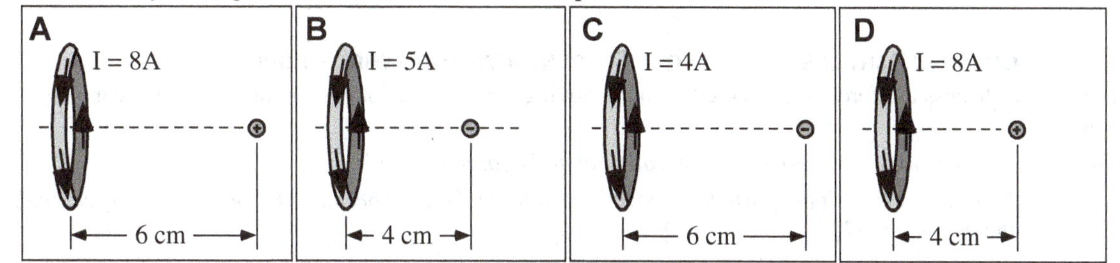

For these situations, draw the direction of the magnetic force exerted on the charged particle and on the current loop in the chart below.

	Direction of force on the current loop	Direction of force on the particle
Case A		
Case B		
Case C		
Case D		

Explain your reasoning.

D3-RT35: Parallel Current–Carrying Wires—Magnetic Force on Wire

In each case below, two very long, straight wires are parallel to each other.

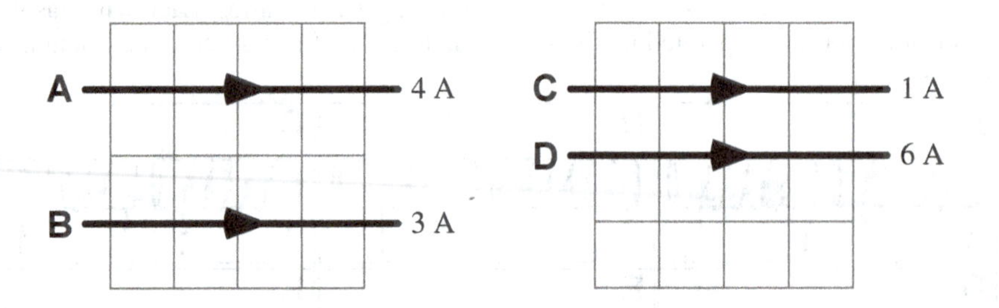

Rank the magnitude of the magnetic force on each of the labeled wires.

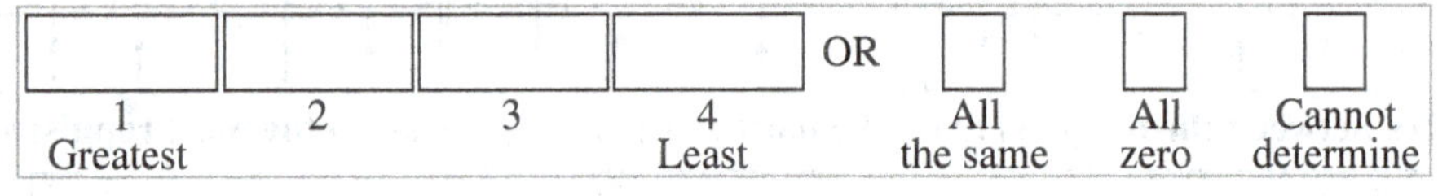

Explain your reasoning.

D3-SCT36: Charged Particle and Straight Current–Carrying Wire—Force

Three students discuss the force on a charged particle moving parallel to a long, straight wire that is carrying a current.

Amalia: *"If the velocity and current are parallel, then the force is zero."*

Brenda: *"There is a force if the particle is moving parallel to the wire because the velocity is perpendicular to the magnetic field."*

Carlos: *"As long as the particle is moving near the wire, the particle will experience a force."*

With which of these students is do you agree?

Amalia _____ Brenda _____ Carlos _____ None of them _____

Explain your reasoning.

D3-LMCT37: THREE PARALLEL CURRENT–CARRYING WIRES—MAGNETIC FORCE ON WIRE

Wires *A* and *B* have 4 A currents coming out of the page, and wire *C* has a 4 A current going into the page. All the wires are parallel and equally spaced.

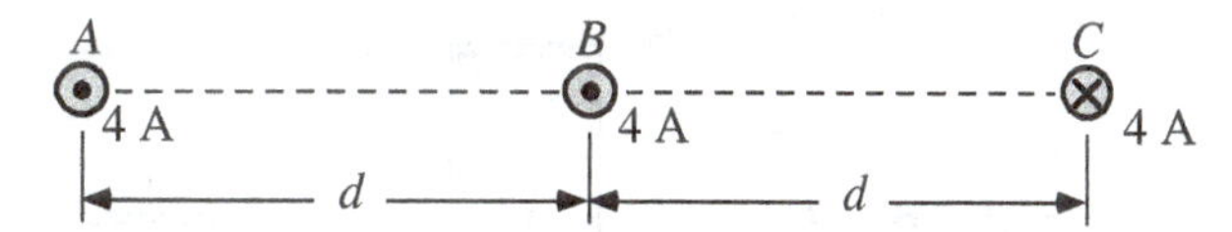

A number of changes to this initial situation are described in (a)–(f) below. Select from choices (i)–(v) how each change will affect the magnetic force on the center wire *B*.

This change will:

(i) **increase** the magnitude of the force without changing its direction.
(ii) **decrease** the magnitude of the force without changing its direction.
(iii) **reverse** the direction of the force without changing its magnitude.
(iv) **change** both the magnitude and direction of the force.
(v) have **no effect** on the magnetic force.

Each change below refers to the initial situation described above:

(a) The current in wire *B* is reversed. _____
Explain your reasoning.

(b) The currents in all three wires are doubled. _____
Explain your reasoning.

(c) The currents in the wires *A* and *C* are both reversed. _____
Explain your reasoning.

(d) The current in wire *B* is reduced. _____
Explain your reasoning.

(e) The currents in the wires *A* and *C* are reversed and cut in half._____
Explain your reasoning.

(f) Both wires *A* and *C* are moved the same distance so that they are both closer to wire *B*. _____
Explain your reasoning.

D3-RT38: Moving Rectangular Loops in Uniform Magnetic Fields—Current

Six identical rectangular wire loops are moving to the right at the same constant speed. There is a uniform magnetic field coming out of the page in the region enclosed by the dashed line. The rectangular loops are all 5 cm by 10 cm.

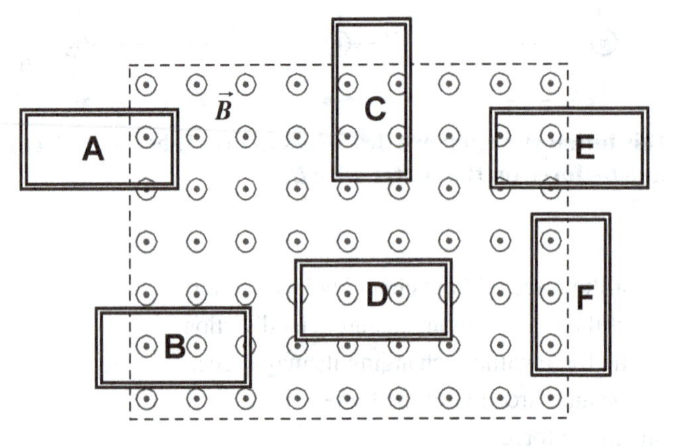

Rank the magnitude of the induced current in the rectangular loops at the instant shown. Assume there is no effect or interaction between the loops.

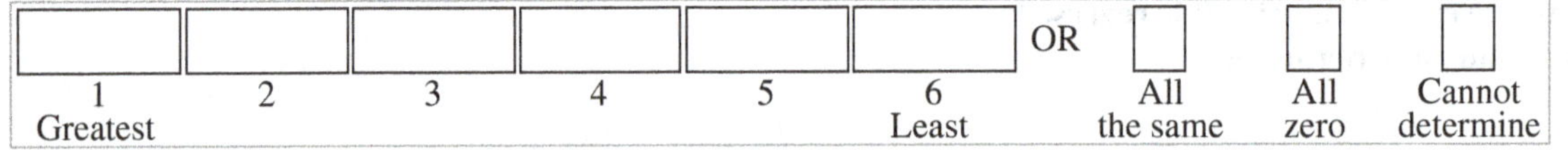

Explain your reasoning.

D3-SCT39: Moving Rectangular Loops in Uniform Magnetic Fields—Current

Three students are discussing a rectangular wire loop moving at a constant speed as it enters a region in which there is a uniform magnetic field perpendicular to the plane of the loop. The sides of the rectangular wire loop are perpendicular or parallel to the leading edge of the magnetic field.

Allison: *"The current will increase as more of the loop gets into the field since there will be more magnetic flux inside the loop."*

Blanca: *"I think the current in the wire loop will start out big and then decrease as the loop moves into the field region since less of the loop will be outside of the field."*

Chithra: *"No, the current in the wire loop will be constant from the time the loop starts into the field region until it is fully into the field region. Then the current will go to zero."*

With which of these students do you agree?

Allison _____ Blanca _____ Chithra _____ None of these students _____

Explain your reasoning.

D3-RT40: CHANGING CURRENT IN LONG WIRE—BULB BRIGHTNESS IN NEARBY LOOP

In each case, a long, straight wire is placed next to a circular wire loop that has a small light bulb in it. The currents in the long wires are changing at uniform rates. The initial current, the final current, and the time interval during which the change occurs are given. The bulbs and wire loops are all identical, and the straight wires are all the same distance from the wire loops.

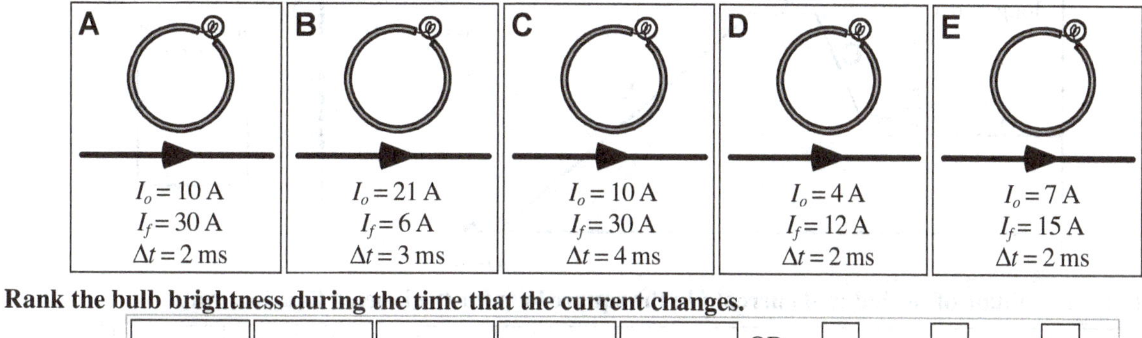

Rank the bulb brightness during the time that the current changes.

☐	☐	☐	☐	☐	OR	☐	☐	☐
1 Greatest	2	3	4	5 Least		All the same	All zero	Cannot determine

Explain your reasoning.

D3-SCT41: CHANGING CURRENT IN LONG WIRE—BULB BRIGHTNESS IN NEARBY LOOP

A circular loop of wire with a small bulb in it is placed beside a long straight current-carrying wire. In both cases below, these loops are the same distance away from the current-carrying wire. Bulb A is brighter than Bulb B. The wire loops, bulbs, and long straight wires are identical for the two situations.

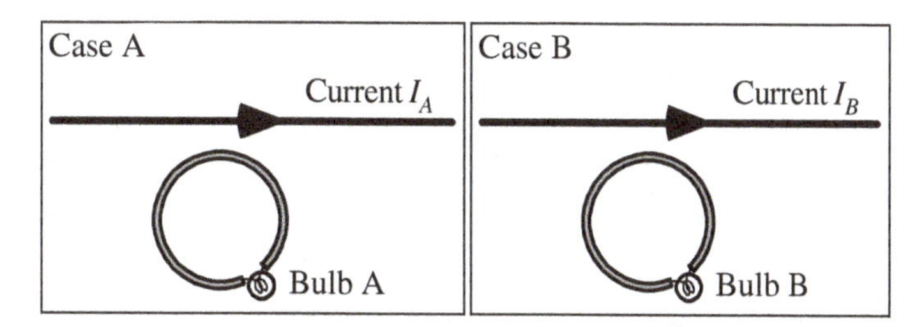

Three students discussing this arrangement contend:

Adela: *"Bulb A is brighter than Bulb B because the long wire next to the brighter bulb has a larger current in it."*

Bryce: *"No, Bulb A is brighter than Bulb B because the current in the long wire next to it is increasing at a faster rate than the current in the other wire."*

Consuelo: *"We don't know that. The current in the long wire must be changing at a faster rate, but it could also be decreasing."*

With which of these students do you agree?

Adela _____ Bryce _____ Consuelo _____ None of these students _____

Explain your reasoning.

D3-RT42: Current in Wire Time Graph—Induced Current in Nearby Loop

The current in a long wire changes with time as indicated in the graph below. A square wire loop is placed near the wire as shown in the diagram.

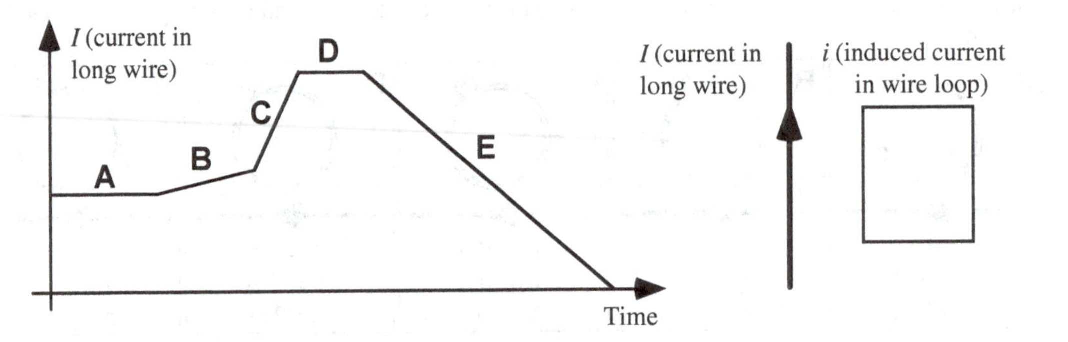

Rank the magnitude of the induced current *i* in the square loop for the labeled time intervals.

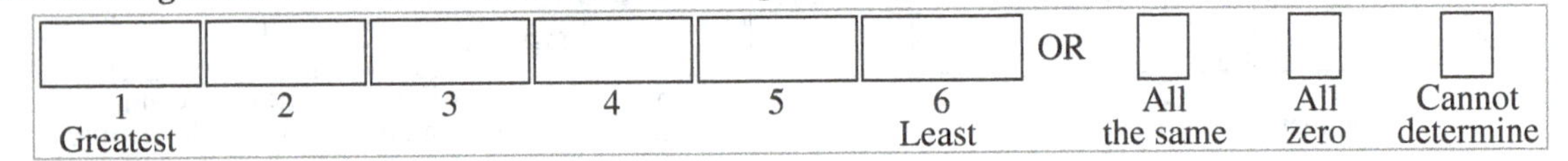

Explain your reasoning.

D3-RT43: WIRE LOOPS AND MOVING MAGNETS—LOOP MOTION

In each case, a permanent magnet is moving toward a circular wire loop that is fixed in place. All of the wire loops are identical, but the wire in Case D has a small gap. The magnets are all identical, but they are approaching the loops at different speeds and with different poles facing the loops. At the instant shown, all of the magnets are the same distance from the wire loops.

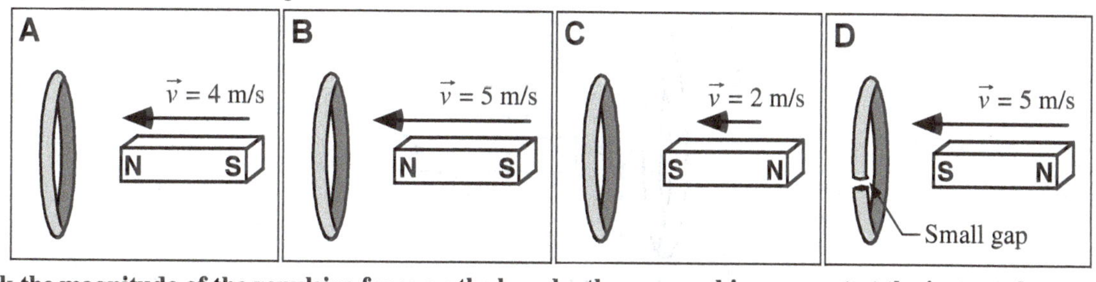

Rank the magnitude of the repulsive force on the loop by the approaching magnet at the instant shown.

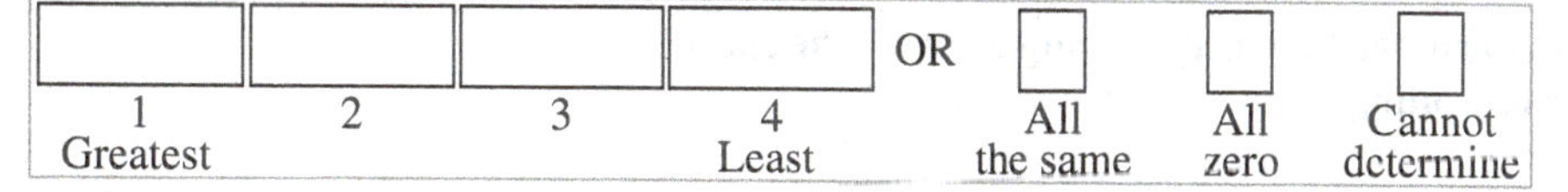

Explain your reasoning.

D3-QRT44: Wire Loops and Moving Magnets—Motion of the System

A circular loop of wire is suspended from a thread so that it hangs freely. A permanent bar magnet is moved toward the center of the wire loop as shown.

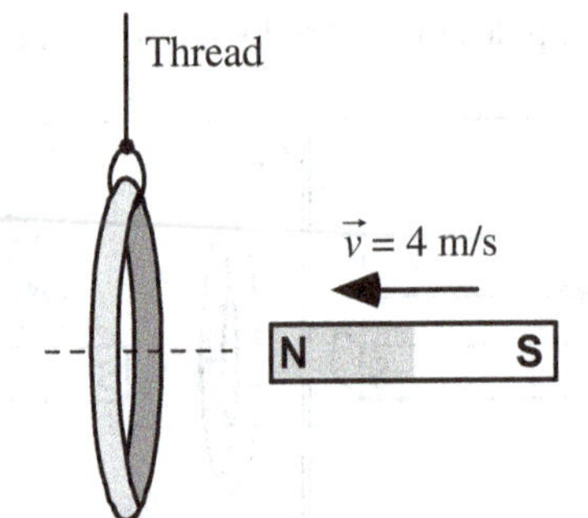

Describe how each of the following changes affects this system.

(a) The magnet is moved toward the loop at twice the speed.
Explain your reasoning.

(b) A small gap is cut in the wire loop.
Explain your reasoning.

(c) The south pole of the magnet is on the side of the magnet closer to the loop.
Explain your reasoning.

(d) The north pole is moving toward the loop from the left at the same speed.
Explain your reasoning.

(d) The strength of the magnet is increased.
Explain your reasoning.

(e) The magnet is moving away from the loop at the same speed.
Explain your reasoning.

E1 WAVES

E1-RT01: WAVES—WAVELENGTH

The drawings represent snapshots taken of waves traveling to the right along strings. The grids shown in the background are identical. The waves all have the same speed, but their amplitudes vary.

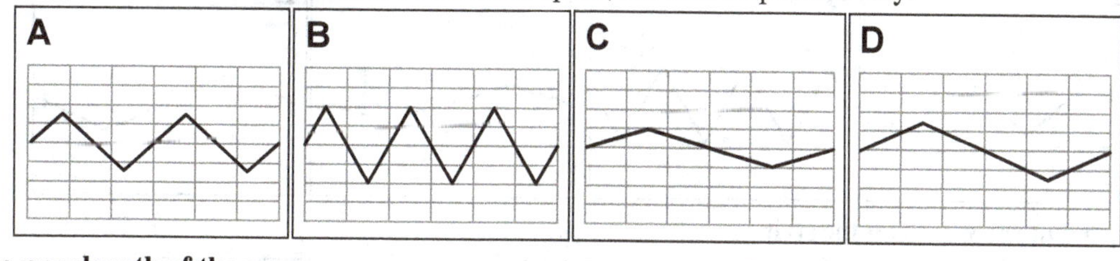

Rank the wavelength of the waves.

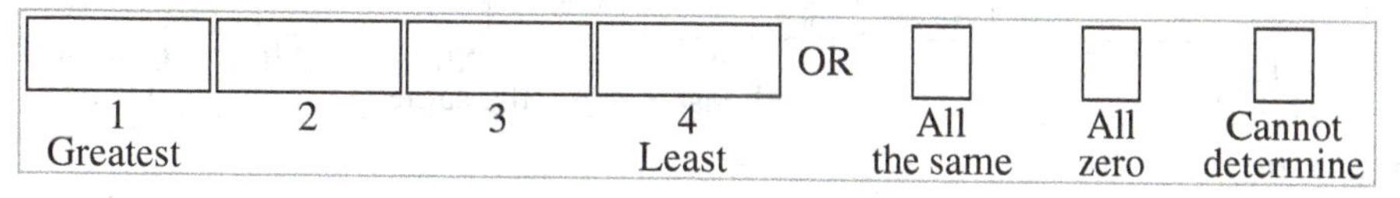

Explain your reasoning.

E1-RT02: WAVES—FREQUENCY

The drawings represent snapshots taken of waves traveling to the right along strings. The grids shown in the background are identical. The waves all have the same speed, but their amplitudes and wavelengths vary.

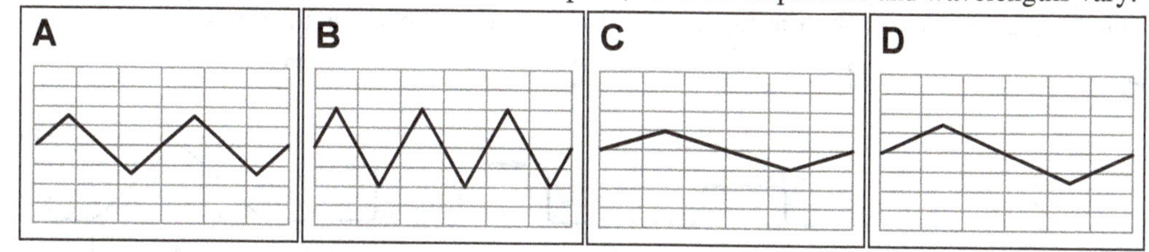

Rank the frequency of the waves.

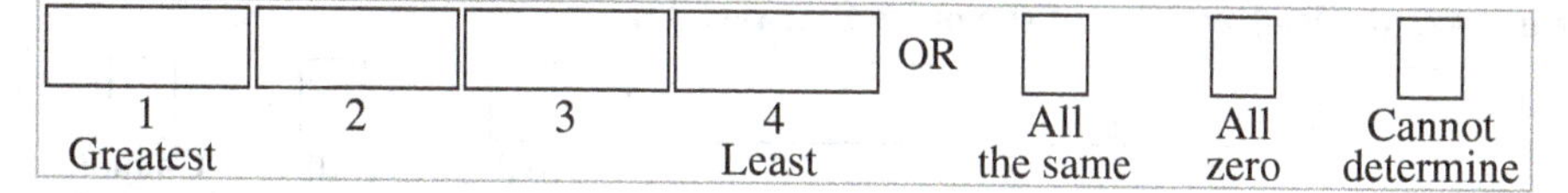

Explain your reasoning.

TIPERs

E1-RT03: WAVES WITH SAME FREQUENCY—WAVE SPEED

The drawings represent snapshots taken of waves traveling to the right along strings. The grids shown in the background are identical. The waves all have the same frequency, but their amplitudes and wavelengths vary.

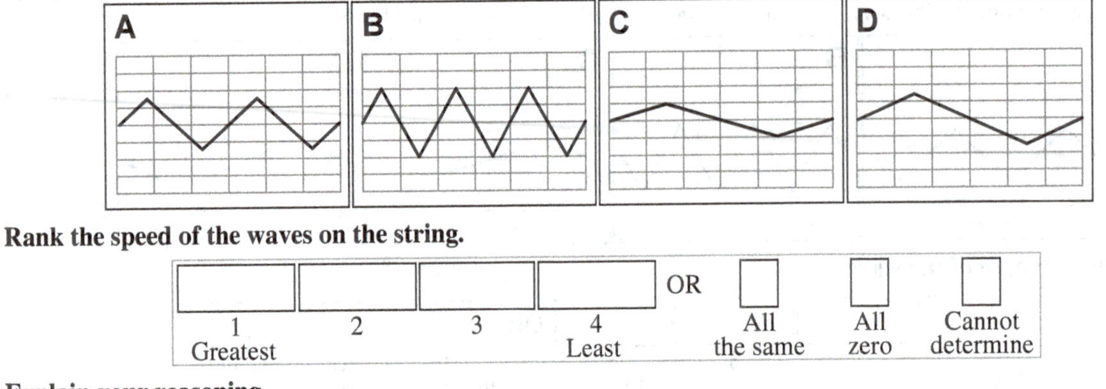

Rank the speed of the waves on the string.

1 Greatest	2	3	4 Least	OR	All the same	All zero	Cannot determine

Explain your reasoning.

E1-RT04: WAVE PULSES—LEADING EDGE TIME TO TRAVEL

The drawings represent snapshots taken of waves traveling along a rope. The grids shown in the background are identical. These pulses, which vary in amplitude, are all sent down identical ropes under equal tension. The ropes are all the same length, and there is no distortion of the pulses as they travel down the ropes.

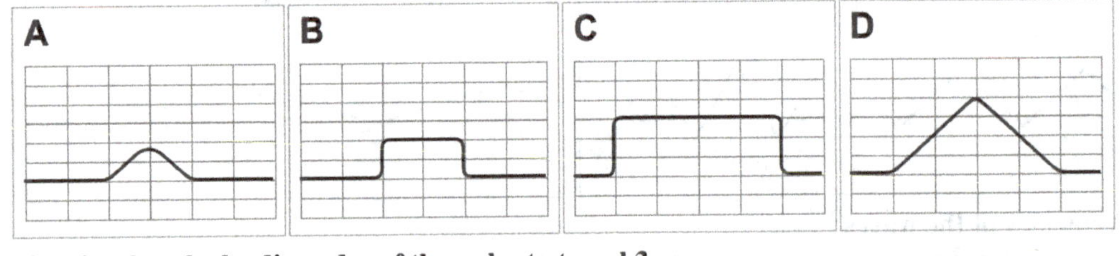

Rank the time it takes the leading edge of the pulse to travel 3 m.

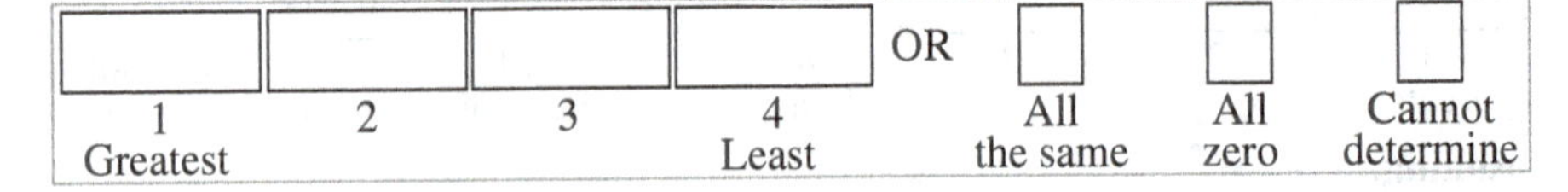

Explain your reasoning.

E1-RT05: PAIRS OF TRANSVERSE WAVES—SUPERPOSITION

Rectangular transverse wave pulses are traveling toward each other along a string. The grids shown in the background are identical, and the pulses vary in height and length. The pulses will meet and interact soon after they are in the positions shown.

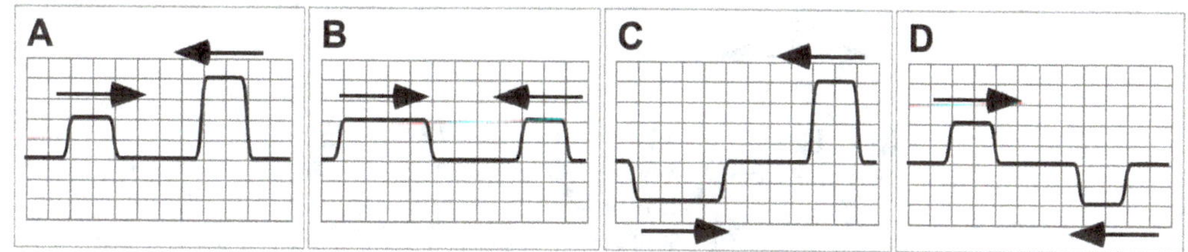

Rank the maximum amplitude of the string at the instant that the positions of the centers of the two pulses coincide.

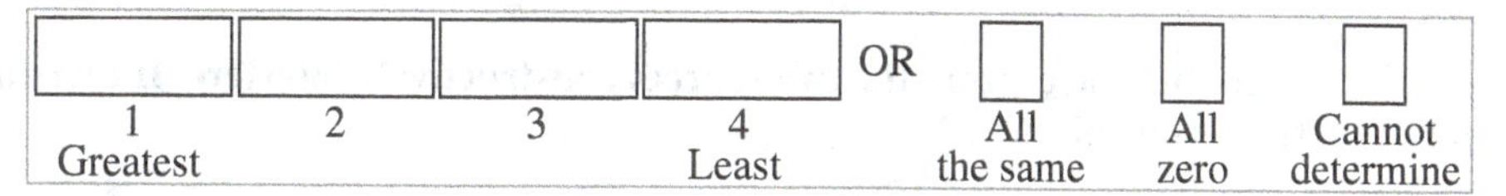

Explain your reasoning.

E1-RT06: WAVE SOURCES SEPARATED BY ONE WAVELENGTH—INTENSITY

Two identical point sources are generating waves with the same frequency and amplitude. The two sources are in phase with each other, so the two sources generate wave crests at the same instant. The wavelength of the waves is equal to the distance between the two sources.

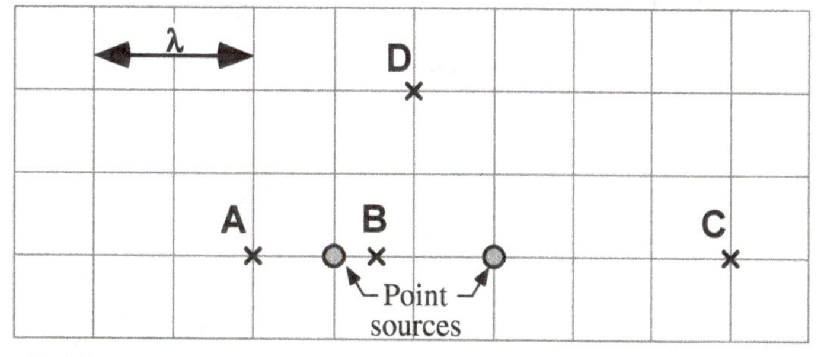

Rank the maximum amplitude of the wave at the labeled points.

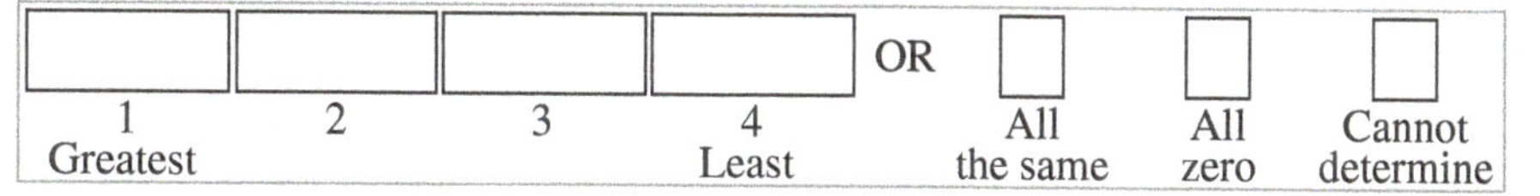

Explain your reasoning.

E1-QRT07: Wave Sources Separated by One Wavelength I—Interference

Two identical point sources are generating waves with the same frequency and amplitude. The two sources are in phase with each other, so the two sources generate wave crests at the same instant. The wavelength of the waves is equal to the distance between the two sources.

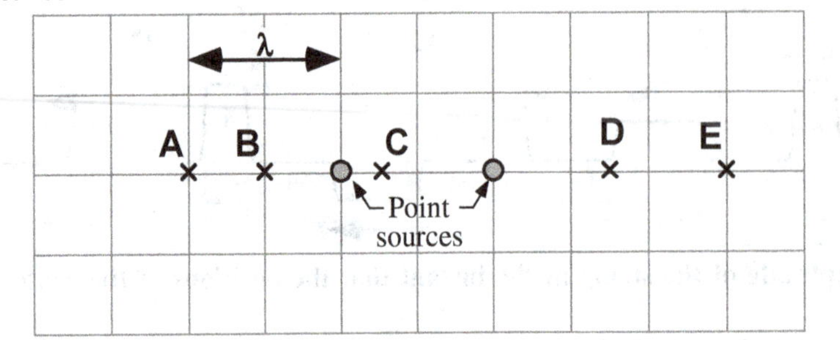

List all the labeled points where the waves from the two sources constructively interfere. If there are no such points, indicate that by stating "none of them." _____

Explain your reasoning.

List all the labeled points where the waves from the two sources destructively interfere. If there are no such points, indicate that by stating "none of them." _____

Explain your reasoning.

E1-QRT08: WAVE SOURCES SEPARATED BY ONE AND ONE-HALF WAVELENGTHS—INTERFERENCE

Two identical point sources are generating waves with the same frequency and amplitude. The two sources are in phase with each other, so the two sources generate wave crests at the same instant. The distance between the two sources is equal to one and one-half times the wavelength, or $1.5\,\lambda$.

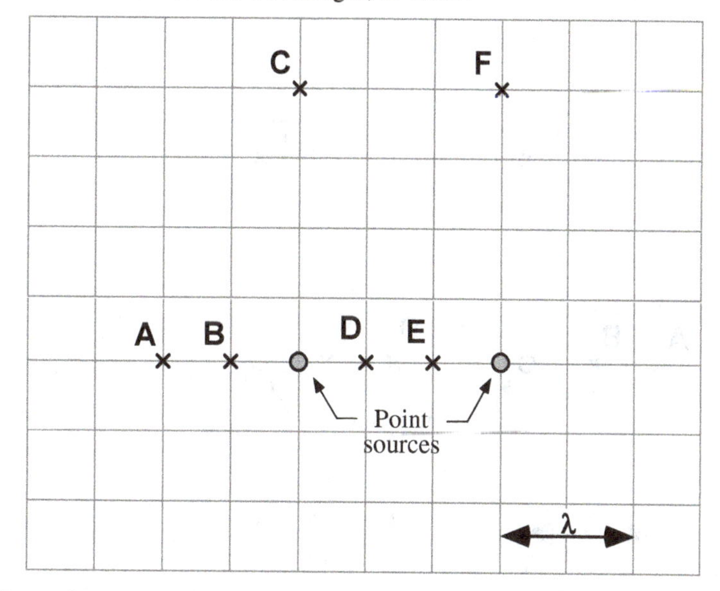

List all the labeled points where the waves from the two sources constructively interfere. If there are no such points, indicate that by stating "none of them." _____

Explain your reasoning.

List all the labeled points where the waves from the two sources destructively interfere. If there are no such points, indicate that by stating "none of them." _____

Explain your reasoning.

E1-QRT09: Wave Sources Separated by Two Wavelengths—Interference

Two identical point sources are generating waves with the same frequency and amplitude. The two sources are in phase with each other, so the two sources generate wave crests at the same instant. The distance between the two sources is equal to two wavelengths, or 2 λ .

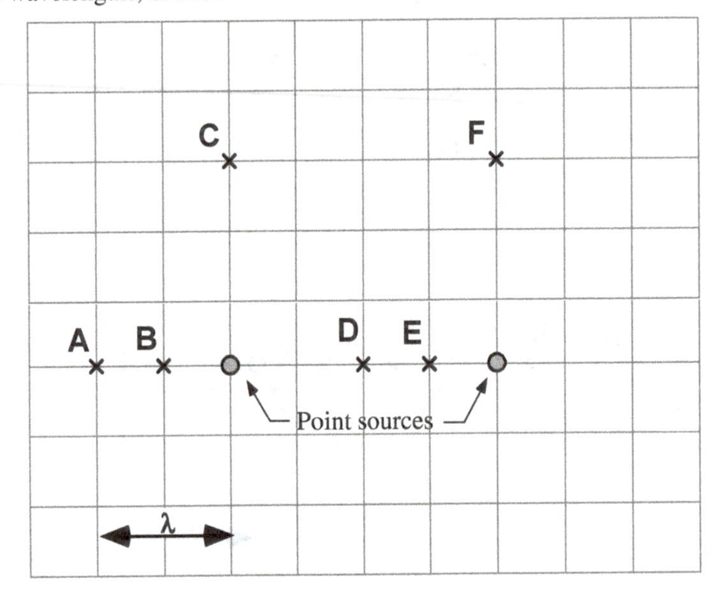

List all the labeled points where the waves from the two sources constructively interfere. If there are no such points, indicate that by stating "none of them." _____

Explain your reasoning.

List all the labeled points where the waves from the two sources destructively interfere. If there are no such points, indicate that by stating "none of them." _____

Explain your reasoning.

E1-QRT10: WAVE SOURCES SEPARATED BY ONE WAVELENGTH II—INTERFERENCE

Two identical point sources are generating waves with the same frequency and amplitude. The two sources are *out of phase* with each other, so at the instant that one source is creating a crest, the other source is creating a trough. The wavelength of the waves is equal to the distance between the two sources.

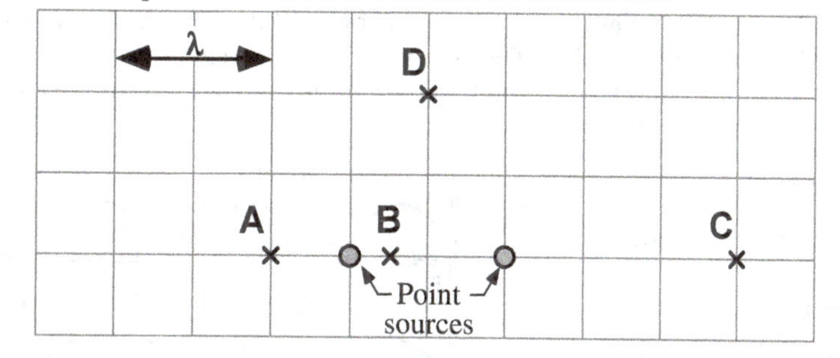

List all the labeled points where the waves from the two sources constructively interfere. If there are no such points, indicate that by stating "none of them." _____

Explain your reasoning.

List all the labeled points where the waves from the two sources destructively interfere. If there are no such points, indicate that by stating "none of them." _____

Explain your reasoning.

E1-RT11: Standing Waves—Frequency

A string is stretched so that it is under tension and is tied at both ends so that the endpoints don't move. A mechanical oscillator then vibrates the string so that a standing wave is created. The dark line in each diagram represents a snapshot of a string at an instant in time when the amplitude of the standing wave is a maximum. The lighter lines represent the string at other times during a complete cycle. All of the strings are identical except for their lengths, and all strings have the same tension. The number of nodes and antinodes in each standing wave is different. The lengths of the strings (L) and the amplitudes at the antinodes (A) are given in each figure.

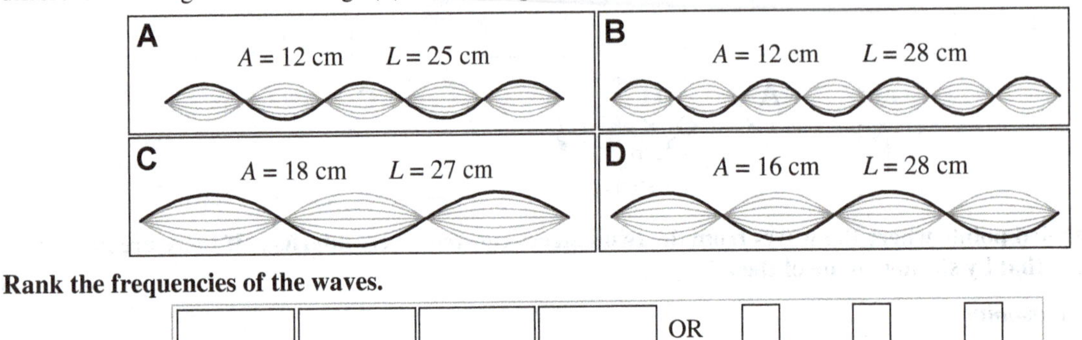

Rank the frequencies of the waves.

1 Greatest	2	3	4 Least	OR	All the same	All zero	Cannot determine

Explain your reasoning.

E1-RT12: STANDING WAVES—WAVELENGTH

A string is stretched so that it is under tension and is tied at both ends so that the endpoints don't move. A mechanical oscillator then vibrates the string so that a standing wave is created. The dark line in each diagram represents a snapshot of a string at an instant in time when the amplitude of the standing wave is a maximum. The lighter lines represent the string at other times during a complete cycle. All of the strings are identical except for their lengths, and all strings have the same tension. The number of nodes and antinodes in each standing wave is different. The lengths of the strings (L) and the amplitudes at the antinodes (A) are given in each figure.

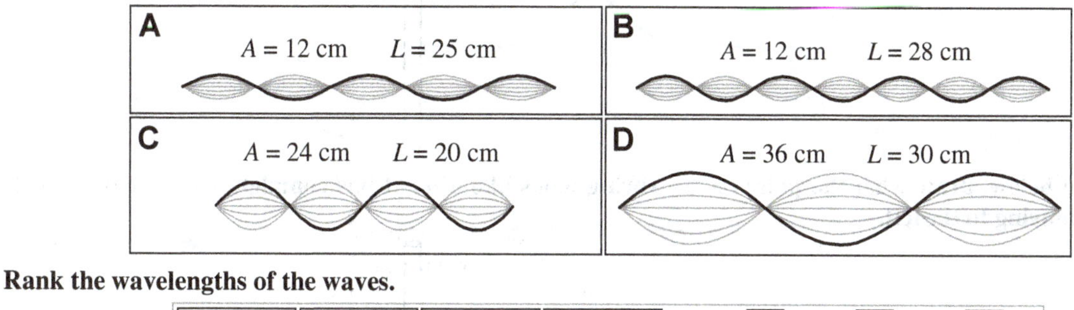

Rank the wavelengths of the waves.

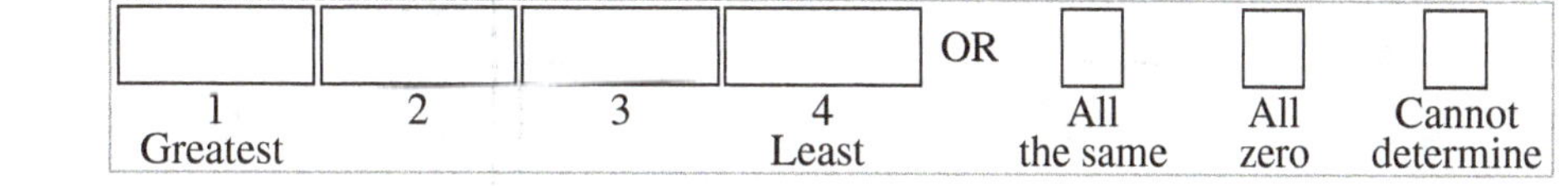

Explain your reasoning.

E1-RT13: STANDING WAVES SYSTEMS—WAVE SPEED

A string is stretched so that it is under tension and is tied at both ends so that the endpoints don't move. A mechanical oscillator then vibrates the string so that a standing wave is created. The dark line in each diagram represents a snapshot of a string at an instant in time when the amplitude of the standing wave is a maximum. The lighter lines represent the string at other times during a complete cycle. All of the strings have the same length but may not have the same mass. The number of nodes and antinodes in the standing wave is the same in Cases A and D. The tensions in the strings (T) and the standing wave frequencies (f) are given in each figure.

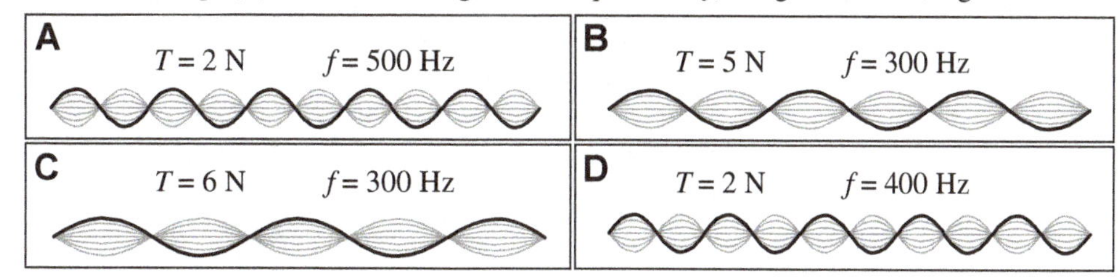

Rank the speeds of the waves in the strings.

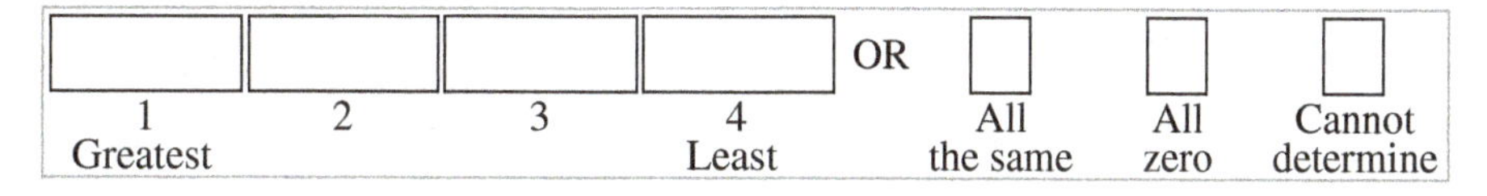

Explain your reasoning.

E1-QRT14: WAVE PULSE ON HORIZONTAL SPRING WITH FIXED END—REFLECTION SHAPE

A long spring is firmly connected to a stationary metal rod at one end. A student holding the other end moves her hand rapidly up and down to create a pulse with the shape shown in the figure. The pulse moves along the taut spring toward the rod.

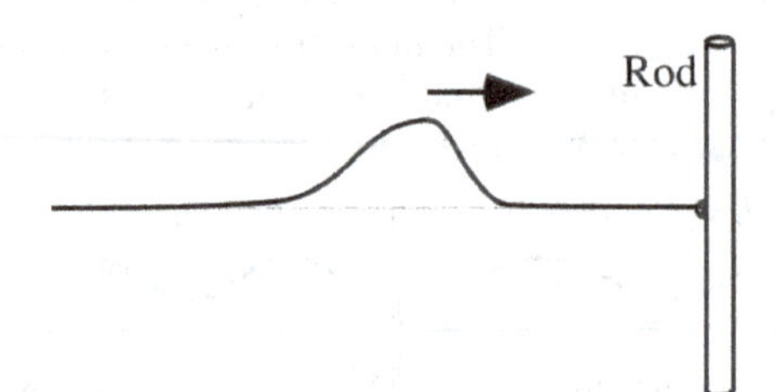

In the space below, draw what the pulse on the spring looks like after it has completely reflected from the wall and is moving to the left.

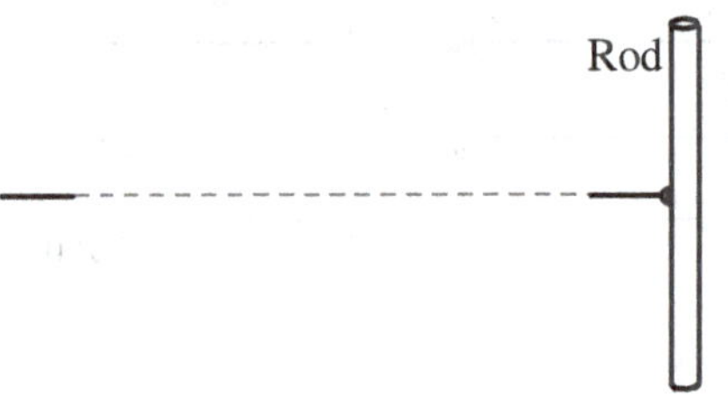

Explain your reasoning.

E1-QRT15: WAVE PULSE ON HORIZONTAL SPRING WITH FREE END—REFLECTION SHAPE

A long spring is connected to a loop that passes around a stationary metal rod at one end. The loop is free to move vertically without friction along the rod. A student holding the other end moves her hand rapidly up and down to create a pulse with the shape shown in the figure. The pulse moves along the taut spring toward the rod.

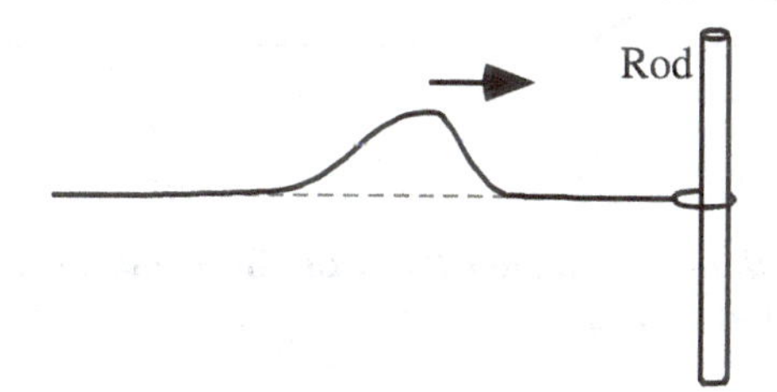

In the space below, draw what the pulse on the spring looks like after it has completely reflected from the wall and is moving to the left.

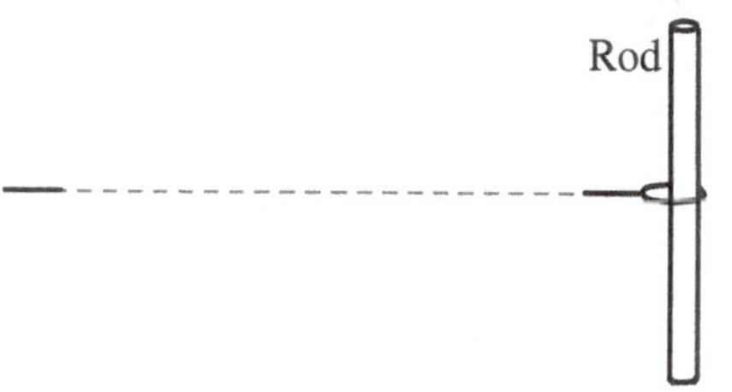

Explain your reasoning.

E1-CT16: WAVE PULSES TRAVELING TOWARD EACH OTHER—SPEED

Two pulses travel toward each other along a long stretched spring as shown. Pulse A is wider than pulse B, but not as high.

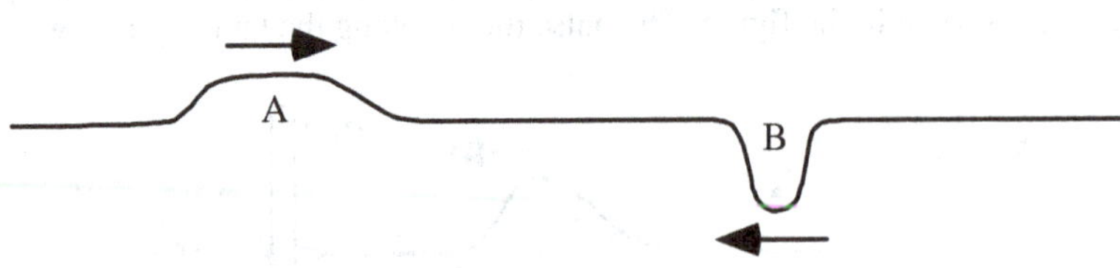

Is the speed of pulse A (i) *larger than,* (ii) *smaller than,* or (iii) *equal to* the speed of pulse B? If there is not enough information to tell, state that explicitly. _____
Explain your reasoning.

E1-WWT17: TWO WAVE PULSES INTERACTING—IMPACT

A student states:

"If two wave pulses traveling in opposite directions along the same string meet, they reflect from one another and go back the way they came from."

What, if anything, is wrong with this statement? If something is wrong, identify it and explain how to correct it. If this statement is correct, explain why.

E1-QRT18: WAVE PULSES TRAVELING TOWARD EACH OTHER—SHAPE AND DIRECTION

Two pulses travel toward each other along a long, stretched spring. The speed of each pulse is 100 cm/s. Each square on the grid is 1 cm long, as can be seen on the ruler markings below the grid. The left pulse is about 6 cm wide, and the right pulse is about 2 cm wide and twice as tall as the left pulse. The figure below represents a snapshot of the spring taken at time $t = 0$.

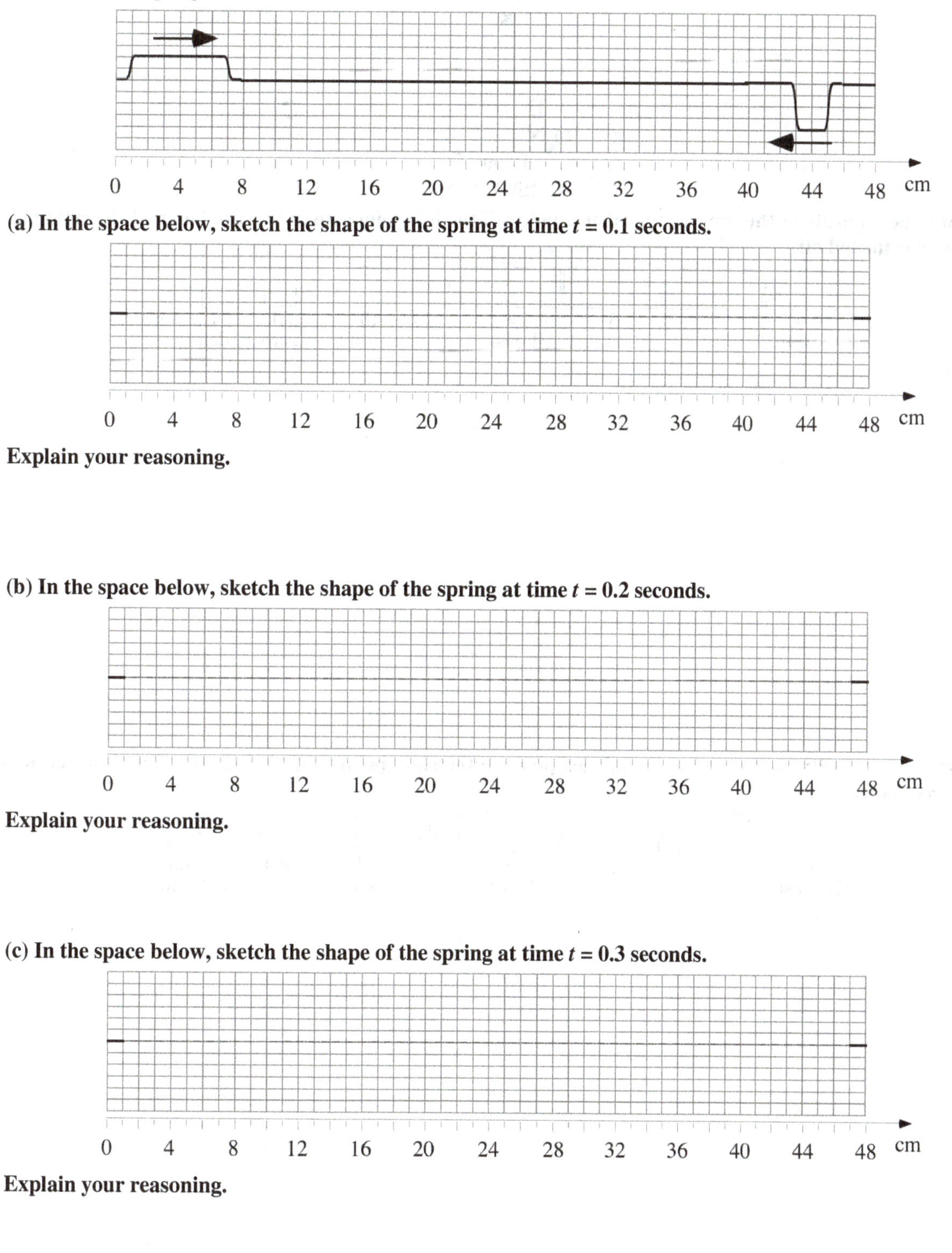

(a) In the space below, sketch the shape of the spring at time $t = 0.1$ seconds.

Explain your reasoning.

(b) In the space below, sketch the shape of the spring at time $t = 0.2$ seconds.

Explain your reasoning.

(c) In the space below, sketch the shape of the spring at time $t = 0.3$ seconds.

Explain your reasoning.

E1-RT19: Wave Source—Intensity at Various Locations

Two point sources can generate waves with the same frequency and amplitude. The sources can be turned on and off individually.

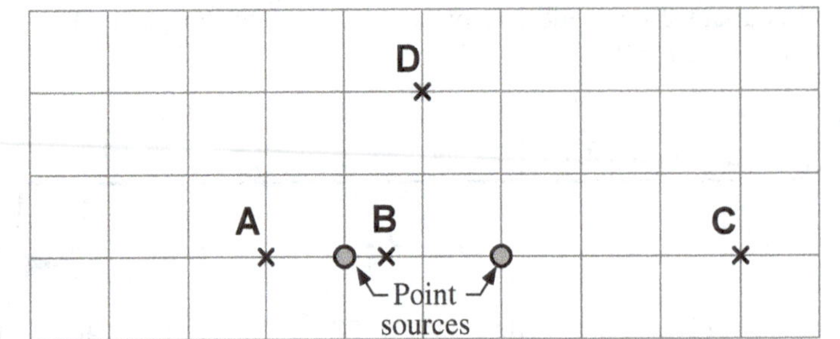

(a) Rank the intensity of the wave at the labeled points if the left point source is turned on and the right point source is turned off.

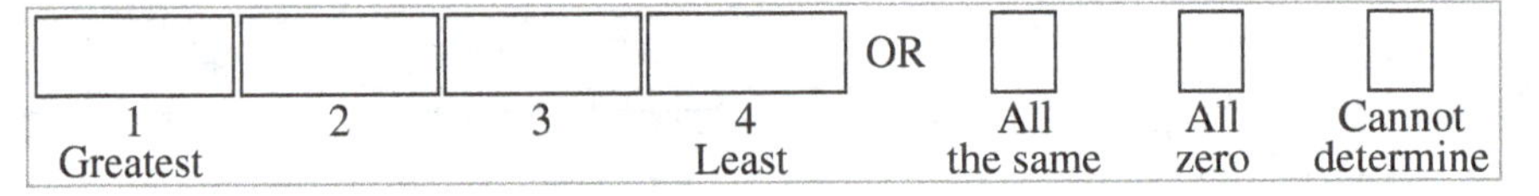

Explain.

(b) Rank the intensity of the wave at the labeled points if the right point source is turned on and the left point source is turned off.

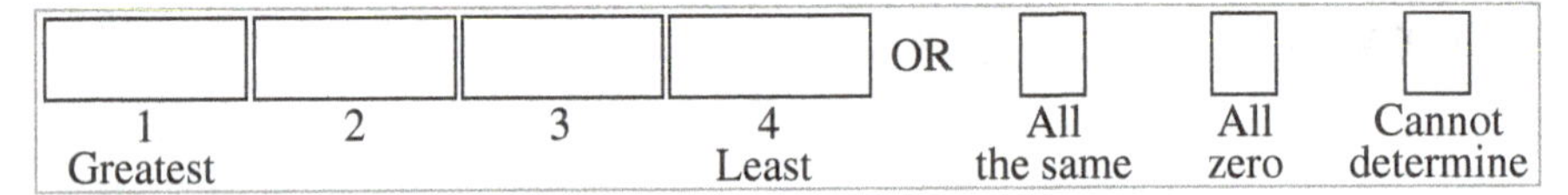

Explain.

E2 OPTICS

E2-QRT01: COLORED BULBS NEAR A DOWEL—SHADOWS

A red light bulb and a blue light bulb are placed near a wall as shown in the top-view diagram. A cylindrical dowel is placed between the bulbs and the wall, creating shadows. Where light from both bulbs hits the wall, the wall will appear magenta in color. Where no light from either bulb hits the wall, the wall will appear black.

(a) Draw rays to determine where light from each bulb hits the wall.

(b) Indicate where the wall appears magenta, where it appears black, where it appears red, and where it appears blue.

Explain your reasoning.

Wall

Blue bulb

Red bulb

Dowel

E2-SCT02: PLANE MIRROR I—OBSERVER LOCATION

A small light bulb is placed near a plane mirror. The locations of five observers are marked, labeled *A*–*E* as shown. Three students are discussing which observers will see an image of the bulb:

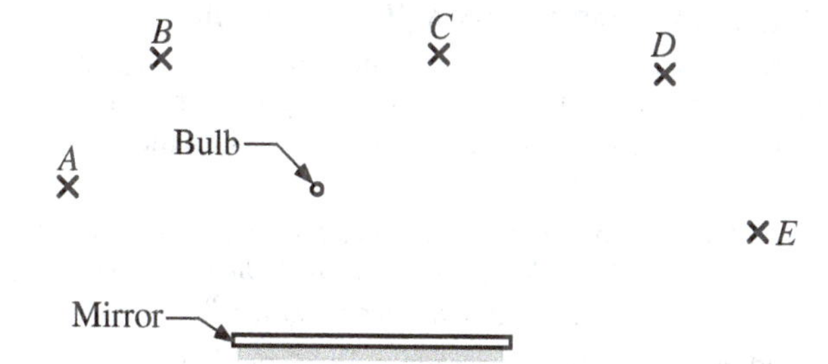

Agustin: *"The image of the bulb will be formed the same distance behind the mirror as the bulb is in front. All of the observers can see the mirror, so they will be able to see the image."*

Bruno: *"The images formed by a plane mirror are virtual images—none of the observers will be able to see them because they aren't real."*

Carol: *"There will be an image of the bulb, because rays from the bulb hit the mirror and bounce off. But only observer C will see the image, because only C is in front of the mirror."*

With which, if any, of these students do you agree?

Agustin _____ Bruno _____ Carol _____ None of them _____

Explain your reasoning.

E2-SCT03: Plane Mirror II—Observer Location

A small light bulb is placed near a plane mirror. The locations of five observers are marked, labeled *A–E* as shown. Three students are discussing which observers will see an image of the bulb:

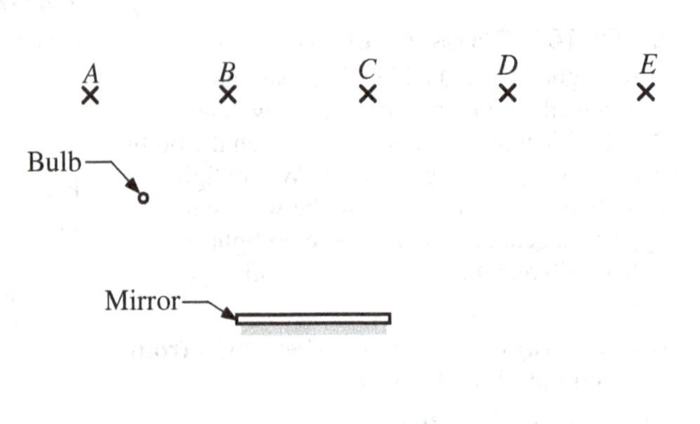

Ashley: *"None of them will see an image. The bulb isn't in front of the mirror, so no image is formed."*

Briar: *"Even though the bulb isn't in front of the mirror, rays from the bulb still hit the mirror and bounce off. They bounce off in the direction of observers D and E, so these two will see the image."*

Capria: *"The image is in the mirror. As long as you are in front of the mirror you will see it. Observer C will see the image, but the rest of the observers aren't in front of the mirror."*

With which, if any, of these students do you agree?

Ashley _____ Briar _____ Capria _____ None of them _____

Explain your reasoning.

E2-SCT04: Shadows on a Wall—Color

A student stands in front of a white wall in a room that is dark except for the light from two small light bulbs, one red and one green. In the top-view at right, four locations on the wall are labeled *A–D*. Three students are discussing the color of the wall at the labeled points:

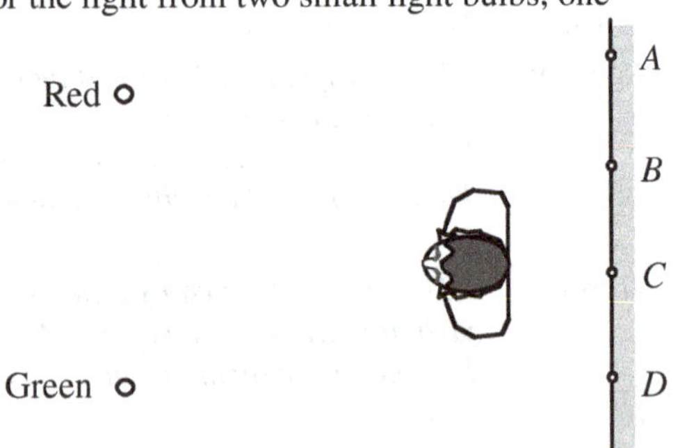

Anuradha: *"No light from either bulb will reach point C, so I think that point is going to be black or very dark. The wall there will have no color."*

Brandon: *"The wall at point A will get light from both bulbs. The color of the wall there will be a mix of red and green light—sort of yellow."*

Carlos: *"Point B will get light from the red bulb but not from the green one, so it will be red. The opposite happens at point D: Light reaches it from the green bulb but not the red one, so it will appear green."*

With which, if any, of these students do you agree?

Anuradha _____ Brandon _____ Carlos _____ None of them _____

Explain your reasoning.

E2-QRT05: COLORED BULBS NEAR A PLANE MIRROR—OBSERVER LOCATIONS

A red light bulb and a blue light bulb are placed near a plane mirror as shown.

(a) A person in front of the mirror can see the image of both bulbs.

Where is this person located? In the diagram, shade the region.

Explain your reasoning.

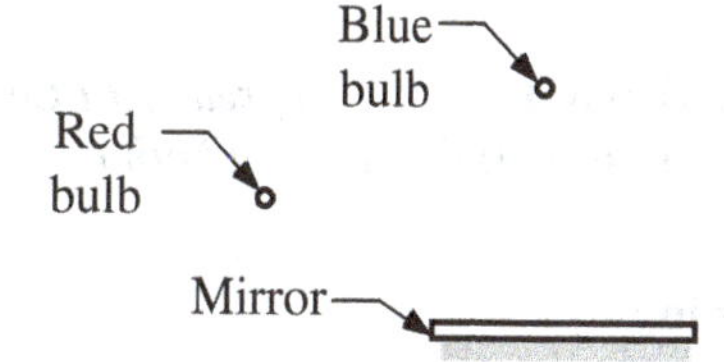

(b) A person in front of the mirror can see the image of the red bulb but cannot see the image of the blue bulb.

Where is this person located? In the diagram, shade the region.

Explain your reasoning.

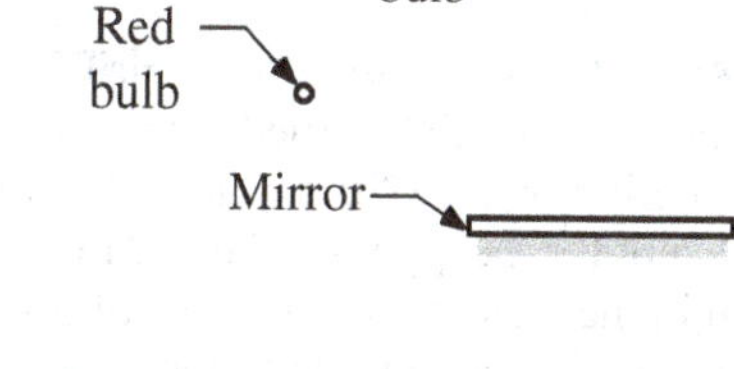

E2-CT06: LIGHT FROM A SMALL BULB—HEIGHT OF SHADOW

In both cases shown, a small light bulb illuminates a wall. A shadow is created on the wall of a rod placed between the wall and the bulb. The two cases are identical except that in Case B there is a glass block between the rod and the wall.

(a) Is the height of the shadow of the rod on the wall (i) *greater in Case A*, (ii) *greater in Case B*, or (iii) *the same in both cases*? _____
Explain your reasoning.

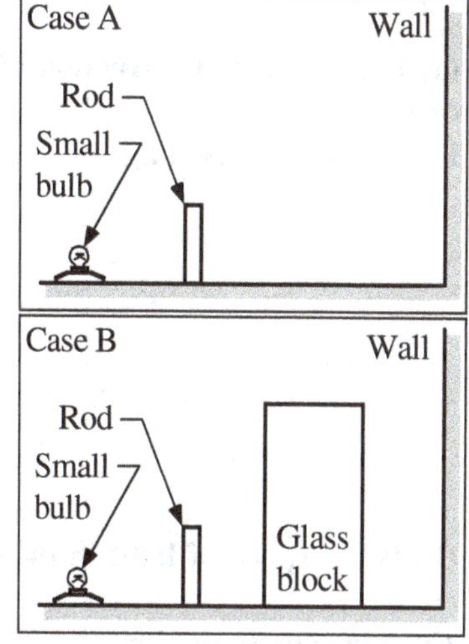

(b) If the entire arrangement from Case B—bulb, rod, glass block, and wall—were submerged in water, would the height of the shadow on the wall (i) *increase,* (ii) *decrease,* or (iii) *remain the same?* _____
Explain your reasoning.

E2-CT07: Light Rays Bent at a Surface—Index of Refraction

In both cases shown, a light ray traveling in water bends at the surface of a cube. The cases are identical except that the cube in Case A is made of a different material than the cube in Case B.

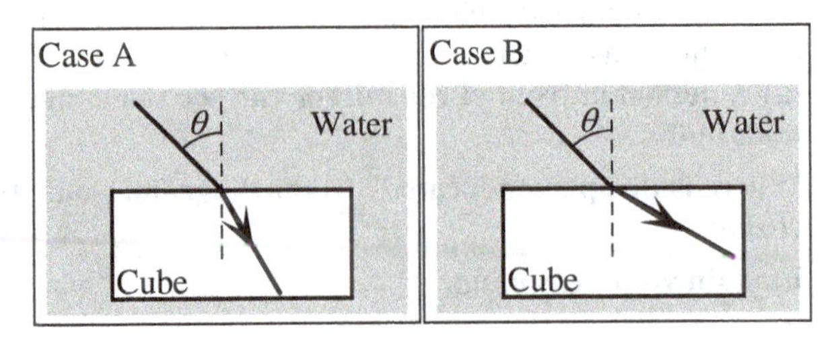

Is the index of refraction of the cube (i) ***greater in Case A*****, (ii)** ***greater in Case B*****, or (iii)** ***the same in both cases?***

_____

Explain your reasoning.

E2-CT08: Water and Oil in an Aquarium—Index of Refraction of Oil

In both cases, an aquarium is partially filled with water, and a layer of oil is floating on top of the water. The two cases are identical except for the type of oil used. In Case A, a laser beam in the water totally reflects off of the oil/water boundary as shown. In Case B, the same laser beam refracts at the oil/water boundary as shown.

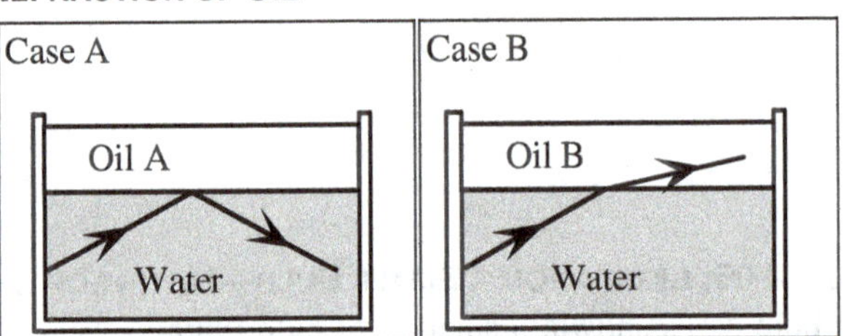

(a) Is the index of refraction of the oil (i) ***greater in Case A*****, (ii)** ***greater in Case B*****, or (iii)** ***the same in both cases?*** _____

Explain your reasoning.

(b) Is the speed of light in the oil (i) ***greater in Case A*****, (ii)** ***greater in Case B*****, or (iii)** ***the same in both cases?*** _____

Explain your reasoning.

E2-CT09: BENDING OF LASER BEAM IN AIR—ANGLE BENT

A laser beam traveling in air enters water at an angle of 60° with respect to the surface and is bent in water to an angle of φ from the surface as shown in Case A. (Note that the drawings may or may not be accurately portraying the situation.)

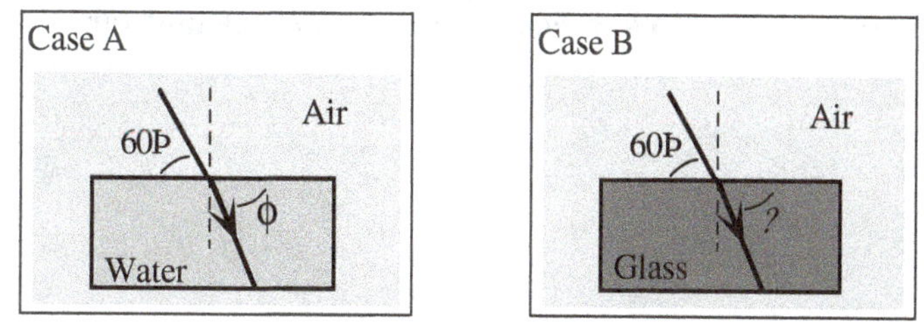

If the water is replaced by glass, is the angle with respect to the surface that the laser beam is bent (i) *greater than*, (ii) *smaller than*, or (iii) *equal to* φ? _____

Explain your reasoning.

E2-CT10: BENDING OF LASER BEAM IN GLASS—ANGLE BENT

A laser beam is traveling in water. This beam strikes a water-glass surface at an angle of 30.0° from the glass surface and is at an angle of β with respect to the surface, as shown in Case A. (Note that the drawings may or may not be accurately portraying the situation.)

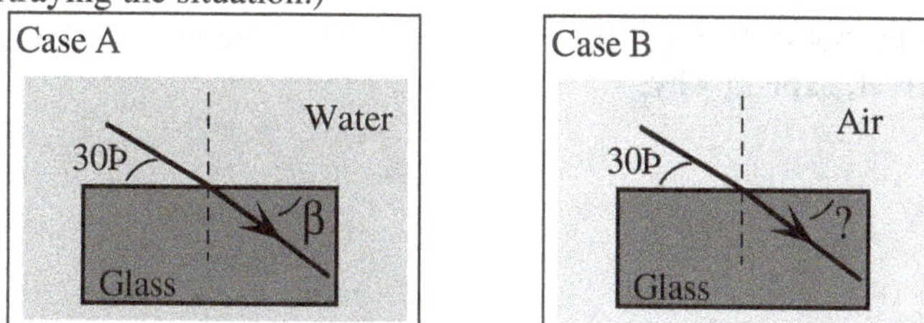

If the water is replaced by air, the angle that the laser beam is bent in the glass from the surface is (i) *greater than*, (ii) *smaller than*, or (iii) *equal to* β? _____

Explain your reasoning.

TIPERs

E2-CT11: Laser Light Pulse Traveling through Slab—Time

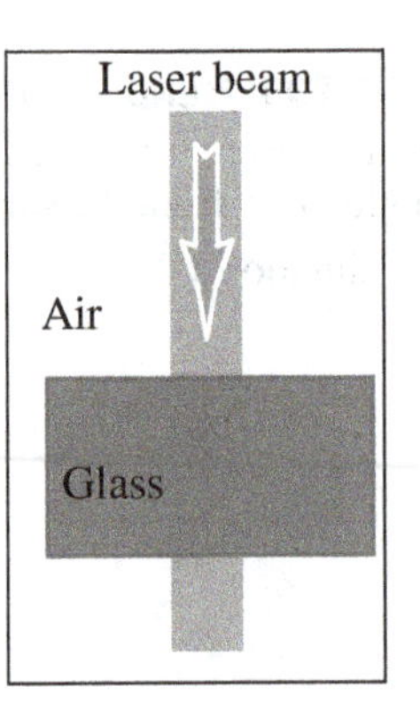

It has been noted that red light bends less than blue light in glass when it enters the glass at an angle.

Is the time that it takes for a pulse of red light to pass perpendicularly through a slab of glass surrounded by air (i) *greater than*, (ii) *less than*, or (iii) *equal to* the time for a blue light pulse? _____

Explain your reasoning.

E2-WWT12: Moving Candle Away from a Lens—Focal Length

A student thinking about a candle in front of a converging lens states:

"I noticed that if I move a candle back away from a converging lens that this results in a larger image being produced. This is due to the change in the focal length of the lens. I think it got shorter."

What, if anything, is wrong with this statement? If something is wrong, identify and explain how to correct all errors. If this statement is correct, explain why.

E2-CT13: OBJECTS INSIDE FOCAL LENGTH OF CURVED MIRRORS—IMAGE DISTANCE

In the two situations shown, the mirrors have the same focal lengths, and the object distance from the mirror to the arrow is the same. In both cases, the object distance is less than the focal distance.

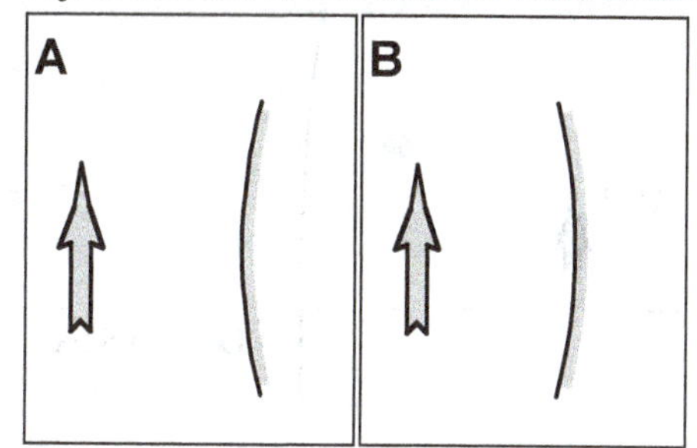

Will the image distance for Case A be (i) *greater than,* (ii) *less than,* or (iii) *equal to* the image distance for Case B? _____

Explain your reasoning.

E2-WBT14: OBJECT AND INVERTED IMAGE FOR A CONVERGING MIRROR—FOCAL POINT

An object is placed in front of a converging mirror. An inverted image of the object is formed at the location shown.

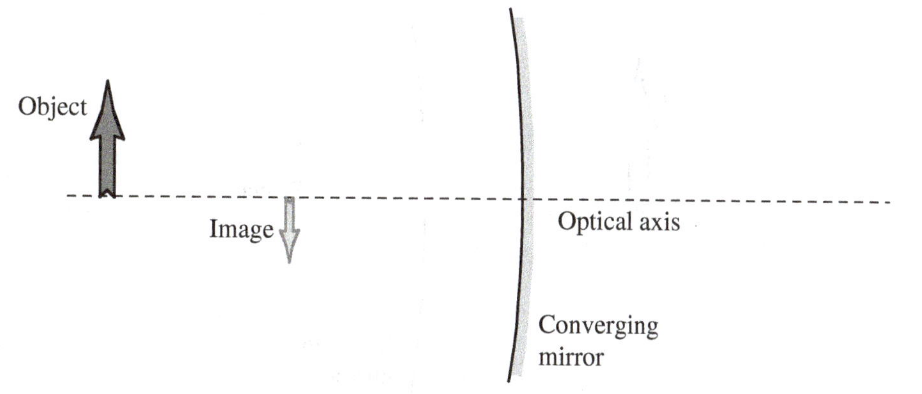

Based on the image and object locations above, find the focal point for this mirror.

Explain your reasoning.

E2-WBT15: Object and Upright Image for a Converging Mirror—Focal Point

An object is placed in front of a converging mirror. An upright image of the object is formed behind the mirror at the location shown.

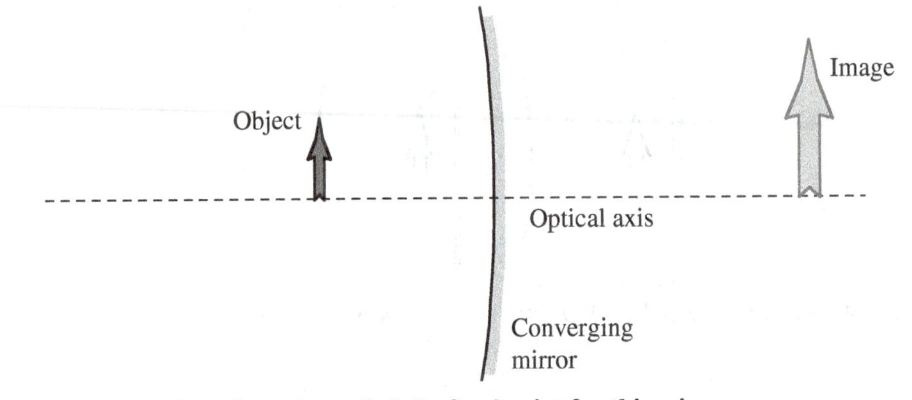

Based on the image and object locations above, find the focal point for this mirror. Explain your reasoning.

E2-WBT16: Object and Upright Image for a Diverging Mirror—Focal Point

An object is placed in front of a diverging mirror. An upright image of the object is formed behind the mirror at the location shown.

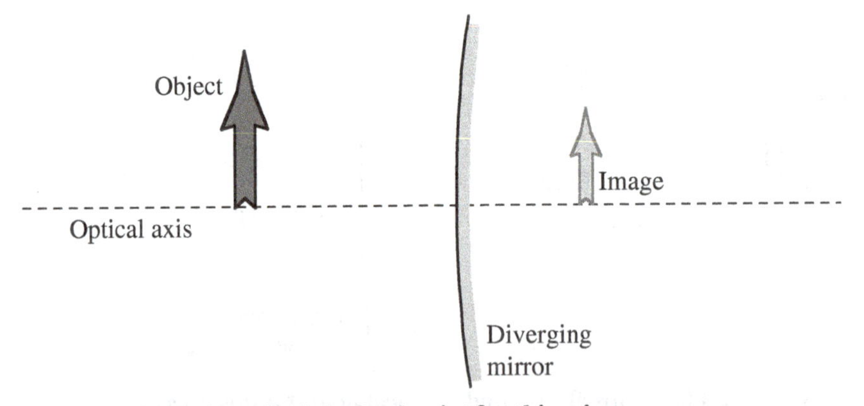

Based on the image and object locations, find the focal point for this mirror. Explain your reasoning.

E2-WBT17: IMAGE FOR A CONVERGING MIRROR—OBJECT

An arrow is placed in front of a converging mirror. An upright image of the arrow is formed to the right of the mirror at the location shown.

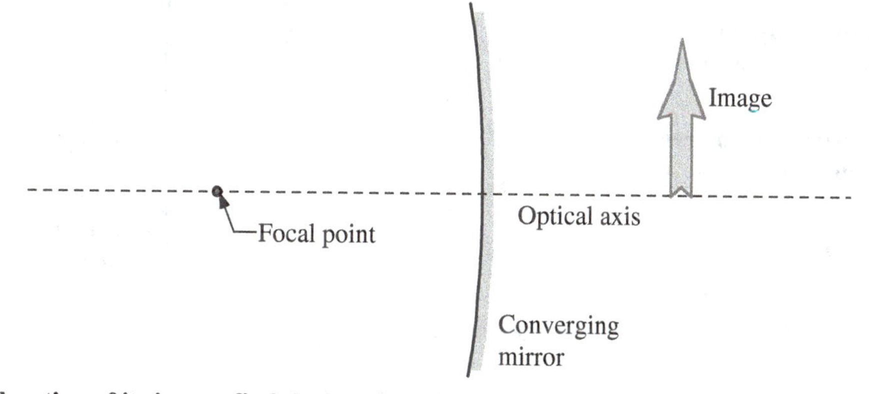

Based on the location of its image, find the location of the arrow.
Explain your reasoning.

E2-WBT18: IMAGE FOR A DIVERGING MIRROR—OBJECT

An arrow is placed in front of a diverging mirror. An upright image of the arrow is formed to the right of the mirror at the location shown.

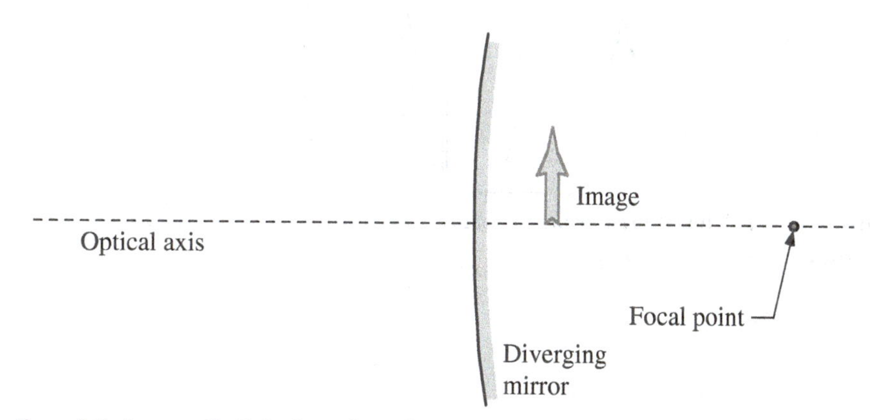

Based on the location of its image, find the location of the arrow.
Explain your reasoning.

E2-QRT19: Bulbs near a Converging Mirror—Observer Location

A small bulb is placed in front of a converging mirror as shown.

In the diagram, (a) draw rays to determine the location of the image of the light bulb.

(b) Then determine where an observer must be located to see the image of the bulb.

Explain your reasoning.

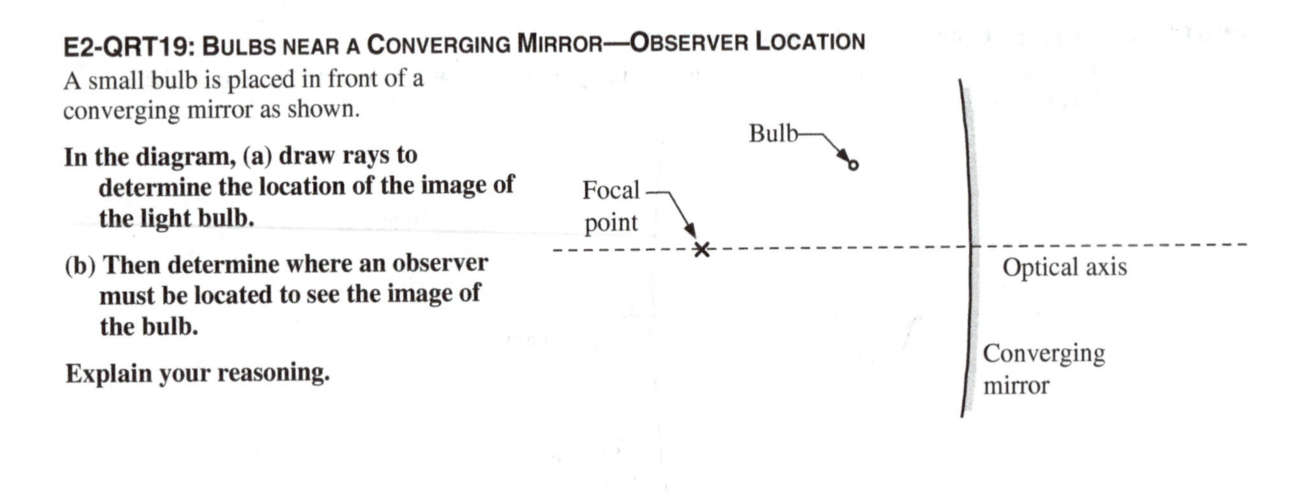

E2-CT20: Objects Outside the Focal Length of Mirrors—Image Distance

In the two situations shown, the mirrors have the same focal lengths, and the object distance from the mirror to the arrow is the same. In both cases, the object distance is greater than the focal distance.

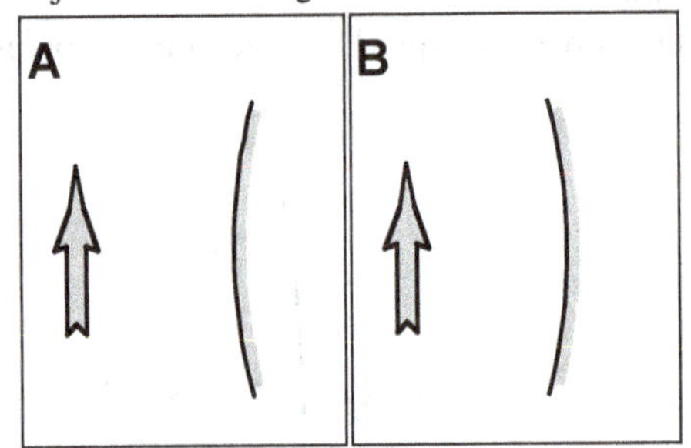

Will the image distance for A be (i) *greater than*, (ii) *less than*, or (iii) *equal to* the image distance for B? _____
Explain your reasoning.

E2-QRT21: IMAGE AND OBJECT LOCATIONS RELATIVE TO MIRRORS—IMAGE TYPES

Consider the following image positions for the mirrors arranged as shown with the objects placed on the left side.

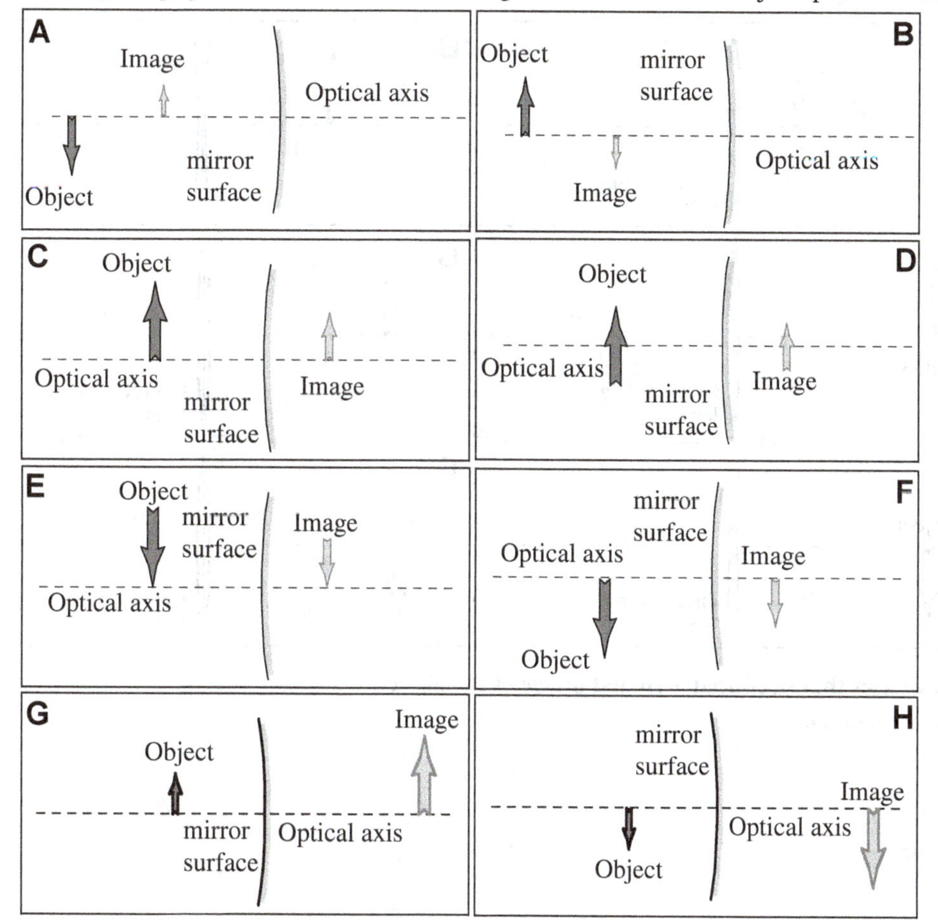

(a) List all the cases that produced a virtual image of the object: _____
Explain your reasoning.

(b) List all the cases that produced an inverted image of the object: _____
Explain your reasoning.

(c) List all the cases that produced a reduced size image of the object: _____
Explain your reasoning.

E2-QRT22: IMAGE AND OBJECT LOCATIONS RELATIVE TO LENS—IMAGE TYPES

Consider the following image positions for the lenses arranged as shown with the objects placed on the left side.

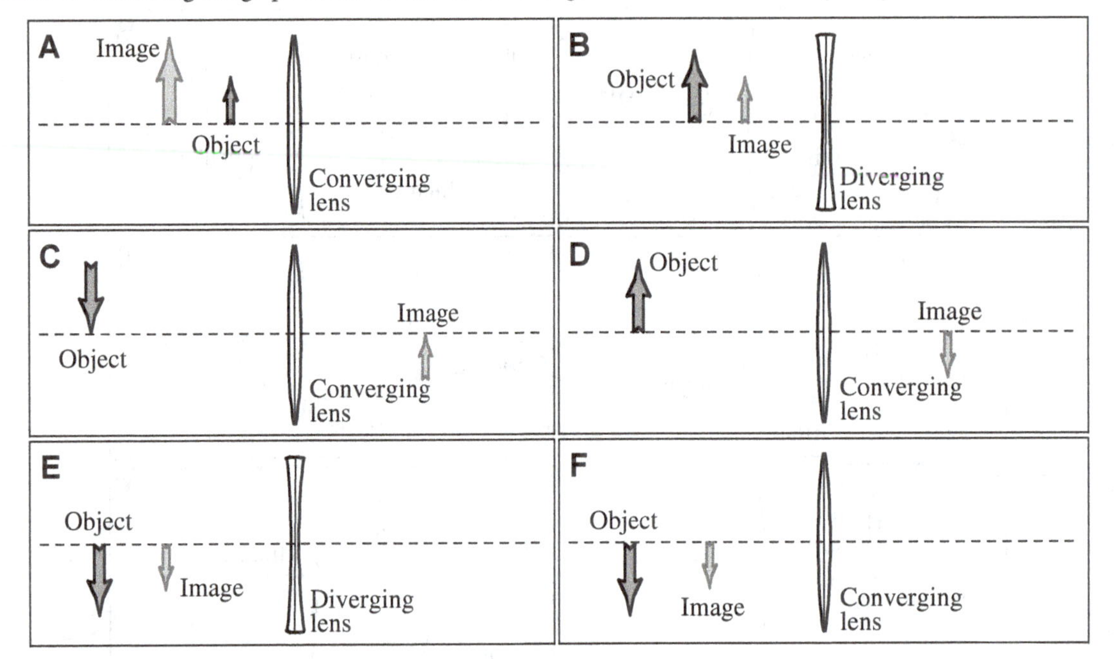

(a) List all the cases that produced a virtual image of the object: _____
Explain your reasoning.

(b) List all the cases that produced an inverted image of the object: _____
Explain your reasoning.

(c) List all the cases that produced a reduced size image of the object: _____
Explain your reasoning.

E2-QRT23: IMAGE AND OBJECT LOCATIONS RELATIVE TO MIRROR—IMAGE TYPE

A student determines the image positions shown for the three arrangements shown below, with the objects placed on the left side of the mirror.

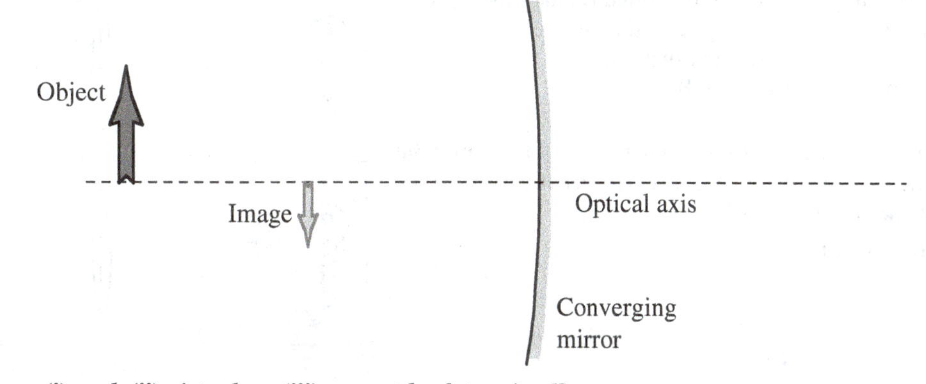

(a) Is this image (i) *real,* (ii) *virtual,* or (iii) *cannot be determined*? _____
Explain your reasonig.

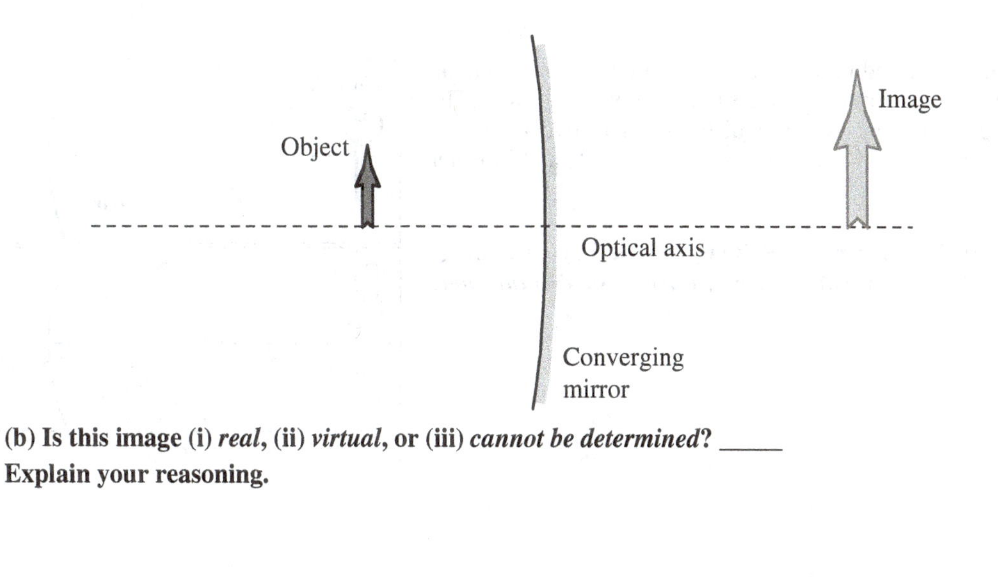

(b) Is this image (i) *real,* (ii) *virtual,* or (iii) *cannot be determined*? _____
Explain your reasoning.

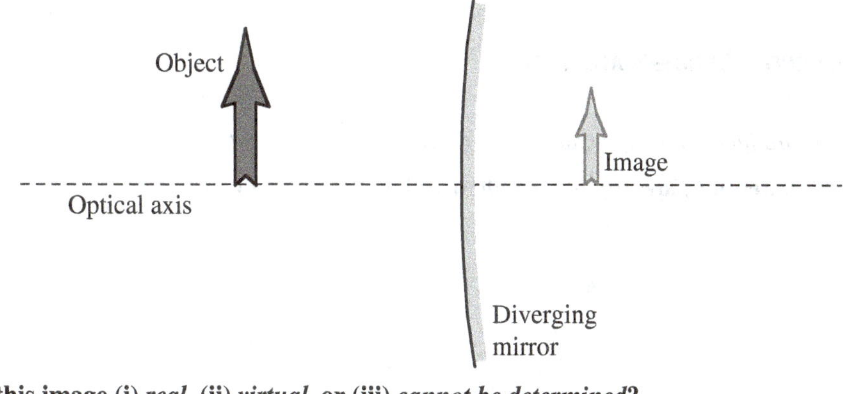

(c) Is this image (i) *real,* (ii) *virtual,* or (iii) *cannot be determined*? _____
Explain your reasoning.

E2-CT24: Image Formed in Air and in Water—Distance from Lens or Mirror

(a) In each case, a key is placed in front of a converging lens so that an inverted image of the key is formed on the other side of the lens. (The object distance is greater than the focal distance.) The two cases are identical except that the key and lens are in air in Case A and in water in Case B.

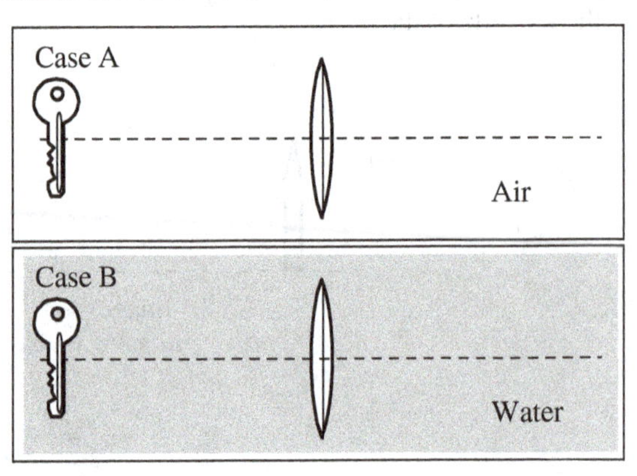

Is the distance from the lens to the image of the key (the image distance) (i) ***greater in Case A*****, (ii)** ***greater in Case B*****, or (iii)** ***the same in both cases*****?** _____

Explain your reasoning.

(b) In each case, a key is placed in front of a converging mirror so that an inverted image of the key is formed on the same side of the mirror. (The object distance is greater than the focal distance.) The two cases are identical except that the key and mirror are in air in Case A, and in water in Case B.

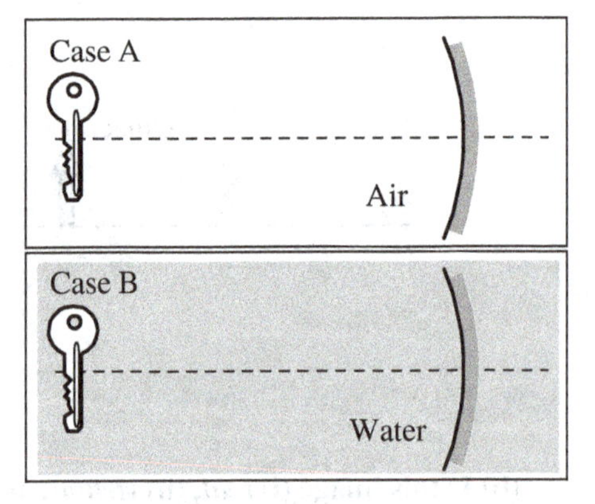

Is the distance from the mirror to the image of the key (the image distance) (i) ***greater in Case A*****, (ii)** ***greater in Case B*****, or (iii)** ***the same in both cases*****?** _____

Explain your reasoning.

E2-TT25: Positive Focal Length Lens—Image Type

A student makes the following contention:

"If a positive focal length lens produces an image that is larger than the object, then the image will be virtual."

There is a problem with this contention; identify the problem and explain how to correct it.

E2-SCT26: REMOVING A CONVERGING LENS—IMAGE

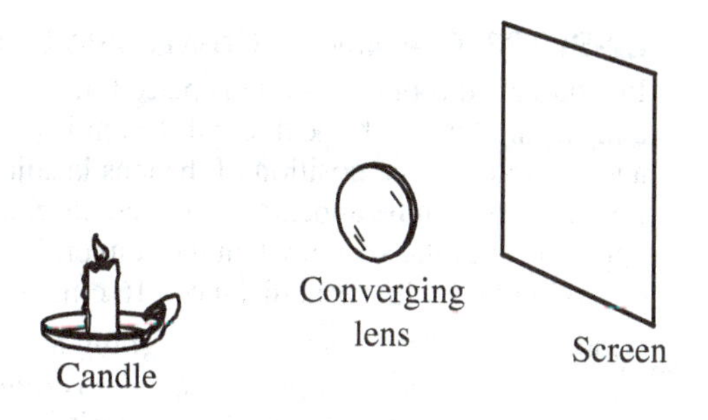

In a darkened room, a candle is placed in front of a white screen, and a converging lens with focal distance 16 cm is placed between the candle and the screen. The position of the lens is adjusted until an inverted image of the candle appears on the screen. The lens is then removed. Three students are discussing the appearance of the screen after the lens is removed:

Adan: *"The lens was bending the light to form the image, and when you remove it, all that's left is the light from the candle lighting up the screen. But there won't be an image of the candle anymore."*

Belmiro: *"The lens was turning the image of the candle upside down, and when you remove it, the image of the candle will be right side up. The light from the tip of the flame now travels in a straight line to the top part of the screen, and the light from the bottom of the candle travels a straight line to the bottom of the screen."*

Chanthana: *"I agree that you can still get an image, and it will be right side up because the lens isn't turning it upside down any more. But I think without the lens to bend the rays, the image will be a lot farther away. Without moving the screen you'll get a fuzzy image of the candle, and you have to move the screen away from the candle to see an upright image of the candle."*

With which, if any, of these students do you agree?

Adan _____ Belmiro _____ Chanthana _____ None of them _____

Explain your reasoning.

E2-SCT27: BLOCKED LENS—IMAGE

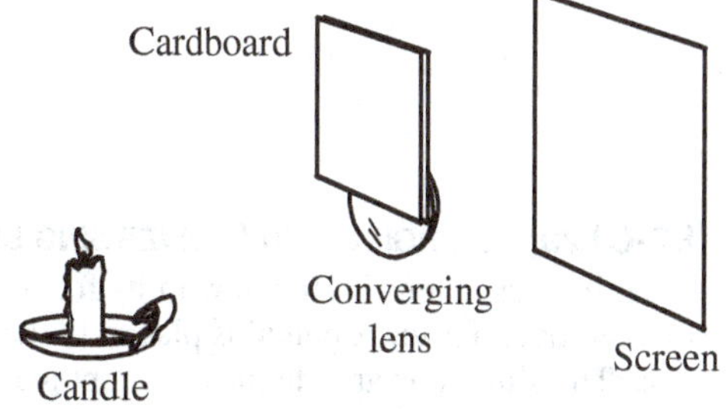

In a darkened room, a candle is placed in front of a white screen, and a converging lens is placed between the candle and the screen. The position of the lens is adjusted until an inverted image of the candle appears on the screen. A piece of cardboard is then placed in front of the lens so that it covers the top half of the lens. Four students are discussing what will happen to the image of the candle when the cardboard is placed in front of the lens:

Ajit: *"No image of the candle will form. The image location is determined by the principal rays, and in this case the ray from the candle parallel to the optical axis has been blocked."*

Binh: *"I think the image doesn't change at all. The bottom half of the lens still works like it did before, and it focuses the light from the candle onto the screen. The image might be dimmer, though, because there is less total light converging to form the image."*

Colette: *"Only the image from the top half of the candle has been blocked. The light from the bottom half still forms an image. What you'll see on the screen is an image of the candle holder and the bottom half of the candle, but not of the flame."*

Diamontina: *"I think you'll only see an image of the top half of the candle. The lens inverts the image, so when you block the top half of the lens, you block the image of the bottom half of the candle."*

With which, if any, of these students do you agree?

Ajit _____ Binh _____ Colette _____ Diamontina _____ None of them _____

Explain your reasoning.

E2-SCT28: Replacing a Converging Lens with a Diverging Lens—Image

In a darkened room, a candle is placed in front of a white screen, and a converging lens with focal length 16 cm is placed between the candle and the screen. The position of the lens is adjusted until an inverted image of the candle appears on the screen. Students are discussing the appearance of the screen when the converging lens is replaced with a diverging lens with focal distance –16 cm:

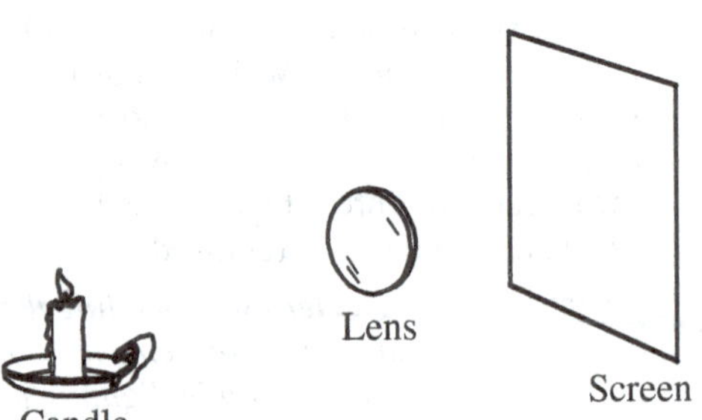

Alonzo: *"The same image is going to form, except this time it will be right side up. Converging lenses invert images, and diverging lenses don't."*

Bonifacio: *"The image will still form, at the same place because the focal distances are the same, but it will be smaller. The image from a diverging lens is always smaller than the object, whereas the image from a converging lens could be smaller or larger."*

Carlita: *"The image is going to be formed on the same side of the screen as the candle, because the focal distance is negative. Also, the negative sign makes the image noninverted instead of inverted. If we move the screen to the front of the lens we'll see the image."*

Dominic: *"I think Carlita is right that the image is on the same side of the lens as the candle, but I don't think you'll see it if you put the screen there."*

With which, if any, of these students do you agree?

Alonzo _____ Bonifacio _____ Carlita _____ Dominic _____ None of them _____

Explain your reasoning.

E2-CT29: Diverging and Converging Lenses—Distance from Object to Image

In both cases, a lens is shown with its focal points marked over a grid. A pencil is placed to the left of the lens. The situations are identical, except that the lens in Case A is a diverging lens and the lens in Case B is a converging lens.

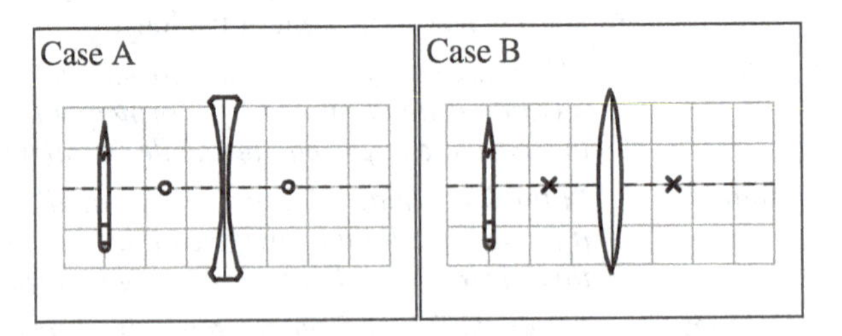

Is the distance from the pencil to the image of the pencil (i) *greater in Case A*, (ii) *greater in Case B*, or (iii) *the same in both cases*? _____

Explain your reasoning.

E2-CT30: CONVERGING LENSES—FOCAL DISTANCE

In both cases, an object is placed 20 cm from a converging lens. The image position in Case A is to the left of the lens (and also to the left of the object), while in Case B the image position is to the right of the lens.

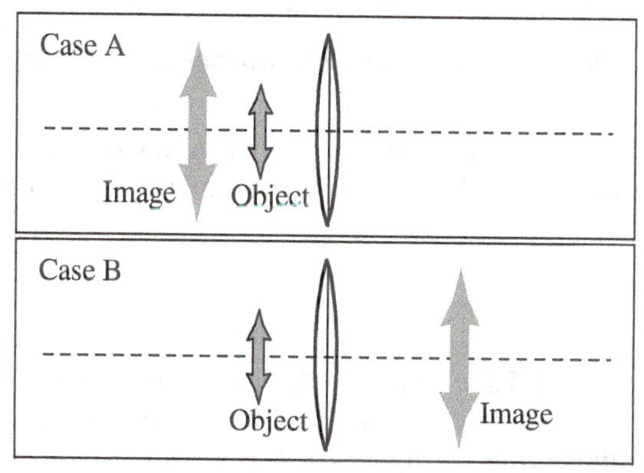

Is the focal distance of the lens (i) *greater in Case A*, (ii) *greater in Case B*, or (iii) *the same in both cases*? _____

Explain your reasoning.

E2-RT31: LENSES AND MIRRORS—FOCAL LENGTH IN WATER

The figures below show converging (A) and diverging (B) thin lenses as well as diverging (C) and converging (D) mirrors. The lenses are made of glass with an index of refraction of 1.6. All lenses and mirrors have the same magnitude focal length when placed in air (n = 1).

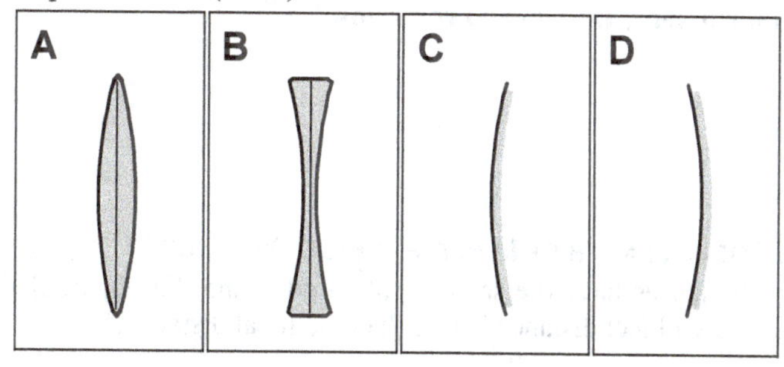

Rank the focal lengths of the mirrors and lenses when they are placed in water (n = 1.33).

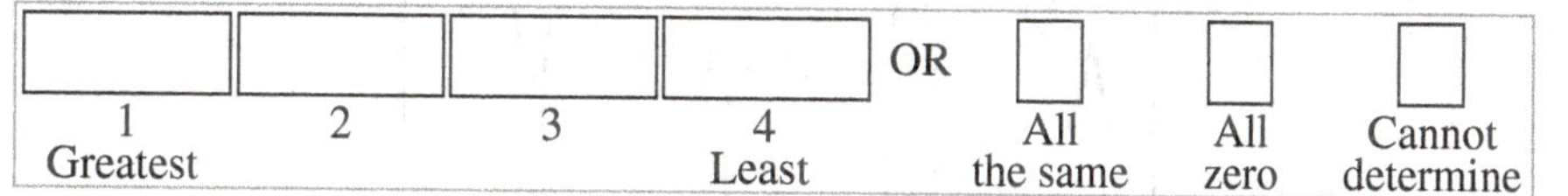

Explain your reasoning.

TIPERs

E2-WWT32: Single Lens—Focal Length

A student makes the following contention:

"If a single lens can only produce images that are smaller than the objects involved, then the lens is a negative focal length lens."

What, if anything, is wrong with this statement? If something is wrong, identify it and explain how to correct all errors. If this statement is correct, explain why.

E2-QRT33: Rays Emerging from a Lens I—Lens and Image Type

The vertical rectangle in the figure represents a thin lens. The dashed line is the optical axis of that lens. Two of the three principal light rays are shown emerging from the lens. These light rays originated at a single point on an object placed to the left of the lens.

(a) What type of lens is this, and how do you know?

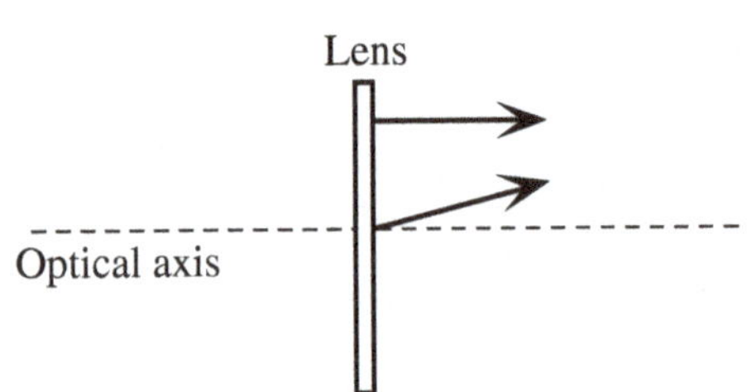

(b) Draw the third ray that is commonly used to find the image.

(c) What type of image will form here, and how do you know?

E2-CT34: Objects Inside Focal Length of Lenses—Image Distance I

In the two situations shown, the lenses have the same focal lengths, and the object distance from the lens to the arrow is the same. In both cases the object distance is less than the focal distance.

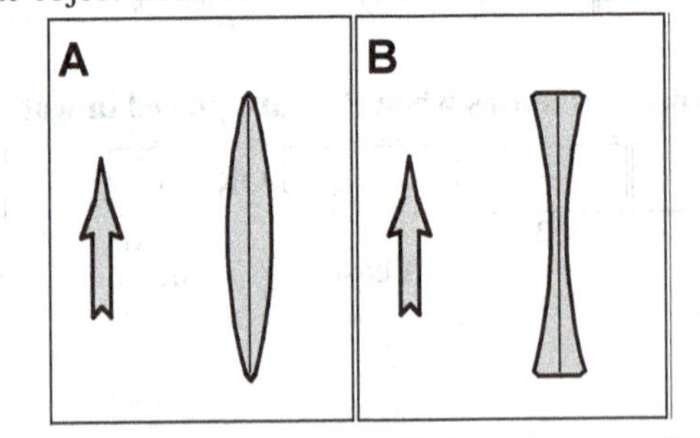

Will the image distance for Case A be (i) *greater than*, (ii) *less than*, or (iii) *equal* to the image distance for Case B? _____

Explain your reasoning.

E2-QRT35: RED AND BLUE LIGHTS IN FRONT OF LENS—IMAGE ON SCREEN

Light from two small bulbs, one red and one blue, passes through a lens to form an image of the bulbs on a screen. Without changing the location of the bulbs, the lens, or the screen, a piece of cardboard is placed so that it covers half of the lens, as shown in the side view diagram.

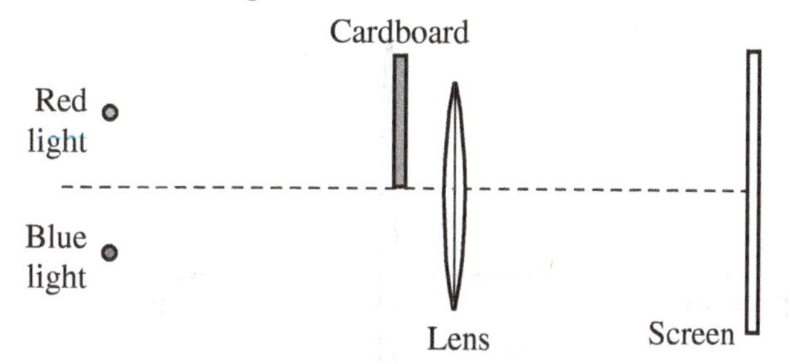

Which choice below best represents what will appear on the screen?

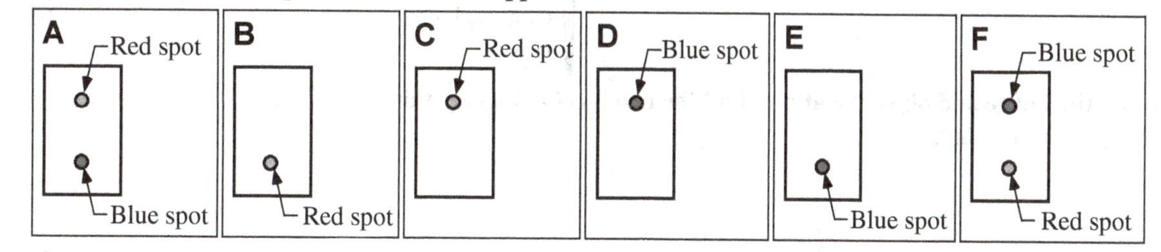

Explain your reasoning.

E2-WBT36: OBJECT AND UPRIGHT IMAGE FOR A DIVERGING LENS—FOCAL POINT

An object is placed in front of a diverging lens. An upright image of the object is formed to the left of the lens at the location shown.

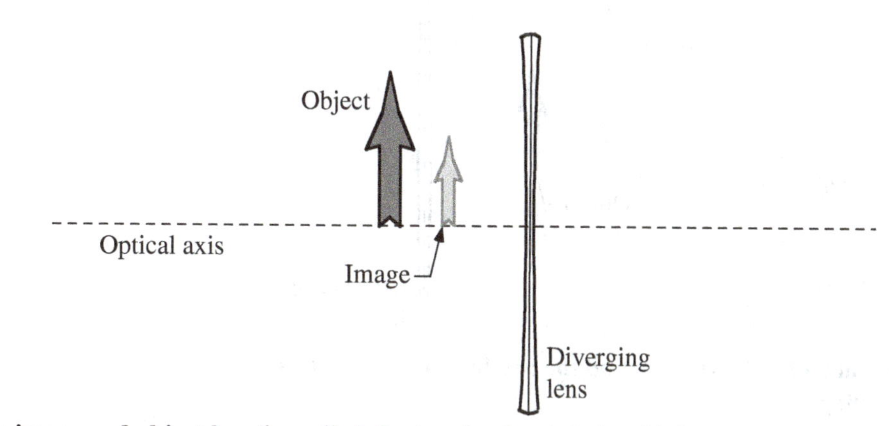

Based on the image and object locations, find the two focal points for this lens.
Explain your reasoning.

E2-WBT37: Object and Inverted Image for a Converging Lens—Focal Point

An object is placed in front of a converging lens. An inverted image of the object is formed to the right of the lens at the location shown.

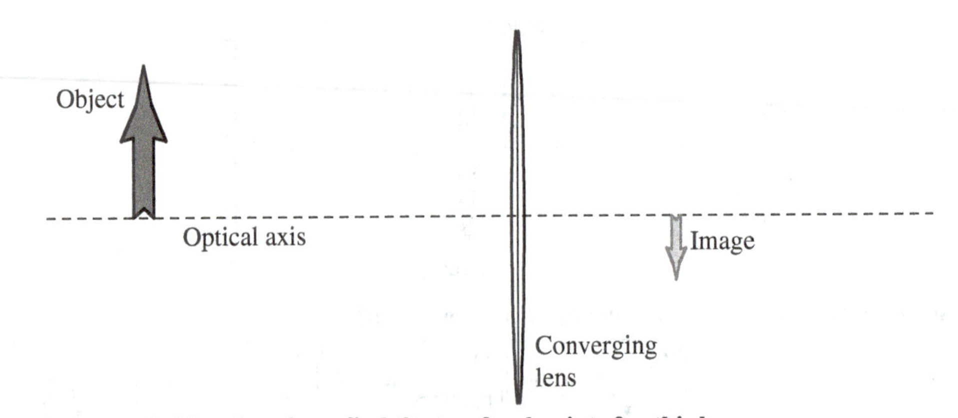

Based on the image and object locations, find the two focal points for this lens.
Explain your reasoning.

E2-WBT38: Object and Upright Image for a Converging Lens—Focal Point

An object is placed in front of a converging lens. An upright image of the object is formed to the left of the lens at the location shown.

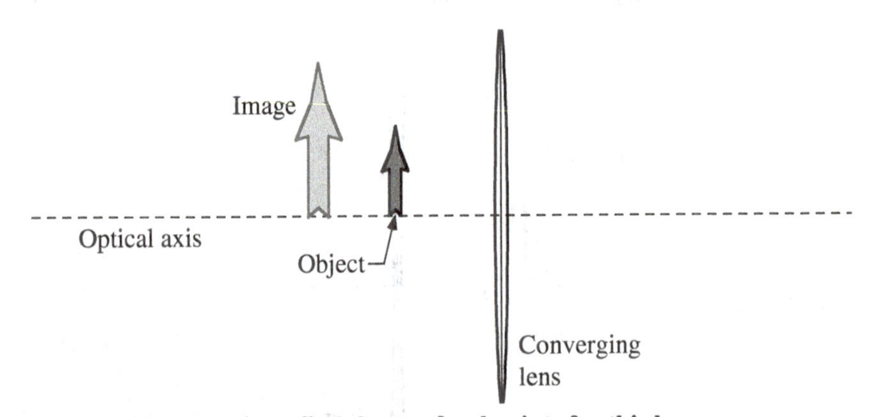

Based on the image and object locations, find the two focal points for this lens.
Explain your reasoning.

E2-WBT39: IMAGE FOR A CONVERGING LENS—OBJECT

An arrow is placed to the left of a converging lens. An inverted image of the arrow is formed to the right of the mirror at the location shown.

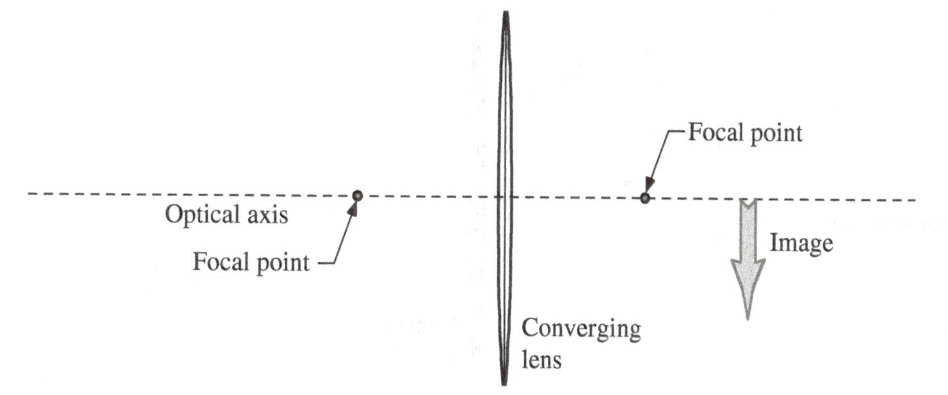

Based on the location of its image, find the location of the arrow.
Explain your reasoning.

E2-CT40: OBJECTS OUTSIDE FOCAL LENGTH OF LENSES—IMAGE DISTANCE

In the two situations shown, the lenses both have the same focal lengths but opposite signs, and the object distance from the lens to the arrow is the same. In both cases the object distance is greater than the focal distance.

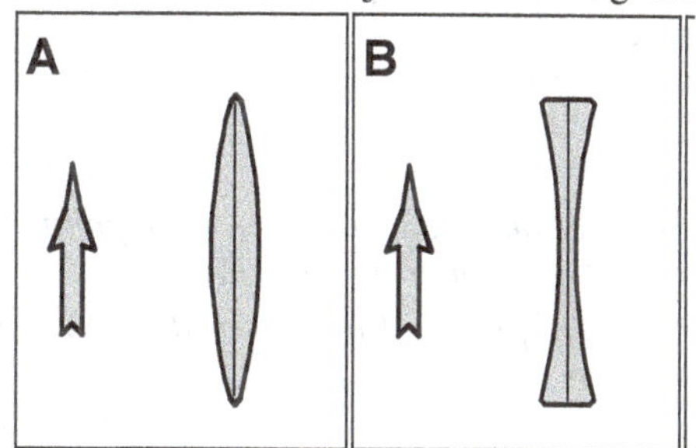

Will the image distance for Case A be (i) *greater than*, (ii) *less than*, or (iii) *equal to* the image distance for Case B? _____
Explain your reasoning.

E2-WBT41: Image for a Converging Lens—Object

An arrow is placed to the left of a converging lens. An inverted image of the arrow is formed to the right of the mirror at the location shown.

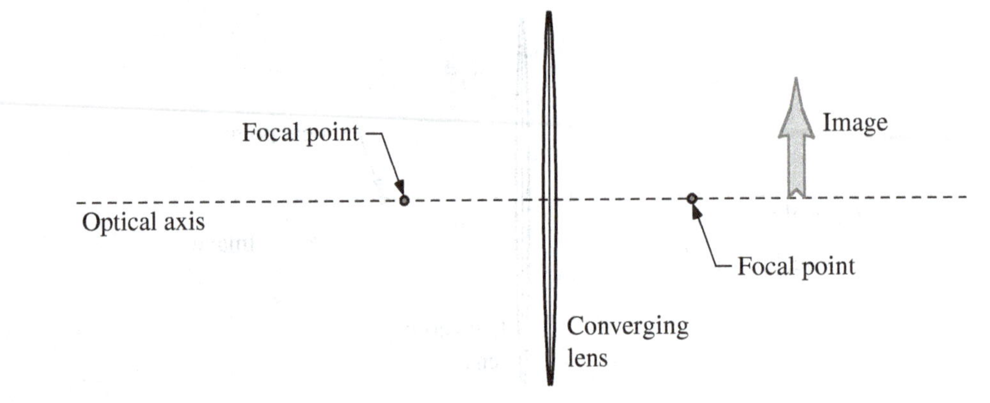

Based on the location of its image, find the location of the arrow.
Explain your reasoning.

E2-SCT42: Emerging Rays—Type of Lens

The vertical rectangle in the figure represents a thin lens. The dashed line is the optical axis of that lens. Two light rays are shown emerging from the lens. These light rays originated at a single point on an object placed to the left of the lens. Three students discussing this figure make the following contentions.

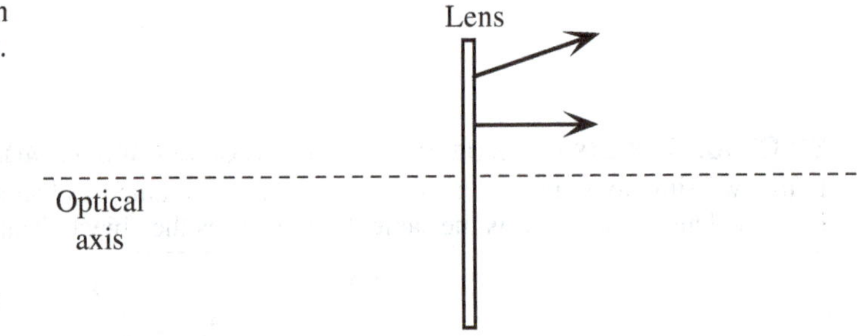

Anna: *"I think this has to be a diverging (negative) lens since the two rays are spreading out after passing through the lens."*

Brandon: *"I don't think we can say this since an object inside the focal point of a converging (positive) lens would also have rays that spread apart after passing through the lens."*

Carlos: *"Well, if these are two of the three principal rays, then this has to be a negative lens, but if they are other rays I'm not sure we can say."*

With which, if any, of these students do you agree?

Anna _____ Brandon _____ Carlos _____ None of them _____

Explain your reasoning.

E2-QRT43: Image and Object Locations Relative to Lenses—Image Type

A student determines the image positions shown for the three setups below with the objects placed on the left side of the lens.

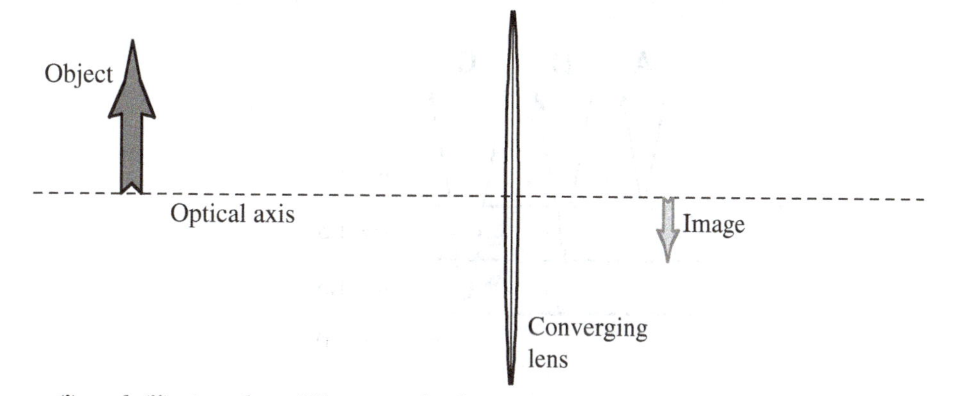

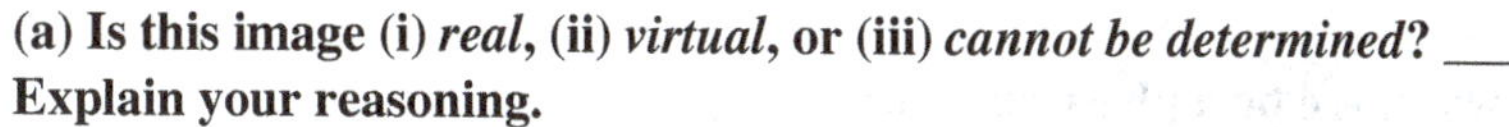

(a) Is this image (i) *real*, (ii) *virtual*, or (iii) *cannot be determined*? _____
Explain your reasoning.

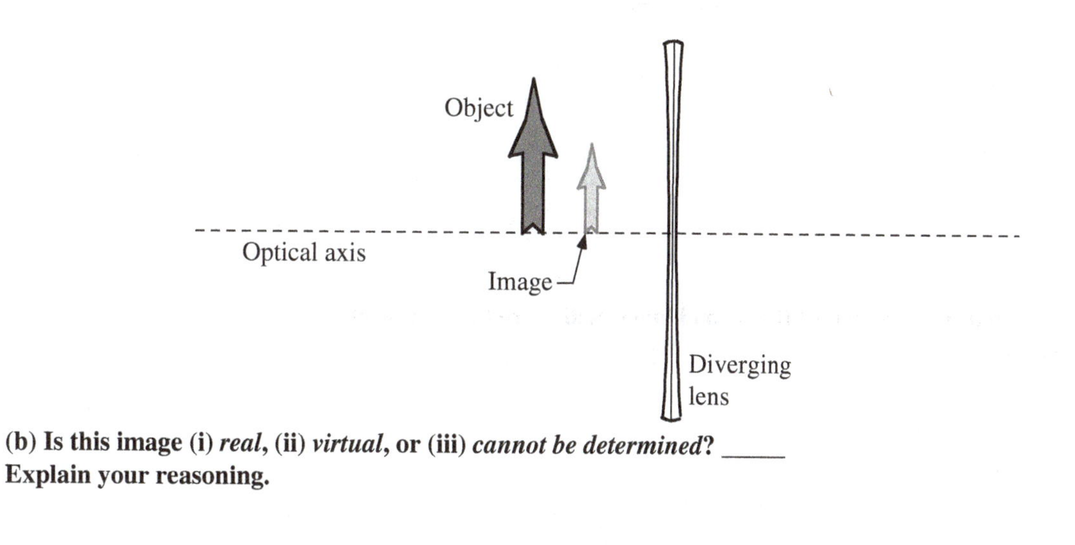

(b) Is this image (i) *real*, (ii) *virtual*, or (iii) *cannot be determined*? _____
Explain your reasoning.

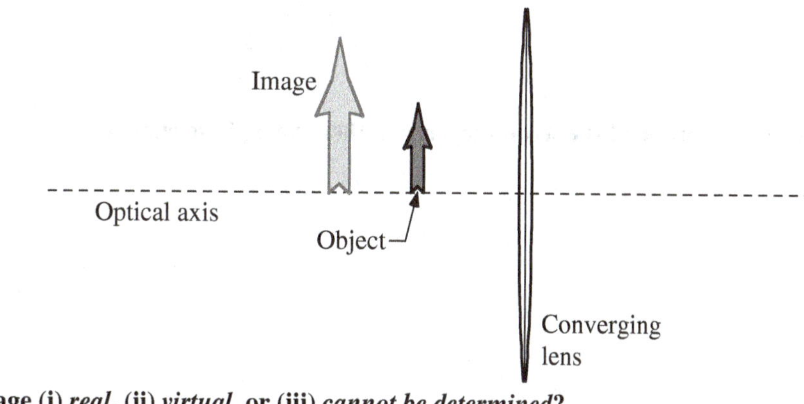

(c) Is this image (i) *real*, (ii) *virtual*, or (iii) *cannot be determined*? _____
Explain your reasoning.

E2-QRT44: Thin Films in Air—Phase Changes

A beam of light is reflected from a thin film made of two layers of different materials. The figure shows what happens to three of the incident rays. The light is initially traveling in air.

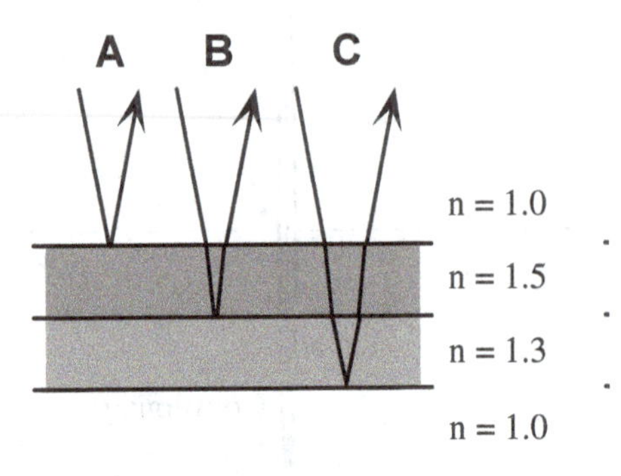

When ray A is reflected from the top layer, will there be a phase change?
Explain your reasoning.

When ray B is reflected from the top of the second layer, will there be a phase change?
Explain your reasoning.

When ray C is reflected from the bottom of the second layer, will there be a phase change?
Explain your reasoning.

E2-QRT45: THIN FILMS IMMERSED IN WATER—PHASE CHANGES

In each case, a light ray is reflected from a thin film made of two layers of different materials immersed in water with a refraction index of 1.3. In all three cases, the incident ray is in water.

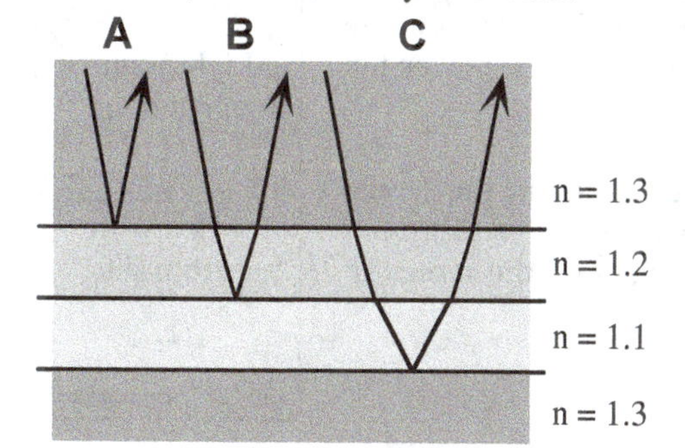

When ray A is reflected from the top layer, will there be a phase change?
Explain your reasoning.

When ray B is reflected from the top of the second layer, will there be a phase change?
Explain your reasoning.

When ray C is reflected from the bottom of the second layer, will there be a phase change?
Explain your reasoning.

E2-CT46: Two Lasers and Two Slits—Bright Fringe Separation

In each case, a beam of light from a laser passes through very narrow slits cut in a mask. The light then hits a distant screen, creating an interference pattern. The two cases are identical except that in Case A the light beam is red, and in Case B it is blue.

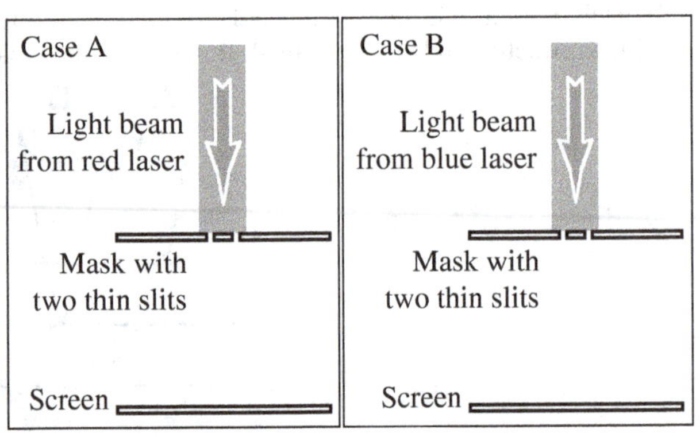

Will the distance between bright fringes on the screen in Case A be (i) *greater than*, (ii) *less than*, or (iii) *equal to* the distance between bright fringes on the screen in Case B? _____

Explain your reasoning.

E3 SOUND

E3-RT01: POLICE CAR AND MOTORCYCLE—SIREN FREQUENCY

A police car with a 600 Hz siren is traveling along the same street as a motorcycle. The velocities of the two vehicles and the distance between them are given in each figure.

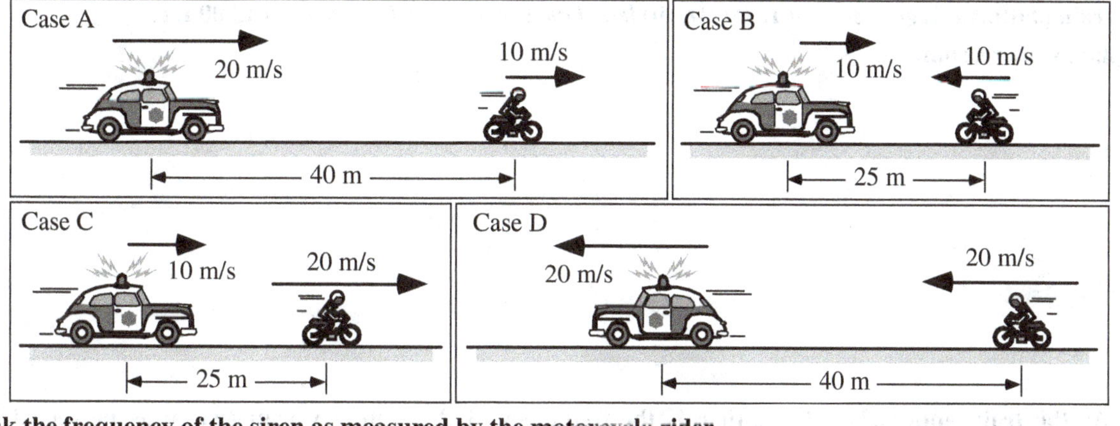

Rank the frequency of the siren as measured by the motorcycle rider.

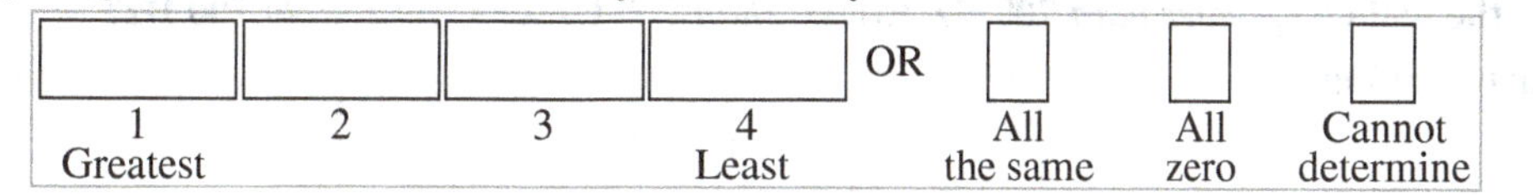

Explain your reasoning.

E3-SCT02: TRAIN APPROACHING OBSERVER AT STATION—FREQUENCY HEARD

A train approaches a station at a constant speed, sounding its whistle continuously. Three students are discussing what an observer standing at a station would hear as the train is approaching:

Anish: *"The train is not accelerating or decelerating. I think the observer will hear a constant pitch that matches the pitch of the whistle."*

Brooke: *"Even if it isn't accelerating, the observer will hear a higher pitch than the whistle actually emits since the train is moving toward the observer. I agree that the observer will hear a constant pitch."*

Cruz: *"I agree with Brooke that the observer will hear a higher pitch, but I think the observer will also hear the frequency increase constantly as the train gets closer and closer."*

With which, if any, of these students do you agree?

Anish _____ Brooke _____ Cruz _____ None of them ______

Explain your reasoning.

TIPERs

E3-CT03 Moving Train Whistle and Stationary Speaker—Perceived Frequency

A train with a 1,000 Hz whistle passes by a train station at a constant speed of 30 m/s. A speaker on the station platform emits a 700 Hz warning siren as the train approaches.

(a) As the train approaches the station, is the frequency of the train whistle as perceived by an observer on the train platform (i) *greater than* 1,000 Hz, (ii) *less than* 1,000 Hz, or (iii) *equal* to 1,000 Hz? _____

Explain your reasoning.

(b) As the train approaches the station, is the frequency of the station warning siren as perceived by a passenger on the train (i) *greater than* 700 Hz, (ii) *less than* 700 Hz, or (iii) *equal to* 700 Hz? _____

Explain your reasoning.

(c) As the train moves away from the station, is the frequency of the train whistle as perceived by an observer on the train platform (i) *greater than* 1,000 Hz, (ii) *less than* 1,000 Hz, or (iii) *equal to* 1,000 Hz? _____

Explain your reasoning.

(d) As the train moves away from the station, is the frequency of the train whistle as perceived by a passenger on the train (i) *greater than* 1,000 Hz, (ii) *less than* 1,000 Hz, or (iii) *equal to* 1,000 Hz? _____

Explain your reasoning.

E3-QRT04: VIBRATING GUITAR STRING AND TUNING FORK—FREQUENCY

A guitar string is strummed near a tuning fork that has a frequency of 512 Hz. Initially, the guitar and tuning fork together create a sound wave with a beat frequency of 5 Hz. The tension in the guitar string is then increased, after which the guitar and tuning fork together create a sound wave with a beat frequency of 4 Hz.

(a) Before the tension in the guitar string is increased, is the frequency of the guitar string (i) *greater than* 512 Hz, (ii) *less than* 512 Hz, or (iii) *equal to* 512 Hz? If it is not possible to determine how the guitar frequency compared to 512 Hz, state that explicitly.

Explain your reasoning.

(b) After the tension in the guitar string is increased, is the frequency of the guitar string (i) *greater than* 512 Hz, (ii) *less than* 512 Hz, or (iii) *equal to* 512 Hz? If it is not possible to determine how the guitar frequency compares to 512 Hz, state that explicitly.

Explain your reasoning.

E3-WWT05: SOUND LEVELS AND DISTANCE—COMPARISON

A student learns that the intensity of sound diminishes as the square of the distance from the source. He states:

"If a sound from a horn is 40 dB at a distance of 3 meters from the source, then it will be only 20 dB at a distance of 6 meters."

What, if anything, is wrong with this statement? If something is wrong, identify it and explain how to correct it. If this statement is correct, explain why.

TIPERs

E3-WWT06: Sound Wave Velocity and Frequency—Distance

A student states:

"The distance between a compression and the next compression for a sound wave with a velocity of 800 ft/s and a frequency of 400 Hz is 2 feet. If the wave had a higher frequency, it would travel faster."

What, if anything, is wrong with this statement? If something is wrong, identify it and explain how to correct it. If this statement is correct, explain why.

E3-SCT07: Clarinet and Saxophone Playing Same Note—Difference

Three students are discussing why a clarinet and saxophone playing the same note can be distinguished from one another. They state:

Anisha: *"I think this is due to the difference in pitch between the two."*

Blanca: *"If they are playing the same note that means they have the same pitch. I think the difference in the way the sound is created results in differences in the velocity of the sound waves, and our ear detects these differences."*

Cristobal: *"I think the real difference is that even though the frequencies are the same, the shapes of the waves are different, and our ear detects this."*

Dawn: *"Neither instrument is really playing a single pure tone—for example, they each are also emitting sound an octave above the note they are playing. It's the differences in these overtones that allow us to tell the difference."*

With which, if any, of these students do you agree?

Anisha _____ Blanca _____ Cristobal _____ Dawn _____ None of them _____

Explain your reasoning.

E3-RT08: TWO DIFFERENT WAVE SOURCES—BEAT FREQUENCY AT VARIOUS LOCATIONS

The point source on the left is generating a 600 Hz sound wave. The point source on the right is creating a 608 Hz sound wave.

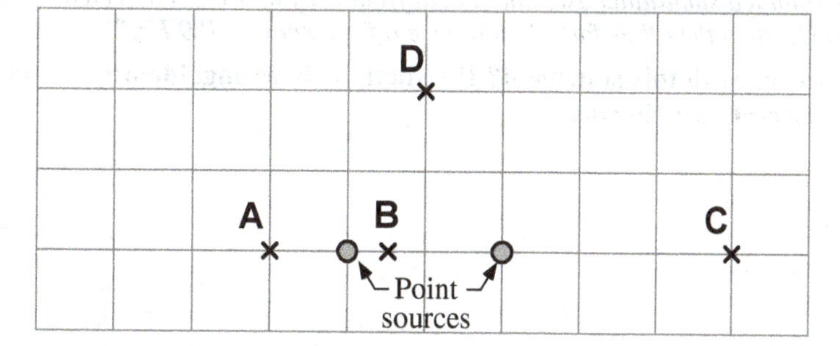

Rank the beat frequency at the labeled points.

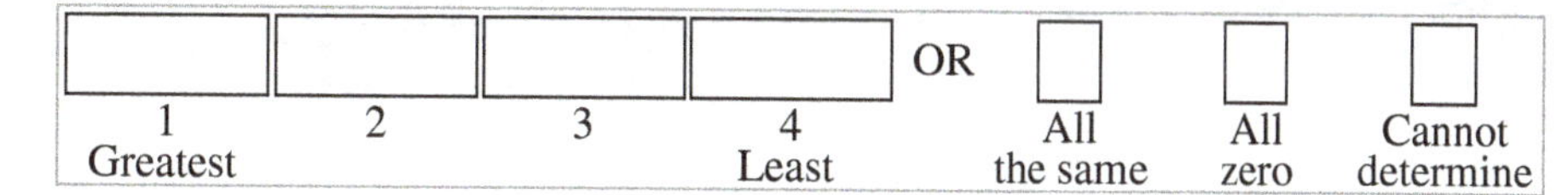

Explain your reasoning.

E3-WWT09: Two Tuning Forks Producing Beats—Frequency

A student states:

"Two tuning forks are sounded simultaneously and a beat frequency of 9 Hz is detected. If one of the tuning forks has a frequency of 480 Hz, then the other fork should have a frequency of 489 Hz."

What, if anything, is wrong with this statement? If something is wrong, identify it and explain how to correct it. If this statement is correct, explain why.

E3-CT10: Three-Dimensional Locations near a Sound Source—Loudness

A point sound source emits sound waves in every direction. The point source is located at the origin, and points *A–E* are located on the corners of an imaginary cube, as shown.

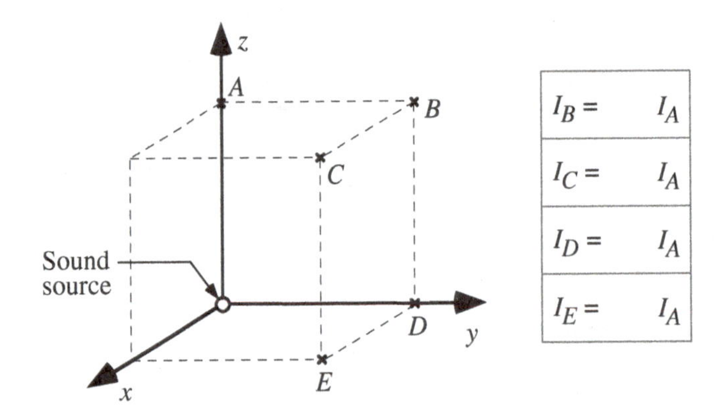

$I_B =$	I_A
$I_C =$	I_A
$I_D =$	I_A
$I_E =$	I_A

If the intensity (loudness) at the point labeled *A* is I_A, in the table to the right of the figure express the intensity at the other labeled points in terms of I_A.
Explain your reasoning.

E3-WWT11: TROMBONE WITH SOAP BUBBLE SOLUTION—TIME TO BURST BUBBLE

A student dips the bell of her trombone into a soap bubble, and makes a prediction about the soap film covering the bell:

"The speed of sound is about 340 meters per second. The air from the trombone will push the film outward pretty quickly. It will only take a fraction of a second for the film to burst after I start playing."

What, if anything, is wrong with this statement? If something is wrong, identify it and explain how to correct it. If this statement is correct, explain why.

E3-RT12: SOUND WAVES WITH SAME WAVELENGTH—WAVE SPEED

Sound waves, all with the same wavelength, are traveling in the same direction through a pool of water. The graphs show the amplitude of the waves as a function of their position. All graphs have the same scale.

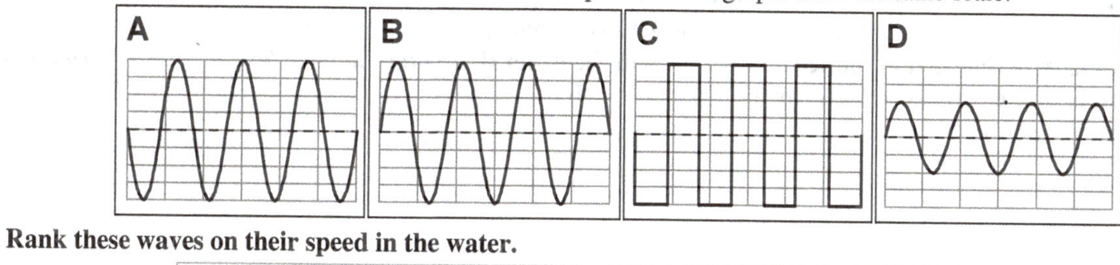

Rank these waves on their speed in the water.

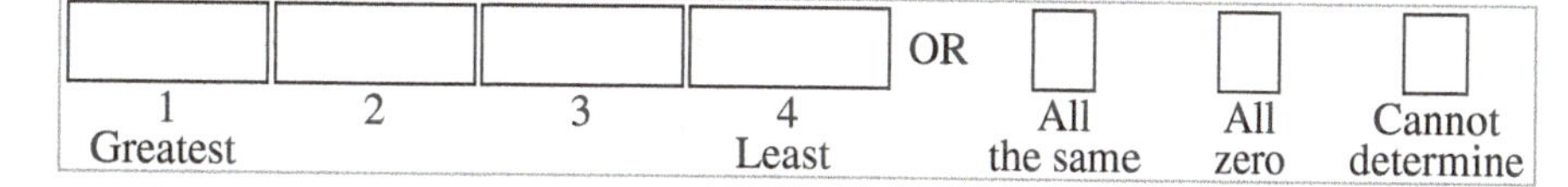

Explain your reasoning.

TIPERs

E3-CT13: Dropped Speaker—Tone in Water

A waterproof siren is emitting sound at a single frequency. The siren is dropped from air into water.

Is the frequency of the waves produced by the siren in the water (i) *greater than*, (ii) *less than*, or (iii) *equal to* the frequency of the waves produced by the siren in the air? _____

Explain your reasoning.

E3-CT14: Dropped Siren—Wavelength in Water

A waterproof siren is emitting sound at a single frequency. The siren is dropped from air into water.

Is the wavelength of the waves produced by the siren in the water (i) *greater than*, (ii) *less than*, or (iii) *equal to* the wavelength of the waves produced by the siren in the air? _____

Explain your reasoning.

E3-WWT15: PIPE ORGAN—FUNDAMENTAL FREQUENCY

After reading that the speed of sound increases as the temperature increases, a student states:

"That's why an instrument needs to be retuned when it gets warm. From the wave equation, when the velocity increases, the frequency also has to increase, since the frequency is proportional to the velocity. Without retuning, the instrument is going to sound sharp."

What, if anything, is wrong with this statement? If something is wrong, identify it and explain how to correct it. If this statement is correct, explain why.

E3-CRT16: PIPE CLOSED AT ONE END—SOUND FREQUENCY, WAVELENGTH, AND VELOCITY

A pipe of length 80 cm is closed at one end and open at the other. Sound is created in the pipe at four different frequencies. The diagram shows the location of nodes (N) and antinodes (A) in the pipe for the four different modes. The table to the right has an entry for wave speed of the second overtone.

Mode	Frequency	Wavelength	Wave speed
Fundamental			
First overtone (Second harmonic)			
Second overtone (Third harmonic)			340 m/s
Third overtone (Fourth harmonic)			

80 cm

Complete the table of frequencies, wavelengths, and wave speeds for the four modes.

Explain your reasoning.

E3-CRT17: PIPE OPEN AT BOTH ENDS—SOUND FREQUENCY, WAVELENGTH, AND VELOCITY

A pipe of length L is open at both ends. Sound is created in the pipe at four different frequencies. The diagram shows the location of nodes (N) and antinodes (A) in the pipe for the four different modes. The table to the right has an entry for wave speed of the first overtone, and an entry for the wavelength of the second overtone.

Use the given information to find the length L of the pipe. Then complete the table of frequencies, wavelengths, and wave speeds for the four modes.

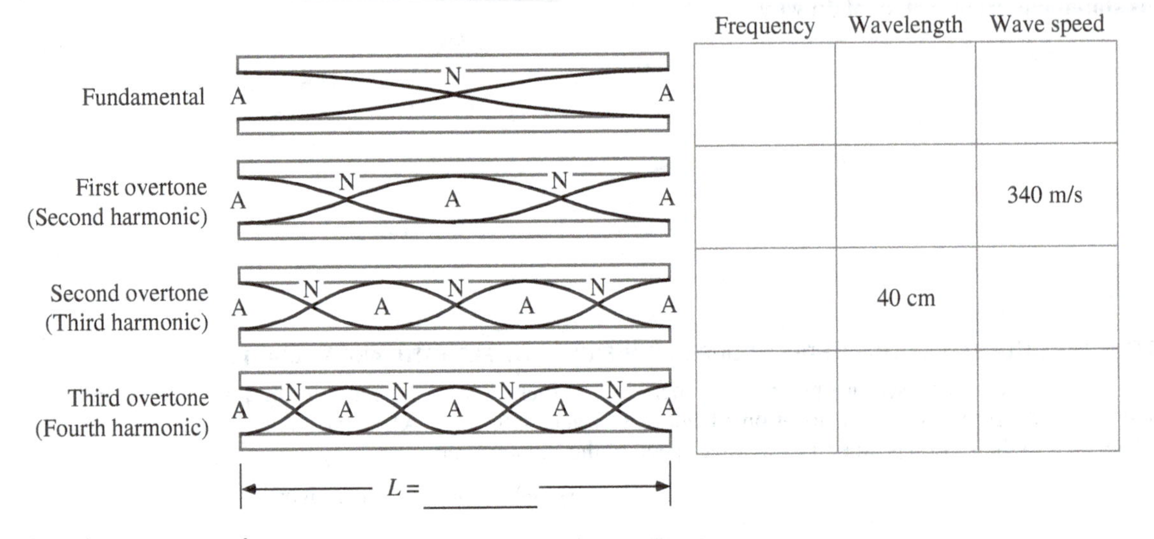

Frequency	Wavelength	Wave speed
		340 m/s
	40 cm	

Explain your reasoning.

F MODERN PHYSICS

F1-SCT01: RADIATION FROM HEATED OBJECTS—COLOR EMITTED

A 1-kg ceramic cube and a 1-kg iron cube both have holes drilled into their sides. When the cubes are heated to the same high temperature, light is emitted through the holes from the interior of the cubes. Three students make the following contentions about the color of the emitted light:

Arkady: *"The color of the light depends on what color the cubes were before they were heated. Most ceramic is white to start with, and iron is dull gray. After they are heated, the ceramic will glow with white light, and the iron will be a dull red like an electric stove element."*

Brigid: *"It takes less energy to heat a metal to a high temperature than it does to heat an equal amount of ceramic to the same temperature. Since less energy was put into the metal, it will glow less brightly and will be a duller red than the ceramic, which will be whiter in color."*

Chen: *"If the cubes are at the same temperature, the molecules in the interior have the same average kinetic energies, and they will emit the same color of light."*

With which, if any, of these students do you agree?

Arkady _____ Brigid _____ Chen _____ None of them _____

Explain your reasoning.

F1-CT02: ULTRAVIOLET LIGHT ON METAL DISCS—WORK FUNCTION

A beam of ultraviolet light shines on a metal disc, causing electrons to be emitted from the disc. The two cases are identical, except that the metals are different and the emitted electrons have a higher maximum speed in Case A than in Case B.

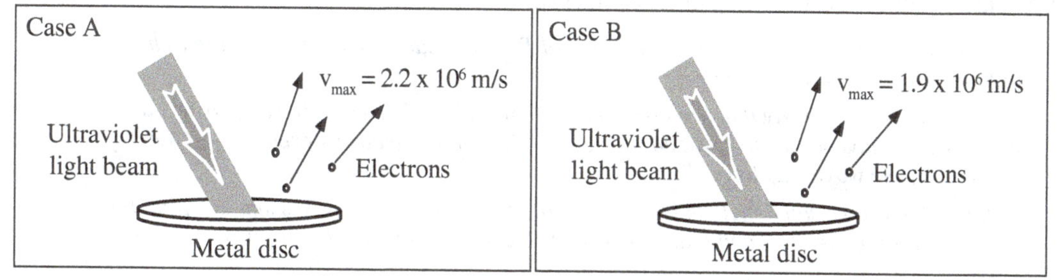

Is the work function of the metal (i) *greater in Case A*, (ii) *greater in Case B*, or (iii) *the same in both cases*? _____

Explain your reasoning.

F1-CT03: Two Illuminated Metals—Work Function

A beam of ultraviolet light shines on a metal disc, causing electrons to be emitted from the disc. The wavelength of the beam is the same in both cases, but the intensities are different. More electrons are ejected from metal A than from metal B, but the electrons ejected from metal B have a higher maximum speed.

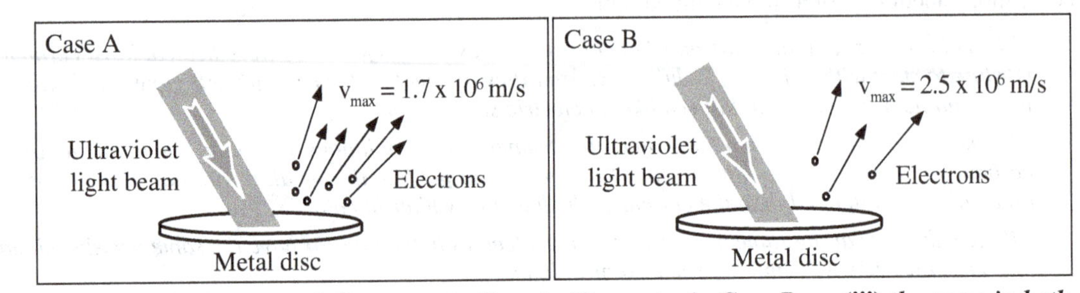

Is the work function of the metal (i) *greater in Case A*, (ii) *greater in Case B*, or (iii) *the same in both cases*? _____

Explain your reasoning.

F1-SCT04: Photoelectric Effect Investigations—More Electrons

In two experiments, electromagnetic waves are used to eject electrons from a metal. The electromagnetic waves have a longer wavelength in experiment A than in experiment B. More electrons were ejected from metal B than from metal A. Three students are discussing the experiments:

Arturo: *"Since more electrons were ejected from metal B, that means the intensity of the light used in that investigation was higher."*

Bonifacio: *"I don't think we can say that for sure. Since the wavelength used in B was shorter, those waves would have more energy, and they could eject more electrons even though the intensity of the wave was lower."*

Carla: *"I think that all we can conclude is that the work function for metal B is smaller than the work function of metal A, and that is why more electrons were ejected from B."*

With which, if any, of these students do you agree?

Arturo _____ Bonifacio _____ Carla _____ None of them _____

Explain your reasoning.

F1-RT05: ULTRAVIOLET LIGHT INCIDENT ON NICKEL—NUMBER OF EJECTED ELECTRONS

A nickel disc emits electrons when it is illuminated with a beam of ultraviolet light. The frequency of the light and the intensity of the light beam are given for each case.

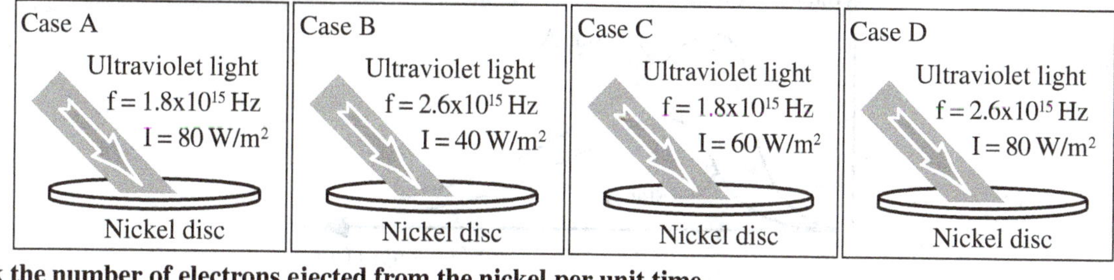

Rank the number of electrons ejected from the nickel per unit time.

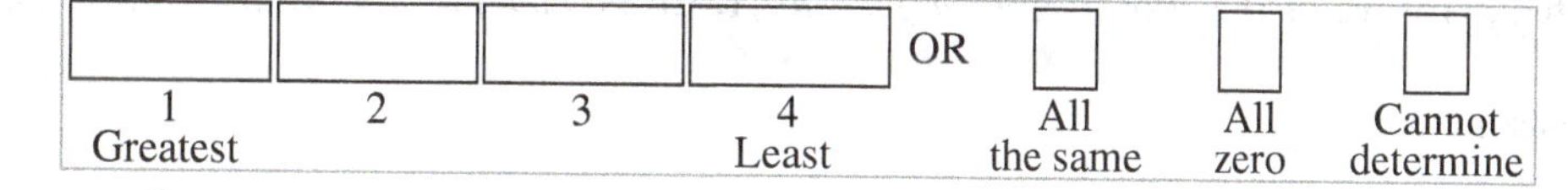

Explain your reasoning.

F1-RT06: ULTRAVIOLET LIGHT INCIDENT ON NICKEL—EJECTED ELECTRON SPEED

A nickel disc emits electrons when it is illuminated with a beam of ultraviolet light. The frequency of the light and the intensity of the light beam are given for each case.

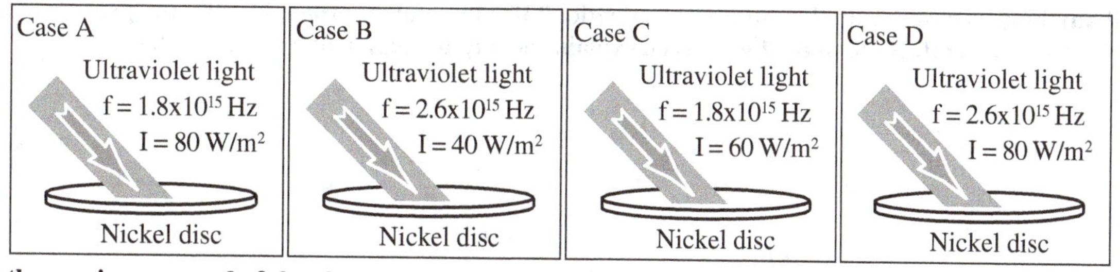

Rank the maximum speed of the electrons ejected from nickel.

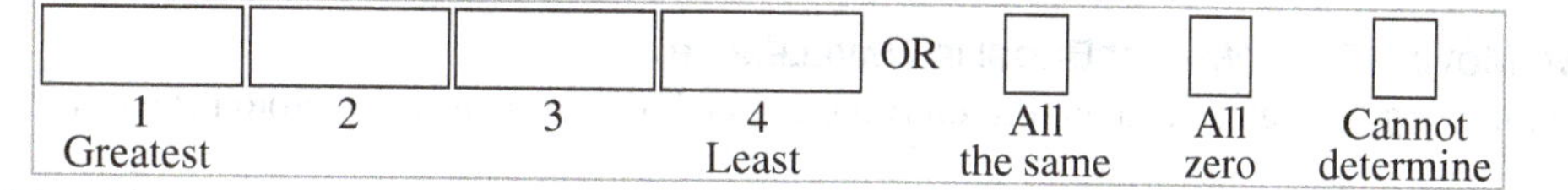

Explain your reasoning.

TIPERs

F1-CT07: BLACKBODY CURVES OF STARS—SURFACE TEMPERATURE

A graph of intensity of emitted radiation as a function of wavelength (a blackbody curve) is shown for two stars.

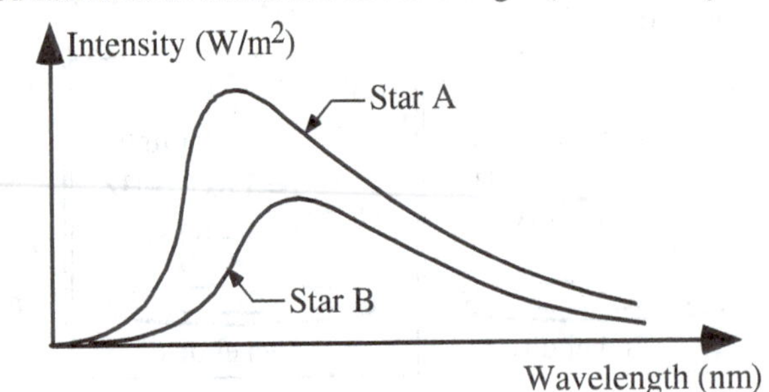

Is the surface temperature of the star (i) *greater for star A*, (ii) *greater for star B*, or (iii) *the same for both stars*? _____

Explain your reasoning.

F1-WWT08: HIGH-FREQUENCY X-RAYS—ENERGETIC PHOTONS

A student contends:

"Higher frequency X-rays contain more energetic photons than lower frequency electromagnetic waves because higher frequency waves have more energy."

What, if anything, is wrong with this student's contention? If something is wrong, identify the problem and explain how to correct it. If the student is correct, explain the physics supporting his/her statement.

F1-CT09: TWO MOVING PROTONS—DEBROGLIE WAVELENGTH

Two protons are moving through a vacuum. Proton A has a speed of 4×10^4 m/s, and proton B has a speed of 9×10^5 m/s.

Will the deBroglie wavelength for proton A be (a) *greater than*, (b) *less than*, or (c) *equal to* the deBroglie wavelength of proton B? _____

Explain your reasoning.

F1-SCT10: HEISENBERG UNCERTAINTY PRINCIPLE—POSITION AND SPEED

Three students talking about the Heisenberg Uncertainty Principle make the following contentions:

Arno: *"The Heisenberg Uncertainty Principle says that you can't measure both the position and speed of an electron at the same time."*

Bobbie: *"I don't think it means you can't measure them at the same time. It just means that if you gain precision in measuring one quantity, you lose precision in measuring the other quantity for the same point in time."*

Carissa: *"I think it is possible to become more precise in measuring both quantities, but only up to a point. Once the product of position and momentum are equal to h over four pi, one quantity has to go up when the other goes down."*

With which, if any, of these students do you agree?

Arno _____ Bobbie _____ Carissa _____ None of them _____

Explain your reasoning.

F2-RT11: ENERGY LEVEL TRANSITIONS—EMISSION FREQUENCY

Shown are four energy levels for an atom along with six possible transitions between pairs of energy levels. Adjacent horizontal lines (light gray or dark) are separated by the same energy difference.

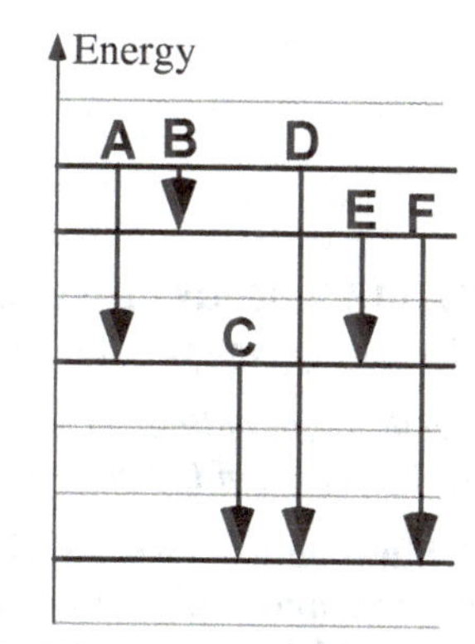

Rank the frequency of the emitted photons for the labeled transitions.

						OR			
1 Greatest	2	3	4	5	6 Least		All the same	All zero	Cannot determine

Explain your reasoning.

TIPERs

F2-RT12: Energy Level Transitions—Emission Wavelength

Shown are four energy levels for an atom along with six possible transitions between pairs of energy levels. Adjacent horizontal lines (light gray or dark) are separated by the same energy difference.

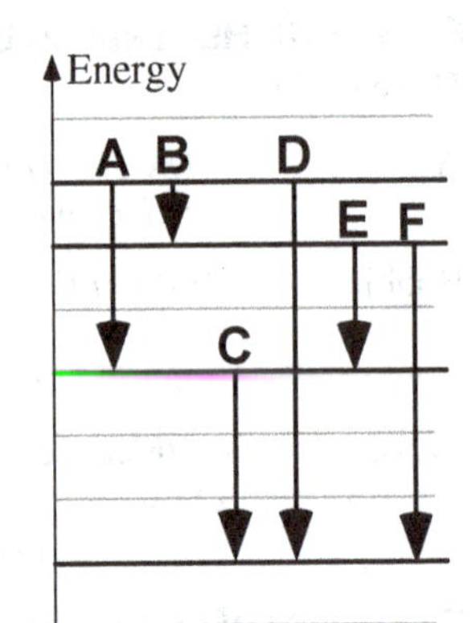

Rank the wavelength of the emitted photons for the labeled transitions.

☐	☐	☐	☐	☐	☐	OR	☐	☐	☐
1 Greatest	2	3	4	5	6 Least		All the same	All zero	Cannot determine

Explain your reasoning.

F2-WWT13: Energy Level Diagram—Wavelength

A student comparing two transitions on an energy level diagram contends:

"The wavelength of light emitted in transition A will be shorter than the wavelength of light emitted in transition B, because transition A starts from a higher energy level."

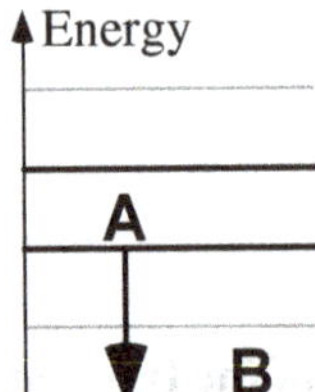

What, if anything, is wrong with this student's contention? If something is wrong, identify the problem and explain how to correct it. If the student is correct, explain the physics supporting his/her statement.

F3-CT14: CARBON ISOTOPES— PROTONS, NEUTRONS, AND ELECTRONS

A carbon-14 atom has 6 electrons, 6 protons, and 8 neutrons.

(a) Will an atom of carbon-11 have (i) *more electrons*, (ii) *fewer electrons*, (iii) or *the same number of electrons* as an atom of carbon-14? _____

Explain your reasoning.

(b) Will an atom of carbon-11 have (i) *more protons*, (ii) *fewer protons*, or (iii) *the same number of protons* as an atom of carbon-14? _____

Explain your reasoning.

(c) Will an atom of carbon-11 have (i) *more neutrons*, (ii) *fewer neutrons*, or (iii) *the same number of neutrons* as an atom of carbon-14? _____

Explain your reasoning.

F3-CT15: TWO RADIOACTIVE SAMPLES—MASS REMAINING

The masses of samples of radioactive elements are measured and then measured again 12 hours later. The initial mass for two samples of different elements are given, along with the half-life of each element.

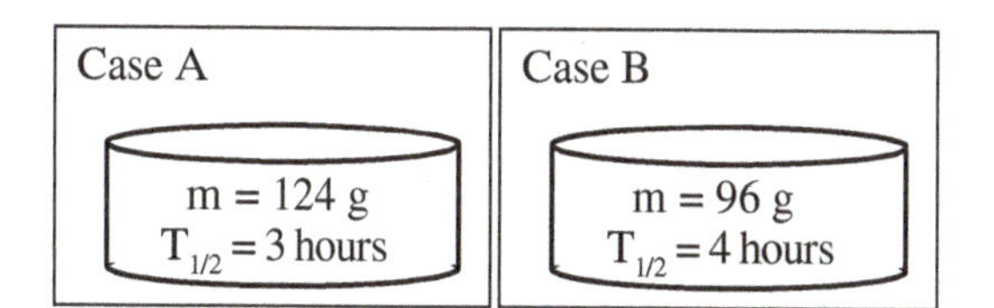

After 12 hours, will the mass of the sample in Case A be (i) *greater than*, (ii) *less than*, or (iii) *equal to* the mass of the sample in Case B? _____

Explain your reasoning.